Springer Series in Optical Sciences

Volume 207

Springer Series in Optical Sciences

The Springer Series in Optical Sciences, under the leadership of Editor-in-Chief William T. Rhodes, Georgia Institute of Technology, USA, provides an expanding selection of research monographs in all major areas of optics: lasers and quantum optics, ultrafast phenomena, optical spectroscopy techniques, optoelectronics, quantum information, information optics, applied laser technology, industrial applications, and other topics of contemporary interest.

With this broad coverage of topics, the series is of use to all research scientists and engineers who need up-to-date reference books.

The editors encourage prospective authors to correspond with them in advance of submitting a manuscript. Submission of manuscripts should be made to the Editor-in-Chief or one of the Editors. See also www.springer.com/series/624

More information about this series at http://www.springer.com/series/624

Yan Feng

Editor

Raman Fiber Lasers

 Springer

Editor
Yan Feng
Shanghai Institute of Optics & Fine Mechanics
Chinese Academy of Sciences
Jiading, Shanghai, China

ISSN 0342-4111   ISSN 1556-1534   (electronic)
Springer Series in Optical Sciences
ISBN 978-3-319-87989-5   ISBN 978-3-319-65277-1   (eBook)
DOI 10.1007/978-3-319-65277-1

Printed on acid-free paper

This Springer imprint is published by Springer Nature
The registered company is Springer International Publishing AG
The registered company address is: Gewerbestrasse 11, 6330 Cham, Switzerland

Preface

Only 2 years after the first demonstration of a working laser, stimulated Raman scattering was discovered by accident. The generation of frequency downshifted light was observed within a nitrobenzene Kerr cell of a Q-switched ruby laser. Similar effects were soon found in a large number of liquids, gases, and solids. The study of stimulated Raman scattering in optical fibers, and so Raman fiber lasers, began in the early 1970s. Optical fibers, as one-dimensional waveguides, are excellent media for nonlinear optical processes like stimulated Raman scattering.

Raman fiber amplifiers had been investigated intensively by the telecom community in the 1990s. In the past decade, with the development of diode and rare-earth-doped fiber lasers, Raman fiber lasers have advanced tremendously in many aspects. It is gradually becoming a versatile laser technology for generating high-power light sources covering spectral range from visible to mid-infrared. Raman fiber lasers and amplifiers are already applied in the fields of telecommunication, astronomy, atom physics, laser spectroscopy, environmental sensing, laser medicine, etc. Meanwhile, as a relatively less-explored laser technology, many new and interesting developments have emerged in recent years.

The remarkable progress and increasing importance of Raman fiber lasers triggered the writing of this book in order to offer readers a comprehensive overview of the state of the art of the technology. A wonderful group of authors, who are leading researchers on each of their topics, have teamed up in preparing the book to accomplish this goal. Chapter 1 explains the technical innovation in the power scaling progress of Raman fiber lasers and future prospects. Chapter 2 focuses on the cascaded operation of Raman fiber lasers to extend the wavelength range. Chapter 3 reviews the state of the art of mid-infrared Raman fiber lasers. Chapter 4 is devoted to infrared supercontinuum sources in which stimulated Raman scattering is a key mechanism. Chapter 5 discusses the specialty optical fibers for Raman fiber lasers. Chapter 6 presents distributed feedback Raman fiber lasers for single-mode operation. And Chap. 7 is a detailed review of random distributed feedback Raman fiber lasers, which are an interesting new development of the field. Numerous applications of Raman fiber lasers are outlined within the chapters.

I am very excited about the result of these collective efforts. It was no easy task for each author, since writing such extensive chapters is very time-consuming. I am fully aware that there are some interesting topics regretfully not included in the present edition due to the tight schedule of candidate authors, for example, mode-locked Raman fiber lasers and gas-filled hollow core fiber Raman lasers. Nevertheless, I believe this text is the most contemporary, informative, and insightful reference and the only book dedicated to Raman fiber lasers today. Of course, your feedback on the book is gratefully welcomed. Feedback can be sent to me via email (mail@yanfeng.org). I expect new developments in the future, and updated reviews will be necessary.

I appreciate all the contributors for their enthusiasm and hard work in writing the chapters. Special thanks go to the editorial staff of Springer Nature for their patience and support.

Shanghai, China   Yan Feng

Contents

Contributors

Sofia R. Abdullina Institute of Automation and Electrometry SB RAS, Novosibirsk, Russia

Kazi S. Abedin OFS Labs, OFS Fitel, LLC, Somerset, NJ, USA

Sergey A. Babin Institute of Automation and Electrometry SB RAS, Novosibirsk, Russia

Novosibirsk State University, Novosibirsk, Russia

Martin Bernier Center for Optics, Photonics and Lasers (COPL), Université Laval, QC, Canada

Dmitry V. Churkin Novosibirsk State University, Novosibirsk, Russia

Yan Feng Shanghai Institute of Optics & Fine Mechanics, Chinese Academy of Sciences, Jiading, Shanghai, China

Vincent Fortin Center for Optics, Photonics and Lasers (COPL), Université Laval, QC, Canada

Mohammed N. Islam Department of Electrical Engineering and Computer Science, University of Michigan, Ann Arbor, MI, USA

Department of Internal Medicine, Division of Cardiovascular Medicine, University of Michigan Medical School, Ann Arbor, MI, USA

Omni Sciences, Inc., Ann Arbor, MI, USA

Sergey I. Kablukov Institute of Automation and Electrometry SB RAS, Novosibirsk, Russia

Tristan Kremp OFS Labs, OFS Fitel, LLC, Somerset, NJ, USA

Alexey G. Kuznetsov Institute of Automation and Electrometry SB RAS, Novosibirsk, Russia

Ivan A. Lobach Institute of Automation and Electrometry SB RAS, Novosibirsk, Russia

Jeffrey W. Nicholson OFS Laboratories, Somerset, NJ, USA

Evgeniy V. Podivilov Institute of Automation and Electrometry SB RAS, Novosibirsk, Russia

Novosibirsk State University, Novosibirsk, Russia

Guanshi Qin Jilin University, Changchun, China

V. R. Supradeepa Centre for Nano Science and Engineering, Indian Institute of Science, Bangalore, India

Sergei K. Turitsyn Aston Institute of Photonic Technologies, Aston University, Birmingham, UK

Novosibirsk State University, Novosibirsk, Russia

Réal Vallée Center for Optics, Photonics and Lasers (COPL), Université Laval, QC, Canada

Ilya D. Vatnik Institute of Automation and Electrometry SB RAS, Novosibirsk, Russia

Novosibirsk State University, Novosibirsk, Russia

Paul S. Westbrook OFS Labs, OFS Fitel, LLC, Somerset, NJ, USA

Lei Zhang Shanghai Institute of Optics & Fine Mechanics, Chinese Academy of Sciences, Jiading, Shanghai, China

Ekaterina A. Zlobina Institute of Automation and Electrometry SB RAS, Novosibirsk, Russia

Chapter 1
High Power Raman Fiber Lasers

Yan Feng and Lei Zhang

1.1 Introduction

Fiber lasers have drawn great attention in the past decades due to robustness, high efficiency, high beam quality, and high power scaling capability [1–3]. The rapid advancement of fiber laser technology was largely benefited from the development of optical fiber and fiber components in telecom industry and high-power laser diodes. The concept of cladding pumping has played a crucial role in power scaling of fiber lasers. It allows efficient absorption of high power but low beam quality laser diodes in gain fibers with 10s-μm-sized core doped with rare-earth ions. The small core size is necessary to have a laser output with good beam quality.

Among different rare-earth-doped fiber lasers, the most established one is Yb-doped fiber lasers. It has advantages of simple energy level structure, low quantum defect, and convenient pump wavelength at 900–1000 nm where high-power laser diodes are well developed. Yb-doped fiber laser at 1 μm has reached 20 kW in nearly diffraction-limited beam quality and 100s kW in multimode output available commercially [3]. However, the emission bands of rare-earth-doped fiber lasers with over 100 W output are narrow and isolated at 1, 1.5, and 2 μm for Yb, Er, and Tm, respectively. But there are plenty of demands in scientific research and biomedical applications on lasers at special wavelengths which cannot be covered by rare-earth-doped fiber lasers.

Various nonlinear frequency conversion technologies can be applied to extend the spectral range of fiber lasers, for example, Raman scattering, parametric process based on four-wave mixing, frequency doubling, supercontinuum generation in optical fibers, and parametric conversion, second-harmonic generation, sum frequency

Y. Feng (✉) • L. Zhang
Shanghai Institute of Optics & Fine Mechanics, Chinese Academy of Sciences,
Jiading, Shanghai 201800, China
e-mail: mail@yanfeng.org

© Springer International Publishing AG 2017   1
Y. Feng (ed.), *Raman Fiber Lasers*, Springer Series in Optical Sciences 207,
DOI 10.1007/978-3-319-65277-1_1

mixing in external nonlinear optical crystals, etc. Among them, Raman fiber devices, in which stimulated Raman scattering (SRS) provides gain, is particularly interesting.

In Raman fiber devices, the frequency conversion is achieved inside the fiber, which allows the laser system in a robust all-fiber configuration. Parametric amplification is also possible inside optical fiber, but it has strict requirement on phase matching between the pump and signal lasers. Parametric sources can be generated only with specially designed dispersion-engineered fibers and within a limited spectral range [4, 5]. In contrast, SRS has no requirement on phase matching (naturally phase matched in other words) between pump and converted laser. Therefore, Raman fiber lasers can work with common optical fibers [6]. Providing appropriate pump laser, Raman fiber laser can be generated at any wavelength within the transparent window of optical fibers. Supercontinuum generation as a result of multiple nonlinearities in optical fibers can cover very wide spectral range from ultraviolet to mid-infrared [7, 8]. However, the spectral intensity is low so that supercontinuum source is not suitable for many applications. Raman fiber lasers and amplifiers are the only proven technology to generate high-power wavelength-agile fiber sources [9].

In recent years, Raman fiber lasers have advanced quickly. Figure 1.1 summarizes the state of art of Raman fiber sources at near- and mid-infrared spectral range. At ~1.1 μm, kilowatt output has been reported [10, 11]. The power decreases exponentially with increasing wavelength, which is mainly due to the availability of pump sources and suitable Raman gain fibers. Within the 1~2 μm wavelength range, the best gain medium is silica fiber which is well developed and optically and mechanically strong [12–14]. The silica fibers can be used to efficiently generate Raman lasers with wavelength as long as 2.5 μm [15–17]. But at even longer

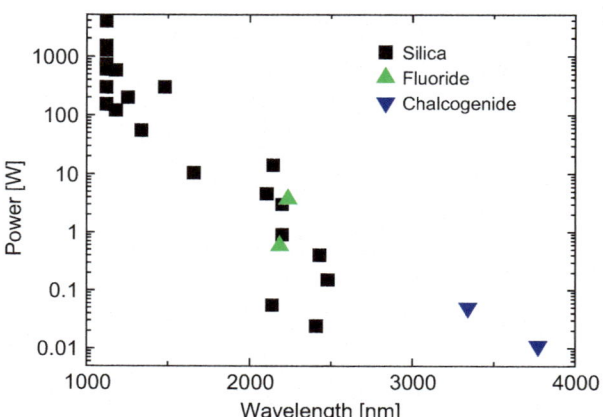

Fig. 1.1 The state of art of Raman fiber devices

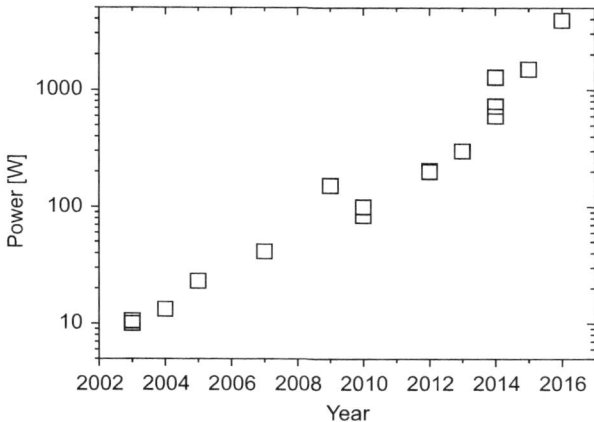

Fig. 1.2 Power evolution of Raman fiber laser and amplifier in the past decade

wavelengths, silica fiber is extremely lossy. Other optical fibers, such as fluoride, tellurite, and chalcogenide fibers, which are transparent at mid-infrared, could be better gain media. The development of mid-infrared Raman fiber lasers is discussed in Chap. 3.

Figure 1.2 summarizes the power scaling of Raman fiber lasers in the past decade, which shows that the power of Raman fiber laser or amplifier progresses exponentially. All these data points are Raman fiber devices pumped by Yb fiber lasers, because Yb fiber lasers had lead the power scaling of fiber lasers. When one recalls the development of nearly diffraction-limited Yb fiber lasers in the past decade [3], it is found that the power increases exponentially as well and always approximately one order of magnitude higher than that of Raman fiber lasers. The rare-earth-doped fiber lasers, as convenient pump sources, enable the development of Raman fiber lasers. But Raman fiber devices face unique challenges for power scaling. The technical innovations which enabled the power scaling are detailed in this chapter.

In this chapter, the developments in the last decade on power scaling of Raman fiber lasers are reviewed. The Raman fiber sources have utilized schemes of simple laser oscillator, master oscillator power amplifier, and pump and signal integrated amplifier to achieve higher output. Motivated by various scientific applications, high-power narrow-linewidth Raman fiber amplifiers have advanced greatly as well. In addition, emerging new Raman fiber laser schemes for high-power operation are described as well, for example, cladding pumping, direct diode pumping, Raman beam combining, etc. Finally, the power scaling potential of Raman fiber lasers is discussed.

1.2 Raman Fiber Oscillators

The simplest Raman fiber laser consists of a fiber laser pump source, a highly reflective fiber Bragg grating (FBG), a piece of Raman gain fiber, and a FBG output coupler. All of them can be fuse-spliced together to form an all-fiber-connected Raman laser.

Figure 1.3a shows a schematic diagram of a typical Raman fiber oscillator [18], which generates 1120 nm laser pumped by a 1070 nm Yb fiber laser. The wavelength difference between the pump and signal laser is determined by the Raman spectrum of the gain fiber. The Raman gain peak is at 440 cm^{-1} for silica fiber [6]. So the optimum pump wavelength is near 1067 nm. But the Raman peak in silica fiber is broad, and a small deviation from the optimum wavelength makes little difference. At the output end, a 1070/1120 nm wavelength division multiplexer (WDM) is added, which is used to remove the residual 1070 nm pump light from the output. Between the high-reflector FBG and the pump fiber laser, two additional 1070/1120 nm WDMs are inserted to prevent the backward 1120 nm laser from the Raman oscillator entering the pump fiber laser.

Since a highly reflective FBG has been used as a rear mirror, why is the laser leakage toward the pump laser still a trouble? Standard FBGs have nanometer-wide reflection band. Raman fiber lasers tend to broaden at high power and can be significantly wider than the FBG bandwidth. Then the high-reflecting FBG could not block the intra-cavity backward propagating light, and laser leakage outside the FBG bandwidth could be an important issue. It not only reduces the laser

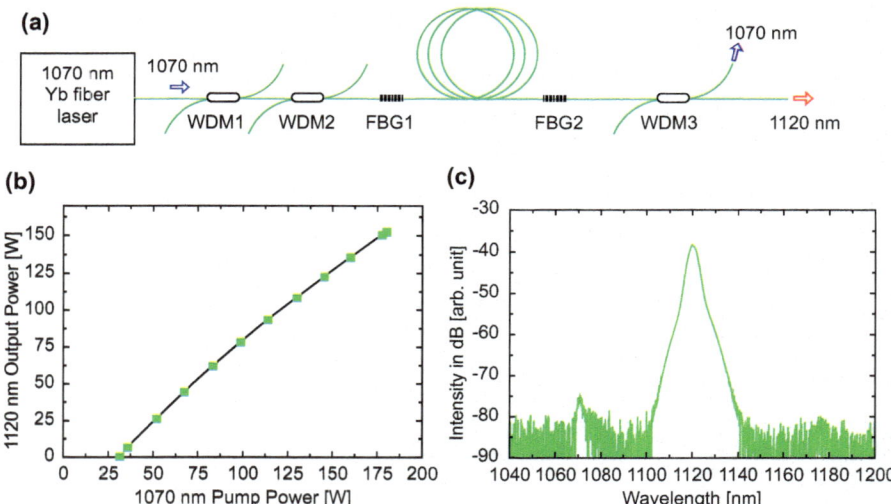

Fig. 1.3 (a) Schematic diagram of a typical Raman fiber oscillator; (b) 1120 nm output power as a function of 1070 nm pump power, where the 1070 nm contamination is already excluded; (c) an 1120 nm laser output spectrum at full power (After [18])

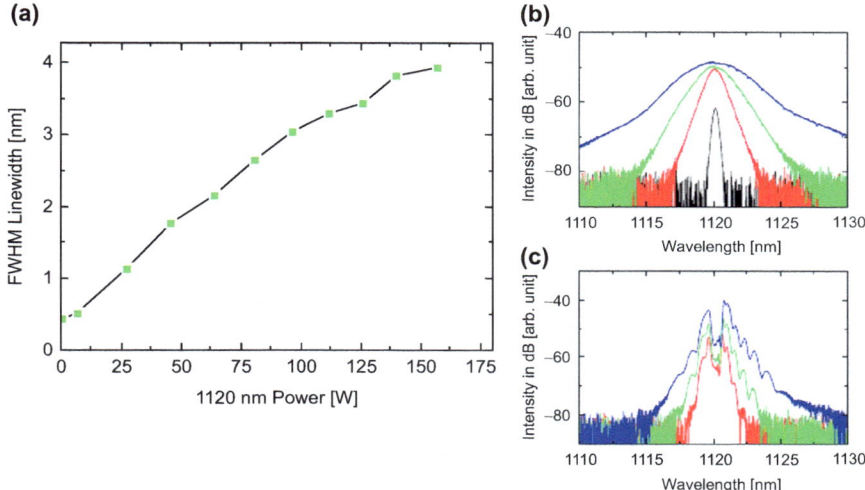

Fig. 1.4 FWHM linewidth as a function of Raman laser output (**a**), the spectra of the laser output (**b**) and laser leakage through the high reflector (**c**) at different power levels (After [18])

efficiency for the higher cavity loss but also influences the pump laser performance. The physical mechanism of the linewidth broadening was found to be four-wave mixing between the numerous longitudinal modes of the long fiber laser cavity [19]. Linewidth broadening exists in rare-earth-doped fiber lasers as well. But it is more significant in Raman fiber lasers, because usually longer gain fiber length is used.

Linewidth broadening became the main limiting factor for power scaling of Raman fiber oscillators. Broadband chirped FBGs were proposed to be used as rear high reflector against the laser linewidth broadening [20, 21]. But the broader the FBG reflection spectrum is, the lossier the FBG is.

In 2009, a 150 W Raman fiber oscillator was reported [18], which was the first Raman fiber laser with over 100 W output. Raman conversion efficiency as high as 85% was achieved with a piece of 30-m-long standard single-mode fiber (HI1060) as gain medium. The length of the gain fiber was chosen as short as possible to avoid the second-order Raman Stokes. In the laser setup, the rear FBG has a bandwidth of 1.05 nm and a peak reflection of 99.7%. The output FBG has a bandwidth of 0.55 nm and a peak reflection of 8.9%. With increasing pump power, the laser linewidth broadens significantly, up to 4 nm at the highest power, as seen in Fig. 1.4a, b. Figure 1.4c shows the spectra of laser leakage through the high reflector. Central dips are seen in the spectra, which is the stop band of the high-reflecting FBG.

The effective reflectivities of two FBGs are expected to be much smaller than the specified peak values, due to the spectral broadening of the circulating light. The exact linewidth broadening is difficult to simulate. Instead, the effective reflectivities can be numerically simulated by fitting the data of 1120 nm laser output, 1120 nm laser leakage through the rear FBG, and unused pump with a classical Raman fiber laser model, which is a system of first-order coupled ordinary differential equations:

$$\frac{\partial P}{\partial z} = -g_r \frac{\upsilon_P}{\upsilon_R} P \left(I^+ + I^- + 2h\upsilon_P B \right) - \alpha_P P$$

$$\frac{\partial I^+}{\partial z} = g_r P \left(I^+ + h\upsilon_R B \right) - \alpha_R I^+ \tag{1.1}$$

$$\frac{\partial I^-}{\partial z} = -g_r P \left(I^- + h\upsilon_R B \right) + \alpha_R I^-$$

where P, I^+, and I^- stand for the power of the pump, the forward propagating signal light, and the backward propagating signal light, respectively. g_r, α_P, and α_R are the Raman gain coefficient, pump, and Raman signal loss coefficients, respectively. $2h\upsilon_P B$ and $h\upsilon_R B$ stand for the contribution of spontaneous Raman emission, where h is Planck constant, υ_P and υ_R are the frequencies of the pump and Stokes emission, and B is a Boltzmann factor, which is very close to 1 in this case. The contribution of the spontaneous terms is important only when the laser is below and near threshold. The boundary conditions (Eq. 1.2) are

$$P(0) = P_0$$

$$I^+(0) = I^-(0)R_A \tag{1.2}$$

$$I^-(L) = I^+(L)R_B$$

where L is the fiber length of the oscillator; the 0 and L in the brackets represent the positions of the FBG1 and FBG2, respectively; R_A and R_B are the effective reflectivities of the FBG1 and FBG2, respectively; and P_0 is the input pump power.

It was found that the effective reflectivity of the nominally highly reflective FBG1 drops from 99.7% to 81.5% at full power, while the effective reflectivity of FBG2 drops from 8.9% to 0.97%. In spite of the low reflectivity of the rear FBG (81.5%), the leaked laser power is only 3.2 W, so the directionality of the laser is still as high as 49:1. The directionality of the laser outputs from a cavity is determined by the reflectivities of two end mirrors. As long as the ratio of the reflectivities of the rear mirror to the outcoupling mirror remains high, good directionality can be guaranteed.

The signal and pump power distributions along the fiber are calculated for understanding the laser power evolution inside the laser cavity. As shown in Fig. 1.5, at full power, only 1.54 W of laser power is fed back to cavity at the outcoupling FBG2, which is amplified to 17.3 W at FBG1, 14.1 W of which is reflected and amplified to 158.7 W at the output end of the laser. The majority of the laser power is generated in the forward direction. For this reason, the loss through the rear FBG1 is low even when the effective reflectivity is reduced to 81.5%. Similarly, the line broadening in backward trip (toward FBG1) is estimated to about 1.5 times, from an initial 0.55 nm defined by the linewidth of FBG2 spectrum. The line broadening in the forward direction is about 4.6 times (3.9 nm/0.84 nm).

Therefore, commonly available FBGs ($\leq$1 nm) are sufficient to overcome the light leakage problem of Raman fiber oscillators, providing that the bandwidth of the

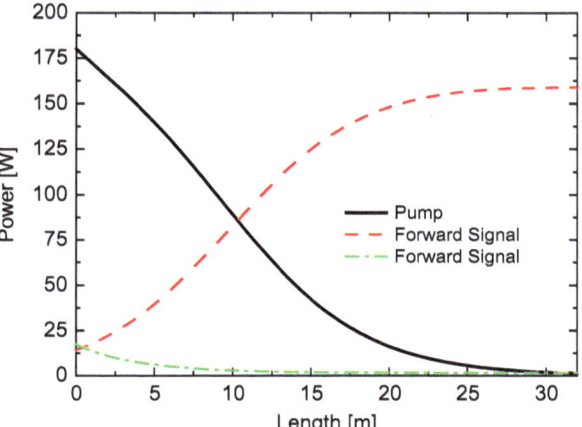

Fig. 1.5 Calculated signal and pump power distribution along the fiber at full power (After [18])

output FBG is much narrower than that of the rear FBG. In such a design, the ratio of the effective reflectivities of the rear and output FBGs becomes extremely large at high power because of the linewidth broadening. Inside the resonator, the backward propagating light is much lower than the forward light. The laser is basically a double-pass amplifier, which minimizes the light leakage. The idea can be further developed with amplifier configuration, with which laser leakage problem can be completely avoided.

For Raman fiber laser study, cascaded Raman shifting is always a topic of great interest to produce laser at otherwise unreachable wavelengths [15, 22, 23]. Generation of fiber laser at 1480 nm is a typical application, which is required for inband pumping of 1.55 μm Er-doped fiber lasers and amplifiers. There are consistent researches on the topic [12, 22, 23]. In 2010, cascaded Raman fiber laser with up to 81 W output at 1480 nm was reported [23]. The pump source is a 162 W Yb fiber laser at 1117 nm. The wavelength shift is achieved using nested cavities of FBGs at 1175, 1239, 1310, 1390, and 1480 nm. An optical conversion efficiency of 50% is achieved even after five Stokes scattering. A long-period grating is used to isolate the Yb-doped fiber amplifier with the cascaded Raman oscillator by introducing 20 dB attenuation at the Stokes wavelength. A specially designed Raman fiber with a long-wavelength fundamental mode cutoff enables efficient multiple Stokes scattering from 1117 to 1480 nm, while preventing further unwanted scattering to 1590 nm. A detailed review on cascaded Raman fiber laser is included in Chap. 2.

1.3   Raman Fiber Amplifiers

It was later realized that master oscillator power amplifier (MOPA) is a better configuration also for Raman fiber devices to further scale the output power, as proven for all other types of lasers. A Raman MOPA consists of a Raman fiber master oscillator and a Raman fiber power amplifier. With a MOPA structure, the power requirement is decoupled with various other performance aspects in wavelength, linewidth, pulse width etc., which decreases the difficulty. It reduces the requirement on thermal and optical damage threshold of fiber optical components, such as FBG, WDM, etc. For Raman fiber MOPA, there is an additional advantage compared with a Raman fiber laser: the backward laser leakage problem associated with linewidth broadening at high power is completely avoided.

1.3.1   Forward-Pumped Raman Fiber Amplifier

Forward-pumped Raman fiber amplifier was implemented by researchers developing high-power 1480 nm Raman fiber laser, which improved the conversion efficiency and boosted the output power significantly. In the first report [24], up to 200 W Raman fiber laser at 1480 nm was achieved with an optical efficiency of 65% (quantum-limited efficiency is ~75%), although five orders of Raman scattering is required from 1117 to 1480 nm. After further optimization, the 1480 nm power was improved to 301 W [12], comparable to the record power levels achieved with rare-earth-doped fiber lasers in the 1.5 micron wavelength region.

Figure 1.6 is the schematic diagram of fiber MOPA. A low-power cascaded Raman fiber laser is combined with a high-power 1117 nm Yb fiber laser via a WDM into a piece of Raman fiber. The low-power Raman fiber laser contains light at each Raman Stokes wavelength of 1175, 1239, 1310, 1390, and 1480 nm. So in the Raman fiber, the high-power 1117 nm laser is converted to next Raman Stokes cascadedly. The Raman fiber has a core with W-shaped index profile, whose attenuation increases steeply after 1500 nm by design. Therefore, the next Raman scattering to 1590 nm is suppressed. The detailed design and optimization process can be found in Chap. 2.

The key component in this configuration is the WDM which combines the high-power pump fiber laser and the multiple-wavelength seed laser. The WDM is a single-mode component, in which the laser propagates in the fiber core. High-power WDM is a challenge to manufacture. The WDM could be a technical bottleneck for further power scaling of Raman fiber amplifiers.

1.3.2   Integrated Rare-Earth and Raman Fiber Amplifiers

A major challenge for power scaling of Raman fiber amplifiers is to combine high-power pump laser with seed laser. As seen in previous section, WDMs are used for the combination in standard configuration. Recent reports had proven that a few 100 W throughput is feasible for WDMs [12, 25, 26]. But further power scaling to kW level is not yet demonstrated. On the other hand, pump and signal combiners have been widely used in high-power rare-earth-doped fiber lasers [1].

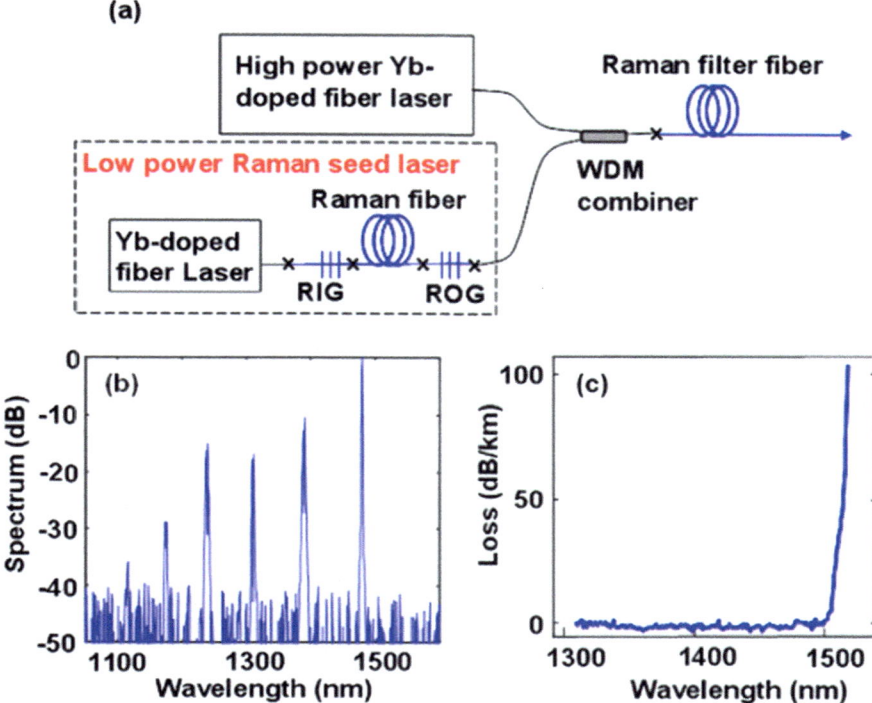

Fig. 1.6 (**a**) Schematic diagram of the cascaded Raman fiber amplifier; (**b**) the output spectrum of the multiple-wavelength Raman seed laser; (**c**) the loss spectrum of the Raman filter fiber (After [12])

In these components, the pump fibers are multimode, which allows higher power input. Power scaling of single-mode Yb-doped fiber amplifier to 20 kW has been demonstrated with pump and signal combiner.

Integrating rare-earth and Raman fiber devices was proposed as a way to scale the Raman fiber amplifier to over kilowatt [27]. To do this, standard high-power Yb-doped fiber amplifiers are seeded with multiple lasers, whose wavelength separations are close to the Raman shift. The seed laser with the shortest wavelength is at the middle of the Yb gain spectrum, which gets amplified efficiently. The laser power is then transferred to the longer wavelengths successively in the following optical fiber by stimulated Raman scattering. The most important improvement in the architecture is the elimination of the WDM that has been used in almost all high-power core-pumped Raman fiber laser. Here the Raman seed lasers and pump laser are propagated and amplified in the core of the same fiber.

A proof-of-principle experiment was carried out first [27]; the all-fiber-connected setup is illustrated in Fig. 1.7. The seed laser is a linearly polarized 1120 nm Raman fiber laser, which emits the residual 1070 nm pump laser as well. The power

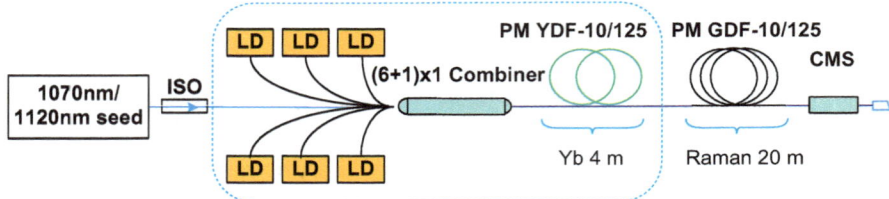

Fig. 1.7 Schematic diagram of an integrated Yb-Raman fiber amplifier (After [27])

ratio of the two wavelengths varies with the total output power. The seed light is coupled into the amplifier with a polarization-maintaining $(6 + 1) \times 1$ pump and signal combiner. Other ends of the combiner (105/125 μm fiber) are connected to six 976 nm laser diodes. The measured available pump power is 390 W after the combiner. The gain fiber is 4 m of PM double-clad Yb-doped fiber (YDF) with a core diameter of 10 μm, a numerical aperture of 0.075, a cladding diameter of 125 μm, and a nominal cladding absorption of 4.8 dB/m at 976 nm. A piece of 20-m-long germanium-doped fiber (GDF) with matching parameters is spliced after the YDF as a Raman converter. A cladding mode stripper (CMS) is used to remove the residual pump light. The output fiber is cleaved at an angle of 8° to suppress the parasitic oscillation.

With a maximum seed laser power of 39.2 W, in which 94% of the output is at 1120 nm, the YDF part of the amplifier was studied at first. A 1 m GDF is spliced to the end of YDF as delivery fiber. The YDF amplifier has higher gain at 1070 nm than 1120 nm; therefore the ratio of 1120 nm laser decreases to 51.1% at the end of the amplifier. After that, the 1 m GDF delivery fiber is replaced by 20 m GDF to convert the 1070 nm laser to 1120 nm. A maximum output of 301 W is achieved with the integrated Yb-Raman fiber amplifier, which is limited by the available pump power. The output spectra are depicted in Fig. 1.8. The 1120 nm power ratio is calculated to be 98.5%. The second-order Raman Stokes at 1180 nm was observed; however, it remains trivial (<0.05%). Note that when the wavelength conversion is over several Stokes components, the higher-order Stokes could cause incomplete conversion between the pump light and final Stokes. But it can be mitigated by optimizing the power of seed lasers at different Stokes wavelength and the Raman gain fiber length or even using filter fiber [12]. The optical efficiency reaches 70% from 976 nm, which is higher than the usual 1120 nm Raman lasers. The linewidth of the 1120 nm laser broadens from 1.6 nm (seed linewidth) to 3.3 nm (amplifier linewidth at an output power of 300 W), and the spectral broadening is mainly due to the four-wave mixing between numerous longitudinal modes associated with a long fiber [28]. The polarization extinction ratio of the laser was measured to be 18 dB at an output power of 50 W. Measurement at higher power was limited by the power handling of the setup. The time domain characteristic of the laser was examined with a high-speed oscilloscope (1GHz bandwidth), and no sign of self-pulsing is observed.

Fig. 1.8 Spectra of the dual-wavelength seed laser, the YDF part of the amplifier, and the YRFA

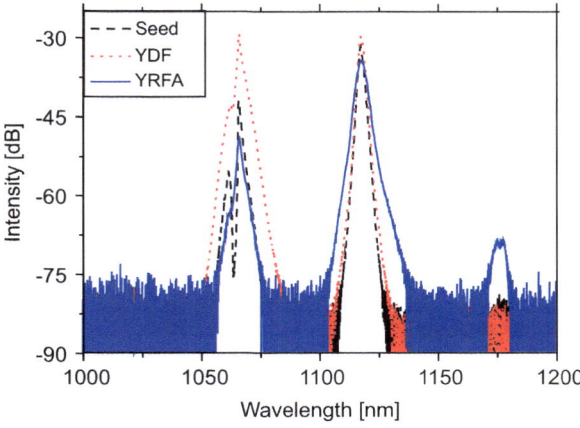

Fig. 1.9 Simulated 976 nm, 1070 nm, and 1120 nm laser power distribution along the fiber within an YRFA

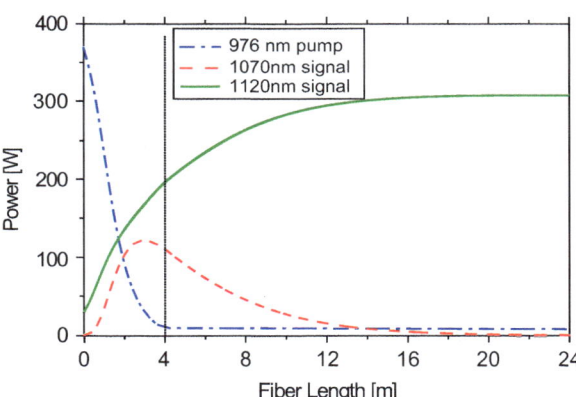

The YRFA can be simulated with a standard differential equation model. The fiber amplifier can be naturally divided into two parts: the 4-m-long YDF amplifier and the 20-m-long GDF Raman amplifier. In the first part, both ytterbium and Raman amplifications exist. In the second part, only stimulated Raman scattering takes place. The signal and pump power distributions along the fiber were then calculated to understand the laser power evolution inside the YRFA. Figure 1.9 shows the results for a YRFA with a dual-wavelength seed laser at 1070 and 1120 nm (2 W and 38 W, respectively) at full pump power. In the first 2.7 m of YDF, both the 1070 and 1120 nm lasers are amplified. The 1070 nm laser reaches a maximum and starts to roll over, because the Raman conversion from 1070 to 1120 nm becomes significant. At the end of YDF, the power of 1070 and 1120 nm laser are 111 and 196 W, respectively. When the dual-wavelength laser propagates along the 20 m GDF, Raman shift continues. At the output end, the power ratio of 1120 nm laser is calculated to be over 99%. The simulation fits well with the experiments.

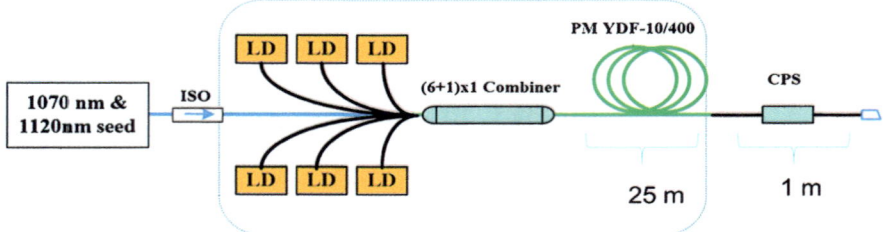

Fig. 1.10 Schematic diagram of a high-power integrated Yb-Raman fiber amplifier, where no extra Raman fiber is needed in the amplifier since the Raman gain in the Yb-doped gain fiber is already high enough

Since ytterbium MOPA can now produce kilowatts or even tens of kilowatts output [3], power scaling of the proposed YRFA to more than kilowatt level is rather straightforward by replacing the master oscillator of the Yb fiber MOPA with a dual-wavelength laser. The proposed YRFA architecture may also be used for cascaded Raman fiber laser generation, by using a multiwavelength seed laser whose wavelengths are separated by the Raman shift of the fiber.

As proofs of the power scalability of the proposed architecture, experiments with gain fibers of larger cladding were carried out. Fibers with larger cladding allow higher power input of diode laser pump. With gain fibers of 400 μm cladding and 10 μm core, up to 580 W single-mode linearly polarized Raman fiber laser at 1120 nm was achieved with an optical efficiency of 70% [29]. The 1120 nm laser purity at maximum power is 95%. The experimental setup is shown in Fig. 1.10. The 10 μm core fiber guarantees diffraction-limited output. The large cladding-to-core area ratio of 40 reduces the absorption coefficient of Yb-doped fiber. Therefore, Yb-doped fiber 25 m long was used in the experiments. With such long gain fiber, no extra Raman fiber after the Yb gain fiber is necessary, because the Raman conversion is already sufficient within the Yb gain fiber.

Furthermore, with gain fibers of 400 μm cladding and 20 μm core, a 1.3 kW Raman fiber laser at 1120 nm was generated with an optical efficiency of 70%, which is the first report of Raman fiber laser with over kilowatt output [10]. A 1.54 kW Yb-doped fiber master oscillator power amplifier at 1080 nm is seeded again with an 1120 nm fiber laser. The amplified 1080 nm laser is Raman shifted to 1120 nm in a subsequent piece of Raman gain fiber. Here at full power, there is still about 190 W residual 1080 nm pump laser which could not be Raman shifted to 1120 nm, as seen in Fig. 1.11. In fact, the conversion ratio saturates already below 1 kW. It was found that most of the residual 1080 nm laser was in the cladding mode. The low-NA cladding mode light leaks through the CMS between the YDF and Raman fiber and cannot be converted to 1120 nm without overlap with the 1120 nm signal laser inside the fiber core. Shortly after that, the output from such Yb-Raman hybrid fiber amplifier was increased to 1.5 kW with higher pump power [11].

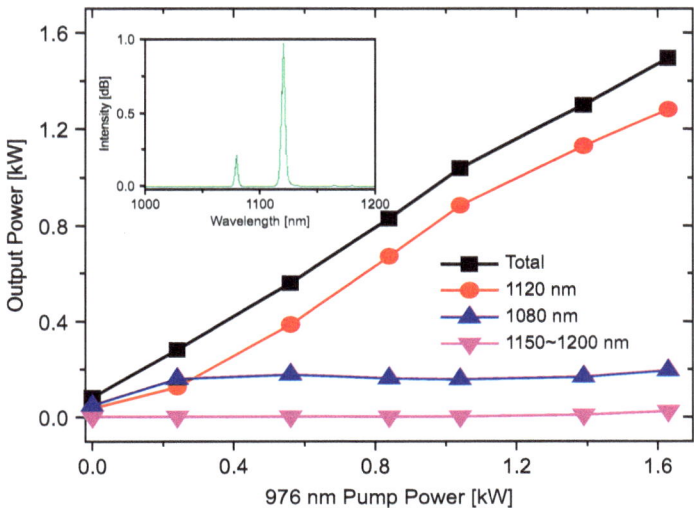

Fig. 1.11 The output power of 1080 nm, 1120 nm, amplified spontaneous Raman emission from 1150 to 1200 nm, and total output power from YRFA as a function of the pump power. Inset, the output spectrum in linear scale at the maximum power

Cascaded Raman fiber laser generation with the YRFA architecture was investigated as well. Preliminary numerical simulation showed that kilowatt-level Raman fiber laser covering 1.1 ∼ 2 μm is feasible with the YRFA architecture and a state-of-the-art high-power Yb fiber MOPA laser. For example, with a 1070 nm Yb fiber laser as pump source, efficient generation of 2066 nm Raman fiber laser seems possible with silica fiber after ten cascaded Raman scattering. Of course, to realize multiple-order cascaded Raman fiber amplifiers, one may encounter problems of significant linewidth broadening and four-wave mixing between the Raman Stokes, etc. Tailoring of fiber dispersion and power distribution in the multiple-wavelength seed laser will be necessary.

The idea was also applied to mid-infrared wavelength range. By integrating Tm and Raman fiber amplifiers, 14.3 W output at 2147 nm was reported [30]. Since all current high-power rare-earth-doped fiber lasers have a master oscillator power amplifier scheme, the proposed architecture can be applied conveniently and further scales the Raman fiber laser output to 10 kW level and extends the spectral coverage.

Most recently, a 1123 nm Raman fiber laser with output as high as 3.89 kW was reported [31]. The laser consists of a single-wavelength (1070 nm) seed and one-stage bidirectional 976 nm laser diodes pumped Yb-doped fiber amplifier. There is no external seeding at 1123 nm in the laser system. The 1123 nm laser action is probably initiated from amplified spontaneous emission of the Yb gain fiber and amplified by stimulated Raman scattering in the Yb fiber amplifier due to the extremely high laser intensity.

1.3.3   Narrow-Linewidth Raman Fiber Amplifiers

The obvious advantage of Raman fiber lasers is wavelength agility. So Raman fiber lasers and amplifiers can be used to generate light sources for a variety of applications, which are not possible with other fiber laser technology. However, many wavelength-sensitive applications also require narrow-linewidth laser. In some cases, further nonlinear frequency conversion processes, such as second-harmonic generation and sum frequency mixing, are necessary to achieve the target wavelength. These processes also require narrow-linewidth sources.

Narrow-linewidth single-longitudinal-mode Raman fiber laser has been reported with distributed feedback (DFB) configuration [32, 33], which will be discussed in Chap. 6. The power scalability of the DFB Raman fiber laser is limited. Direct generation of high-power narrow-linewidth laser from usual Raman fiber oscillators formed by two FBG reflectors is difficult. Raman gain coefficient is relatively low as compared with rare-earth gain, so Raman fiber lasers usually have longer fiber cavity. Nonlinear processes, especially four-wave mixing, would broaden the laser linewidth in square-root law when the laser power is increased [19, 34]. Nevertheless, high-power narrow-linewidth lasers can be generated by fiber Raman amplification of single-frequency seed lasers. The seed lasers are usually DFB or external-cavity diode lasers, which also can emit over an ultrawide spectral range. The combination of low-power diode seed laser and Raman fiber amplifier has proved to be a versatile laser technology.

The development in narrow-linewidth Raman fiber amplifiers has been driven by a particular application in astronomy, called sodium laser guide star adaptive optics [35, 36]. For this application, a high-power narrow-linewidth laser at 1178 nm is needed in order to generate laser at 589 nm by frequency doubling. The 589 nm laser is used at astronomical telescope to excite a layer of sodium atoms at an altitude of $90 \sim 100$ km, to generate an artificial "guide star." With the help of it, the image distortion induced by atmosphere turbulence can be compensated with adaptive optics technology [37]. The laser guide star technology is essential for large-aperture telescopes to approach the designed angle resolution.

In early days, Rhodamine dye lasers at 589 nm were used for sodium guide star. However, dye lasers are difficult to maintain. The sum frequency mixing of two Nd:YAG laser lines at 1064 nm and 1319 nm happens to be 589 nm. High-power 589 nm Nd:YAG lasers with different formats have been demonstrated. However, they were still too bulky and complicated for operation in remote astronomical sites. Therefore, fiber-based sodium guide star lasers have been pursued. Among different technical paths, Raman fiber amplifier-based guide star laser is most advanced at present.

Figure 1.12 shows a typical 1178 nm narrow-linewidth Raman fiber amplifier configuration. Backward pumping is chosen to preserve narrow linewidth, since dramatic linewidth broadening can be observed in forward-pumped Raman fiber amplifiers. The pump wavelength 1120 nm is selected to match the Raman shift in silica fiber. A 1120/1178 nm WDM is used to couple the pump light into the

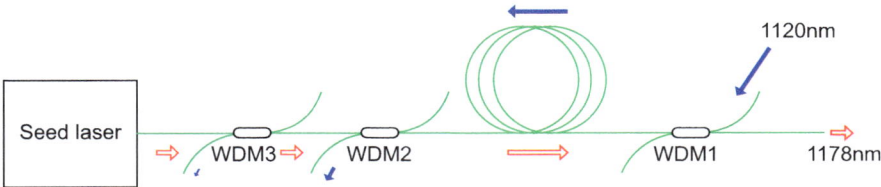

Fig. 1.12 Typical experimental setup of a 1178 nm Raman fiber amplifier pumped by a 1120 nm fiber laser

backward-pumped single-mode Raman fiber amplifier. Two more 1120/1178 nm WDMs are inserted between the seed laser and amplifier to extract the remaining pump laser and protect the seed laser.

Stimulated Brillouin scattering (SBS) is the main limiting factor in high-power narrow-linewidth Raman fiber amplifiers, because the Brillouin gain coefficient is two orders of magnitude higher than the Raman gain coefficient [6]. As long as the SBS reaches threshold, the laser would be downshifted and propagates backwardly toward the master laser. It not only limits the amplification but also unstabilizes the amplifier. SBS and SRS are two major nonlinear effects in optical fiber, which have similar origin. Both are from photon and phonon interaction in the medium. Therefore, it is challenging to suppress SBS while maintaining SRS. Common methods for suppressing SBS, like increasing the core diameter, shortening the fiber length, etc., do not work well. However, there is difference between the two in that optical phonons participate in SRS, while acoustic phonons participate in SBS. SBS is more sensitive to material composition, temperature, strain, fiber geometry, etc. So there are methods to suppressing SBS in narrow-linewidth Raman fiber amplifiers.

Brillouin gain coefficient is two orders of magnitude higher than the Raman gain coefficient in optical fiber, but 10% efficiency amplification of a MHz linewidth 1178 nm laser was reported in the first demonstration of multiwatt amplifier [38]. The reason behind is that the power distribution of the laser inside the Raman fiber amplifier is uneven, close to exponential in an backward-pumped far-to-saturation condition. Most laser power is generated in a short piece of fiber at the end of amplifier. Therefore, the effective fiber length for SBS process is much shorter than the amplifier fiber length. For the 150 m amplifier in reference [38], the effective fiber length for SBS process is only about 20 m. Detailed numerical simulation shows that the SBS-limited output is proportional to the pump power [39]. One can increase the amplifier output by raising the pump power.

But to improve the amplifier efficiency, SBS has to be well suppressed. One way to suppress the SBS is to broaden the seed laser linewidth [38], which however is not wanted for many applications. Another way is to broaden the SBS gain spectrum, which includes many methods.

Acoustically tailored fibers had been explored to reduce the Brillouin gain, while maintaining or even enhancing Raman gain [40, 41]. It is achieved through the manipulation of the concentration of dopants in the core as a function of position in

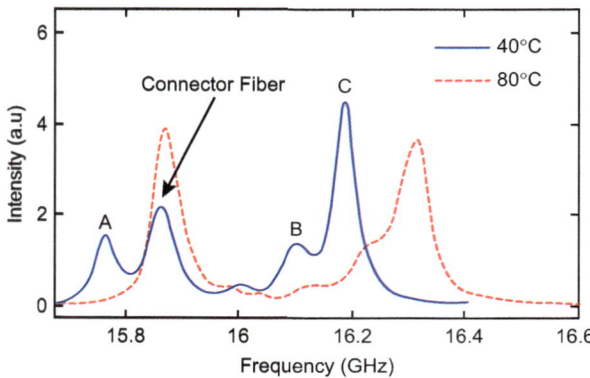

Fig. 1.13 Brillouin gain spectra of an acoustically tailored fiber at 40 °C and 80 °C. There are multiple peaks associated with the fiber as well as one peak due to the connector fiber

the transverse direction. A nonuniform acoustic index of refraction is obtained by varying the concentrations of fluorine and aluminum in the core. This can reduce the acousto-optical interaction and mitigate SBS. Figure 1.13 shows the Brillouin gain spectrum of the fabricated fiber. There are three peaks associated with the acoustically tailored fiber. Ideally, these peaks should be of equal height for best SBS mitigation. But due to the small size of the core ($\sim$6 μm), the ideal design was not achieved.

With the fabricated acoustically tailored fiber, up to 22 W single-frequency 1178 nm output was obtained in a backward-pumped two-stage amplifier with an optical efficiency of about 25% [41]. The same group of researchers also reported suppression of SBS by applying thermal gradients along the gain fiber and achieved 1.5 times improvement in a single-stage amplifier experiment [42]. Similarly, cascading multiple pieces of different fibers and multiple isolated amplifier stages can be used to increase the Brillouin threshold [38, 41]. But the scalability of all these methods is limited.

The most successful method of suppressing SBS in narrow-linewidth Raman fiber amplifiers until now is to apply longitudinally varied strain along the gain fiber. The strain method has been studied since the 1990s in the telecom community [43, 44]. The strain introduces a proportional shift of SBS gain spectrum. Therefore, with a strain distribution, SBS light from different portion of the gain fiber is spectrally isolated and could not get amplified efficiently in other portions of the fiber. By this way, the effect of SBS can be mitigated. Application of the technique to narrow-linewidth Raman fiber amplifiers has been investigated intensively since 2003 [39, 45–47] and described in these literatures [26, 48, 49].

In high-gain Raman fiber amplifiers, the power varies along the gain fiber steeply. The strain distribution has to be carefully designed according to the power distribution to achieve desirable SBS suppression. Numerical model was developed to simulate narrow-band RFA [48]. The SBS spectrum and power variation along

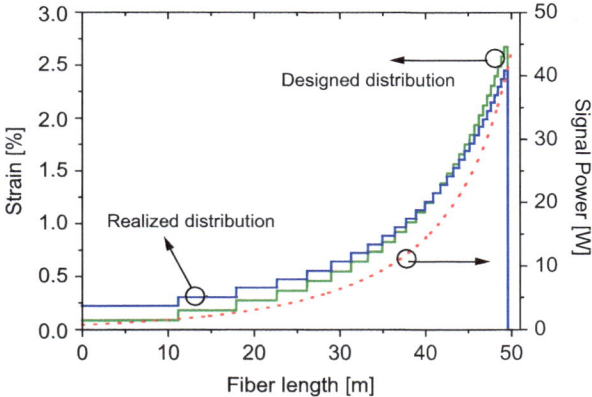

Fig. 1.14 Calculated signal power evolution (*dotted*), the designed strain distribution (solid, *green*), and the applied strain distribution (solid, *blue*) along the fiber (From [48])

the fiber were included in the model. Although continuous variation of strain could be better for SBS suppression, stepwise strain distribution was simulated in the model, since this is easier to implement in practice. At each strain step, SBS light is generated at different frequencies because of the strain-dependent SBS gain spectrum. With a chosen number of strain steps, the best SBS suppression relates to the case where each strain step sees the same SBS light generation at different frequency shifts. This can be achieved by adjusting the individual step lengths, since SBS gain is an integration along the fiber length.

Figure 1.14 shows typical design and realization of stepwise strain distribution for a Raman fiber amplifier [48], in which a 0.8 W seed laser is amplified to about 45 W with 30 strain steps in a 50-m-long gain fiber. Also shown in the figure is the laser power distribution along the amplifier. The fiber length for different strain steps varies significantly. The length of fiber at the highest level of strain is only 0.4 m, while the longest step is about 12 m. Precise design of strain distribution is found to be crucial for effective SBS suppression. This is very different from the suppression of SBS in passive delivery fiber, where uniform strain steps are close to the optimum.

With the strain methods, the highest conversion efficiency reported is 52% [48]. In that work, up to 44 W, 1 MHz linewidth, 1178 nm CW laser is obtained by Raman amplification of a distributed feedback diode laser in a variably strained polarization-maintaining fiber pumped by a linearly polarized 1120 nm fiber laser. Thirty strain steps were applied in the amplifier, and a 20 times reduction in the effective stimulated Brillouin scattering coefficient was achieved.

By improving the power of the 1120 nm pump laser, an even higher-power 1178 nm narrow-linewidth Raman fiber amplifier was demonstrated [26, 29]. It can work in both pulsed and CW format. Therefore, it offers much flexibility for astronomical application. In the pulsed case, the 1178 nm FRA produced

Fig. 1.15 1178 nm peak
power and backward light
power as functions of the
1120 nm pump peak power
(After [26])

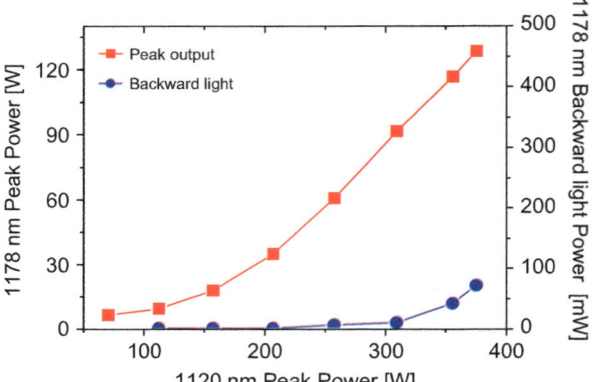

Fig. 1.15 1178 nm peak power and backward light power as functions of the 1120 nm pump peak power (After [26])

square-shaped pulses with tunable repetition rate (500 Hz to 10 kHz) and duration
(1 ms to 30 μs), while the peak power remain constant with a record peak power
of 120 W. The peak power and backward light power as functions of the 1120 nm
pump peak power are shown in Fig. 1.15. In the CW case, the same laser setup
produced up to 84 W 1178 nm output. These results represent the highest output
reported for single-frequency Raman fiber amplifiers.

The RFA output spectrum and backward light spectrum are shown in Fig. 1.16.
Although a spectral pedestal was observed in the output, the signal to noise ratio
is over 50 dB at maximum power. SBS is well suppressed and does not reach
its threshold even at maximum pump power. The backward propagating light
increases slowly, which is mainly the residual 1120 nm pump laser and amplified
spontaneous Raman emission, as shown in Fig. 1.16. The fraction of 1178 nm light
in the backward propagating light is calculated by spectral integration and is found
to be small. To further scale the output power, the primary challenge is still the
mitigation of SBS. In addition, the power-handling capability of the WDMs used in
the Raman fiber amplifier is a technical limiting factor.

For application to laser guide star, the high-power narrow-linewidth 1178 nm
Raman fiber laser should be converted to 589 nm by second-harmonic generation.
It can be done by single pass in periodically poled nonlinear crystals [50–52].
However, the conversion efficiency is below 30% in the investigated power level.
And practical source by this method remains below 10 W for the reliability of
the periodically poled nonlinear crystals. Practical 589 nm guide star lasers can
be generated in LBO crystal, which has much higher laser damage threshold. LBO
crystal has lower nonlinear coefficient; therefore it is placed in an enhancement
cavity.

Figure 1.17 illustrates the setup for the resonant frequency doubling. A noncriti-
cally phase-matched LBO crystal is placed between two curved mirrors. The cavity
length is designed and precisely adjusted to be 157 mm, ensuring a free spectral
range of 1.71 GHz. The cavity is locked to the laser using the well-established
Pound-Drever-Hall (PDH) method. The 1178 nm laser contains a main frequency

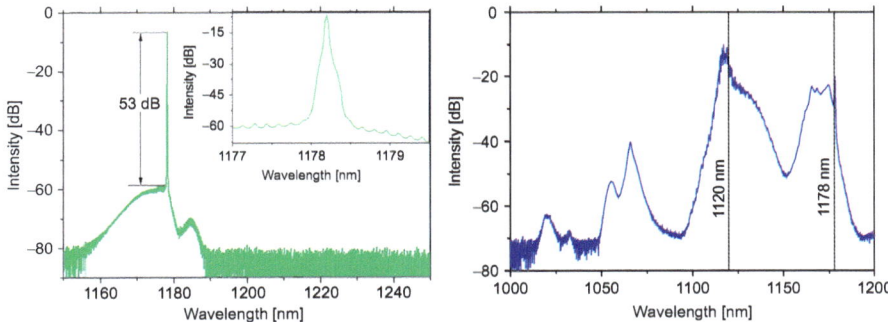

Fig. 1.16 (*Left*) Spectrum of the 1178 nm RFA output at full output power, inset, and spectrum with a narrow wavelength range from 1177 to 1079.5 nm. (*Right*) Typical backward propagating light spectrum

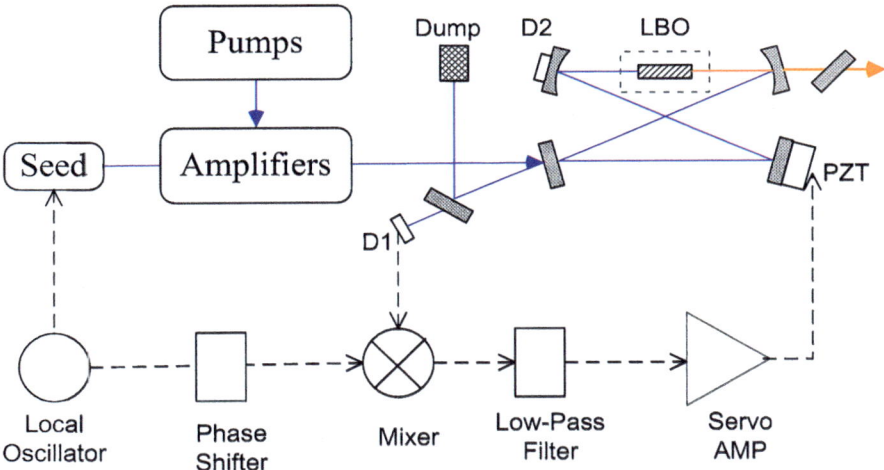

Fig. 1.17 Diagram of frequency doubling in an enhancement cavity with active cavity locking

and two symmetrical sidebands 1.71 GHz away, which is generated by phase modulating the seed laser. In this case, the main frequency and sidebands of the 1178 nm laser could resonate in the cavity simultaneously. The second harmonics of the main frequency and the sum frequency mixing of the main frequency and the sidebands are the dominant frequency-converted output. So the generated 589 nm laser has a main frequency and two sidebands 1.71 GHz away. Figure 1.18 shows the output spectrum of the 589 nm laser measured with a Fabry-Perot interferometer. The blue-shifted sideband will be used for repumping the sodium atoms to increase the return flux from the sodium guide star [53]. The amplitude of repumping light can be continuously adjusted with respect to the main line by changing the modulation intensity. For laser guide star application, the required power ratio is

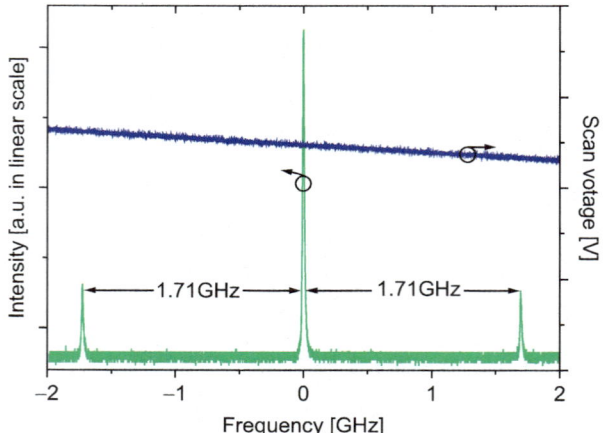

Fig. 1.18 Spectrum of the 589 nm laser with repumping light measured with a Fabry-Perot interferometer

around 10–20%. And the frequency shift of sidebands can be tuned as well by changing the modulation frequency. From the same figure, the linewidth of the 589 nm laser can be read at the same time, which is 13 MHz limited by the resolution of the FPI. The real linewidth should be well below 10 MHz, since the DFB diode laser has a linewidth of $\sim$1 MHz and no spectral broadening could be observed after amplification.

By optimizing the input mirror reflectivity and mode-matching optics, the narrow-linewidth 1178 nm laser can be coupled into enhancement cavity efficiently after cavity-length locking. Maximum 57 W 589 nm laser has been reported with a conversion efficiency of 80% [26]. Twenty-five watt class sodium guide star laser has been prototyped and being used in sodium LIDAR application.

The Raman fiber amplifier technology stands out among all possible guide star laser technologies at present. It is the only type of sodium guide star laser that has been fully engineered to product level [54, 55]. Twenty-two watt class laser devices have been successfully installed at Very Large Telescope of ESO [56] and Keck Observatory in Hawaii [57]. A photo for first light for the Four Laser Guide Star Facility on ESO's Very Large Telescope is shown in Fig. 1.19. Furthermore, all next-generation extremely laser telescope (ELT) projects have chosen the Raman fiber amplifier-based guide star laser as their baseline approach [59].

As mentioned before, the main advantage of Raman fiber devices is the wavelength agility. Raman fiber laser and amplifier can operate at any wavelength where the optical gain fiber is transparent, providing appropriate pump laser. Therefore, the developed technology of single-frequency Raman fiber amplifier can be applied to many other applications which require high-power laser at specific wavelength not convenient for rare-earth-doped fiber lasers.

Fig. 1.19 First light for the
Four Laser Guide Star
Facility on ESO's Very Large
Telescope (Credit ESO [58])

For example, at a slightly different wavelength of 588 nm, which is generated by frequency doubling of a 1176 nm single-frequency Raman fiber amplifier, it can be used for precision spectroscopic measurement of helium [60]. A few-watt narrow-linewidth Raman fiber amplifier at 1.27 μm was reported for remote sensing measurements of atmospheric oxygen (O_2). The amplifier uses phosphosilicate fiber as Raman gain fiber so that it can be converted from 1.08 μm to 1.27 μm with a single Raman shift. A high-power, single-frequency, quasi-continuous-wave 1336 nm Raman fiber amplifier was developed for future application in laser cooling of $^{27}Al^+$ after eighth harmonic generation in KBBF crystal [61]. The pump laser is a 1256 nm ytterbium-Raman integrated fiber amplifier with phosphosilicate fiber as Raman gain fiber. The 1336 nm amplifier produces square-shaped pulses with tunable repetition rate and duration. The peak power is as high as 53 W which remains constant during the tuning. A polarization extinction ratio of >25 dB is achieved due to the all-polarization-maintaining fiber configuration. These are just examples of current applications for single-frequency Raman fiber amplifiers. More applications are expected due to the wavelength versatility of Raman fiber devices.

1.4 Cladding Pumping

Cladding pumping technique has revolutionized rare-earth-doped fiber laser technology in power scaling. A natural progression of Raman fiber devices is cladding pumping as well for further power increase. Cladding-pumped Raman fiber amplifier was first discussed by Nilsson et al. [62]. While core-pumped Raman fiber laser shifts the wavelength but has no improvement in brightness, the cladding-pumped Raman fiber laser emits light with higher brightness than the pump laser, which is an important property of usual lasers.

To gain a high brightness enhancement, big cladding-to-core area ratio is preferred. However, if the cladding-to-core area ratio is too large, the intensity generated in the core can greatly exceed that in the cladding long before the pump laser is depleted. It leads to the generation of parasitic second-order Stokes light in the core, limiting the conversion from pump light to first Stokes light [63]. Figure 1.20 shows calculated conversion efficiency of a single-pass co-pumped double-cladding Raman fiber amplifier with different inner-cladding-to-core area ratio [64]. With increasing inner-cladding-to-core area ratio, the conversion efficiency decreases quickly. To obtain a favorable conversion efficiency, the area ratio between the inner cladding and core has to be sufficiently small, for example, less than $\sim$8 [65, 66], which then limits the actual enhancement of brightness.

One solution is to design a specialty fiber which attenuates light at the second Raman Stokes while transmitting the first Stokes light. W-type fiber, which has sharp spectral characteristics, was designed to improve the allowable cladding-to-core area ratio. Parasitic Raman conversion to the second Stokes in W-type fiber can be suppressed by bending loss, while the loss at the first Stokes is kept low. The ratio between the inner-cladding area and the effective area of the Stokes wave can be improved, up to 40 was shown to be feasible [63, 67]. With this method, up to

Fig. 1.20 Conversion efficiency of a single-pass co-pumped double-cladding Raman fiber amplifier with different inner-cladding-to-core area ratio [64]

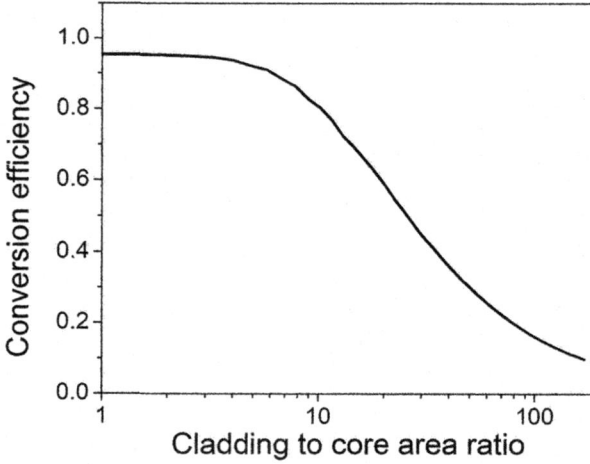

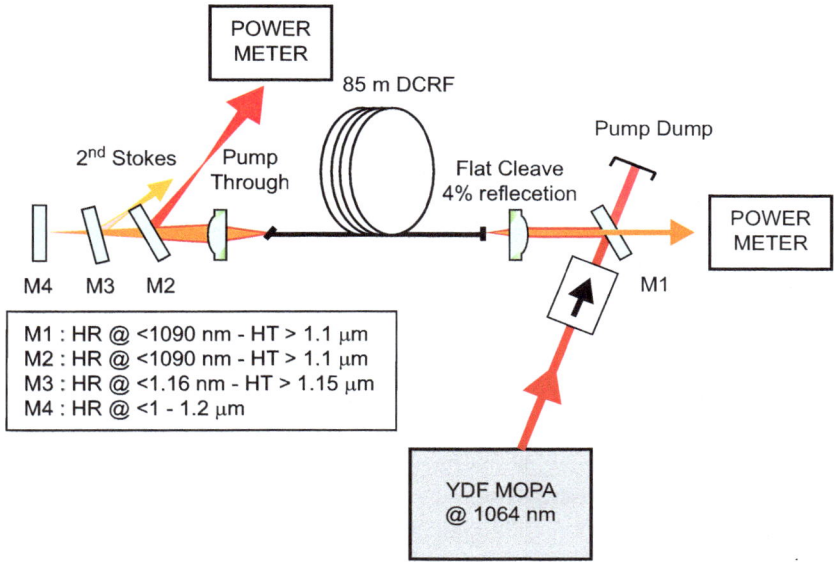

Fig. 1.21 Configuration of a cladding-pumped Raman fiber laser (After [68])

100 W Raman fiber laser had been demonstrated [68]. The experimental setup is shown in Fig. 1.21. In pulsed format, up to 1 mJ output had been reported [66].

Cascaded-cladding-pumped cascaded Raman fiber lasers or amplifiers with multiple-clad fiber were also proposed as gain medium to overcome the inner-cladding-to-core area ratio restriction [64]. The low brightness pump light is coupled into the second cladding (first inner cladding) of the fiber, and Raman Stokes light is generated in the third cladding. Similarly, the generated Stokes light will act as pump light and generate the higher-order Stokes light in the fourth cladding. The processes happen cascadedly until the single-mode laser is generated in the fiber core. By designing the neighboring claddings with small area ratio (e.g., less than 8), parasitic Stokes laser could be suppressed. Therefore, the restriction on the cladding-to-core area ratio can be removed by introducing the intermediate claddings.

Theoretical model of cascaded-cladding-pumped cascaded Raman fiber amplifier was developed to verify the idea [64]. In the simulation, a Ge-doped silica fiber with first inner-cladding diameter of 105 μm and core diameter of 10 μm was investigated, which matches commercial fiber and guarantees single-mode output from the core. The pump laser was assumed to be a long-pulsed (ns or longer) Yb fiber laser at 1070 nm with a peak power of 5 kW. The simulation shows that a four-cladding fiber is optimum. A maximum output power of 3.0 kW at 1246 nm third-order Raman Stokes can be achieved with a conversion efficiency of 59.6%. The optimal intermediate cladding diameters are 25 and 52 μm. Compared to double-clad case, the brightness enhancement increases from 650 to 2800. The numerical simulation proved that the multiple-cladding fiber scheme can indeed improve the conversion efficiency and brightness enhancement. But the concept is not yet experimentally demonstrated.

Cladding pumping is an important technique for future power scaling of Raman fiber sources, so it deserves further investigation. A related technique is to use multimode gain fiber to have higher power and to use the self-beam-cleaning effect to obtain brightness enhancement. Self-beam-cleaning effect is routinely observed in Raman lasers with graded-index fiber as gain medium. It is due to the better overlap of Raman light in the lower-order modes with the pump lasers [69] and also due to the Raman gain distribution profile in the graded-index fiber. The multiple-cladding fiber described in previous paragraph is in fact a stepwise version of graded-index fiber, which is believed to be more robust in enhancing the brightness of input pump laser via stimulated Raman scattering.

1.5 Direct Diode Pumping

In most high-power Raman fiber lasers already reported, the pump sources are rare-earth-doped fiber lasers or solid-state lasers. These sources are themselves complicated laser devices pumped by laser diodes. Direct diode pumping of Raman fiber lasers could reduce the complexity substantially and improve the robustness of the laser system. But Raman lasers require highly bright pump sources, which makes direct diode pumping challenging. For single-mode laser diodes, the maximum output power is on the watt level. Directly diode-pumped Raman fiber amplifiers had been employed in telecom community [70]. But direct diode-pumped Raman fiber lasers at other wavelength region and with more than 1 W output were not reported until a few years ago. Driving by the demands in high-power Yb fiber lasers, the brightness of commercial laser diodes has reached a point that high-power direct diode-pumped Raman fiber laser becomes feasible.

In 2013, a CW Raman fiber laser below 1 μm was reported with direct pumping by a high-power multimode laser diode at 938 nm [71]. The laser cavity is formed with a 4.5-km-long multimode graded-index fiber by a normally cleaved fiber end and a highly reflective fiber Bragg grating. The experimental setup is shown in Fig. 1.22. A 3 W low-index transverse mode laser was generated at the first Stokes

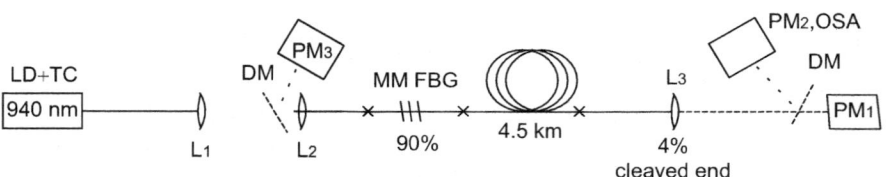

Fig. 1.22 Scheme of the experimental setup for the first diode-pumped Raman fiber laser operating below 1 μm (After [70])

wavelength of ~980 nm with 45 W coupled pump power, limited by the onset of the second Stokes wave at 1025 nm. This demonstration is the first report of diode-pumped Raman fiber laser operating below 1 μm. A similar setup was used to demonstrate a diode-pumped random Raman fiber laser. Point-like reflections were removed in the setup by omitting the FBG and cleaving the fiber ends with angle >15°. Random lasing was achieved with Rayleigh backscattering in the long optical fiber as the necessary feedback. A 0.5 W output at 980 nm was generated with 4.5 times lower divergence as compared to the pump beam [72].

Another group of researchers had demonstrated directly diode-pumped Raman fiber laser at an even shorter wavelength of 835 nm [73]. At this wavelength the background loss is about 10 times higher that at 1550 nm, so diode laser with higher brightness is required. To overcome this, a high-brightness broad-stripe diode laser at 806 nm was pulsed to generate multiwatt peak-power pump pulses of 50–100 ns duration and synchronously pumped a Raman fiber laser. A slope efficiency of 65% was achieved in a 600-m-long ultrahigh NA fiber with watt-level threshold in a ring cavity. The average output power was 1.4 mW, primarily limited by the pump driver. In another report, the same group of authors described a Raman fiber laser that is pumped directly by spectrally combined high-power multimode laser diodes at ~975 nm and emit at ~1019 nm [74]. With a commercial multimode graded-index fiber, 20 W output power was reached with a record slope efficiency of 80% (optical efficiency 50%). Directly diode cladding-pumped Raman fiber laser was also demonstrated with an in-house double-clad fiber (inner-cladding diameter 38 μm, core diameter 14.6 μm). The beam quality was improved to $M^2 = 1.9$, albeit with lower output power and slope efficiency due to higher fiber loss.

In 2016, the output power from directly diode-pumped Raman fiber laser was scaled to 80 W with a higher power laser diode module at 976 nm [75]. The 80 W 1020 nm output was obtained at CW pumping with an optical-to-optical efficiency of 53%. When working quasi-CW, at a duty cycle of 30%, 85 W of peak power was produced with an efficiency of 60%. The experimental setup is illustrated in Fig. 1.23. A commercial graded-index core fiber acts as the Raman fiber in a power

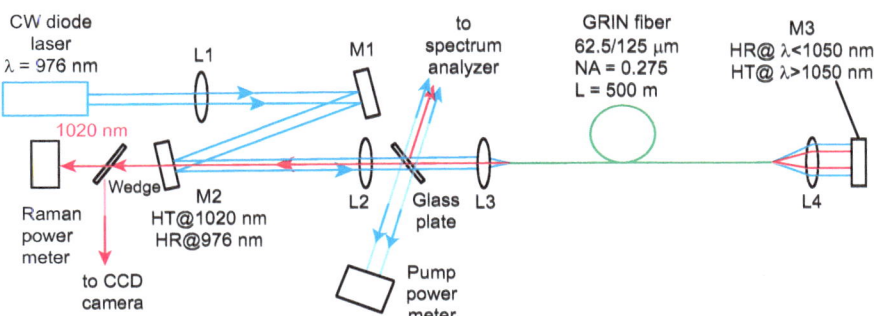

Fig. 1.23 Experimental setup for the directly diode-pumped Raman fiber laser with 80 W output [75]

oscillator configuration, which includes spectral selection to prevent generation of the second Stokes. A brightness enhancement factor of 7 for Raman output with respect to pump diode is attained in the experiment. By polarization combining two laser diodes, up to 154 W Raman laser output was recently demonstrated in a similar setup with an optical efficiency of 65% [76].

1.6   Raman Fiber Device for Beam Combining

Power scaling of fiber lasers is limited by optical nonlinearity, optical damage, and thermal issues [77, 78]. A variety of beam-combining technologies have been investigated for further power scaling. Among them, spectral beam combining proves to be successful. Due to limited gain bandwidth of rare-earth-doped fiber, diffraction optics such as gratings is used to combine multiple beams. However, the number of beams that can be combined is limited by the gain bandwidth divided by spectral separation of beams, which is determined by the dispersion capability of the gratings. And the linewidth of individual beams has to be sufficiently narrow compared to the spectral separation of beams. Otherwise, the quality of beam after the grating will be a problem, because a broad-linewidth beam will have additional angular spread in the output beam. However, high-power narrow-linewidth fiber laser remains a technical challenge due to stimulated Brillouin scattering and linewidth broadening issue.

Raman fiber laser and amplifier are known to be wavelength agile. Therefore, Raman fiber devices are naturally fitted for spectral beam combining [29]. Figure 1.24a illustrates the spectral coverage of cascaded Raman scattering pumped by a 1070 nm Yb fiber laser. Within ten Raman shifts, laser at any wavelength between 1 and 2 μm can be generated in principle. Wavelength tuning of a continuous wave fiber laser covering 1–2 μm has been demonstrated recently with a random Raman fiber laser configuration [79]. High-power cascaded Raman fiber amplification should be feasible. Of course, careful design of the cascaded Raman fiber laser configuration and gain fiber itself will be required to achieve the goal. Nevertheless, even with only one Raman shift, the spectral coverage of high-power fiber laser is doubled as compared with Yb-doped fiber laser alone.

The ultra-broadband tunability of Raman fiber laser offers a new way to spectrally combine laser beams: dichroic-mirror-based spectral beam combination. For example, 15 fiber lasers with spectral separation of 20 nm can be generated with cascaded Raman conversion (from 1030 to 1310 nm). Because of the wide spectral separation of 20 nm, the laser beams can be combined with common dielectric bandpass filters. A schematic diagram of the dichroic-mirror-based spectral beam combining is illustrated in Fig. 1.24b. Here, the dielectric-coated mirrors, which are easy to manufacture, are the combining devices instead of the gratings. Interference effect is used to combine beams at different wavelengths, instead of diffraction effect. There is no diffraction-induced angular spreading effect in the output with

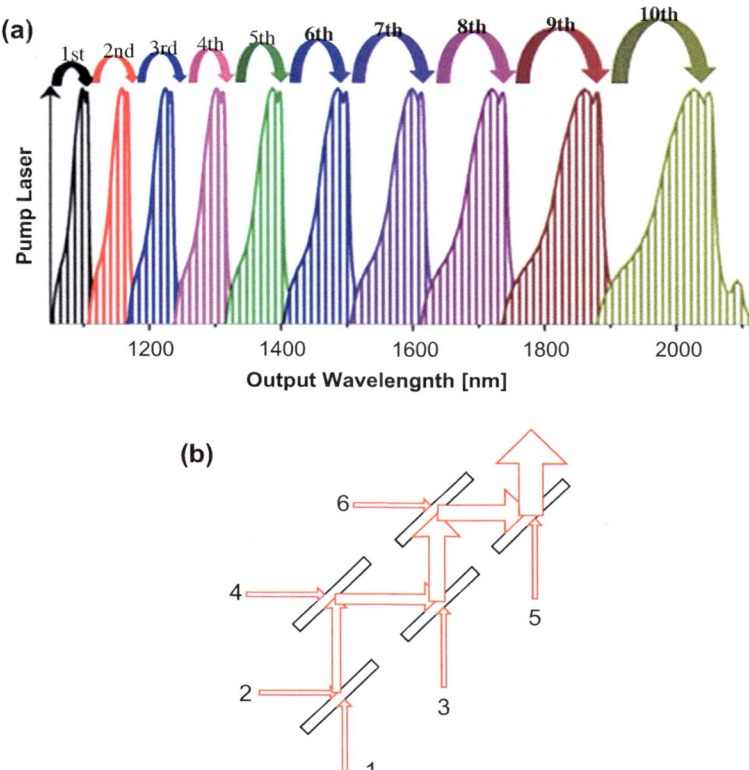

Fig. 1.24 (**a**) Spectral coverage of cascaded Raman scattering pumped by a 1070 nm Yb fiber laser and (**b**) a schematic diagram of interference-based spectral beam combining

dichroic mirror. Thus, the requirement on the linewidth of individual beams is relaxed with this method. This is important for spectral beam combining, because it is much easier to obtain higher-power fiber lasers with broader linewidth.

The nonlinear optical effect of stimulated Raman scattering can be utilized for combining multiple laser beams. A Raman fiber laser or amplifier can be pumped by multiple lasers and outputs a single beam. A simple realization of the concept is illustrated in Fig. 1.25. Multiple multimode fiber lasers and a seed Raman laser of high beam quality are combined together into a single multimode Raman fiber. The Raman fiber could be graded-index fiber or multiple-cladding fiber. The combined pump lasers can be converted to a single Raman laser beam. The beam quality of the combined Raman laser can be much higher than that of the multiple pump lasers, due to the Raman self-beam-cleaning effect with graded-index fiber or the cladding-pumping mechanism. The number of combined beams is scalable with the diameter of the combining fiber. This "Raman beam combining" method has advantages

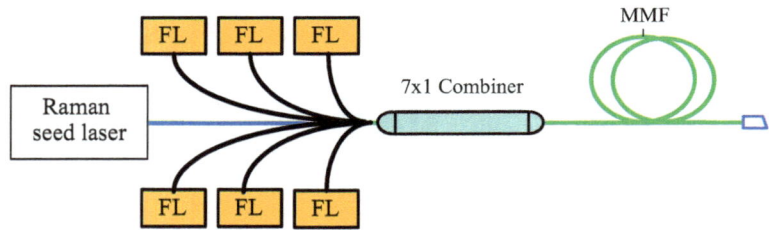

Fig. 1.25 A schematic diagram of Raman beam combining

compared with other methods. As a laser-pumped laser, it does not require phase locking as in coherent beam combining. The combined laser beam has a single wavelength, comparing to spectral beam combining.

Indeed, the scalability of Raman beam combining will be limited by thermal effect, optical nonlinearity, and optical damage finally, as in common rare-earth-doped fiber lasers [77]. In the following section, some intrinsic advantages of Raman fiber devices for high-power operation are discussed.

1.7 Power Scaling Potential

Following the rapid development of rare-earth-doped fiber laser, especially Yb-doped fiber laser, the power scaling capability had been analyzed systematically [77, 78, 80]. Power scaling of fiber lasers is limited by thermal issues, nonlinear optical effects, optical damage, and pump coupling limits. A few tens of kilowatts was predicted depending on the different restriction on fiber core diameter. Raman fiber lasers are limited by the same issues, but they have some intrinsic advantages for high-power operation [29].

In Raman fiber lasers, thermal effect is less a problem because the quantum defect is small. For silica fiber Raman laser pumped at 1070 nm and operating at peak Raman gain (1123 nm), the quantum defect is only about 4.7%. This can only be realized with tandem pumping in Yb fiber laser. The quantum defect can be reduced further by forcing the Raman fiber laser to work at smaller wavelength shift, since the Raman gain spectrum is broadband.

In Raman fiber lasers, the gain medium is passive fiber without doping of rare-earth or other active ions. Therefore, it is free of photo darkening and more resistant to optical damage. Without active dopants, the refractive index of the core material can be controlled more precisely. Thus, fibers with extremely small NA can be fabricated, which means extremely large mode area.

For Raman fiber lasers, the pump coupling limit is almost removed, since high-brightness fiber lasers are used as pump sources. Moreover, Raman fiber devices allow cascaded tandem pumping, so laser power can be combined successively until reaching the other limits.

With these intrinsic advantages, higher-power scalability than Yb-doped fiber laser may be expected for Raman fiber lasers. Indeed, higher power output with cladding-pumped solid-core photonic bandgap Raman fiber amplifiers is theoretically predicted [81]. But experimental demonstration is yet to be achieved. As in rare-earth-doped fiber lasers, modal instability is expected in high-power Raman fiber lasers or amplifiers [82].

1.8 Summary

During the past 10 years, the output power of Raman fiber lasers has advanced significantly with two orders of magnitude, from 10s watts to 1000s watts. The output of simple FBG formed Raman fiber oscillator is limited, because of the nonlinear linewidth broadening at high power. Master oscillator power amplifier scheme was used to solve the technical issues induced by linewidth broadening. Furthermore, pump and signal integrated amplifier scheme was utilized to solve the challenge in combining high-power pump and signal light and achieved more than kilowatt output. With this scheme, 10 kW Raman fiber laser is achievable by small modification to currently available high-power Yb fiber amplifiers. And by cascaded Raman shifting, high power fiber lasers at extended wavelength region can be obtained.

Many scientific and high-tech applications require high-power narrow-linewidth laser at specific wavelengths, which cannot be reached with common laser technology. In this context, high-power narrow-linewidth Raman fiber amplifiers have been studied and advanced greatly in the past decade. It appears Raman fiber amplifier seeded by narrow-linewidth diode laser is a versatile solution.

High-power Raman fiber lasers have already been applied in laser guide star adaptive optics; LIDAR; pumping Er, Tm, and Ho lasers, biomedicine; etc. Yet there are still many developments in schemes for high-power operation. Cladding pumping is pursued to have brightness enhancement and scale output power. Direct diode pumping is actively researched in recent years to reduce the complexity and improve the robustness of the Raman fiber lasers. Furthermore, Raman fiber devices are proposed for laser beam combination.

As the only proven technology for generating high-power wavelength-agile fiber sources, the interest in Raman fiber lasers is growing in the scientific and industrial community. Considering the advantages of Raman fiber lasers in low quantum defect, high-quality gain fiber, and tandem pumping capability, further power scaling is expected in the future.

References

1. Richardson, D.J., Nilsson, J., Clarkson, W.A.: High power fiber lasers: current status and future perspectives [Invited]. J. Opt. Soc. Am. B. **27**, B63–B92 (2010)
2. Limpert, J., Roser, F., Klingebiel, S., Schreiber, T., Wirth, C., Peschel, T., Eberhardt, R., Tiinnermann, A.: The rising power of fiber lasers and amplifiers. IEEE J. Sel. Top. Quantum Electron. **13**, 537–545 (2007)
3. Jauregui, C., Limpert, J., Tünnermann, A.: High-power fibre lasers. Nat. Photonics. **7**, 861–867 (2013)
4. Zlobina, E.A., Kablukov, S.I., Babin, S.A.: Tunable CW all-fiber optical parametric oscillator operating below 1 μm. Opt. Express. **21**, 6777–6782 (2013)
5. Murray, R.T., Kelleher, E.J.R., Popov, S.V., Mussot, A., Kudlinski, A., Taylor, J.R.: Widely tunable polarization maintaining photonic crystal fiber based parametric wavelength conversion. Opt. Express. **21**, 15826–15833 (2013)
6. Agrawal, G.P.: Nonlinear Fiber Optics. Academic Press, San Diego (2007)
7. Dudley, J.M., Genty, G., Coen, S.: Supercontinuum generation in photonic crystal fiber. Rev. Mod. Phys. **78**, 1135 (2006)
8. Dudley, J.M., Taylor, J.R.: Supercontinuum Generation in Optical Fibers. Cambridge University Press, Cambridge/New York (2010)
9. Supradeepa, V.R., Feng, Y., Nicholson, J.W.: Raman fiber lasers. J. Opt. **19**, 23001 (2017)
10. Zhang, L., Liu, C., Jiang, H., Qi, Y., He, B., Zhou, J., Gu, X., Feng, Y.: Kilowatt Ytterbium-Raman fiber laser. Opt. Express. **22**, 18483 (2014)
11. Zhang, H., Tao, R., Zhou, P., Wang, X., Xu, X.: 1.5 kW Yb-Raman combined nonlinear fiber amplifier at 1120 nm. IEEE Photon. Technol. Lett. **27**, 1–1 (2014)
12. Supradeepa, V.R., Nicholson, J.W.: Power scaling of high-efficiency 1.5 μm cascaded Raman fiber lasers. Opt. Lett. **38**, 2538 (2013)
13. Zhang, L., Jiang, H., Yang, X., Pan, W., Feng, Y.: Ultra-wide wavelength tuning of a cascaded Raman random fiber laser. Opt. Lett. **41**, 215 (2016)
14. Dianov, E.M.: Advances in Raman fibers. J. Light. Technol. **20**, 1457 (2002)
15. Rakich, P.T., Fink, Y., Soljačić, M.: Efficient mid-IR spectral generation via spontaneous fifth-order cascaded-Raman amplification in silica fibers. Opt. Lett. **33**, 1690 (2008)
16. Jiang, H., Zhang, L., Feng, Y.: Silica-based fiber Raman laser at > 2.4 μm. Opt. Lett. **40**, 3249 (2015)
17. Dianov, E.M., Bufetov, I.A., Mashinsky, V.M., Neustruev, V.B., Medvedkov, O.I., Shubin, A.V., Mel'kumov, M.A., Gur'yanov, A.N., Khopin, V.F., Yashkov, M.V.: Raman fibre lasers emitting at a wavelength above 2 μm. Quantum Electron. **34**, 695–697 (2004)
18. Feng, Y., Taylor, L.R., Calia, D.B.: 150 W highly-efficient Raman fiber laser. Opt. Express. **17**, 23678–23683 (2009)
19. Babin, S.A., Churkin, D.V., Ismagulov, A.E., Kablukov, S.I., Podivilov, E.V.: Four-wave-mixing-induced turbulent spectral broadening in a long Raman fiber laser. J. Opt. Soc. Am. B. **24**, 1729–1738 (2007)
20. Vallée, R., Bélanger, E., Déry, B., Bernier, M., Faucher, D.: Highly efficient and high-power Raman fiber laser based on broadband chirped fiber Bragg gratings. J. Light. Technol. **24**(12), 5039–5043 (2006)
21. Babin, S.A., Churkin, D.V., Podivilov, E.V.: Intensity interactions in cascades of a two-stage Raman fiber laser. Opt. commun. **226**, 329 (2003)
22. Emori, Y., Tanaka, K., Headley, C., Fujisaki, A.: High-power Cascaded Raman Fiber Laser with 41-W output power at 1480-nm band. In: 2007 Conference on Lasers and Electro-Optics (CLEO) (IEEE, 2007), pp. 1–2
23. Nicholson, J.W., Yan, M.F., Wisk, P., Fleming, J., DiMarcello, F., Monberg, E., Taunay, T., Headley, C., DiGiovanni, D.J.: Raman fiber laser with 81 W output power at 1480 nm. Opt. Lett. **35**, 3069–3071 (2010)

24. Supradeepa, V.R., Nichsolson, J.W., Headley, C.E., Yan, M.F., Palsdottir, B., Jakobsen, D.: A high efficiency architecture for cascaded Raman fiber lasers. Opt. Express. **21**, 7148–7155 (2013)
25. Heeg, T.H.T., Ttenhues, C.H.O., Ayinc, H.A.S., Eumann, J.Ö.R.G.N., Racht, D.I.K.: Core-pumped single-frequency fiber amplifier with an output power of 158 W. Opt. Lett. **41**, 9–12 (2016)
26. Zhang, L., Jiang, H., Cui, S., Hu, J., Feng, Y.: Versatile Raman fiber laser for sodium laser guide star. Laser Photon. Rev. **8**, 889–895 (2014)
27. Zhang, L., Jiang, H., Cui, S., Feng, Y.: Integrated ytterbium-Raman fiber amplifier. Opt. Lett. **39**, 1933 (2014)
28. Suret, P., Randoux, S.: Influence of spectral broadening on steady characteristics of Raman fiber lasers: from experiments to questions about validity of usual models. Opt. Commun. **237**, 201–212 (2004)
29. Feng, Y., Zhang, L., Jiang, H.: Power scaling of Raman fiber lasers. In: Shaw, L.B. (ed.) Spie Lase, pp. 93440U. International Society for Optics and Photonics (2015)
30. Liu, J., Tan, F., Shi, H., Wang, P.: High-power operation of silica-based Raman fiber amplifier at 2147 nm. Opt. Express. **22**, 28383–28389 (2014)
31. Xiao, Q., Yan, P., Li, D., Sun, J., Wang, X., Huang, Y., Gong, M.: Bidirectional pumped high power Raman fiber laser. Opt. Express. **24**, 6758–6768 (2016)
32. Shi, J., Alam, S., Ibsen, M.: Highly efficient Raman distributed feedback fibre lasers. Opt. Express. **20**, 5082–5091 (2012)
33. Westbrook, P.S., Abedin, K.S., Nicholson, J.W., Kremp, T., Porque, J.: Raman fiber distributed feedback lasers. Opt. Lett. **36**, 2895–2897 (2011)
34. Bouteiller, J.C.: Spectral modeling of Raman fiber lasers. Photonics Technol. Lett. IEEE. **15**, 1698–1700 (2003)
35. Taylor, L.R., Feng, Y., Calia, D.B.: 50W CW visible laser source at 589nm obtained via frequency doubling of three coherently combined narrow-band Raman fibre amplifiers. Opt. Express. **18**, 8540–8555 (2010)
36. Feng, Y., Taylor, L.R., Calia, D.B.: 25 W Raman-fiber-amplifier-based 589 nm laser for laser guide star. Opt. Express. **17**, 19021–19026 (2009)
37. Ageorges, N., Dainty, C. (eds.): Laser guide star adaptive optics for astronomy. NATO Adv. Study Inst. **551**, (1997)
38. Feng, Y., Taylor, L., Calia, D.B.: Multiwatts narrow linewidth fiber Raman amplifiers. Opt. Express. **16**, 10927–10932 (2008)
39. Vergien, C., Dajani, I., Zeringue, C.: Theoretical analysis of single-frequency Raman fiber amplifier system operating at 1178nm. Opt. Express. **18**, 26214–26228 (2010)
40. Robin, C., Dajani, I.: Acoustically segmented photonic crystal fiber for single-frequency high-power laser applications. Opt. Lett. **36**, 2641–2643 (2011)
41. Dajani, I., Vergien, C., Robin, C., Ward, B.: Investigations of single-frequency Raman fiber amplifiers operating at 1178 nm. Opt. Express. **21**, 12038–12052 (2013)
42. Vergien, C., Dajani, I., Robin, C.: 18 W single-stage single-frequency acoustically tailored Raman fiber amplifier. Opt. Lett. **37**, 1766–1768 (2012)
43. Yoshizawa, N., Imai, T.: Stimulated Brillouin scattering suppression by means of applying strain distribution to fiber with cabling. Light. Technol. J. **11**, 1518–1522 (1993)
44. Boggio, J.M.C., Marconi, J.D., Fragnito, H.L.: Experimental and numerical investigation of the SBS-threshold increase in an optical fiber by applying strain distributions. J. Light. Technol. **23**, 3808 (2005)
45. Engelbrecht, R., Bayer, M., Schmidt, L.P.: Numerical calculation of stimulated Brillouin scattering and its suppression in Raman fiber amplifiers. In: CLEO/Europe pp. 641. 2003 Conference on (2003)
46. Engelbrecht, R., Mueller, M., Schmauss, B.: SBS shaping and suppression by arbitrary strain distributions realized by a fiber coiling machine. In 2009 IEEE/LEOS Winter Topicals Meeting Series, pp. 248–249. IEEE, New York (2009)

47. Leng, J., Chen, S., Wu, W., Hou, J., Xu, X.: Analysis and simulation of single-frequency Raman fiber amplifiers. Opt. Commun. **284**, 2997–3003 (2011)
48. Zhang, L., Hu, J., Wang, J., Feng, Y.: Stimulated-Brillouin-scattering-suppressed high-power single-frequency polarization-maintaining Raman fiber amplifier with longitudinally varied strain for laser guide star. Opt. Lett. **37**, 4796–4798 (2012)
49. Engelbrecht, R.: Analysis of SBS gain shaping and threshold increase by arbitrary strain distributions. J. Light. Technol. **32**, 1–1 (2014)
50. Zhang, L., Yuan, Y., Liu, Y., Wang, J., Hu, J., Lu, X., Feng, Y., Zhu, S.: 589 nm laser generation by frequency doubling of a single-frequency Raman fiber amplifier in PPSLT. Appl. Optics. **52**, 1636–1640 (2013)
51. Taylor, L., Feng, Y., Calia, D.B., Hackenberg, W.: Multi-watt 589-nm Na D/sub 2/−line generation via frequency doubling of a Raman fiber amplifier: a source for LGS-assisted AO. Proc. SPIE - Int. Soc. Opt. Eng. **6272**, 627249 (2006)
52. Georgiev, D., Gapontsev, V.P., Dronov, A.G., Vyatkin, M.Y., Rulkov, A.B., Popov, S.V., Taylor, J.R.: Watts-level frequency doubling of a narrow line linearly polarized Raman fiber laser to 589 nm. Opt. Express. **13**, 6772–6776 (2005)
53. Holzlöhner, R., Rochester, S.M., Bonaccini Calia, D., Budker, D., Higbie, J.M., Hackenberg, W.: Optimization of cw sodium laser guide star efficiency. Astron. Astrophys. **510**, A20 (2010)
54. Kaenders, W.G., Friedenauer, A., Karpov, V., Protopopov, V., Clements, W.R.L., Taylor, L.R., Feng, Y., Calia, D.B., Ernstberger, B.: Diode-seeded fiber-based sodium laser guide stars ready for deployment. In: Ellerbroek, B.L., Hart, M., Hubin, N., Wizinowich, P.L. (eds.) Adaptive Optics Systems Ii, vol. 7736. Spie-Int Soc Optical Engineering. Bellingham (2010)
55. Enderlein, M., Friedenauer, A., Schwerdt, R., Rehme, P., Wei, D., Karpov, V., Ernstberger, B., Leisching, P., Clements, W.R.L., Kaenders, W.G.: Series production of next-generation guide-star lasers at TOPTICA and MPBC. In: Marchetti, E., Close, L.M., Véran, J.P. (eds.) SPIE Astronomical Telescopes C Instrumentation, pp. 914807. International Society for Optics and Photonics. SPIE, Bellingham (2014)
56. First Light of New Laser at Paranal | ESO. http://www.eso.org/public/announcements/ann15034/
57. $4 Million Laser Marks Ground Zero for Adaptive Optics Science W. M. Keck Observatory. http://www.keckobservatory.org/recent/entry/4_million_laser_marks_ground_zero_for_adaptive_optics_science
58. Four Lasers Over Paranal. http://www.eso.org/public/news/eso1613/
59. Enderlein, M., Kaenders, W.G.: Sodium guide star (R) evolution. Opt. Photonik. **11**, 31–35 (2016)
60. Luo, P.-L., Hu, J., Feng, Y., Wang, L.-B., Shy, J.-T.: Doppler-free intermodulated fluorescence spectroscopy of 4He 23P–31,3D transitions at 588 nm with a 1-W compact laser system. Appl. Phys. B Lasers Opt. **120**, 279–284 (2015)
61. Zhang, L., Jiang, H., Yang, X., Gu, X., Feng, Y.: High-power single-frequency 1336 nm Raman fiber amplifier. J. Light. Technol. **34**, 4907–4911 (2016)
62. Nilsson, J., Sahu, J.K., Jang, J.N., Selvas, R., Hanna, D.C., Grudinin, A.B.: Cladding-pumped Raman fiber amplifier. In: Optical Amplifiers and Their Applications, pp. PD2. OSA (2002)
63. Heebner, J.E., Sridharan, A.K., Dawson, J.W., Messerly, M.J., Pax, P.H., Shverdin, M.Y., Beach, R.J., Barty, C.P.J.: High brightness, quantum-defect-limited conversion efficiency in cladding-pumped Raman fiber amplifiers and oscillators. Opt. Express. **18**, 14705–14716 (2010)
64. Jiang, H., Zhang, L., Feng, Y.: Cascaded-cladding-pumped cascaded Raman fiber amplifier. Opt. Express. **23**, 13947 (2015)
65. Ji, J., Codemard, C.A., Sahu, J.K., Nilsson, J.: Design, performance, and limitations of fibers for cladding-pumped Raman lasers. Opt. Fiber Technol. **16**, 428–441 (2010)
66. Ji, J.: Cladding-pumped Raman fibre laser sources. University of Southampton (2011)
67. Junhua, J., Codemard, C.A., Nilsson, J.: Analysis of spectral bendloss filtering in a cladding-pumped W-type fiber Raman amplifier. J. Light. Technol. **28**, 2179–2186 (2010)

68. Codemard, C.A., Ji, J., Sahu, J.K., Nilsson, J.: 100 W CW cladding-pumped Raman fiber laser at 1120 nm. Proc. SPIE - Int. Soc. Opt. Eng. **7580**, 75801N (7) (2010)
69. Terry, N.B., Alley, T.G., Russell, T.H.: An explanation of SRS beam cleanup in graded-index fibers and the absence of SRS beam cleanup in step-index fibers. Opt. Express. **15**, 17509–17519 (2007)
70. Namiki, S., Emori, Y.: Ultrabroad-band Raman amplifiers pumped and gain-equalized by wavelength-division-multiplexed high-power laser diodes. IEEE J. Sel. Top. Quantum Electron. **7**, 3–16 (2001)
71. Kablukov, S.I., Dontsova, E.I., Zlobina, E.A., Nemov, I.N., Vlasov, A.A., Babin, S.A.: An LD-pumped Raman fiber laser operating below 1 μm. Laser Phys. Lett. **10**, 85103 (2013)
72. Babin, S.A., Dontsova, E.I., Kablukov, S.I.: Random fiber laser directly pumped by a high-power laser diode. Opt. Lett. **38**, 3301 (2013)
73. Yao, T., Nilsson, J.: 835 nm fiber Raman laser pulse pumped by a multimode laser diode at 806 nm. J. Opt. Soc. Am. B. **31**, 882 (2014)
74. Yao, T., Harish, A., Sahu, J., Nilsson, J.: High-power continuous-wave directly-diode-pumped fiber Raman lasers. Appl. Sci. **5**, 1323–1336 (2015)
75. Glick, Y., Fromzel, V., Zhang, J., Dahan, A., Ter-Gabrielyan, N., Pattnaik, R.K., Dubinskii, M.: High power, high efficiency diode pumped Raman fiber laser. Laser Phys. Lett. **13**, 65101 (2016)
76. Glick, Y., Fromzel, V., Zhang, J., Ter-Gabrielyan, N., Dubinskii, M.: High-efficiency, 154 W CW, diode-pumped Raman fiber laser with brightness enhancement. Appl. Optics. **56**, B97 (2017)
77. Dawson, J.W., Messerly, M.J., Beach, R.J., Shverdin, M.Y., Stappaerts, E.A., Sridharan, A.K., Pax, P.H., Heebner, J.E., Siders, C.W., Barty, C.P.J.: Analysis of the scalability of diffraction-limited fiber lasers and amplifiers to high average power. Opt. Express. **16**, 13240–13266 (2008)
78. Zhu, J., Zhou, P., Ma, Y., Xu, X., Liu, Z.: Power scaling analysis of tandem-pumped Yb-doped fiber lasers and amplifiers. Opt. Express. **19**, 18645–18654 (2011)
79. Zhang, L., Jiang, H., Yang, X., Pan, W., Cui, S., Feng, Y.: Nearly-octave wavelength tuning of a continuous wave fiber laser. Sci. Rep. **7**, 42611 (2017)
80. Dawsona, J.W., Messerlya, M.J., Heebnera, J.E., Paxa, P.H., Sridharana, A.K., Bullingtona, A.L., Beacha, R.J., Sidersa, C.W., Bartya, C.P.J., Dubinskiib, M.: Power scaling analysis of fiber lasers and amplifiers based on non-silica materials. In: Proc. of SPIE, vol. 7686, pp. 768611
81. Ward, B.: Solid-core photonic bandgap fibers for cladding-pumped Raman amplification. Opt. Express. **19**, 11852–11866 (2011)
82. Naderi, S., Dajani, I., Grosek, J., Madden, T.: Theoretical and numerical treatment of modal instability in high-power core and cladding-pumped Raman fiber amplifiers. Opt. Express. **24**, 16550 (2016)

Chapter 2
Cascaded Raman Fiber Lasers

V. R. Supradeepa and Jeffrey W. Nicholson

2.1 Introduction

In the last decade, there has been a tremendous development in high-power fiber lasers with output powers exceeding several kWs from a single fiber [1]. The primary technology responsible for this has been ytterbium-doped fiber lasers. This is due to several inherent material advantages in ytterbium-doped lasers such as low quantum defect and the ability to dope ytterbium into fibers at high concentrations without losing efficiency. However, the same level of performance has not been obtained with other rare-earth dopants. This has resulted in power scaling of fiber lasers limited to the narrow emission region of ytterbium. This is a significant limitation since many applications rely upon the diversity of emission wavelengths. Figure 2.1 shows the status of power scaling of the primary fiber laser technologies of today. Corresponding to the rare-earth dopant, each laser has a specific emission window. The window of emission wavelengths from ytterbium-doped fiber lasers is limited to a small band (1050–1120 nm). In this region, over 10 kW of power has been demonstrated in power combined systems, and in more conventional modules, it is now common to have over 3 kW of power. Other dopants such as thulium/holmium (1900–2100 nm) and erbium and erbium-ytterbium (1530–1590 nm) partly complement the emission range. Thulium lasers have achieved kW class power levels [2], and the erbium-based lasers are limited to a few 100 W [3]. Power scaling aspects as well as other important laser parameters such as efficiency and reliability of these lasers have significantly lagged ytterbium-doped fiber lasers.

V.R. Supradeepa (✉)
Centre for Nano Science and Engineering, Indian Institute of Science, Bangalore 560012, India
e-mail: supradeepa@cense.iisc.ernet.in

J.W. Nicholson
OFS Laboratories, 19 Schoolhouse Road, Somerset, NJ 08873, USA
e-mail: jwn@ofsoptics.com

© Springer International Publishing AG 2017
Y. Feng (ed.), *Raman Fiber Lasers*, Springer Series in Optical Sciences 207,
DOI 10.1007/978-3-319-65277-1_2

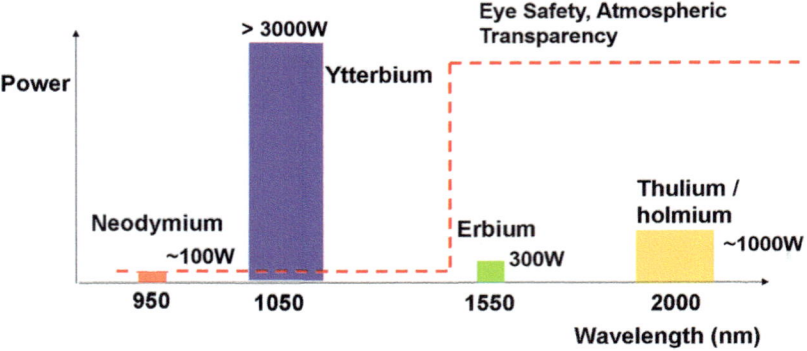

Fig. 2.1 Status of power scaling of various fiber laser technologies

The reasons for this include higher quantum defect, difficulty in enhanced doping, quenching, and ion pair formation.

In addition, there are substantial white spaces in the wavelength spectrum where no suitable rare-earth dopant is available which can function at high powers. This is a significant constraint since several desirable attributes for high-power applications are available at alternate wavelength bands. One such important attribute is eye safety. The maximal permissible exposure to the human eye at the 1.5 micron wavelength is several orders (three to four orders) of magnitude higher (depending on the pulse width) than the 1 micron region [4]. Another important application involving lasers would be free space propagation. This can be in the context of free space communications, LIDAR, sensing, or directed energy. For these applications, again, 1 micron region is not the best suited since the propagation losses in the atmosphere are substantially higher compared to other bands at longer wavelengths [5]. Other applications include acting as pump sources to other nonlinearity-constrained lasers such as pulsed lasers or narrow-linewidth lasers. These pump sources need to be in high brightness while having emission windows in the absorption region of the lasers which can be in the existing white spaces [6]. For the reasons mentioned above, it is imperative to develop lasers which can provide high optical powers, in a scalable fashion, in a variety of wavelength bands.

Cascaded Raman fiber lasers are an agile, fiber laser technology that are scalable and provide high optical powers at various wavelength bands [7]. Figure 2.2 shows the schematic of a cascaded Raman fiber laser. The fundamental operating principle is to wavelength convert available high-power fiber lasers to inaccessible wavelengths using stimulated Raman scattering in optical fibers. However, since one Raman shift is insufficient to vary the wavelength substantially, a series of cascaded Raman shifts is utilized, thus, the name cascaded Raman fiber lasers. Frequently, ytterbium lasers serve as the primary input sources to cascaded Raman fiber lasers. This is anticipated from our previous discussions which pointed to ytterbium fiber lasers as the only source which has achieved tremendous power scaling. Recently, the use of thulium lasers as the starting wavelength has also started. In this chapter,

Fig. 2.2 Schematic of cascaded Raman fiber lasers utilizing cascaded Raman conversion

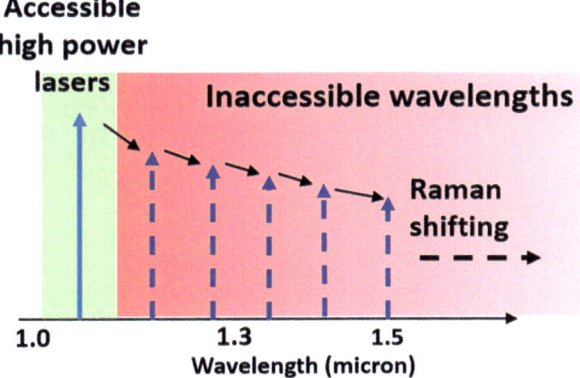

we will largely confine to work based on wavelength conversion of ytterbium lasers, though the general principles are equally applicable to any starting wavelength. There has been extensive development in cascaded Raman lasers in the decade. This includes new laser architectures, scaling output power, improving efficiency, enhancing wavelength diversity, and improving spectral quality. In this chapter, we will first look into the general principles of Raman lasers followed by some of these recent developments.

2.2 Conventional Implementation of Cascaded Raman Fiber Lasers

Figure 2.3 shows the schematic of a conventional implementation of cascaded Raman fiber lasers. An ytterbium-doped fiber laser pumps a Raman conversion module referred to as a cascaded Raman resonator. A cascaded Raman resonator has several nested cavities comprised of fiber Bragg gratings at each of the intermediate wavelengths, initial and final wavelengths. Often, all the gratings of one side are fabricated on a single piece of fiber to minimize splice losses, and these are collectively referred to as the Raman input grating (RIG) and Raman output grating (ROG) sets. Between the grating sets, a low effective area (high nonlinearity) fiber, frequently referred to as Raman fiber, is utilized. Intermediate wavelengths in the cascaded resonator are selected at the peak of the Raman gain of the wavelength preceding it. The cascade of Raman conversion is terminated with a low reflectivity output coupler in the Raman output grating side. At sufficient input powers, the output will primarily be comprised of the final wavelength with a small fraction of light in all the intermediate wavelengths.

In the example shown in Fig. 2.3, a 1117-nm-ytterbium-doped fiber laser is used as the pump source, and it converts light to an output wavelength of 1480 nm. Representative applications for high-power 1480 nm include the use as pump

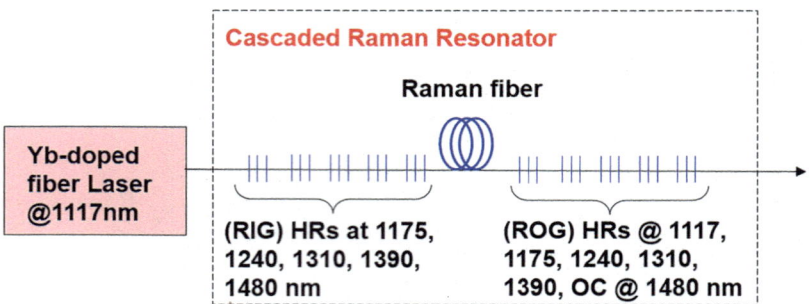

Fig. 2.3 Schematic of a cascaded Raman fiber laser converting 1117–1480 nm (five Raman shifts) (RIG, Raman input grating set; ROG, Raman output grating set)

sources for core pumping of erbium-doped fiber amplifiers. We will discuss this application in more detail at a later stage. In addition, the laser becomes significantly more eye safe and becomes very attractive for a variety of laser applications involving people.

The three key components in a cascaded Raman resonator are the Raman fibers which create the Raman gain, Raman input and output grating sets, and at higher powers, components for isolation between the rare-earth-doped fiber laser and the Raman convertor. The necessity for the first two components is anticipated; however, the third component becomes necessary as we scale power. Fundamentally, the cascaded Raman laser involves two cavities, one from the ytterbium-doped fiber laser and the other from the cascaded Raman resonator. As we will discuss later, beyond a threshold power, coupling between the two can result in destabilization of the laser. This deleterious effect is overcome with the use of isolation components. In later sections, we will discuss modified architectures for cascaded Raman conversion which can achieve higher efficiency than the conventional implementation while simultaneously providing higher reliability. The three key components discussed here, however, continue to be the case even for the modified architectures.

2.2.1 Raman Fibers

Optical fibers utilized to provide Raman gain are characterized by a frequency shift associated with the Raman process and a need to satisfy the following properties – high Raman gain, low loss necessary to sustain long propagation distances, and the ability to minimize other nonlinear effects such as four-wave mixing, supercontinuum generation, etc. These aspects necessitate the following requirements from the optical fiber used for such a purpose:

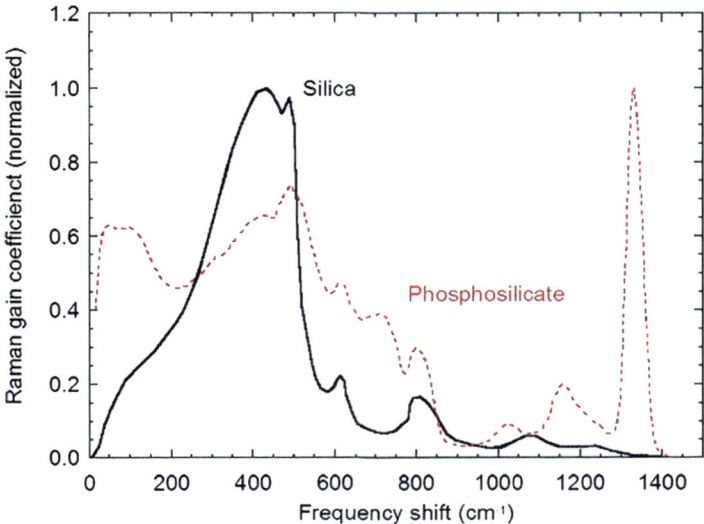

Fig. 2.4 Raman gain coefficients in conventional silica fibers and phosphosilicate fibers (From Ref. [8])

- Frequency shift – This refers to the frequency difference between the pump wavelength and the Raman-shifted component. This is characterized by the peak Raman gain as a function of frequency offset from the carrier and primarily depends on the material composition. Silica-based fibers have a frequency shift of ~13 THz. An alternative fiber such as phosphosilicate fiber has a peak gain close to ~40 THz. Having a larger Raman shift is often a desirable attribute since this necessitates fewer shifts to reach the final wavelength. A comparison between the Raman gain coefficients of conventional silica fibers and phosphosilicate fibers is shown in Fig. 2.4. However, though the peak of Raman gain might be at a specific offset, at higher powers, lasing can occur at all frequencies with sufficient Raman gain. Phosphosilicate fiber-based lasers, thus, have not been used at higher powers since, as seen from the gain plot, there is substantial gain even at much lower frequency offsets.
- Effective area and Raman gain coefficient – Since stimulated Raman scattering is an intensity-dependent effect, Raman fibers tend to have lower effective area. Lower effective area is achieved by utilizing a smaller core and maintaining robust single-modedness of the core; this also requires higher numerical apertures. This is an added advantage since higher numerical apertures are achieved by enhanced Germania doping, and this also enhances the Raman gain coefficient compared to pure silica fibers. Conventionally, fibers used in Raman fiber lasers have an effective area between 10 and 15 micron^2 at 1100 nm. This however is not a strict guideline since, given the sufficient power, the mode-field area

requirement of the fiber can be relaxed. However, smaller fibers work better for the following dispersion aspect.

- Dispersion – A high-power laser propagating in a fiber in the anomalous dispersion region can generate a continuous wave supercontinuum [9]. As we will discuss this in later sections, this occurs due to combination of modulational instability, four-wave mixing, and a variety of other nonlinear effects. Even in the absence of supercontinuum generation, in the case of low dispersion (anomalous or normal), since four-wave mixing is better phase-matched, substantial line broadening can occur [10]. These effects are highly counterproductive to the cascaded Raman conversion. In the case of supercontinuum generation, the Raman cascade is lost. In the case of substantial line broadening, the efficiency of conversion to the next Raman shift reduces, and this manifests as a lower fraction of the power in the final wavelength or line-broadened output wavelength or inability of the laser to reach the final wavelength. It is preferable, thus, for Raman fibers to have a large normal dispersion. Conventional Raman fibers [11] have a dispersion coefficient, as high as -80 ps/nm/km in the wavelength region of interest. This is naturally enabled using small effective area fibers with higher-core numerical aperture. Conventional transmission-type silica fibers achieve zero dispersion at $\sim$1300 nm which sets a limit to the amount of wavelength conversion. However, the small effective fibers, owing to a greater waveguide contribution to their dispersion behavior, can have zero-dispersion wavelengths at much longer wavelengths.
- Loss – Raman fibers typically have comparable loss to standard single-mode fibers used in optical communications. For analyzing performance of Raman lasers, a better figure of merit would be 1/(loss × effective area) since the net Raman gain increases with reduced area necessitating a shorter length of Raman fiber needed for the conversion. As the loss numbers are similar, the figure of merit for Raman fibers is much higher. This discussion, however, is primarily limited to wavelength's windows where silica as a material is very transparent. However, it is very lossy in the mid-IR region and thus cannot be used for mid-IR Raman lasers. Currently, Raman lasers operate in two main wavelength bands. In the near-IR region, silica fibers are primarily used. In the mid-IR region, fluoride or chalcogenide fibers are used [12, 13].

2.2.2 Cascaded Raman Resonator Gratings

Raman grating sets are like Bragg gratings used in fiber lasers [14]. Specific details for Raman resonators are:

- Fibers – As it was discussed previously, the fibers used for Raman conversion tend to be a small effective area. If gratings are fabricated in standard fibers, the splice losses between the two fibers can result in reduction of efficiency. Thus, the Raman gratings are fabricated in the same type of fiber used to provide the Raman gain.

- Gratings in sets – Raman grating sets include multiple gratings, and with independent fabrication of each grating, efficiency would be compromised due to multiple splices. Thus, all gratings needed for the Raman resonator on one side are fabricated together on a single fiber. Losses can be further reduced by having the Raman input and output grating sets fabricated directly on the Raman gain fiber to make it a splice-less monolithic entity.
- Power handling – The process for grating fabrication should be capable in handling high optical powers. Fiber fuse [15] is a dangerous aftereffect of faulty gratings.

2.2.3 Components for Isolation

A cascaded Raman fiber laser as discussed before is a configuration involving two cavities – one for the rare-earth-doped fiber laser and the other for the cascaded Raman resonator. In such circumstances, it is possible to expect instabilities due to coupling between the cavities. However, as with most instabilities, until a certain power threshold is reached, there is no issue. When operated at lower-power levels (10s of W), the lasers operate well [16, 17]. It was observed in [18] that even at 40 W output power, a 1480-nm Raman laser could operate stably without isolation between the cavities.

The stability of the system at higher-power levels degrades due to coupling between the cavities which occurs in the form of feedback from the Raman cavity into the rare-earth-doped fiber laser [19, 20]. It is observed that feedback at a wavelength one Raman shift away from the input wavelength back into the rare-earth-doped laser leads to lasing at this wavelength as well in the rare-earth cavity. This initially causes temporal fluctuations [20] and, if pushed further, leads to complete laser failure.

For reliable, high-power operation, the rare-earth fiber laser cavity should be isolated from the backward propagating Raman-shifted wavelengths. An optical isolator in principle could be used. However, fiber-coupled optical isolators currently have power handling significantly lower than the output power of fiber lasers. The maximum power handling of commercially available, fiber-coupled isolators is currently around 50 W, while ytterbium-doped fiber lasers have already crossed a kW. In addition, the isolator must provide low loss at the signal wavelength and high isolation at a wavelength one Raman shift away. Such broadband performance is still lacking in current high-power isolators.

Fortunately, a frequency-selective element can be used to isolate the rare-earth laser cavity from the cascaded Raman cavity. A fused fiber wavelength division multiplexer has been found to be the most effective. Fused biconical tapers [21, 22] can provide wavelength-selective directional couplers with very low loss, an important characteristic for high-power fiber laser operation. Single-mode WDMs provide low-loss throughput for the high-power rare-earth laser output to the Raman laser cavities, while diverting the backward propagating Raman-shifted light away from the rare-earth cavity. These WDMs have been demonstrated to isolate the rare-earth cavity from the backward Stokes lasing to provide stable operation [19] and have been successfully operated with more than 400 W of input power [23].

Another option for wavelength-selective isolation of the backward Stokes is UV-inscribed gratings. A long-period grating (LPG) provides phase matching between modes in an optical fiber. To create loss (filtering) at a particular wavelength, phase matching can be achieved between the fundamental mode and a cladding mode at that wavelength, which can then be stripped out to provide a narrowband rejection filter [24]. LPGs have low insertion loss and high (20 dB) in-band rejection making them suitable for inter-cavity isolation. However, unlike WDMs which remove rejected light through a single-mode fiber, an LPG dumps rejected light into the fiber cladding, requiring careful thermal management at high-power operation. LPGs have also been demonstrated successfully as a means for isolating the ytterbium laser from the backward Stokes light [20].

Tilted fiber Bragg gratings (FBGs) are another potential possibility for isolating the rare-earth and Raman cavities. Unlike conventional FBGs, the index modulation that forms the grating is written at an angle with respect to the fiber core [14, 25]. In tilted FBGs, the reflected wavelengths are ejected out of the side of the fiber, rather than backward in the optical core. While tilted FBGs have not yet been used for cavity isolation in Raman fiber lasers, they have been demonstrated to remove other unwanted Stokes lines from a high-power Raman fiber laser [26].

It is worth noting that in architectures (described in more detail in later sections) that do not have a Raman cavity, such as cascaded Raman amplifiers [23], the isolation is not strictly necessary, and these architectures are inherently more stable.

2.3   Design of Conventional Cascaded Raman Fiber Lasers

2.3.1   Numerical Modeling of Cascaded Raman Fiber Lasers

In a cascaded Raman resonator, there is light at each of the intermediate Raman-shifted wavelengths and in the input and the final wavelength. Using a subscript to denote the forward or backward nature, the equation for cascaded Raman lasers can be written as [16, 27]:

$$
\frac{dP_k^{b,f}}{dz} = \pm \alpha_k P_k^{b,f} \mp \sum_{j<k} \frac{g_R(j,k)}{A_{\mathrm{eff}}^j} \left(P_j^f + P_j^b \right) P_k^{b,f} \pm \sum_{j>k} \frac{\upsilon_k g_R(k,j)}{\upsilon_j A_{\mathrm{eff}}^j} \left(P_j^f + P_j^b \right) P_k^{b,f}
$$
$$
\mp \sum_{j<k} \frac{g_R(j,k)}{A_{\mathrm{eff}}^j} h\upsilon_k \Delta\upsilon\,(j) \left(P_j^f + P_j^b \right) \pm \sum_{j>k} \frac{\upsilon_k g_R(k,j)}{\upsilon_j A_{\mathrm{eff}}^j} h\upsilon_j \Delta\upsilon\,(j) P_j^{b,f}
$$

Here, $g_R(j,k)$ is the Raman gain coefficient which connects the 'j'th and 'k'th components. In conventional cascaded Raman lasers, 'j' and 'k' differ by one (i.e.,

Raman gain for components separated by more than one Raman shift is negligible). A^j_{eff} is the effective area for the 'j'th component and α_k is the fiber loss coefficient for the 'k'th component. The transmission spectrum of the fiber can be incorporated through the fiber loss term. $\Delta\upsilon$ is the bandwidth of spontaneous emission needed to start the process and υ is the frequency of each wave. The above equations are generic and they incorporate both the forward and backward components. To solve the above equations, we need the boundary conditions. For cascaded Raman resonators, the boundary conditions can be written as:

$$P^f_k(0) = R^f_k P^b_k(0), P^b_k(L) = R^b_k P^f_k(L)$$

Here, "k" represents the Stokes component with $k = 1$ being the pump and in an "n" Raman cascade the "nth" component will be the final signal. "L" is the length of the Raman fiber. "R" is the reflectivity provided by the fiber Bragg gratings. In a conventional cascaded Raman laser, all the reflectivities are close to 1 except for the two following components:

$$R^f_1 = 0, \quad R^b_n = R_{oc}$$

The first condition indicates that the pump is coupled straight into the cavity (there is no grating at the pump wavelength on the input side), and the second condition indicates that the final Stokes component has a lower reflectivity output coupler (OC). The output power and efficiency is given by:

$$P_{\text{out}} = (1 - R_{oc}) P^f_n(L)$$

$$\eta = \frac{P_{\text{out}}}{P_{\text{in}}} = \frac{(1 - R_{oc}) P^f_n(L)}{P^f_1(0)}$$

2.3.2 Optimization of Resonator Components

The above equations, in a general case, owing to the presence of both forward and backward components, needs to be solved numerically. The goal of optimization is to maximize the efficiency of the Raman conversion process. This is conventionally done by tuning the following parameters to maximize efficiency:

- Fiber length – The optimal length of fiber necessary primarily depends on two aspects – the input power into the cascaded Raman resonator and the number of Raman shifts in the cascaded Raman resonator. At low powers and several cascades, the length of fiber can run into kms. This indicates the value of having smaller effective area fibers so that the total length can be reduced.
- Grating reflectivity – Since all gratings except one in the cascaded Raman resonator are high reflectivity (HR) gratings which have reflectivities close to 1,

optimization is specifically aimed at the reflectivity of the output coupler (OC). The reflectivity of the output coupler will also closely depend on the length of the fiber, the number of cascades in the Raman laser, and the operating power. Enhanced reflectivity of the output coupler will result in lowering the threshold for the Raman laser. However, this comes at the cost of laser efficiency. Thus, a compromise is usually necessary, as with most lasers, between the threshold and efficiency at high powers. In high-power Raman lasers, it is common to have output coupler reflectivity less than 20%.

- Grating bandwidths – In addition to Raman gain, the long lengths and low effective area of the fiber enhances other nonlinear processes such as four-wave mixing and self-phase modulation. This results in line broadening of each of the components inside the cavity. Unless the bandwidth of each of the gratings is substantial enough, there can be light leakage outside of the band of the gratings which results in incomplete conversion and reduced efficiency. The gratings used in the cascaded Raman resonator routinely have bandwidths well above 1 nm. Wider bandwidths come at the cost of longer storage time of light inside the gratings. This can cause reliability issues and needs to be monitored.

Regarding assembly and operation, a Raman laser has a variety of optical fibers. It can be conventional or large-mode-area fibers in the rare-earth-doped laser, standard fibers in the fused fiber components such as the WDMs used for isolation, and small effective area fibers in the Raman cavity. This results in the need for several splices between components of the cavity. Suboptimal splices can lead to significant reduction in efficiency. Splicing between dissimilar fibers has been extensively researched [28], and sufficient care should be given to optimize all the splices before the laser can be assembled.

2.4   Cascaded Raman Fiber Lasers Using Filter Fibers

In this section, we will describe the first of the recent advances in the technology of cascaded Raman fiber lasers. The advance relates to the issue of enhancement of efficiency which also enables power scaling. A significant limitation of early high-power Raman lasers that used conventional Raman fiber in the cavity was unwanted scattering into the next Raman order. For example, in the case of 1480-nm Raman lasers, the next Raman order occurs at 1583 nm [18]. It was seen that, for a configuration resulting in the optimal efficiency for 1480 nm (the final wavelength), the next unwanted wavelength at 1583 nm was only 15 dB below at 41 W of output power. The 1583-nm peak could be reduced to 30 dB below the 1480 peak, by shortening the fiber length, but with a negative impact on conversion efficiency. Furthermore, as the fiber length was shortened, a smaller fraction of light at the output was contained at 1480 nm, with more residual radiation at the intermediate Stokes wavelengths, ultimately reducing the amount of output power at 1480 nm. On the flip side, if the fiber length were to be further increased, it would have resulted in further growth of the unwanted 1583-nm component, further resulting in the reduction of efficiency.

In order to maintain a long fiber length in the Raman cavity while suppressing scattering to the next Raman order, a Raman filter fiber (RFF) was designed and fabricated. The RFF was a germanosilicate fiber that used a fundamental mode cutoff to achieve distributed loss at wavelengths longer than 1480 nm. Depressed clad fibers with fundamental mode cutoff have provided distributed loss for short wavelength erbium-doped fiber amplifiers [29] as well as for Raman suppression in Yb-doped fiber lasers [30]. The RFF thus provides both cascaded Raman gain up to 1480 nm and distributed loss at longer wavelengths. The wavelength where the loss increases depends on both the scaling of the refractive index as well as the bend diameter. For example, an RFF was drawn to multiple diameters to find the fiber with the optimal cutoff wavelength. The measured loss curves of two different diameter RFFs are shown in Fig. 2.5a, and compared to a conventional Raman gain fiber. At short wavelengths, from 1000 to 1480 nm, the loss of the RFF and conventional Raman fiber were similar, but at long wavelengths, the RFF loss rapidly increased.

Fig. 2.5 (a) Loss of the Raman filter fiber compared to a conventional Raman gain fiber. (b) Calculated and measured dispersion of the Raman filter fiber (From Ref. [20])

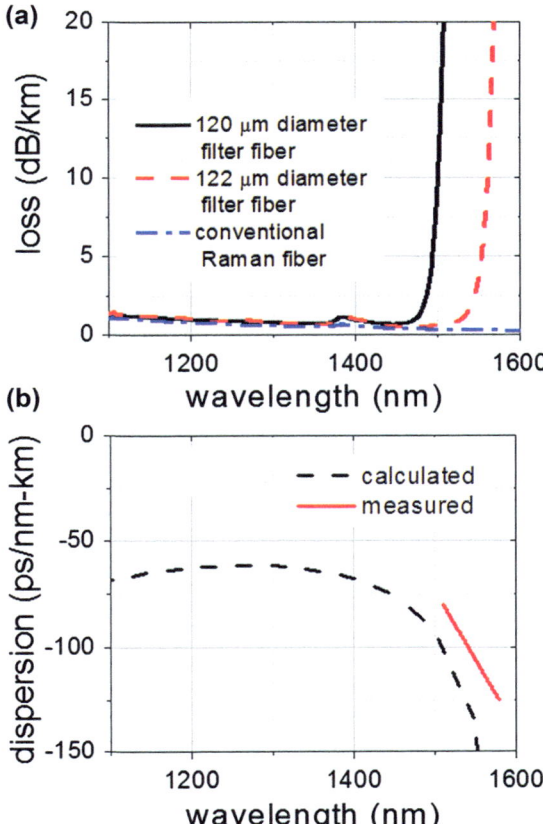

Another consideration which we have discussed before is the dispersion of the Raman gain fiber over the operating wavelength range. If the dispersion of the Raman gain fiber were to become anomalous, modulation instability would lead to supercontinuum generation at high powers [31]. Therefore, the RFF must have normal dispersion over the entire operating range. As an example, a 1480-nm Raman fiber laser pumped by a 1050-nm Yb fiber laser requires normal dispersion from 1050 to 1600 nm. The dispersion calculated for a normal dispersion filter fiber from the measured index profile of the fiber, as well as the measured dispersion, is shown in 5(b). It is seen that the dispersion is strongly normal and well below −50 ps/nm/km across the band. This, as described before, also has the additional benefit of reducing line broadening of all the intermediate Raman components as well as the final wavelength. This has a positive effect on the net efficiency.

The introduction of Raman filter fibers into cascaded Raman lasers has led to considerable increases in output power from 1480-nm Raman fiber lasers [20, 23, 32, 33]. In addition, because the distributed loss suppresses the unwanted scattering to the next Stokes order, the nonlinearity in a Raman cavity with filter fiber can be driven harder than in a comparable cavity based on conventional Raman fiber. As a result, Raman lasers based on filter fiber have higher efficiency and a higher fraction of in-band power compared to Raman lasers based on conventional Raman fiber. For example, for a pump power of 250 W, 81 W of 1480-nm output power was demonstrated [20] for an optical to optical conversion efficiency of 32% from a cavity with Raman filter fiber. In comparison, a laser with conventional Raman fiber required 175 W of pump power for 41 W of 1480-nm output power or 23% conversion efficiency [16]. The increased conversion efficiency could be attributed to the Raman filter fiber, which allowed long lengths of Raman fiber to be used without being limited by scattering to 1590 nm. Using filter fiber in the conventional cascaded Raman laser architecture, power was further scaled to above 100 W (104 W in [32]) with a further enhanced conversion efficiency of 35%. Despite the improvement in efficiency with the use of filter fibers, the efficiency of the cascaded Raman lasers is still substantially lower than the fundamental quantum-limited efficiency. For example, the achieved efficiency of 35% in for a 1480-nm laser pumped using 975-nm diodes [32] is still reasonably lower than the quantum-limited efficiency of $\sim$66%. In the next section, a new architecture would be described which aims at overcoming this gap.

2.5 High-Efficiency Raman Lasers Using Cascaded Raman Amplifiers

Efficiency reduction in conventional cascaded Raman lasers occurs because of the following reasons – propagation losses in the grating sets [34], intra-cavity losses between often dissimilar fibers, transmission loss due to scattering and absorption in the Raman fiber, and enhanced backward and forward light leaking from the cavity

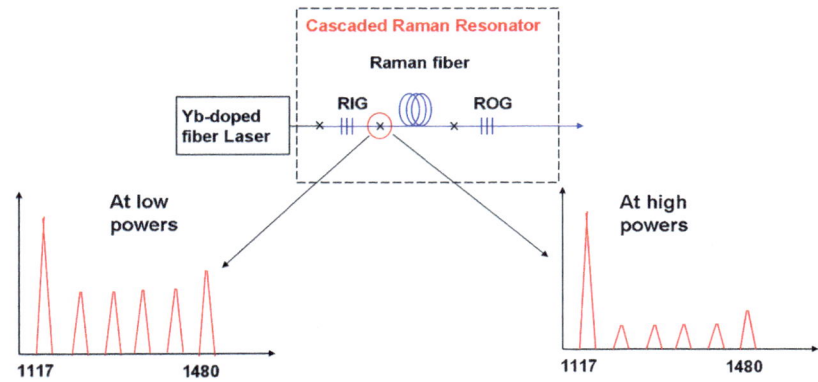

Fig. 2.6 Qualitative representation of spectra for a fifth order cascaded Raman fiber laser at lower and higher powers on the input side

at the intermediate wavelengths due to their bandwidth being higher than the grating bandwidths [16, 27, 35].

Most of the loss mechanisms identified above occur in the cascaded Raman resonator assembly. Figure 2.6 shows a schematic of the spectral content of light at the start of the cascaded Raman resonator. Such a measurement can be done by analyzing the light scattered at the splice spectrally. At low powers, there is a sizable fraction of light at all intermediate Stokes wavelengths. However, at higher powers, this fraction reduces considerably. This suggests that, instead of a resonator, a single pass cascaded Raman amplifier can work equally well for high-power Raman lasers as long as it is sufficiently seeded at all the intermediate Stokes wavelengths. Seed powers at all the intermediate wavelengths reduce the gain requirement, provide wavelength selectivity, and enable preferential forward Raman scattering. Since this method avoids most of the losses inherent to the cascaded Raman resonator assembly, the efficiency and power scaling is anticipated to be substantially better.

For high-power Raman lasers based on the cascaded Raman amplifier, there are two key requirements. First is a multiwavelength source which can simultaneously provide sufficient powers at all the intermediate wavelengths. A lower-power conventional cascaded Raman laser lends itself ideally for this purpose. Light present at the output at all the intermediate wavelengths provides sufficient seed power at the exact required wavelengths. Second is that, scattering of the output wavelength to the next Raman order can be higher in a single pass configuration compared to a cascaded resonator configuration. The use of Raman filter fiber discussed previously eliminates this problem and provides an ideal technique to terminate the cascade of wavelength conversion. Figure 2.7a shows the experimental setup based on the new architecture. A high-power Yb-doped fiber laser is combined with a lower-power Raman seed laser (spectrum shown in Fig. 2.7). This is then sent through gain medium comprising the Raman filter fiber (a representative transmission spectrum is shown in Fig. 2.7c).

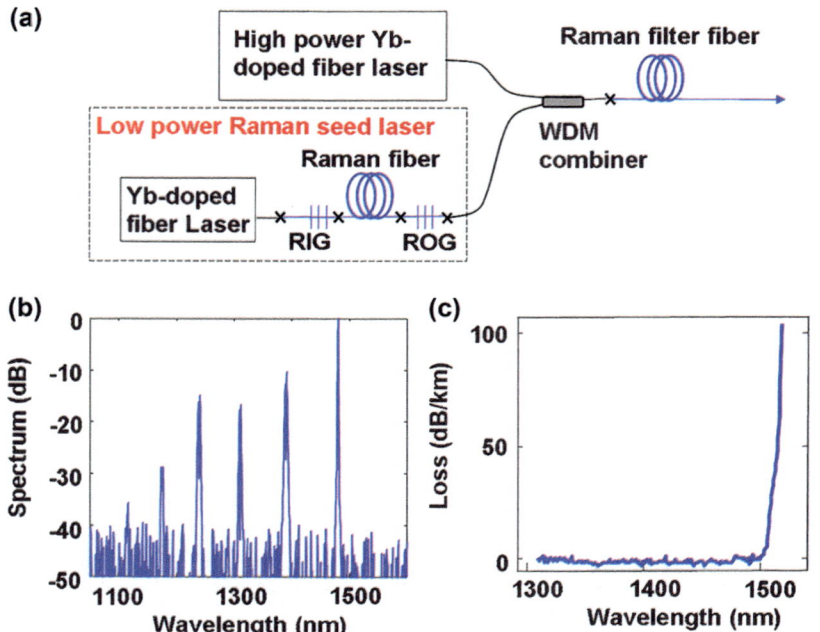

Fig. 2.7 (a) Schematic of the Raman laser based on the cascaded Raman amplifier architecture. (b) Spectrum of the seed source. (c) Transmission spectrum of the Raman filter fiber (From Ref. [23])

Figure 2.8 shows the results from an experiment to utilize the cascaded Raman amplifier architecture to develop a fifth order laser from 1117 to 1480 nm. The total output power and the 1480-nm component as a function of input power at 1117 nm is shown in Fig. 2.8a. An output power of ~301 W at 1480 nm is achieved (limited only by input power) for a total input power of ~470 W (including the Yb-doped fiber laser in the seed source) with a total conversion efficiency of ~64% (1117 nm to 1480 nm, quantum-limited efficiency of 75%). Figure 2.8b shows the measured output spectrum at full power (log scale in the inset). More than 95% of the power is in the 1480-nm band indicating a high level of wavelength conversion, while high suppression of the next Stokes order at 1590 nm is maintained using the filter fiber. The efficiency of ytterbium-doped fiber lasers can be 75% or better. This would result in the net system having an efficiency of close to 50%. This is a substantial improvement over the previous architectures. A comparison to this system would be the most common method to develop high optical powers in the 1.5-micron wavelength region using erbium-ytterbium co-doped fiber lasers. The conversion efficiency in [3] was <25% and decreasing due to parasitic lasing. Thus, compared to alternatives, Raman lasers are already more efficient and even more so when the high-efficiency architecture is considered. The reduction in efficiency below the quantum limit can be accounted for by the efficiency of the ytterbium-

Fig. 2.8 (**a**) Output power and component at 1480 nm for a Raman laser based on a cascaded Raman amplifier. (**b**) Output spectrum (in linear and log scale) (From Ref. [23])

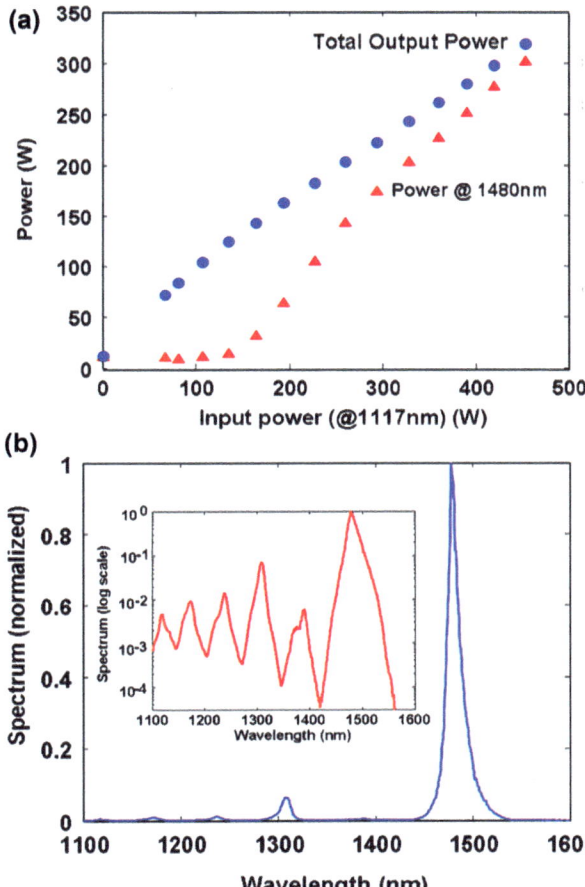

doped laser, loss of the WDM combiner, splices, and residual power in the other Stokes components. This also provides suitable directions to further enhance the performance.

To illustrate the enhancement in efficiency compared to the conventional architecture, the results obtained from this scheme [33] was compared to results in [32] where 104 W of output power was obtained. Compared to results shown in Fig. 2.8, the length of the Raman filter fiber here is longer to achieve higher efficiencies at lower input powers. The power from the seed source in this experiment is ∼3 W at 1480 nm and 4 W in total. Figure 2.9a compares the output power at 1480 nm for the two architectures. Also shown is the expected power assuming quantum-limited conversion. At lower powers, power from the new architecture is low. This is expected since the cascade is not driven all the way to 1480 nm. However, once the threshold to reach this is achieved, power rapidly grows and at full power is significantly higher than that from the conventional cascaded Raman laser. At the

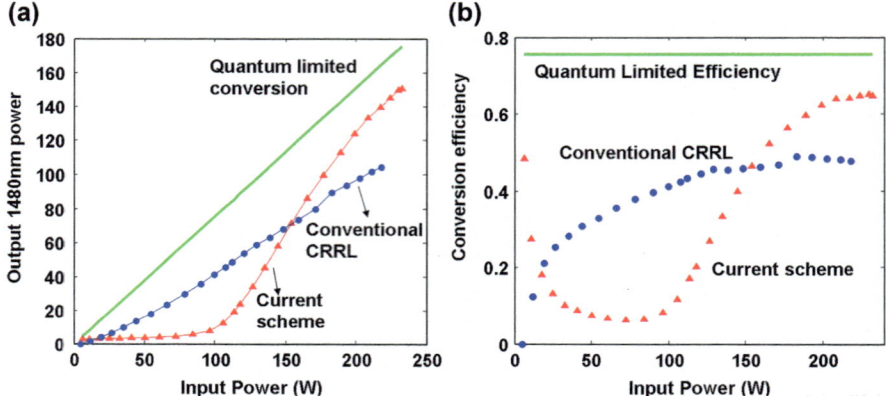

Fig. 2.9 (**a**) Comparison of the input power to the output power for different cascaded Raman laser architectures for conversion from 1117 to 1480 nm. (**b**) Corresponding conversion efficiencies (From Ref. [33])

maximum power point for the conventional Raman laser, the new scheme provides 40% more power. Figure 2.9b shows the corresponding conversion efficiencies. At low power, the current scheme's output is dominated by the seed source and has the corresponding ~48% conversion efficiency. With increasing powers, but prior to threshold, this efficiency drops. Beyond the threshold, however, the efficiency quickly recovers and achieves a maximum of ~65%.

Figure 2.10 shows the behavior of the cascaded Raman amplifier more intuitively. The total output power and components at each intermediate Raman wavelength measured as a function of input power at 1117 nm are plotted. A progressive growth and decay of all the intermediate Stokes components with increasing power is clearly observed. A rapid growth of the final output wavelength is seen beyond a power threshold. The filter fiber ensures there is no further conversion of the output wavelength. At maximum power, a high degree of wavelength conversion with most of the output power being in the final, 1480-nm component is seen. The initial offset of the 1480-nm component (and the total power) is due to the seed source. An interesting behavior with the penultimate Stokes component (1390 nm) is seen which is unlike the previous ones. The preexisting presence of significant power at the output wavelength (1480 nm) manifests as additional loss through stimulated Raman scattering for the 1390-nm component. This creates a more complicated growth and decay condition. This behavior can become the limiting factor on power of the seed source. If the seed source and, thus, the 1480-nm component are too powerful, it can prematurely deplete the 1390-nm component suppressing the power transfer from 1310 nm. This can result in incomplete wavelength conversion and reduced efficiency. These effects as well as the modeling of the cascaded Raman amplifier architecture is discussed in the next section.

Fig. 2.10 Experimental measurement of different components of the Raman cascade as a function of input power (From Ref. [33])

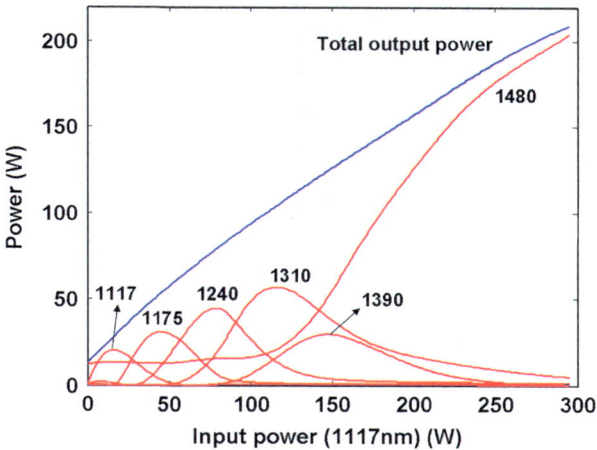

2.6   Modeling and Optimization of Cascaded Raman Amplifier-Based Raman Lasers

In this section, we explore using simulations, increasing the seed power and identifying proper operating conditions for the cascaded Raman amplifier architecture. Let P_i, $i = 0, 1, \ldots, n$ be the power components in each of the wavelengths, P_0 being the pump, P_{n-1} being the signal, and P_n the undesirable wavelength created by further conversion of the signal. From 1117 to 1480 nm, $n = 6$. Since the conversion is seeded at all intermediate wavelengths, we can ignore the amplified spontaneously generated backward components, and the equations governing each Stokes components are

$$\frac{dP_i}{dz} = -\alpha_i P_i + \frac{g_{i-1}}{A_{\text{eff}_i}} P_i P_{i-1} - \left(\frac{\lambda_i}{\lambda_{i-1}}\right) \frac{g_i}{A_{\text{eff}_i}} P_i P_{i+1}$$

where α_i is the loss for each wavelength, A_{eff} is the effective area of the fiber at each of the Stokes wavelengths, g_i is the Raman gain coefficient, and the term $\lambda_i / \lambda_{i-1}$ represents the quantum defect. The above equation has three terms corresponding to linear loss, growth due to gain from previous Stokes wavelength, and loss due to further conversion to the next Stokes wavelength. For the pump P_0, there is no gain component and the equation is modified to

$$\frac{dP_0}{dz} = -\alpha_0 P_0 - \left(\frac{\lambda_1}{\lambda_0}\right) \frac{g_0}{A_{\text{eff}_0}} P_0 P_1$$

And for the final Stokes component which is created by the undesirable Raman conversion of the signal, we can ignore the term governing further conversion to the next Raman order, and the equation is modified to

$$\frac{dP_n}{dz} = -\alpha_n P_n + \frac{g_n}{A_{\text{eff}_n}} P_{n-1} P_n$$

In this work, A_{eff} increases from $12 \mu m^2$ at 1117 nm to $18 \mu m^2$ at 1480 nm. All the wavelength components are unpolarized, and the Raman gain coefficient can be represented as [from 36]

$$g_i = 0.5 \times 10^{-13} \times \frac{1 \mu m}{\lambda_i}$$

Linear loss $\alpha_i, i = 0, 1, \ldots, n-1$ is assumed to be ~ 1 dB/km. This is a conservative estimate for Raman fibers. α_n, the excess loss for the converted signal component introduced by the Raman filter fiber, needs to be high enough to suppress further conversion. From Eq. (3), we see that if

$$\alpha_n > (g_n/A_{\text{eff}_n}) P_{n-1},$$

the spurious component (at 1590 nm) cannot grow. For an output power of 300 W at 1480 nm, this corresponds to ~ 3 dB/m. Filter fibers used for these applications can easily provide the level of loss per unit length necessary to fully suppress parasitic conversion of the signal.

The pump wavelength is provided by the high-power Yb-doped fiber laser, and intermediate and final wavelengths are provided by the cascaded Raman seed laser. In the simulations, as appropriate to the experiment, the high-power Yb-doped fiber laser was assumed to provide up to 450 W of output power at 1117 nm, and the intermediate Stokes components in the seed source is 15 dB below the power at its final wavelength, corresponding to experimental conditions. The simulation results were found to be insensitive to the exact power levels in each of the intermediate Stokes components. The simulations of the cascaded Raman amplifier are simpler compared to the conventional cascaded Raman laser since all the components are forward moving. The set of differential equations can be solved using a simple method such as Euler's method. Starting from the initial values, the derivative can be evaluated at each length step and the components updated for the successive step on a discretized length grid.

Figure 2.11a shows the output power at 1480 nm as the power in the seed source is varied from 0.1 to 100 W for three different lengths of the Raman filter fiber close to the optimal point. Also shown is the optimal conversion in the amplifier given by quantum-limited conversion. As expected, when the seed source power is too low, there is incomplete conversion. This can be compensated to some extent by using longer fiber lengths with the addition of a small increase to linear loss in the fiber. There is interesting behavior observed as the seed source power is increased. The output power shows saturation like behavior. The difference between the optimal power and obtained output progressively increases. This indicates that power scaling cannot be achieved by simply increasing the power in the seed source. Figure 2.11b shows the corresponding plots in terms of conversion efficiency. Beyond the

Fig. 2.11 (a) Output power as a function of total power of the seed source for different lengths, also shown is the ideal output power. (b) Conversion efficiency in the amplifier as a function of seed power

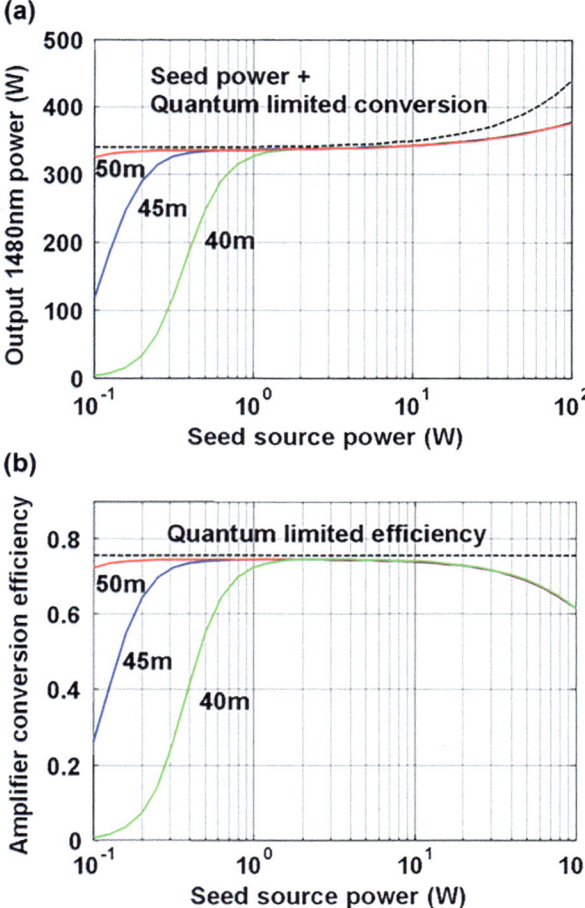

10 W power level in the seed source results in progressively reduced conversion efficiencies and in the 1-10 W level, close to quantum-limited efficiency can be achieved. Thus, having a moderate power seed source in the range of 1–10 W is optimal for achieving high conversion efficiency.

The origins of this behavior can be understood by looking at the growth of individual Stokes components as a function of input power for various seed source powers. Figure 2.12a shows the case for a 45-m-long cascaded Raman amplifier with a seed source power of 10 W. This per simulations shown in Fig. 2.11 should achieve a high conversion efficiency. There is a progressive growth and decay of all the intermediate Stokes components except the 1390-nm component which is slightly suppressed. At maximum power, a high degree of wavelength conversion with most of the power in the final wavelength is achieved. Figure 2.12b shows the situation when the seed source power is increased to 40 W. The growth of individual

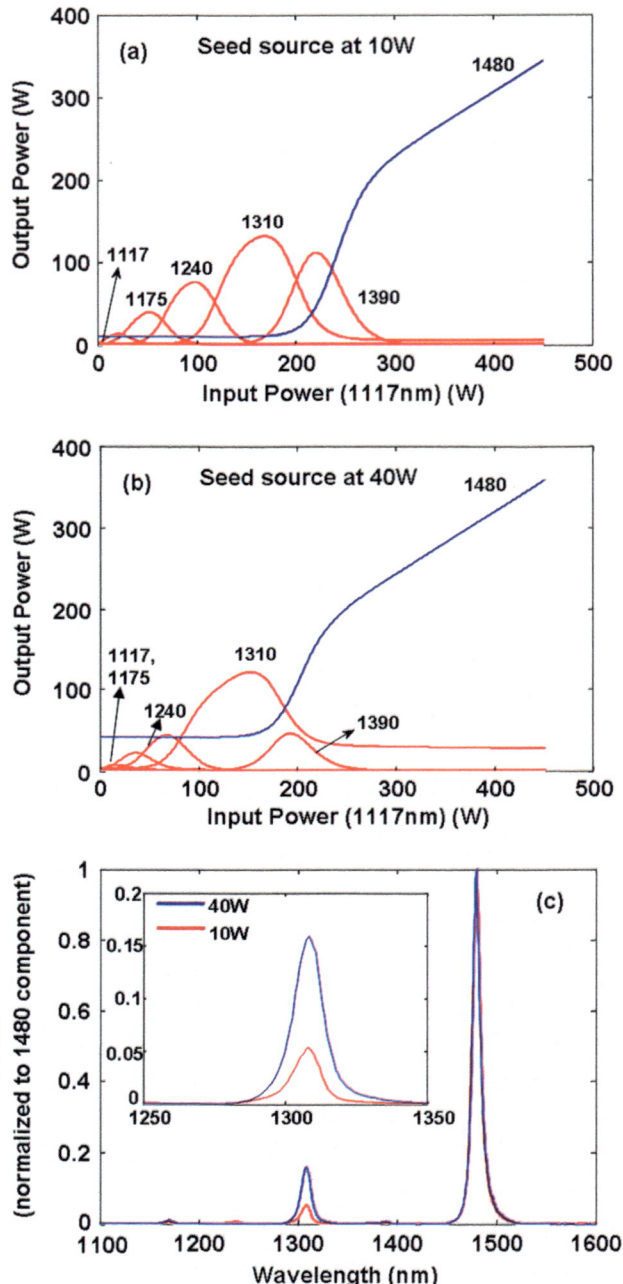

Fig. 2.12 (**a, b**) The evolution of power at all the components as a function of input power for different seed source powers and Raman fiber length of 45 m. (**c**) Experimentally measured spectra at an output power of 200 W at 1480 nm as the seed power is increased from 10 to 40 W for a fiber length of 50 m

components is shifted to lower power owing to higher seed powers. The penultimate component, however, at 1390 nm is further suppressed. This is due to the presence of significant power at 1480 nm from the start. The consequence is incomplete power transfer out of 1310 nm resulting in incomplete wavelength conversion. Figure 2.12c shows experimentally measured spectra at an output power of ∼200 W at 1480 nm (corresponding to an input power of ∼300 W at 1117 nm) as the seed source power is increased from 10 to 40 W. Incomplete power transfer out of 1310 nm is clearly seen as the seed power is increased. The 1310 component relative to the 1480-nm component (zoomed in the inset) nearly triples in height from a ratio of 0.05 to 0.15.

From the simulations and experiments above, to achieve power scaling, increasing the power in the ytterbium-doped fiber laser while maintaining the seed laser at low powers is optimal for conversion efficiency.

2.7   Simplified Architectures for High-Efficiency Cascaded Raman Amplifiers

In the cascaded Raman amplifier architecture described in the previous section, to generate the intermediate Stokes wavelengths, a complete low-power cascaded Raman laser is needed. This necessitates an additional intermediate power fiber laser and associated optics. It is strongly desired to achieve high-efficiency Raman conversion using a single, high-power fiber laser using an all-passive Raman conversion module. This was demonstrated in [37], where such a conversion was achieved and a Raman laser was demonstrated providing 64 W at 1480 nm (pumped at 1117 nm) with a conversion efficiency of ∼60% (corresponding to over 45% efficiency w.r.t 975-nm pumping).

Figure 2.13 shows the schematic of the architecture. The goal is to utilize a single ytterbium-doped fiber laser for both tasks. This is done by tapping out a small fraction from the high-power fiber laser and using that to generate the

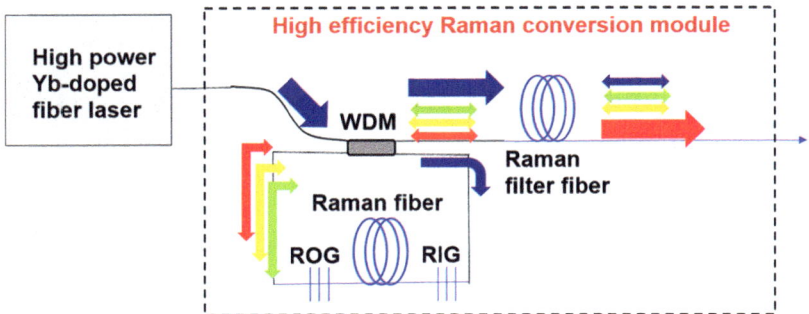

Fig. 2.13 Schematic of the all-passive cascaded Raman amplifier, RIG, ROG, Raman input and output grating sets; WDM, wavelength division multiplexer

intermediate Stokes wavelengths. The generation of seed is carried out as before, by a conventional cascaded Raman resonator which achieves this through leakage at all intermediate wavelengths. Owing to the small power tapped out, the hit on efficiency is not substantial. In addition to the generation of all the intermediate wavelengths, it also needs to be efficiently coupled back with the high-power laser source for cascaded Raman conversion. This necessitates a device which can tap out a small fraction of the incident light while also efficiently coupling back the newly generated wavelength components. Such a functionality is easily achieved with a fused fiber WDM which can multiplex the final and initial wavelengths and operate at high power. A 1117/1480 WDM with an isolation of $\sim$15 dB between the 1117 and 1480 ports was utilized. The isolation parameter results in leakage of light at the 15-dB level ($\sim$3%) to the cross port which is sufficient tap out of power to power the cascaded Raman conversion. The wavelength multiplexing nature enables efficient recoupling back of the generated wavelength components. An additional benefit observed was from the gratings in the Raman resonator which recoupled a substantial fraction of the backward Raman components generated in the amplifier (as shown in Fig. 2.13). Cascaded Raman conversion was achieved in a Raman filter fiber which prevented further Raman conversion of the final wavelength.

Figure 2.14a shows the total output power and the 1480-nm component as a function of input power at 1117 nm. An output power of $\sim$64 W at 1480 nm (limited by input power) for a total input power of $\sim$107 W with a total conversion efficiency of $\sim$60% (1117 nm–1480 nm, $\sim$80% of the quantum-limited efficiency) was obtained. Figure 2.14b shows the measured output spectrum at full power. $\sim$95% of the power is in the 1480-nm band indicating a high level of wavelength conversion, while high suppression of the next Stokes order at 1590 nm is maintained through the use of the filter fiber. The conversion efficiency at lower powers is lower by the necessity of longer Raman fibers. It is anticipated that the conversion efficiency will increase as the output power is scaled.

In summary, this architecture achieves the benefit of the cascaded Raman amplifier architecture, which is to bypass the cascaded Raman resonator in the main optical path while at the same time keeping it comparable to the conventional architecture in terms of simplicity.

2.8 Some Applications of Cascaded Raman Fiber Lasers

2.8.1 Optical Pumping

As described at the start of this chapter, cascaded Raman lasers offer the mechanism to generate high-brightness optical sources across the spectrum. This wavelength versatility means that it can be used for high-power pumping of optical absorptions at wavelengths which are difficult to achieve with other means such as direct diode pumping or rare-earth-doped fiber lasers. Furthermore, the single-mode

Fig. 2.14 (a) Plot of total output power and output power at 1480 nm as a function of input power at 1117 nm. (b) Spectrum of the output in linear scale at maximum power

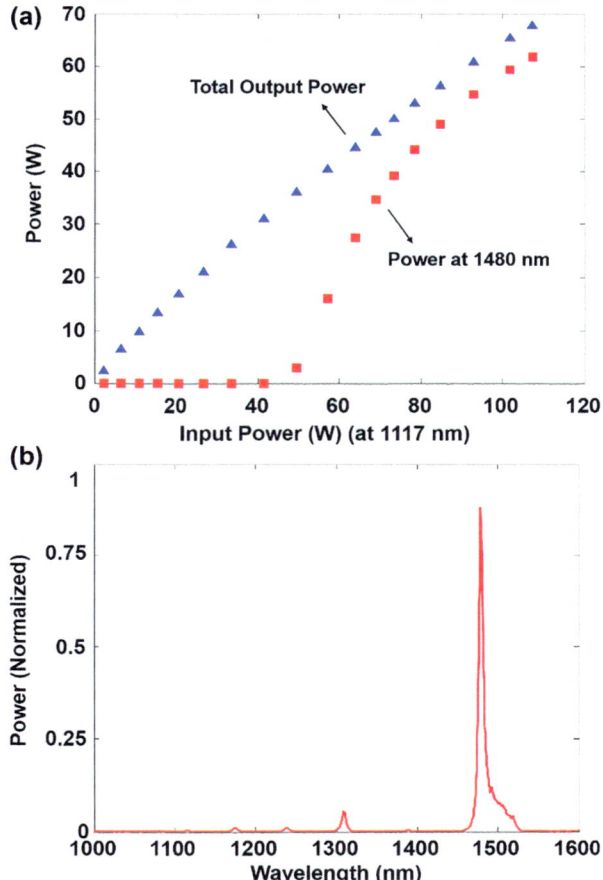

output from the Raman laser can enable core pumping or high-brightness cladding pumping. Core pumping with Raman lasers is especially attractive for nonlinearity-constrained lasers such as pulsed fiber laser systems and narrow-linewidth lasers. Core pumping helps keep the active fiber length short even with relatively low absorption fibers. This ensures that the total nonlinearity (such as modulational instability, stimulated Brillouin scattering, and four-wave mixing) is substantially reduced.

Er-doped fiber lasers have lagged Yb-doped fibers in power scaling. Nevertheless, there is significant interest in high-power erbium-doped fibers due to the relative eye safety of the 1.5 micron wavelength range, compared to operation at 1 micron. Furthermore a window of atmospheric transparency at 1.5 microns makes Er-doped fiber laser attractive for applications involving free space operation, such as remote sensing, LIDAR, and free space communications. The highest power demonstrated to date from a fiber laser at 1.5 μm was from an Er-Yb laser. An output power of 297 W was obtained at 1.56 μm for a pump power of 1.2 kW at 975 nm, but the slope efficiency and further power scaling were limited by parasitic lasing of the Yb ions at 1.06 μm [3]. High-efficiency in-band core pumping of Er-Yb has also been reported using a 1535-nm fiber laser as a pump, but the maximum output power was limited to 18 W [38].

As a result of difficulties in power scaling Er-Yb fibers, there has been recent interest in power scaling of Yb-free Er fiber lasers. One approach to high-power operation of Yb-free Er fiber lasers is direct cladding pumping with multimode 976-nm diode lasers [39–41]. The maximum power achieved to date using direct diode cladding pumping of an Yb-free, Er-doped fiber is 100 W [41]. Another approach is to cladding, in-band pump either Yb-free Er, or Er-Yb fiber, using multimode 15xx pump diodes or fiber lasers [42–44]. One difficulty with diode pumping with 15xx diodes is that they are not as well developed, or high performance, as multimode diodes operating at 976 nm. The highest power reported to date from a single-mode Yb-free Er fiber in-band, cladding pumped with diodes is 88 W [42].

In comparison, a single-transverse-mode Er fiber laser core-pumped by a high-power, 1480-nm Raman fiber laser has also generated over 100 W of output power [45]. A schematic of the laser setup, plot of output power vs. launched pump power, and plot of the output optical spectrum are shown in Fig. 2.15. Because of the core pumping configuration, a standard telecom single-mode Er fiber, OFS MP980, could be used as the active gain medium. Although OFS MP980 has a low Er absorption, a relatively short length of 21 m of fiber could be used due to the core-pumping architecture. Furthermore, the low dopant concentration helps reduce clustering, resulting in the high-slope efficiency of 71% with respect to launched pump power. Finally, using a cascaded Raman pump distributes the heat load due to the quantum defect among multiple stages (Yb laser, cascaded Raman resonator, and Er laser) compared to direct diode pumping of an erbium laser. This heat load distribution helps to ease thermal management.

2.8.1.1   Core Pumping of Pulsed, Er-Doped Amplifiers with High-Power 1480-nm Raman Lasers

While using Raman lasers to pump CW erbium lasers has generated over 100-W output power, Raman lasers are also advantageous for pulsed applications. For example, at moderate power levels, a cascaded Raman fiber laser can provide several Watts of pump power for generating 10s of nanoJoule of pulse energy from an all-normal dispersion, femtosecond mode-locked fiber oscillators using a single-mode Er fiber [46].

To scale pulse energies and peak power significantly beyond what can be achieved with a single-mode fiber, the effective area of the mode in the erbium fiber needs to be increased substantially. In this regime, Raman fiber lasers and the core-pumped architecture have some distinct additional advantages. As mentioned previously, core pumping helps keep the fiber length short, compared to cladding pumping, which is important for increasing nonlinear thresholds such as self-phase

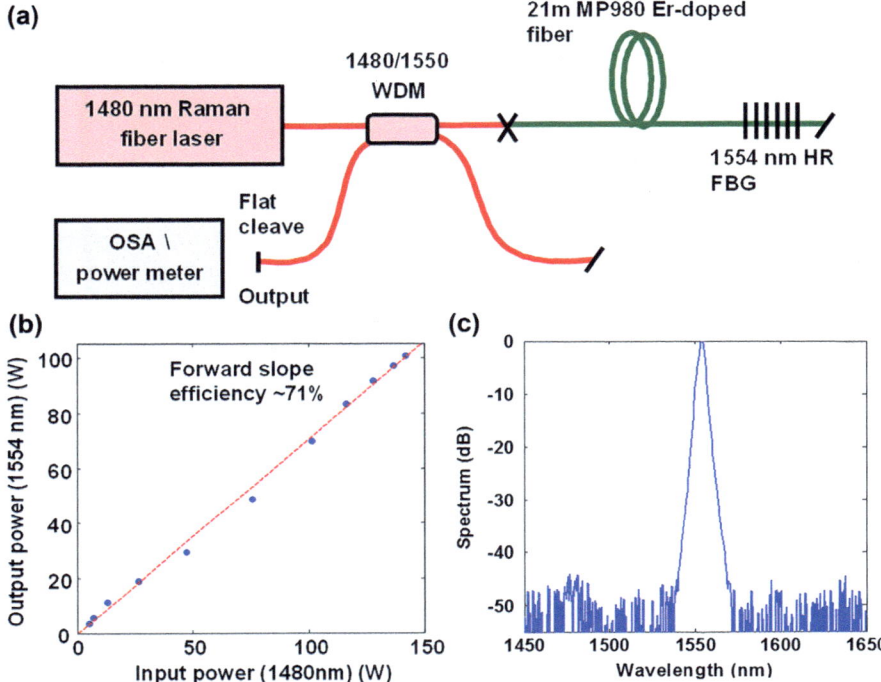

Fig. 2.15 CW erbium fiber laser with over 100 W pumped by a high-power cascaded Raman laser. (**a**) Setup. (**b**) Output power vs. pump power. (**c**) Output optical spectrum (From Ref. [45])

modulation, Brillouin scattering, and Raman scattering. Furthermore, Yb-free, Er fibers more readily lend themselves to large-mode-area (LMA) fibers with diffraction-limited performance [6] as the relatively simple composition aids in fabricating precision refractive index profiles. Finally, in fibers that support multiple modes, but are intended to operate on only a single mode, the high pump/signal overlap in a core-pumped architecture provides suppression of unwanted higher-order modes via differential gain [47, 48]. One negative aspect of core pumping is that experiments and simulations have shown lower efficiency in core-pumped erbium-ytterbium co-doped fiber amplifiers, compared to cladding pump amplifiers, due to a lower pump intensity mitigating pair-induced quenching [49].

A general schematic of a very-large-mode-area (VLMA) amplifier architecture pumped by a 1480-nm Raman fiber laser is shown in Fig. 2.16. In such an amplifier, the pulsed signal and the high-power 1480-nm Raman laser are coupled together with a 1480/1550 single-mode fiber wavelength division multiplexer. Pump and signal are then launched together into the fundamental mode of the VLMA-Er amplifier where amplification takes place.

Direct amplification of picosecond pulses in approximately 3-m-long VLMA amplifier with 1100-μm^2 effective area has been demonstrated to peak powers of

Fig. 2.16 General schematic
of a large-mode-area
erbium-doped amplifier
pumped by a high-power,
1480-nm Raman fiber laser

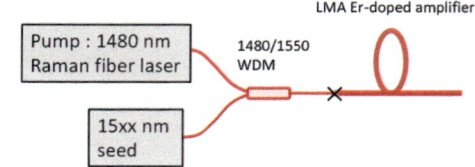

~127 kW [6]. Femtosecond, chirped-pulse fiber amplifiers (CPA) using Raman-pumped VLMA-Er fibers have also been demonstrated. In a germanosilicate 4.5-m-long, VLMA fiber with effective area of ~875-μm^2, 25-μJ, and 800-fs pulses were demonstrated [51]. More recently, using a 28-cm-long, Er-doped phosphate glass with effective area of 2290 μm^2 and pumped by a Raman laser, sub-500-fs pulses with a pulse energy of 915 μJ were demonstrated [52]. The large-mode area of VLMA fiber combined with high-power Raman pumps also makes them well suited to systems that require both high average power and high peak power. The 100-W average power, 10-GHz, 130-femtosecond pulses with diffraction-limited beam quality were generated from a 3-m-long VLMA-Er fiber with effective area of 1100 μm^2 [50]. Optical performance of this system is illustrated in Fig. 2.17.

For further scaling of effective area and pulse peak power beyond what a VLMA fiber can achieve, a specially designed higher-order-mode (HOM) fiber can be used [53]. HOM fibers are few-moded fibers designed to operate on a single higher-order mode, typically the $LP_{0,N}$ mode. In an HOM amplifier, the signal is launched into the fundamental mode of the fiber. A long-period grating (LPG) then provides phase-matched coupling to the desired $LP_{0,N}$ HOM. In the case of an erbium-HOM amplifier pumped by a single-mode Raman laser, both pump and signal are launched into the same HOM [54], provided almost perfect overlap between pump and signal, and conferring the benefits of core pumping discussed above. At the output of the amplifier, the signal can be converted to a diffraction-limited beam either with a second matched LPG or in the case of a high peak power system, a bulk-optic mode converter such as an axicon [55]. Er-doped HOM amplifiers have been demonstrated with effective areas as large as 6000 μm^2 [56]. A schematic of a typical Er-doped HOM amplifier, pumped by a Raman fiber laser is illustrated in Fig. 2.18. An important benefit of HOM fibers is that they are less susceptible to bend-induced area reductions compared to fundamental mode fibers [57].

Comparison between Er-VLMA and Er-HOM amplifiers in femtosecond CPA systems has shown that the achievable pulse energy scales with the effective area of the mode, whether the mode is a fundamental mode or higher-order mode [55]. In a high-power CPA system based on Raman fiber laser pumped, Er-doped higher-order-mode fiber, sub-500-fs with 300-μJ pulses were achieved [58]. This particular system used an LPG for output mode conversion which adds nonlinearity to the overall system. Thus, a system using a bulk-optic axicon for mode conversion would be expected to provide further pulse energy scaling.

The large effective area of the Er-doped HOM fiber is also advantageous for direct amplification of nanosecond pulses up to 700-kW peak power [59]. Finally,

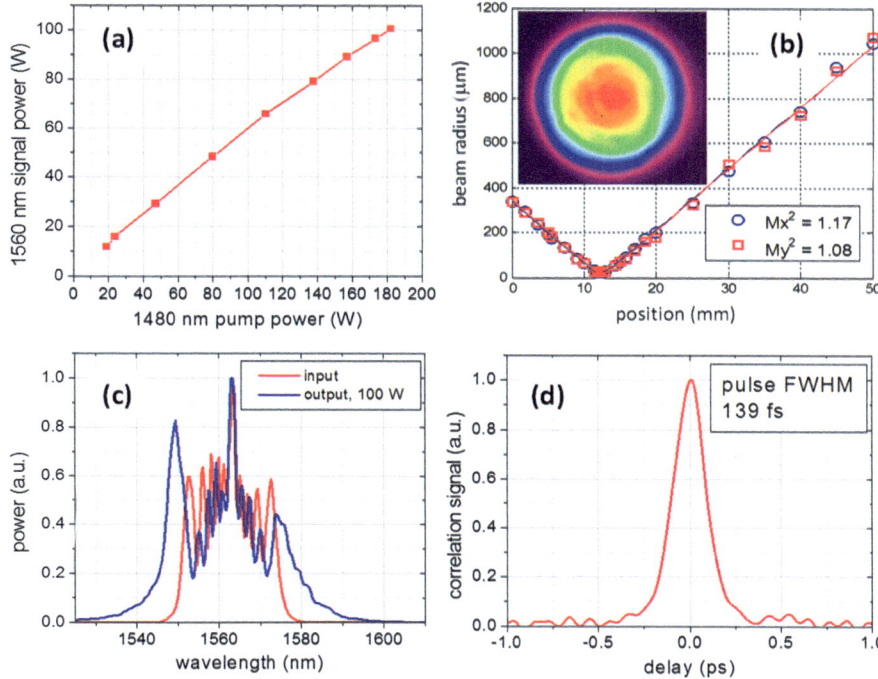

Fig. 2.17 Amplification of 10-GHz, femtosecond pulses in a VLMA-Er amplifier pumped by a high-power 1480-nm Raman laser. (**a**) Output power vs. pump power. (**b**) Beam quality at 100-W output power. (**c**) Amplifier input spectrum and output spectrum at 100-W power. (**d**) Autocorrelation at 100-W output power (From Ref. [50])

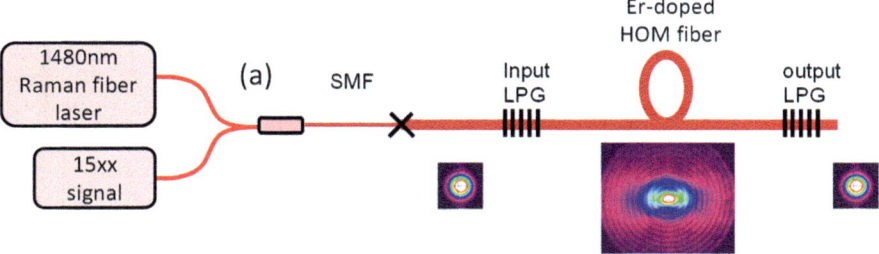

Fig. 2.18 Erbium-doped HOM amplifier pumped by a 1480-nm cascaded Raman fiber laser

the large effective area of the HOM, together with high anomalous dispersion, is ideal for generating self-frequency-shifted solitons in a cascaded Raman laser-pumped amplifier. The 186 nJ, 91-fs solitons, tunable from 1570 to 1620 nm, with a peak power of 1.9 MW were demonstrated from an Er-HOM amplifier pumped by a Raman laser, compared to 17 nJ, 106 fs from a VLMA-Er fiber [60].

2.8.2   Continuous Wave Supercontinuum Generation

Supercontinuum lasers are an important technology which has found applications in a variety of fields [61, 62]. Commonly referred to as "broad as a lamp, bright as a laser," they are lasers which have the bandwidth of a white light source, but the spatial brightness of a laser. These sources are useful for a variety of applications in imaging, test and measurement, sensing, and spectroscopy. High-power fiber supercontinuum lasers are generated by pumping a nonlinear medium with a high-power fiber laser in the anomalous dispersion region close to the zero-dispersion point. Modulation instability coupled with four-wave mixing, soliton generation, and Raman scattering each play a crucial role in the supercontinuum generation [61, 62]. Optical fibers owing to low-loss and long interaction lengths are best suited for supercontinuum generation. However, conventional silica fibers have large normal dispersion in the 1050–1120-nm wavelength region. Since the waveguide component of dispersion is also normal, there is no way of making standard silica fibers have zero dispersion in this band. Thus, high-power supercontinuum generation has been primarily demonstrated using microstructure fibers [63]. With microstructure fibers, an all-fiber architecture is difficult to achieve due to problems with splicing needing free space coupling is used. Further, the technology is still relatively recent and more expensive, and widespread adoption as in the case of silica fibers is not yet there.

Silica fibers on the other hand have anomalous dispersion at wavelengths longer than 1310 nm. In particular, at the 1550-nm region, mature component technology is available due to applications in optical communications. Highly nonlinear fibers [65] with high nonlinear coefficients and zero dispersion are available in this band which is ideal for supercontinuum generation. However, there is a lack of suitable sources to pump these fibers. This is due to the limitations of erbium or erbium-ytterbium co-doped fibers to scale in power while maintaining brightness. Cascaded Raman lasers provide an excellent method to wavelength convert high-power ytterbium-doped fiber lasers to the zero-dispersion region of the highly nonlinear fibers and generate broadband, high-power supercontinua. In [64, 66], a highly nonlinear fiber with zero dispersion at 1480 nm was pumped by a 1117–1480-nm Raman laser to generate over 3 W of power and a 20-dB bandwidth of over 500 nm spanning from 1200 to 1700 nm. Figure 2.19 shows the results for growth of the supercontinuum from [64]. For three different fiber lengths, as the power input to the fiber is enhanced, there is a substantial growth of the supercontinuum. The performance in the normal dispersion region (left side of the pump) in these supercontinuum sources is not at the same level as the anomalous region. It would be interesting going forward to develop Raman-pumped supercontinuum sources which overcome this problem.

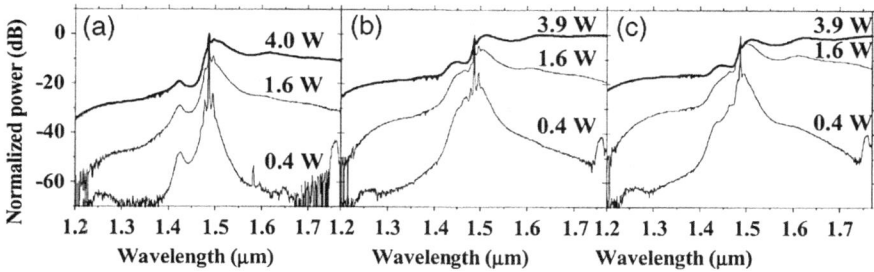

Fig. 2.19 Growth of the CW supercontinuum with increasing power for HNLF lengths of 0.5 km (**a**), 1 km (**b**), and 1.5 km (**c**) (From Ref. [64])

2.9 Summary

In this chapter, we introduced the technology of cascaded Raman fiber lasers which are a wavelength agile, scalable, high-power fiber laser technology. It is anticipated that this agility will further enhance the importance of cascaded Raman lasers in the coming years. Recent developments to enhance efficiency and reliability of these lasers were discussed in detail. A small subset of cascaded Raman laser applications in the areas of laser pumping and NIR supercontinuum generation was also discussed.

References

1. Richardson, D.J., Nilsson, J., Clarkson, W.A.: High power fiber lasers: current status and future perspectives [invited]. J. Opt. Soc. Am. B. **27**, B63–B92 (2010)
2. Ehrenreich, T., Leveille, R., Majid, I., Tankala, K., Rines, G., Moulton, P.: 1-kW, all-glass Tm: fiber laser. Proc. SPIE. **7580**, 758016 (2010)
3. Jeong, Y., Yoo, S., Codemard, C., Nilsson, J., Sahu, J., Payne, D.N., Horley, R., Turner, P.W., Hickey, L., Harker, A., Lovelady, M., Piper, A.: Erbium:ytterbium codoped large-core fiber laser with 297-W continuous-wave output power. IEEE J. Sel. Top. Quantum Electron. **13**, 573 (2007)
4. https://www.lia.orgs/publications/ansi/Z136-1
5. Clark, R.N.: Spectroscopy of Rocks and Minerals, and Principles of Spectroscopy. In: Rencz, A.N. (ed.) Manual of Remote Sensing, vol. 3, Remote Sensing for the Earth Sciences. John Wiley and Sons, New York (1999)
6. Jasapara, J.C., Andrejco, M.J., DeSantolo, A., Yablon, A.D., Varallyay, Z., Nicholson, J.W., Fini, J.M., DiGiovanni, D.J., Headley, C., Monberg, E., Dimarcello, F.V.: Diffraction-limited fundamental mode operation of core-pumped very-large-mode-area Er fiber amplifiers. IEEE J. Sel. Top. Quantum Electron. **15**, 3–11 (2009)
7. Grubb, S.G., Strasser, T., Cheung, W.Y., Reed, W.A., Mizrahi, V., Erdogan, T., Lemaire, P.J., Vengsarkar, A.M., DiGiovanni, D.J., Peckham, D.W., Rockney, B.H.: High-Power 1.48 μm Cascaded Raman Laser in Germanosilicate Fibers. In: Optical Amplifiers and Their Applications, Vol. 18 of 1995 OSA Technical Digest Series. Optical Society of America, Washington, DC (1995). paper SaA4

8. Chen, M., Shirakawa, A., Fan, X., Ueda, K.-i., Olausson, C.B., Lyngsø, J.K., Broeng, J.: Single-frequency ytterbium doped photonic bandgap fiber amplifier at 1178 nm. Opt. Express. **20**, 21044–21052 (2012)
9. Genty, G., Coen, S., Dudley, J.M.: Fiber supercontinuum sources. J. Opt. Soc. Am. B. **24**(8), 1771 (2007)
10. Babin, S.A., Churkin, D.V., Ismagulov, A.E., Kablukov, S.I., Podivilov, E.V.: Four-wave-mixing-induced turbulent spectral broadening in a long Raman fiber laser. J. Opt. Soc. Am. B. **24**, 1729–1738 (2007)
11. http://fiber-optic-catalog.ofsoptics.com/item/optical--fibers/raman-optical-fiber1/raman-fiber
12. Fortin, V., Bernier, M., Faucher, D., Carrier, J., Vallée, R.: 3.7 W fluoride glass Raman fiber laser operating at 2231 nm. Opt. Express. **20**, 19412–19419 (2012)
13. Bernier, M., Fortin, V., El-Amraoui, M., Messaddeq, Y., Vallée, R.: 3.77 μm fiber laser based on cascaded Raman gain in a chalcogenide glass fiber. Opt. Lett. **39**, 2052–2055 (2014)
14. Kashyap, R.: Fiber Bragg Gratings. Academic press, CA (1999)
15. Kashyap, R.: The fiber fuse - from a curious effect to a critical issue: a 25th year retrospective. Opt. Express. **21**, 6422–6441 (2013)
16. Agrawal, G., Headley, C.: Raman amplification in fiber optical communication systems. (Elsevier Academic press, 2005)
17. Dianov, E.M., Proghorov, A.M.: Medium-power CW Raman fiber lasers. IEEE J. Sel. Top. Quantum Electron. **6**, 1022 (2000)
18. Emori, Y., Tanaka, K., Headley, C., Fujisaki, A.: High-Power Cascaded Raman Fiber Laser with 41-W Output Power at 1480-nm Band, In: Conference on Lasers and Electro-Optics/Quantum Electronics and Laser Science Conference and Photonic Applications Systems Technologies, OSA Technical Digest Series (CD) (Optical Society of America, 2007), paper CFI2
19. Feng, Y., Taylor, L.R., Calia, D.B.: 150 W highly-efficient Raman fiber laser. Opt. Express. **17**, 23678–23683 (2009)
20. Nicholson, J.W., Yan, M.F., Wisk, P., Fleming, J., DiMarcello, F., Monberg, E., Taunay, T., Headley, C., DiGiovanni, D.J.: Raman fiber laser with 81 W output power at 1480 nm. Opt. Lett. **35**, 3069–3071 (2010)
21. Kawasaki, B.S., Hill, K.O., Lamont, R.G.: Biconical-taper single-mode fiber coupler. Opt. Lett. **6**, 327–328 (1981)
22. Agrawal, G.P.: Fiber-Optic Communication Systems, 2nd edition. Wiley Interscience, 1997
23. Supradeepa, V.R., Nicholson, J.W.: Power scaling of high-efficiency 1.5 um cascaded Raman fiber lasers. Opt. Lett. **38**, 2538–2541 (2013)
24. Vengsarker, A.M., Lemaire, P.J., Judkins, J.B., Bhatia, V., Erdogan, T., Sipe, J.E.: Long-period fiber gratings as band-rejection filters. IEEE J. LightWave Tech. **14**, 58–65 (1996)
25. Kashyap, R., Wyatt, R., McKee, P.F.: Wavelength flattened saturated erbium amplifier using multiple side-tap Bragg gratings. Electron. Lett. **29**(11), 1025 (1993)
26. Peng, X., Kim, K., Xinhua, G., Mielke, M., Jennings, S., Rider, A., Fisher, N., Woodbridge, T., Dionne, R., Trepanier, F.: Root cause analysis and solution to the degradation of wavelength division multiplexing (WDM) couplers in high power fiber amplifier system. Opt. Express. **21**, 20052–20061 (2013)
27. Rini, M., Cristiani, I., Degiorgio, V.: Numerical modeling and optimization of cascaded CW Raman fiber lasers. IEEE J. Quantum Electron. **36**, 1117–1122 (2000)
28. Yablon, A.D.: Optical Fiber Fusion Splicing. Springer, Heidelberg (2005)
29. Arbore, M., Zhou, Y., Keaton, G., Kane, T.: 36 dB gain in S-band EDFA with distributed ASE suppression. OAA 2002, paper PD4
30. Kim, J., Dupriez, P., Codemard, C., Nilsson, J., Sahu, J.K.: Suppression of stimulated Raman scattering in a high power Yb-doped fiber amplifier using a W-type core with fundamental mode cut-off. Opt. Express. **14**, 5103–5113 (2006)
31. Nicholson, J.W., Abeeluck, A.K., Headley, C., Yan, M.F., Jorgensen, C.G.: Pulsed and continuous-wave supercontinuum generation in highly nonlinear, dispersion-shifted fibers. Applied Physics B. **77**, 211–218 (2003)

32. Supradeepa, V.R., Nicholson, J.W., Headley, C., Lee, Y.W., Palsdottir, B., Jakobsen, D.: Cascaded Raman fiber laser at 1480 nm with output power of 104 W. In: Fiber Lasers IX Technology Systems, and Applications, Proceedings of SPIE vol. 8237, paper 8237-48
33. Supradeepa, V.R., Nicholson, J.W., Headley, C.E., Yan, M.F., Palsdottir, B., Jakobsen, D.: A high efficiency architecture for cascaded Raman fiber lasers. Opt. Express. **2013**(21), 7148–7155 (2013)
34. Reed, W.A., Stentz, A.J., Strasser, T.A.: Article comprising a cascaded Raman fiber laser. U.S. Patent 5,815,518 (1998)
35. Jackson, S.D., Muir, P.H.: Theory and numerical simulation of nth-order cascaded Raman fiber lasers. J. Opt. Soc. Am. B. **18**, 1297–1306 (2001)
36. Agrawal, G.P.: Nonlinear fiber optics. Academic press, Oxford (2007)
37. Supradeepa, V.R.R., Balaswamy, V., Arun, S., Chayran, G.: A simplified architecture for high efficiency cascaded Raman fiber lasers. In: Conference on Lasers and Electro-Optics, OSA Technical Digest (2016) (Optical Society of America, 2016), paper SM2Q.5
38. Lim, E.-L., Alam, S.-u., Richardson, D.J.: Optimizing the pumping configuration for the power scaling of in-band pumped erbium doped fiber amplifiers. Opt. Express. **20**, 13886–13895 (2012)
39. Kuhn, V., Kracht, D., Neumann, J., Weßels, P.: Er-doped photonic crystal fiber amplifier with 70 W of output power. Opt. Lett. **36**, 3030–3032 (2011)
40. Kotov, L.V., Likhachev, M.E., Bubnov, M.M., Medvedkov, O.I., Yashkov, M.V., Guryanov, A.N., Lhermite, J., Février, S., Cormier, E.: 75 W 40% efficiency single-mode all-fiber erbium-doped laser cladding pumped at 976 nm. Opt. Lett. **38**, 2230–2232 (2013)
41. Kotov, L.V., Likhachev, M.E., Bubnov, M.M., Medvedkov, O.I., Yashkov, M.V., Guryanov, A.N., Février, S., Lhermite, J., Cormier, E.: Yb-free Er-doped all-fiber amplifier cladding-pumped at 976 nm with output power in excess of 100 W. Proceedings SPIE 8961, Fiber Lasers XI: Technology, Systems, and Applications, 89610X (March 7, 2014); doi:10.1117/12.2038503
42. Zhang, J., Fromzel, V., Dubinskii, M.: Resonantly cladding-pumped Yb-free Er-doped LMA fiber laser with record high power and efficiency. Opt. Express. **19**, 5574–5578 (2011)
43. Zhang, J., Fromzel, V., Dubinskii, M.: Resonantly (in-band) cladding-pumped Yb-free Er-doped fibre laser with record efficiency. Electron. Lett. **48**(20), 1298–1300 (2012)
44. Jebali, M.A., Maran, J.-N., LaRochelle, S.: 264 W output power at 1585 nm in Er–Yb codoped fiber laser using in-band pumping. Opt. Lett. **39**, 3974–3977 (2014)
45. Supradeepa, V.R., Nicholson, J.W., Feder, K.: Continuous wave Erbium-doped fiber laser with output power of > 100 W at 1550 nm in-band core-pumped by a 1480 nm Raman fiber laser, in CLEO Technical Digest, 2012, paper CM2N.8
46. Ruehl, A., Kuhn, V., Wandt, D., Kracht, D.: Normal dispersion erbium-doped fiber laser with pulse energies above 10 nJ. Opt. Express. **16**, 3130–3135 (2008)
47. Nicholson, J.W., Jasapara, J.C., Desantolo, A., Monberg, E., Dimarcello, F.: Characterizing the modes of a core-pumped, large-mode area Er fiber using spatially and spectrally resolved imaging. CLEO 2009, paper CWD4
48. Várallyay, Z., Jasapara, J.C.: Comparison of amplification in large area fibers using cladding-pump and fundamental-mode core-pump schemes. Opt. Express. **17**, 17242–17252 (2009)
49. Lim, E.-L., Alam, S.-u., Richardson, D.J.: High-energy, in-band pumped erbium doped fiber amplifiers. Opt. Express. **20**, 18803–18818 (2012)
50. Nicholson, J.W., Ahmad, R., De Santolo, A.: 100 W, 10 GHz, femtosecond pulses from a very-large-mode-area Er-doped fiber amplifier. Conference on Lasers and Electro-Optics 2016, paper SM1Q.1
51. Yilmaz, T., Vaissie, L., Akbulut, M., Gaudiosi, D.M., Collura, L., Booth, T.J., Jasapara, J.C., Andrejco, M.J., Yablon, A.D., Headley, C., Di Giovanni, D.J.. Large-mode-area Er-doped fiber chirped-pulse amplification system for high-energy sub-picosecond pulses at 1.55 μm, Proceedings SPIE 6873, Fiber Lasers V: Technology, Systems, and Applications, 68731I (February 25, 2008)

52. Peng, X., Kim, K., Mielke, M., Jennings, S., Masor, G., Stohl, D., Chavez-Pirson, A., Nguyen, D.T., Rhonehouse, D., Zong, J., Churin, D., Peyghambarian, N.: Monolithic fiber chirped pulse amplification system for millijoule femtosecond pulse generation at 1.55 μm. Opt. Express. **22**, 2459–2464 (2014)
53. Ramachandran, S., Fini, J.M., Mermelstein, M., Nicholson, J.W., Ghalmi, S., Yan, M.F.: Ultra-large effective-area, higher-order mode fibers: a new strategy for high-power lasers. Laser Photonic Rev. **2**, 429–448 (2008)
54. Nicholson, J.W., Fini, J.M., DeSantolo, A.M., Monberg, E., DiMarcello, F., Fleming, J., Headley, C., DiGiovanni, D.J., Ghalmi, S., Ramachandran, S.: A higher-order-mode Erbium-doped-fiber amplifier. Opt. Express. **18**, 17651–17657 (2010)
55. Nicholson, J.W., DeSantolo, A., Westbrook, P.S., Windeler, R.S., Kremp, T., Headley, C., DiGiovanni, D.J.: Axicons for mode conversion in high peak power, higher-order mode, fiber amplifiers. Opt. Express. **23**, 33849–33860 (2015)
56. Nicholson, J.W., Fini, J.M., DeSantolo, A.M., Liu, X., Feder, K., Westbrook, P.S., Supradeepa, V.R., Monberg, E., DiMarcello, F., Ortiz, R., Headley, C., DiGiovanni, D.J.: Scaling the effective area of higher-order-mode erbium-doped fiber amplifiers. Opt. Express. **20**, 24575–24584 (2012)
57. Fini, J.M., Ramachandran, S.: Natural bend-distortion immunity of higher-order-mode, large-mode-area fibers. Opt. Lett. **32**(7), 758–750 (2007)
58. Peng, X., Kim, K., Mielke, M., Booth, T., Nicholson, J.W., Fini, J.M., Liu, X., DeSantolo, A., Westbrook, P.S., Windeler, R.S., Monberg, E.M., DiMarcello, F., Headley, C., DiGiovanni, D.J.: Higher-order mode fiber enables high energy chirped-pulse amplification. Opt. Express. **30**, 32411–32416 (2013)
59. Nicholson, J.W., Fini, J.M., DeSantolo, A., Westbrook, P.S., Windeler, R.S., Kremp, T., Headley, C., Giovanni, D.J.I.: High energy pulse amplification in a higher-order mode fiber amplifier with axicon for output mode conversion. In: Conference on Lasers and Electro-Optics (CLEO) 2015 paper STu4L.4
60. Zach, A., Kaenders, W., Nicholson, J.W., Fini, J., DeSantolo, A.: Demonstration of Soliton Self Shifting Employing Er3+ Doped VLMA- and HOM-Fiber Amplifiers. In: Conference on Lasers and Electro-Optics (CLEO) 2015, paper ATu2M.6
61. Dudley, J.M., Genty, G., Coen, S.: Supercontinuum generation in photonic crystal fiber. Rev. Mod. Phys. **78**, 1135–1184 (2006)
62. Dudley, J.M., Taylor, J.R.: Supercontinuum Generation in Optical Fibers. Cambridge University Press, New York (2010)
63. Cumberland, B.A., Travers, J.C., Popov, S.V., Taylor, J.R.: 29 W high power CW supercontinuum source. Opt. Express. **16**, 5954–5962 (2008)
64. Abeeluck, A.K., Headley, C., Jørgensen, C.G.: High-power supercontinuum generation in highly nonlinear, dispersion-shifted fibers by use of a continuous-wave Raman fiber laser. Opt. Lett. **29**, 2163–2165 (2004)
65. Hirano, M., Nakanishi, T., Okuno, T., Onishi, M.: Silica-based highly nonlinear fiber and their application. IEEE J. Sel. Top. Quantum Electron. **15**(1), 103–113 (2009)
66. Abeeluck, A.K., Headley, C.: Continuous-wave pumping in the anomalous- and normal-dispersion regimes of nonlinear fibers for supercontinuum generation. Opt. Lett. **30**, 61–63 (2005)

Chapter 3
Mid-Infrared Raman Fiber Lasers

Vincent Fortin, Martin Bernier, and Réal Vallée

3.1 Introduction

3.1.1 Laser Sources Operating in the MIR

Because they encompass the spectral range of most molecular vibrations, mid-infrared (MIR) sources of coherent radiation are of great importance to both fundamental research and applications. Nowadays, MIR sources essentially belong to three main categories: gas, semiconductors, and solid state, either under bulk or fiber form.

The CO_2 laser has long been considered the golden standard of MIR gas lasers. Relying on an electric discharge optical pumping scheme, it can be operated either in CW or pulsed mode. Generally operated at the wavelength of 10.6 μm, but also available near 9.4 μm, the CO_2 laser can reach record efficiencies as high as 25% [1] and output powers in excess of 100 kW in CW operation [2]. Although intrinsically cumbersome, this laser has reached a high level of maturity and is still one of the most cost-effective laser sources. For this reason, it has been widely used in various application fields ranging from material processing [3] to the biomedical sector [4]. The carbon monoxide (CO) laser is also commercially available at wavelengths ranging between 5 and 6 μm at power levels reaching hundreds of watts. Other more marginal MIR gas laser emissions have also been demonstrated at various wavelengths with other gases: He–Ne (3.39 μm) [1], N_2O (10 μm) [5], and HBr (4 μm) [6].

V. Fortin (✉) • M. Bernier • R. Vallée
Center for Optics, Photonics and Lasers (COPL), Université Laval, Québec City, QC, Canada
e-mail: vincent.fortin@copl.ulaval.ca

© Springer International Publishing AG 2017
Y. Feng (ed.), *Raman Fiber Lasers*, Springer Series in Optical Sciences 207,
DOI 10.1007/978-3-319-65277-1_3

Quantum cascade lasers (QCLs) are semiconductor injection lasers based on intersubband transitions in group III–IV multiple-quantum well structures, which are especially suited for laser emission in the MIR [7, 8]. Originally operated in pulsed mode and at cryogenic temperature, QCLs now stand as very promising practical sources available at room temperature in both CW and pulsed mode at wavelengths ranging in principle between 3 and 25 μm. QCLs are well suited for high-selectivity and high-sensitivity spectroscopic applications, such as trace-gas sensing. For this purpose, they can be operated single mode, with narrow linewidth (distributed feedback cavity) and tunable either over an absorption peak (e.g., with temperature tuning) or over a broader range with an external cavity [8]. Their output power at room temperature is however generally limited to a few watts, or even below the watt level at wavelengths below 4 μm, a spectral region where they are currently less efficient.

Bulk solid-state lasers emitting in the MIR are either based on crystals doped with transition metals or on frequency mixing (i.e., nonlinear) processes. Vibronic solid-state lasers relying on period 4 transition metal active ions currently provide a good coverage of the 2–6 μm spectral region [9]. The Cr^{2+} cation has been efficiently used as dopant in either ZnS/ZnSe or CdSe crystals with spectral coverage of 1.8–3.1 μm and 2.0–3.5 μm, respectively. In addition, power-scaling experiments have achieved up to 140 W (CW) at 2.5 μm in a Cr^{2+}:ZnSe crystal [10]. At longer wavelengths, the Fe^{2+} cation is preferred with a spectral coverage of 3.5–5.5 μm and 4.0–6.5 μm in ZnSe and CdMnTe crystals, respectively. Both Cr^{2+}- and Fe^{2+}-doped crystal lasers are now commercially available [11]. Optical parametric oscillators also represent a convenient alternative for the generation of MIR coherent radiation over a broad spectral range, especially with the advent of new orientation-patterned materials with very broad MIR transmission which allows full coverage of the whole molecular fingerprint region [12]. MIR OPO systems are also available commercially over the 2.2–4.6 μm spectral range with average output powers ranging between 1 and 5 W [13].

Fiber lasers operating in the MIR are under development based either on rare-earth ions or on Raman gain. Now, fiber lasers not only take advantage of the merits of the solid-state laser family to which they belong but also present unique features with regard to their waveguiding nature as well as their ruggedness and isolation from adverse external conditions. This is especially true for fiber lasers operated in the MIR region where ambient molecules lying along a laser cavity optical path can lead to intracavity losses and thus be detrimental to laser operation. Accordingly, there is currently a strong incentive for developing MIR all-fiber lasers adapted to address the various application challenges. The recent availability of low-loss and strongly reflective fiber Bragg gratings (FBGs) in fluoride glass (FG) fibers, photoinscribed with femtosecond pulses, is playing a key role in that regard [14]. Also instrumental is the fact that FGs can be doped efficiently with trivalent lanthanide (rare-earth) cations which actually become glass constituents rather than actual dopants. This is a different situation than with silica where clusters of these ions tend to form at high doping concentrations. For this reason, FG optical fibers with rare-earth concentrations approaching 10 mol% are feasible. Until recently, most rare-earth-doped fluoride fibers belonged to the fluorozirconate family where concentrations of up to 7% were demonstrated. Now, indium fluoride glass bulks were recently demonstrated with holmium concentrations of $\sim 10\%$ [15], a major step toward InF_3-based doped fibers. Finally, another important cornerstone of MIR fiber lasers consists in the development of low-loss double-clad fluoride optical fibers [16] allowing for efficient laser diode clad-pumping.

There are basically four cations allowing for the partial coverage of the 2–4 μm spectral region with rare-earth-doped FG fiber lasers: thulium (Tm^{3+}) with emission bands near 2.0 and 2.3 μm but more importantly erbium (Er^{3+}) with emission around 2.8 and 3.5 μm; holmium (Ho^{3+}) around 2.1, 2.9, 3.2, and 3.9 μm; and finally, dysprosium (Dy^{3+}) around 3.0 μm [17]. However, several of these laser emissions were reported to be inefficient especially as the wavelength is increased because they are quenched by the phonon-mediated non-radiative processes prevailing at room temperature. In fact, the multiphonon decay rate, which is competing with the photon radiative emission, is actually increasing with the temperature. For this reason, the holmium transition at 3.9 μm could only be achieved at 77 K [18]. Another limitation factor for MIR fiber lasers is related to the intrinsic inefficiency of their pumping scheme (quantum defect), which typically relies on near-infrared optical pumps. New scenarios are being explored to increase efficiency and mitigate thermal effects [19]. Table 3.1 presents the highest output powers reported to date from rare-earth-doped FG CW fiber lasers in the 2–4 μm spectral range.

It is noted from Table 3.1 that although only seven atomic transitions are involved, a relatively good coverage of the 2.3–4.0 μm spectral window is obtained. The reason for this arises from the fact that most of these transitions occurring in FGs correspond to bands rather than lines. For instance, the nominal 2.8 μm ($^4I_{11/2} \rightarrow {}^4I_{13/2}$) atomic transition of Er^{3+} was shown to support laser operation over 300 nm, i.e., from 2.7 to 3.01 μm [28, 29]. Similarly, the Er^{3+} ($^4F_{9/2} \rightarrow {}^4I_{9/2}$) transition was demonstrated to be tunable between 3.33 and 3.78 μm [27]. In terms of maximum power, the highest values reported to date were on the order of 30 W

Table 3.1 The highest reported output powers from rare-earth-doped FG CW fiber lasers operating between 2 and 4 μm

Emission wavelength (nm)	Active ion	Output power (W)	Reference
2305	Tm^{3+}	0.15	[20]
2825	Er^{3+}	28	[21]
2940	Ho^{3+}	2.5	[22]
2940	Er^{3+}	30	[23]
3222	Ho^{3+}	0.011	[24]
3265	Dy^{3+}	0.12	[25]
3442	Er^{3+}	1.5	[26]
3470	Er^{3+}	1.47	[27]
3950	Ho^{3+}	0.011	[18]

for the Er^{3+} ($^4I_{11/2} \rightarrow {}^4I_{13/2}$) transition [23]. The watt level has also been reached with the Er^{3+} ($^4F_{9/2} \rightarrow {}^4I_{9/2}$) transition near 3.5 μm, which is promising for the future [25, 26].

Raman fiber lasers (RFLs) complement rare-earth-doped fiber lasers by generating wavelengths outside their electronic emission bands. While RFLs have been used extensively in the near-IR, especially between 1 and 1.5 μm, it has been difficult to extend this technology to wavelengths above 2 μm. In fact, the specialty fibers made of fluoride and chalcogenide glasses first needed to reach the required maturity level, especially in terms of propagation losses and high-power handling. In addition, MIR pump sources such as 2.8 μm Er^{3+}-doped fiber lasers also had to achieve a high level of maturity.

3.1.2   MIR Applications

The MIR spectrum, often labeled the "fingerprint region," hosts the vibrational absorption lines associated to a wide range of chemical species. While near-IR sources typically probe the overtone and combination bands, MIR lasers give direct access to the fundamental absorption bands, which allow high-sensitivity spectroscopic measurements. Possible applications include environmental monitoring, control of manufacturing processes, and noninvasive medical diagnosis [30, 31]. The existence of two MIR atmospheric transmission windows (i.e., at 2.9–5.3 μm and 7.6–16 μm) [30] also enables long-distance remote sensing of different gases. Fiber lasers are especially suited for remote detection applications as they can generate high output powers with a near-perfect beam quality. In the 2.5–4 μm spectral range, where fiber lasers currently operate, one can find the fundamental stretching absorption of O–H ($\sim$2.9–3.0 μm) and C–H bonds ($\sim$3.4 μm) [31, 32]. The C–H resonance is of great interest as it is found in all hydrocarbons [33] as well as many industrial compounds [34, 35]. At slightly longer wavelengths, between 4 and 5 μm, carbon oxides such as CO and CO_2 display strong absorption bands [36]. There is thus a growing demand for high-power sources emitting at longer MIR wavelengths to meet the requirements of novel spectroscopic applications.

Laser surgery is another application field that could benefit from high-power MIR sources. Unlike UV photons, MIR photons do not proceed by photochemical dissociation of tissues but rather by resonant excitation of vibrational transitions of their basic constituents. In the MIR, water and proteins are the main absorbing constituents of soft tissues. As previously mentioned, the strongest resonance in water occurs near 2.94 μm, corresponding to its symmetric stretching vibration. Water deformation vibration also occurs near 6.1 μm which also corresponds to a protein (amide I) resonance. Accordingly, penetration depths of the order of 1 μm and 2.8 μm are measured at 2.94 μm and 6.1 μm, respectively [37]. This is crucial as collateral temperature increase, which is leading to tissue denaturation, is generally inversely proportional to its absorption. Now, thermal issues are especially difficult to handle in soft tissues since their absorption was shown to significantly decrease with temperature, especially at 2.94 μm [38]. Therefore, most of the lasers currently used for biomedical applications, i.e., Ho:YAG ($\lambda = 2.1$ μm), Er:YSGG ($\lambda = 2.79$ μm), Er:YAG ($\lambda = 2.94$ μm), and CO_2 ($\lambda = 10.6$ μm), actually suffer from collateral damage problems to different extents depending on the targeted application. This is complicated by the fact that some of these lasers can only be operated in free-running mode or present very poor spatial beam profiles (e.g., Er:YAG). The regime of operation (i.e., CW vs pulsed vs burst mode) is also of crucial importance since a laser operated with a pulse length shorter than tissue thermal decay may prevent thermal damage but results in mechanical damage due to photoinduced shock wave [39]. Therefore, although several laser systems have been developed for biomedical applications in areas such as urology [40], ophthalmology [41], dentistry [42], neurosurgery [43], and aesthetics [44], more research is required in order to fully exploit the potential of MIR lasers, and this strongly relies on the development of more versatile laser sources, namely, in terms of their operation wavelength, power, and repetition rate/pulse width. In addition, more convenient sources (i.e., more compact and reliable), like Raman gain or rare-earth-doped fiber lasers, are required in order for actual commercial applications to emerge.

Defense and security applications such as directed infrared countermeasures (DIRCM) can also highly benefit from new MIR laser development. These systems generally operate on three spectral bands (2–2.5 μm, 3–3.8 μm, and 3.8–5 μm) since these are the bands used by most heat-seeking guided missiles [45]. Wavelengths above 3.8 μm are typically the most difficult to obtain as there is a general lack of laser sources in this spectral range. Other requirements for the source include a high beam quality, a compact size with a limited number of discrete components, the potential for power scaling (usually >1 W average power [45]), and the ability to modulate its output at high repetition rates (in the 20–50 kHz frequency range). The current systems are mostly based on nonlinear wavelength conversion of solid-state laser sources (e.g., OPOs). Through stimulated Raman scattering, fiber lasers could generate these long wavelengths in a more efficient and much simpler architecture.

3.2 MIR Raman Fiber Laser Components

Raman fiber lasers, especially in the MIR domain, critically depend on their constituents. This section therefore describes each of the components required in a Raman fiber laser setup. We begin by reviewing the infrared glass materials and their properties (i.e., attenuation and Raman gain) and then discuss the fundamental concepts associated with MIR fiber Bragg gratings. Finally, the potential pump lasers and their characteristics are reviewed.

3.2.1 Glass Fiber Compositions

Fibers considered in the design of a laser cavity may be made of various materials: oxide, fluoride (FG), or chalcogenide (ChG) glasses. Oxide glasses, such as silica, are undoubtedly the most commonly used in fiber laser technologies. Standard silicate fibers, which typically contain a small GeO_2 concentration in their core, only allow transmission up to 2.2 μm, thus limiting their use for the development of MIR sources. It was shown that the GeO_2 concentration could be increased to very high levels (>50 mol. %) to slightly extend the glass transmission window toward the MIR but more importantly to increase the Raman gain coefficient. Another very promising class of oxide glasses for efficient Raman sources is the multicomponent tellurite glasses (i.e., made of TeO_2). They can have a broad transmission window combined with a much higher nonlinearity and Raman cross section. Notable compositions include TBZN (TeO_2–Bi_2O_3–ZnO–Na_2O) [46] and TZNL (TeO_2–ZnO–Na_2O– La_2O) [47].

Fluoride glasses (FGs) are part of the halogen family and are generally composed of multiple fluorine compounds in order to ensure their stability. They include fluorozirconate, fluoroaluminate, and fluoroindate glasses primarily composed of ZrF_4, AlF_3, and InF_3, respectively. In the field of optics, ZBLAN (ZrF_4–BaF_2–LaF_3–AlF_3–NaF) is the most widely used fluorozirconate glass due to its good stability and low intrinsic losses up to 4 μm [48]. These glasses generally have a lower damage threshold compared to silicate glasses.

As for chalcogenide glasses (ChGs), they are composed of at least one chalcogen element, either sulfur (S), selenium (Se), or tellurium (Te), combined with another neighboring element from the periodic table, such as gallium (Ga), germanium (Ge), arsenic (As), or antimony (Sb). For fiber drawing, the most common glass compositions are arsenic sulfide (As_2S_3) and arsenic selenide (As_2Se_3). These glasses have significantly higher linear and nonlinear refractive indices compared to those of oxides and FGs. They also suffer from the lowest laser damage threshold.

3.2.1.1 Fiber Fabrication Process and Attenuation Coefficient

The main consideration for selecting a Raman fiber is its optical attenuation at the Stokes emission wavelength. A large part of the attenuation coefficient is set by the fabrication techniques employed (i.e., glass preparation and fiber drawing). Most germanosilicate fibers have very low propagation losses since they are drawn from a preform made by a chemical vapor deposition method (MCVD, for instance). Such method is the best option to avoid glass contamination and core/clad interface defects. For heavily GeO_2-doped silicate (HGDS) fibers, the MCVD process can be significantly challenging due to the large thermal expansion mismatch between SiO_2 and GeO_2 [49]. A dip in the central region of the index profile is often observed in these fibers due to the high vapor pressure of GeO_2 (compared to SiO_2) during the collapsing and sealing steps, resulting in high losses [50]. While the attenuation of commercial SiO_2 fibers currently approaches the theoretical limit, HGDS fibers still exhibit losses that are orders of magnitude higher than the theoretical attenuation of the glass.

Most soft-glass fibers (e.g., fluorides, chalcogenides, and some compositions of oxides such as tellurites) cannot be drawn from a preform fabricated by MCVD because gas precursors are not available for these glass compositions. Fluoride glasses, for instance, are generally drawn from preforms prepared by built-in casting or rotational casting methods [48]. The tendency for these glasses to crystallize combined with their viscosity properties makes the fiber-drawing process especially difficult. In the case of chalcogenides, fibers are currently drawn using the double crucible method [51]. One of the main drawbacks of this technique lies in the core-cladding interface. Indeed, during fusion, inclusions and air bubbles may get trapped in the glass, resulting in an increased level of attenuation. In addition, contaminants are more likely to be encountered. OH- impurities, for instance, result in a strong absorption band near 3 μm and must be reduced to very weak concentrations. In order to achieve an acceptable attenuation level, no effort should be spared in purifying the glass prior to the fiber-drawing process. For all these reasons, the lowest levels of attenuation achieved with specialty soft-glass fibers remain significantly higher than the theoretical limit [52].

Apart from glass purification, preform fabrication, and drawing process, the intrinsic attenuation in the MIR is mainly fixed by the multiphonon absorption limit. It is set by the vibrational energy levels (i.e., phonon energy) associated with atomic and molecular bonds. For example, silica has a maximum phonon energy around $1100 \, cm^{-1}$ [53], which is significantly higher than for soft glasses (e.g., $<600 \, cm^{-1}$ for FGs [53] and $<450 \, cm^{-1}$ for ChGs [54]). As a result, silica is basically opaque at wavelengths longer than 2.4 μm (≥ 1 dB/m). On the other hand, the infrared transmission limit (at 1 dB/m level) of common FG fibers is around 4 μm for AlF_3 [55], 4.2 μm for ZrF_4 [16, 56], and 5.2 μm for InF_3 [57]. As for chalcogenides, they show the broadest transmission window in the mid-infrared, approximately up to 6.5 μm and 9.5 μm, respectively, for As_2S_3 and As_2Se_3 [21]. Figure 3.1 shows the theoretical attenuation for some of the abovementioned glasses. The UV-visible attenuation edge is a result of both Urbach tail and Rayleigh scattering, while the MIR edge is defined by multiphonon absorption.

3.2.1.2 Raman Gain Coefficient and Spectrum

Beyond fiber attenuation, the ideal material for a Raman source should have a high Raman gain coefficient as well as a broad spectrum for tunability. Since materials that are transparent in the mid-infrared are low phonon energy glasses, they often have a narrow Raman spectrum with a small shift at their maximum Raman gain (e.g., see chalcogenides). The Raman gain spectra for silica and for various infrared glasses are plotted in Fig. 3.2.

For pure silica glass, the maximum gain is around 5×10^{-14} m/W (at 2 μm) and is located at a shift of $445 \, cm^{-1}$ (13.4 THz) [60]. This peak corresponds to the

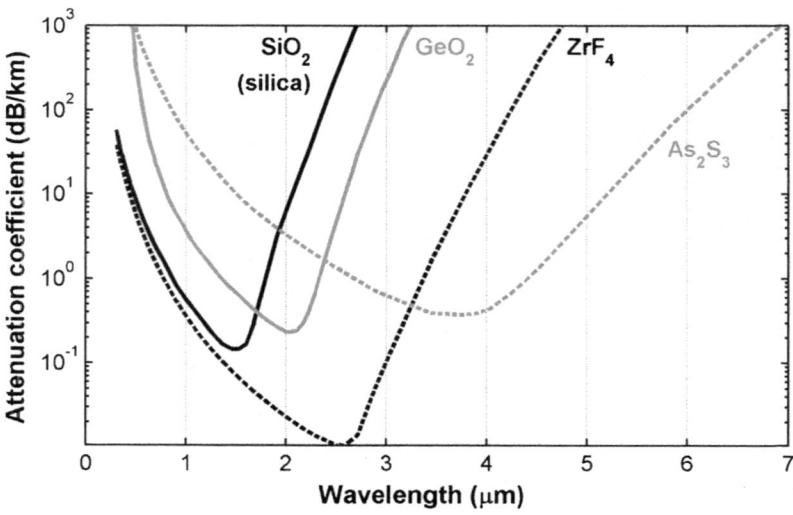

Fig. 3.1 Theoretical attenuation of different glass compositions (SiO₂, GeO₂, ZrF₄-based, and As₂S₃ glasses) [49, 52]

symmetric stretching vibrations of bridging oxygen atoms (bonding SiO_4 tetrahedra together primarily in 6-membered rings [60]). GeO_2 glass on the other hand has a slightly lower Raman frequency shift at 420 cm^{-1} but has a Raman gain coefficient approximately nine times greater than silica [61]. Like silica, the main peak of GeO_2 is also associated to symmetric stretching of bridging oxygens. For germanosilicate fibers, the Raman gain and its spectrum can be predicted with good accuracy for any given GeO_2 concentration and refractive index profile [62].

As seen in Fig. 3.2, tellurite glasses usually have a dual band Raman spectrum. They often show a first band around 400 cm^{-1} and a second one composed of multiple peaks in the 640–740 cm^{-1} range [46, 63]. The second band is commonly stronger, with a Raman gain coefficient more than ten times that of silica glass [46]. With such a broad Raman spectrum, tellurite glasses make excellent candidates for widely tunable Raman lasers and amplifiers [64].

Fluoride glass fibers typically have a Raman gain coefficient comparable to that of silica, but their Raman spectrum is narrower compared to most oxide glasses. In particular, fluorozirconates like ZBLAN show a maximum around 570–590 cm^{-1} (17.1–17.7 THz) and a few lower-frequency secondary peaks [65–67]. The exact position of these latter peaks is strongly dependent on the actual fiber composition. The Raman gain coefficient of the main peak is about 7×10^{-14} m/W at 2 μm [68], which is attributed to the symmetric stretching vibrations of the Zr-F bonds (where the fluorine atom is non-bridging) in the elementary cell consisting of ZrF_6 octahedra [67] or ZrF_8 dodecahedra [66]. Figure 3.3 depicts the structure of ZrF_4-based glasses [66]. Fluoroaluminates are known to have Raman gain maxima around 560 and 620 cm^{-1} (16.8 and 18.6 THz) [69]. As for fluoroindates, their

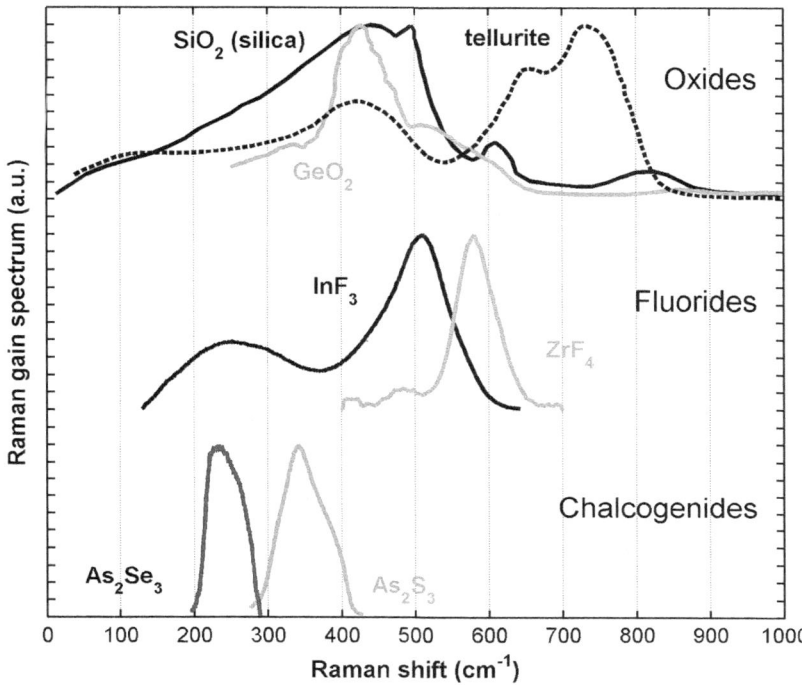

Fig. 3.2 Raman gain spectrum for various fiber compositions. Silica spectrum is from [58], GeO$_2$ from [59], and tellurite (TBZN: TeO$_2$–Bi$_2$O$_3$–ZnO–Na$_2$O) from [46]. FG spectra were recorded with micro Raman spectroscopy system. ChG spectra were evaluated in a pump-probe amplification experiment

Fig. 3.3 Basic structure of zirconium fluoride glass (ZrF$_4$) showing ZrF$_6$ octahedra (Adapted from [66]) (F_b bridging fluorine atom, F_{nb} non-bridging fluorine atom)

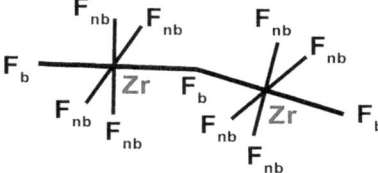

maximum gain is near 510 cm^{-1} (15.3 THz) [70]. To the best of our knowledge, no experimental characterization of the absolute gain coefficient exists for these two FG compositions.

Finally, chalcogenide fibers present a generally higher gain than fluoride fibers but a much weaker spectral shift. Arsenic sulfide (As$_2$S$_3$) and arsenic selenide (As$_2$Se$_3$) have respective maxima at frequency shifts of 340 cm^{-1} (10.2 THz) and 230 cm^{-1} (6.9 THz) [71]. These peaks correspond to the symmetric stretching vibrations of sulfur (or selenium) atoms inside a pyramidal structure (see Fig. 3.4) [72]. Their absolute Raman gain can reach very high values: up to 5.5×10^{-12} m/W

Fig. 3.4 Basic structure of
As$_2$S$_3$ glass showing the
pyramidal structure for three
molecules (I, II, and III) [72]

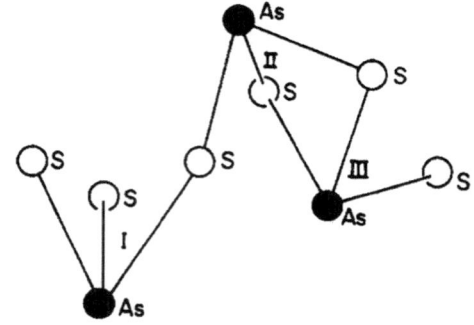

for As$_2$S$_3$ and 2.7 × 10^{-11} m/W for As$_2$Se$_3$, at 2 μm [71, 73]. These values are at
least two orders of magnitude higher compared to silica glass.

Generally speaking, the conversion efficiency from pump to Stokes wavelength
is not only fixed by the material Raman gain; it also depends on the intensity
generated inside the fiber. The effective Raman gain is thus defined as g_R/A_{eff}, where
A_{eff} is the effective area of the LP$_{01}$ mode. The material (intrinsic) Raman gain is
assumed to scale as $1/\lambda$, and the mode area scales approximately as $1/\lambda^2$ if the
core size is properly adjusted to preserve the same V number (see Gaussian beam
approximation for LP$_{01}$ mode [74]). As a result, the effective Raman gain has an
approximate $1/\lambda^3$ dependence. To bypass this major limitation at MIR wavelengths,
fibers that allow a strong modal confinement (i.e., with high NA and small core) are
generally selected.

3.2.1.3 Figure of Merit (FOM)

The characteristics described above allow us to evaluate a figure of merit (FOM)
for each glass. This FOM must take into account the attenuation (α_S) at the Stokes
wavelength as well as the material Raman gain (g_{RP}) at the pump wavelength – i.e.,
the parameters that have the most impact on the system's performances. We thus
simply define the FOM as the ratio of gain (per unit of pump power) and losses
divided by the square of the Stokes wavelength (λ_S), as shown in the following
equation:

$$\text{FOM} = \frac{g_{RP}}{\lambda_S^2 \alpha_S}. \tag{3.1}$$

Table 3.2 lists the values calculated at 1.5, 2.5, and 3.5 μm for common glass
compositions. In these calculations, a realistic attenuation coefficient was assumed
for each fiber, taken either from fiber manufacturer data or recent experimental
papers. The Raman gain coefficients listed in Table 3.2 are valid for two unpolarized
beams (pump and Stokes). For two identically polarized beams, the gain coefficient
should be twice as large.

Table 3.2 Figure of merit (FOM) of various MIR materials used in Raman fiber lasers

Material	Wavelength (λ_S) nm	Raman gain (g_{RP}) 10^{-14} m/W	Attenuation (α_S) m^{-1}	FOM
Silica	1500	3.6 [75]	0.00003 [52]	533.3
Heavily GeO$_2$-doped silicate	1500	31.9 [61]	0.009 [76]	15.8
	2500	19.9	1.04 [77]	0.03
Fluorozirconate	1500	4.9 [68]	0.006 [16]	3.6
	2500	3.1	0.0003	16.5
	3500	2.3	0.06	0.03
As$_2$S$_3$	1500	390 [73]	0.12 [78]	14.4
	2500	242	0.03	12.9
	3500	178	0.006 [73]	24.2
As$_2$Se$_3$	1500	1836 [71]	0.12 [78]	68.0
	2500	1126	0.08	22.5
	3500	822	0.06	11.2

It comes as no surprise that silica is the material of choice for shorter wavelengths (i.e., at 1.5 μm). As the wavelength is increased to 2.5 μm, fluorozirconate, As$_2$S$_3$, and As$_2$Se$_3$ glasses are expected to offer comparable performances. However, at 3.5 μm, As$_2$S$_3$ stands out quite clearly because of its low attenuation in this spectral range. Note that the material's damage threshold was not considered in these calculations. For similar FOMs, glasses having a higher damage threshold are obviously better choices for achieving high powers (e.g., fluorozirconate fibers instead of ChG fibers at 2.5 μm).

3.2.2 Bragg Gratings in MIR Fibers

Fiber Bragg grating (FBG) technology is key for the development of monolithic fiber lasers, generally used as integrated end mirrors in the cavity [79] or in fancier schemes as all-fiber dispersion compensators for short-pulse generation [80]. Since the FBG structure is directly inscribed in the fiber core, the laser cavity is fairly well isolated from external perturbations. This gives to fiber lasers a well-recognized stability and ruggedness that allow for power scaling to the multi-kW regime in both continuous and pulsed operation modes [81]. In this section, we discuss the specific characteristics sought in FBGs used in efficient MIR Raman cavities. Previous studies on the inscription of FBGs in soft glasses are then reviewed, relying either on standard UV-based writing techniques or on the most versatile ultrashort pulse approach.

3.2.2.1   FBG Design for Efficient MIR Raman Cavities

Similarly to fiber laser cavities based on rare-earth gain mediums, the FBG design required to maximize RFL efficiency is usually composed of a highly reflective and broadband input cavity coupler (IC) and an overlapping narrowband output cavity coupler (OC) having a peak reflectivity adapted to maximize RFL performances. Such a cavity design ensures that the signal generated in the laser cavity is efficiently reflected back by the IC and delivered to the actual laser output (i.e., through the OC). The IC targeted design generally has a bandwidth of 3–5 nm and a peak reflectivity in excess of 99.5%. Such design requires the use of a chirped FBG to achieve a large bandwidth from a finite refractive index modulation (Δn_{AC}). The FBG's chirp rate and length are chosen according to the available Δn_{AC}, which can vary significantly from one glass composition to another. An IC FBG length in the range of 1–4 cm is generally a good target to mitigate the writing stability challenges associated with very long FBGs and considering that a permanent Δn_{AC} of the order of 10^{-3} is achievable. In these conditions, an FBG chirp rate in the range of 0.3 and 1.3 nm/cm can meet the design. If the available Δn_{AC} is lower, an increase in the ratio FBG length/chirp rate has to be considered.

The design of the OC is generally much less demanding compared to the IC since the bandwidth–reflectivity product is lower. The OC targeted design generally has a bandwidth of 0.5–1.0 nm and a peak reflectivity in the range 20–80% depending on the Raman cavity design. Such FBG design can generally be achieved with a short-length uniform period FBG even with a relatively low Δn_{AC}. However, the OC Bragg wavelength (λ_B) has to be precisely controlled to match the center of the IC reflectivity spectrum and ensure a good overlap between them. The use of a Gaussian apodization is not necessary but preferred in order to reduce the reflectivity sidebands [82] to ensure a stable laser emission. This is particularly important when a high-reflectivity OC is considered for which the sidebands can have a significant value.

Examples of reference designs for both a highly reflective IC ($R_{max} > 99.5\%$) and OCs of different reflectivities ($R_{max} = 20\%$, 50% and 80%) are presented in Fig. 3.5. We consider an operating wavelength of 2.94 µm and a refractive index of 1.48 (e.g., fluoride glasses). Table 3.3 shows a set of grating parameters (length, chirp rate, and index modulation) that allows to meet RFL requirements in terms of bandwidth and maximum reflectivity. The reflectivity spectra are then computed using the *OptiGrating V4.2.3* software (*Optiwave Inc.*), as shown in Fig. 3.5.

It is important to note that the effective reflectivity (R_{eff}) of FBGs in a Raman cavity can be significantly lower than their maximum value (R_{max}). In fact, intracavity spectral broadening, an effect usually attributed to self-phase modulation and four-wave mixing, is often observed in such cavities since high core intensities are involved. This effect is stronger in highly nonlinear fibers (i.e., chalcogenides) and when the laser is operated close to the zero dispersion wavelength. The effective reflectivity associated with a FBG is given as follows [83]:

Table 3.3 Characteristics of the reference design FBGs ($n_{eff} \approx 1.48$ and $\lambda_{B\ center} \approx 2.94$ μm)

FBG	Length cm	Chirp rate nm/cm	Index mod. (Δn_{AC}) 10^{-4}	Bandwidth nm	Max. reflectivity (R_{max}) %
IC	2.5	3.6	9.25	5 at R ≥ 99.5%	>99.5
OC$_1$	0.6	0	0.20	0.68 (FWHM)	20
OC$_2$	0.6	0	0.36	0.80 (FWHM)	50
OC$_3$	0.6	0	0.59	1.04 (FWHM)	80

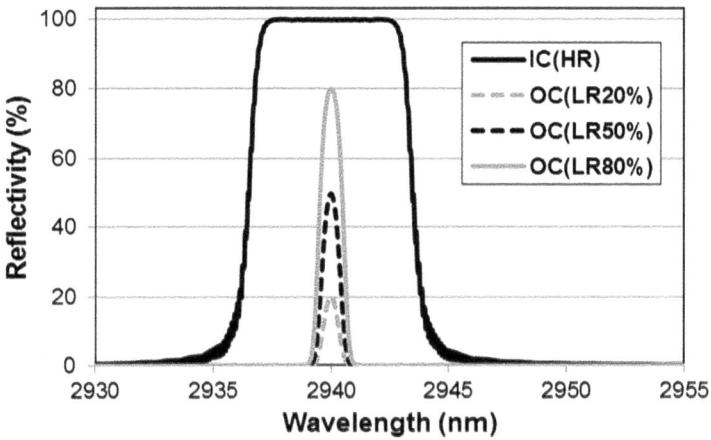

Fig. 3.5 Reflectivity spectra of the reference design for the IC FBG and OC FBGs of 20%, 50%, and 80% peak reflectivity

$$R_{eff} = \frac{P_{reflected}}{P_{incident}} = \frac{\int S_{intracavity}(\lambda)\,R(\lambda)\,d\lambda}{\int S_{intracavity}(\lambda)\,d\lambda} \tag{3.2}$$

where $R(\lambda)$ and $S_{intracavity}(\lambda)$ are, respectively, the FBG and intracavity laser spectra. Spectral broadening affects the laser performances by allowing a significant portion of the intracavity power to leak outside the cavity. It is possible to mitigate this effect by selecting a broadband IC and a narrowband OC (as in the reference design) to ensure that no leaks occur through the input FBG [84]. In these conditions, only the effective reflectivity of the output FBG is reduced, and the degradation of laser performances can be avoided altogether with a careful choice of R_{max} (i.e., higher than the optimal value of the cavity) [85].

As for other considerations, since the Raman gain is generally quite low in the mid-infrared (see Sect. 3.2.1), the losses associated with FBG inscription can be an issue and have to be minimized through the optimization of the inscription process. Minimizing FBG losses of the IC at the pump wavelength is also very critical to ensure efficient pump coupling since RFLs are generally core-pumped with shorter wavelengths and since coupling to cladding modes which overlap the short wavelength region of the FBG is generally a concern for highly reflective and broadband FBGs. Because of the presence of such cladding mode coupling losses, a very important aspect to implement is the orientation of the chirped IC FBG. The shorter Bragg wavelength side of the IC chirped FBG should always be oriented toward the RFL cavity to minimize the influence of such losses in the laser cavity [86].

3.2.2.2 Early Studies of FBG Inscription in Soft Glasses

Fluoride glasses were first studied for FBG writing by using the standard UV laser radiation that coincides with an absorption band of the glass. However, it was shown that only a very weak photosensitivity could be obtained (with a maximum Δn_{AC} of the order of 10^{-6}) using 193 nm radiation [87], whereas an even lower photosensitivity was achieved with 248 nm radiation [88]. Doped fluorozirconate glasses with various rare-earth elements were also studied [88], and only glasses doped with cerium displayed a significant increase in photosensitivity [89, 90]. Such approach was recently pursued, and the possibility of writing high-reflectivity FBGs in cerium–thulium co-doped glass fibers was reported [91]. Unfortunately, the maximum stable Δn_{AC} achieved is of the order of 5×10^{-5} which significantly limits the FBG design and does not currently allow to meet the RFL FBG reference design detailed above.

Various FBG writing techniques were also investigated for optical fibers made of chalcogenide glasses (ChGs). Unfortunately, ChGs do not transmit in the UV region given their low-energy bandgap which prohibits the use of standard UV-based writing techniques for FBG fabrication. Many attempts at writing FBGs in both As_2S_3 and As_2Se_3 fibers were reported using transverse exposure of near- or sub-bandgap CW light. Since As_2S_3 glass has a bandgap of $\sim$620 nm (short wavelength absorption edge at 1 cm^{-1}) [92], the use of a He–Ne laser operating at 633 nm is suitable to access the core's material photosensitivity by transverse exposure. The transverse holographic method was first demonstrated [93] and yields an FBG with a maximum Δn_{AC} of $\sim$5 $\times$ 10^{-5} following an exposure of about 20 min. In subsequent reports, this method was replaced with the phase-mask inscription setup which is more mechanically stable [94, 95]. The spectral responses of the resulting FBGs were relatively poor, and the required exposure time to obtain a significant photosensitivity (max $\Delta n_{AC} \sim 10^{-4}$) was up to an hour long. Similar results were reported in As_2Se_3 glass fibers (short wavelength absorption edge at 1 cm^{-1} of 850 nm) [92] by using sub-bandgap exposure at 785 nm from a semiconductor laser and the phase-mask writing technique [96] with similar mitigated success. More recently, successful attempts at writing FBGs with a large index modulation using even shorter exposure wavelengths of 532 nm for As_2S_3 (max. $\Delta n_{AC} \sim 1.5 \times 10^{-3}$) [97] and 633 nm for As_2Se_3 (max. $\Delta n_{AC} \sim 6 \times 10^{-3}$) [98] were reported with the added requirement of tapering the fibers to reduce the cladding diameter to access the core's photosensitivity. Such results demonstrate the strong photosensitivity of

ChGs with the potential to meet the RFL FBG reference design detailed above by using exposure to sub-bandgap radiation while also revealing the limitation of using the side exposure in the linear absorption regime with standard (i.e., not tapered) fibers.

3.2.2.3 FBG Inscription with Ultrashort Pulses

For most MIR-transmitting glasses, the common FBG writing technologies discussed above cannot meet the requirements of the reference design. A new approach to photosensitivity based on the nonlinear absorption of ultrashort pulses was demonstrated in the late 1990s and led to major advances in the processing of transparent materials [99]. It was originally demonstrated by Davis et al. [100] that a refractive index change can be induced inside various bulk materials including silica and fluoride glasses by focusing femtosecond pulses at 800 nm. Such new photosensitivity technique was adapted to FBG writing soon after with the demonstration of FBG inscription in standard germanosilicate fibers using the phase-mask technique [101] without the need for photosensitization with hydrogen or deuterium to reach a high level of photosensitivity (max. $\Delta n_{AC} \sim 2 \times 10^{-3}$). In addition, since most of the fiber jackets are transparent at a wavelength of 800 nm, such technique was recently reported to efficiently write FBGs through the polymer jacket of standard germanosilicate fibers [102]. This feature eliminates the need for the cumbersome stripping/recoating of the fiber associated with the traditional FBG writing process, which is particularly important in the context of processing mid-infrared transmitting fibers known to be mechanically weak. Most of the mid-infrared materials, such as rare-earth-doped and undoped fluoride glass fibers [14] and As_2S_3 ChG fibers [103], were tested successfully for FBG writing with this new approach. Note that the physical properties of MIR glasses, which significantly differ from those of silica glasses, were found to lead to thermomechanical instabilities when exposed to femtosecond pulses at high fluence. A special control over laser exposure conditions was found necessary in order to ensure an exploitable photosensitivity (i.e., $\Delta n_{AC} \sim 10^{-3}$) [104]. Such control was refined during the following years to keep step with the ongoing development of mid-infrared RFLs and rare-earth-doped fiber lasers.

A schematic of the experimental setup used to write FBGs using femtosecond pulses at 800 nm and the phase-mask technique is shown in Fig. 3.6. The incident femtosecond pulsetrain having a pulse duration of 35 fs at 1 kHz repetition rate is provided by a Ti-sapphire laser (Coherent Legend-HE). Such pulses are focused along the fiber using a cylindrical lens, through the phase-mask that splits the beam in several diffraction orders. The phase-mask diffraction efficiency is generally optimized to diffract most of the energy in the ± 1 orders. These two diffracted beams interfere together after the phase mask. Such interference pattern is then imprinted along the fiber core placed at the focal plane of the cylindrical lens to construct the Bragg grating along a certain length of the fiber. To increase the FBG

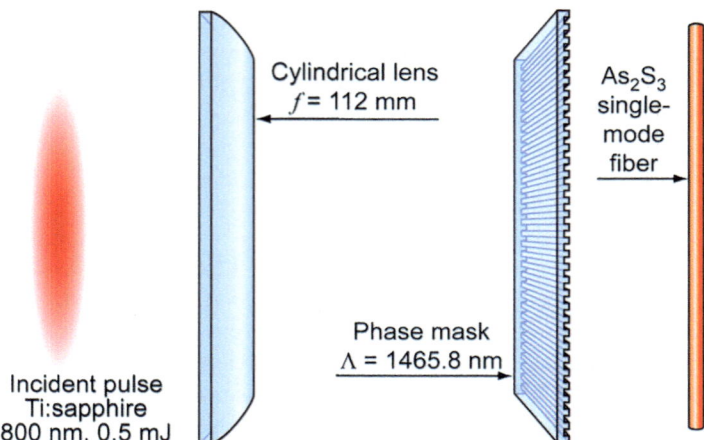

Fig. 3.6 Schematic of an FBG writing setup using femtosecond pulses and a phase mask (From [105])

length, the focused beam is translated parallel to the fiber axis, while the fiber and the phase mask are held in close proximity.

As an example of real components fabricated using this technique, a pair of IC and OC FBGs was written in an undoped fluoride glass fiber (15/250 μm, NA = 0.125, *Le Verre Fluoré inc.*). The IC design was based on the use of a chirped phase mask having a central period of $\Lambda_{PM} = 1980.5$ nm leading to a central Bragg wavelength of 2936 nm ($\lambda_B = 2 \cdot n_{eff} \cdot \Lambda_{FBG}$, with $n_{eff} = 1.4825$ and $\Lambda_{FBG} = \Lambda_{PM}/2$). The phase mask had a chirp rate of 1.6 nm/cm, leading to an FBG chirp rate half of that value (i.e., 0.8 nm/cm). The beam was scanned during the IC fabrication over a length of 2.0 cm to reach the desired bandwidth of ≥ 5 nm (5.4 nm at $R \geq 99.5\%$), and the laser fluence was adjusted to reach the desired peak reflectivity. The resulting IC FBG transmission spectrum is shown in Fig. 3.7 (black curve) from which we can infer through modeling a $\Delta n_{AC} \approx 1.8 \times 10^{-3}$. Note that cladding modes are quite evident in the transmission spectrum of the IC FBG, where the loss band is overlapping the short wavelength side of the FBG reflectivity spectrum and extends down to 2750 nm with significant losses (i.e., 10–50%). As mentioned previously, the IC FBG should be properly oriented (i.e., shorter Bragg wavelengths toward the cavity) to reduce the impact of such losses [86].

The narrowband OC FBG was fabricated with a uniform phase mask having a period of 1983 nm, leading to a Bragg wavelength of 2939 nm, a few nm above the IC central bandwidth. A controlled strain was applied during the OC FBG fabrication in order to reach a final Bragg wavelength (i.e., after releasing the strain) of 2937 nm, well centered in the IC bandwidth. The writing fluence was adjusted to reach the targeted reflectivity of 20%, and the beam was scanned over a length of 0.5 cm to provide the targeted bandwidth of 0.8 nm. Given the low targeted reflectivity, the apodization of the grating was not implemented for this OC. The

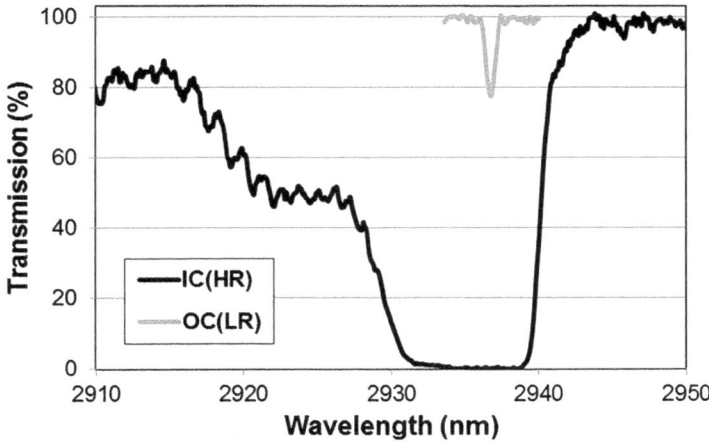

Fig. 3.7 Transmission spectra of a pair of IC and OC FBGs inscribed in fluoride glass fibers with the reference design

resulting OC FBG transmission spectrum is shown in Fig. 3.7 (gray curve) from which we can infer through modeling a $\Delta n_{AC} \approx 1.3 \times 10^{-4}$. Both FBGs were thermally annealed at 120 °C during 30 min after their inscription to ensure a long-term stability of their spectral response.

Another example of real components fabricated using the same technique is a pair of IC and OC FBGs written in an As_2S_3 ChG fiber (4/145 μm, NA = 0.36, *COPL/CorActive High-Tech Inc.*). The IC design was also based on the use of a chirped phase mask having a central period of $\Lambda_{PM} = 1422$ nm leading to a central Bragg wavelength of 3338 nm ($n_{eff} = 2.3474$). The phase mask had a chirp rate of 1.3 nm/cm, leading to an FBG chirp rate of 0.65 nm/cm. The beam was scanned during the IC fabrication over a length of 2.5 cm to reach the desired bandwidth of ≥ 5 nm (8 nm at $R \geq 99.5\%$). The laser fluence was adjusted to maximize the index modulation, which explains the significantly larger bandwidth than the target since ChG glasses are known to be less thermally stable than fluoride glasses given their lower Tg (180 °C for As_2S_3 vs 360 °C for zirconium fluoride glass). The resulting IC FBG transmission spectrum is shown in Fig. 3.8 (black curve) from which we can infer through modeling a large $\Delta n_{AC} \approx 7.5 \times 10^{-3}$.

The narrowband OC FBG was fabricated with a uniform phase mask having a period of 1424 nm, leading to a Bragg wavelength of 3342 nm, a few nm above the IC central bandwidth. A controlled strain was also applied during the OC FBG fabrication in order to reach a final Bragg wavelength (i.e., after releasing the strain) of 3338.8 nm, well centered in the IC bandwidth. The writing fluence was adjusted to reach the targeted reflectivity of 80%, and a large Gaussian beam of 10 mm diameter (at $1/e^2$) was directly used to provide the targeted FWHM bandwidth of 0.5 nm while providing the Gaussian uncompensated apodization profile. The resulting OC FBG transmission spectrum is shown in Fig. 3.8 (gray curve) from

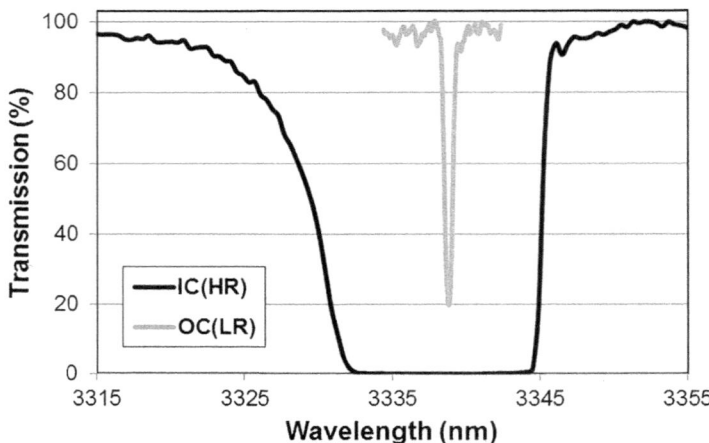

Fig. 3.8 Transmission spectra of a pair of IC and OC FBGs inscribed in As_2S_3 ChG fibers with the reference design

which we can infer through modeling a $\Delta n_{AC} \approx 6.5 \times 10^{-4}$. In this case, only the OC was thermally annealed at 100 °C during 60 min after its inscription to ensure a midterm stability of its spectral response. The thermal stability of FBGs written in As_2S_3 ChG fibers is still under investigation as well as FBGs written in As_2Se_3 ChG fibers that were not yet reported by using the femtosecond technique.

3.2.3 Optical Pump for MIR Raman Emission

The laser pump is a critical component in high-power MIR Raman laser sources. Above all requirements, the basic one of providing several watts of power at a wavelength above 2 μm is rather challenging. Although a few different laser types can reach such a power level in this spectral range (see Sect. 3.1.1), rare-earth-doped fiber lasers undoubtedly possess the best affinity with RFLs and are usually preferred. Indeed, when the pump active fiber and the Raman fiber have a similar geometry, it is possible to splice them (either mechanically or by fusion) to improve the robustness and the stability of the system. In cases where a large mode mismatch between the fibers exists, rare-earth-doped fiber lasers still facilitate efficient free-space coupling into the Raman cavity due to their excellent beam quality.

As discussed in Sect. 3.2.1 on MIR fibers, the effective Raman gain is subject to a $1/\lambda^3$ decrease as we raise the wavelength of the pump source. One strategy to ease the high pump power requirement is to nest the Raman oscillator inside the pump cavity [85]. Then, if the pump cavity is made of high reflectors, the intracavity power builds up quickly and the Raman threshold is readily achieved. This strategy is readily implemented when a rare-earth-doped fiber laser is selected as a pump source.

Two rare-earth-ion laser transitions are especially interesting for pumping Raman cavities in the MIR: the $^3F_4 \rightarrow {}^3H_6$ transition of Tm^{3+} ions near 2 μm and the $^4I_{11/2} \rightarrow {}^4I_{13/2}$ transition of Er^{3+} ions in the neighborhood of 2.8 μm. These transitions can be used to pump Raman cavities between 2 and 3 μm and between 3 and 4 μm, respectively. Alternatively, the 3.5 μm erbium transition in FGs ($^4F_{9/2} \rightarrow {}^4I_{9/2}$) can also be used to pump Raman cavities when aiming for even longer wavelengths. Since Raman fiber lasers mostly benefit from CW or long-pulse pumping (>ns), the description of the different pump sources is confined hereafter to these operation regimes.

3.2.3.1 Tm-Doped Fiber Laser Near 2 μm

The $^3F_4 \rightarrow {}^3H_6$ transition of thulium can generate a high-power laser emission between 1.9 and 2.1 μm. Several pumping schemes exist to stimulate this transition, namely, around 790 nm, 1210 nm, and 1630 nm [106]. Generally, 790 nm pumping is preferred as this allows powerful AlGaAs laser diodes to be used. Even with an unfavorable quantum defect (790/2000 ≈ 0.4), the 790 nm pumping efficiency remains high because of the cross relaxation process ($^3H_4, {}^3H_6 \rightarrow {}^3F_4, {}^3F_4$) that transfers two ions to the upper level of the laser transition [see CR in Fig. 3.9a] [108]. This effect is enhanced in fibers with higher doping levels and by using a 790 nm pump wavelength (compared to other pump schemes). Nonetheless, greater efficiency can still be achieved with intraband pumping around 1630 nm. However, since high-power laser diodes are not available at this nonstandard wavelength, other fiber lasers must be used as pump sources (e.g., Raman or Er^{3+}-doped).

The best results to date were obtained with silica fibers as host material. For continuous wave operation (CW), Tm^{3+} laser systems reaching 900 W laser at 2040 nm [shown in Fig. 3.9b] [107] and over one kilowatt at 2045 nm [109, 110] were reported. The broad gain generated by the Tm^{3+} ion can also be used advantageously in the development of tunable sources. For instance, a 300 W source with a 150 nm tunable range (1890–2050 nm) was recently reported based on an all-fiber architecture [111]. CW Tm^{3+}:silica fiber lasers operating at around 2 μm are currently commercialized by several manufacturers [11, 112, 113].

Q-switching of the $^3F_4 \rightarrow {}^3H_6$ transition has also been studied intensively [114–116] using both passive and active switching mechanisms. Nowadays, the most common approach to reach high average and peak powers is the master oscillator power amplifier (MOPA) configuration. The seed source for the power amplifier is typically made of a low-power version of a Q-switched Tm fiber laser. This approach has recently led to the generation of mJ nanosecond pulses with an average output power greater than 100 W [116].

Figure 3.10 illustrates the potential spectral coverage of Raman cavities made of various glass compositions and pumped with a Tm^{3+}:SiO$_2$ fiber laser. To set the limits of the Stokes signal range, the pump wavelength was assumed to be tunable between 1.9 and 2.1 μm. The width of the Raman gain spectrum (at FWHM) for each of the five common fiber materials was also considered. It appears from this

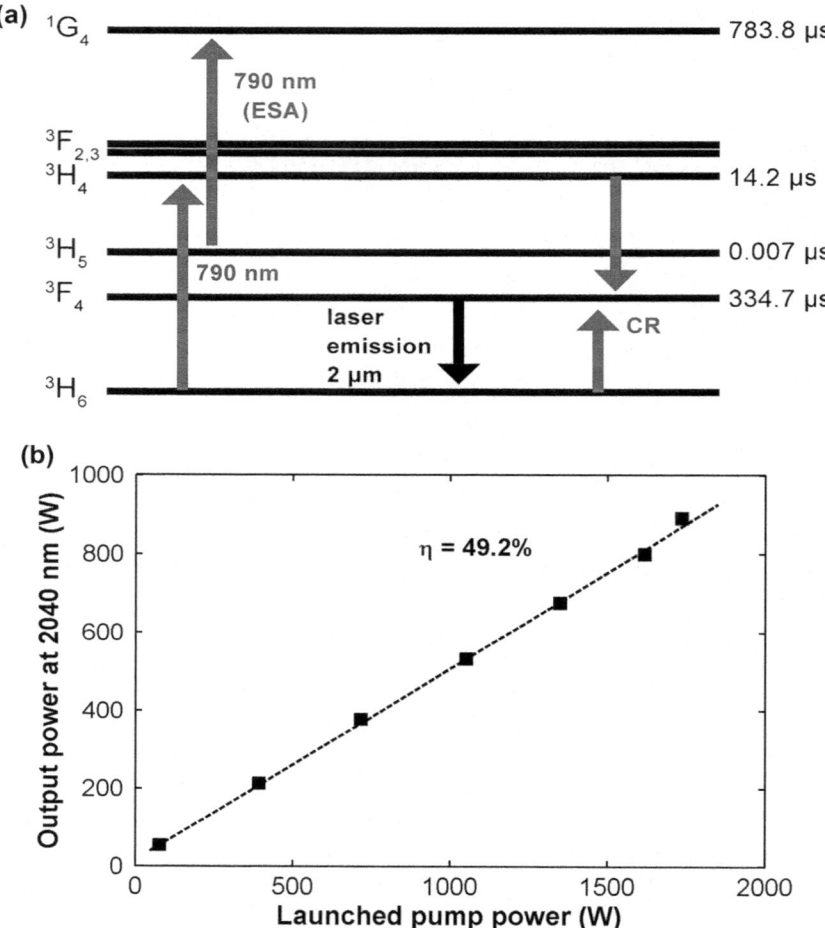

Fig. 3.9 (a) Simplified energy diagram for Tm^{3+} ions in silica glass (From [106]). (b) Output power at 2040 nm as a function of the launched pump power for a kW-class Tm^{3+}:silica fiber laser (Adapted from [107])

diagram that a fairly good coverage of the 2–3 μm spectral range is in principle possible, especially with tellurite glass, based on the first- and second-order Stokes shifts.

3.2.3.2 Er-Doped Fiber Lasers Near 2.8 and 3.4 μm

Erbium-doped fiber lasers emitting on the $^4I_{11/2} \rightarrow {}^4I_{13/2}$ transition have achieved the highest powers near 3 μm, mainly because they can be pumped with powerful

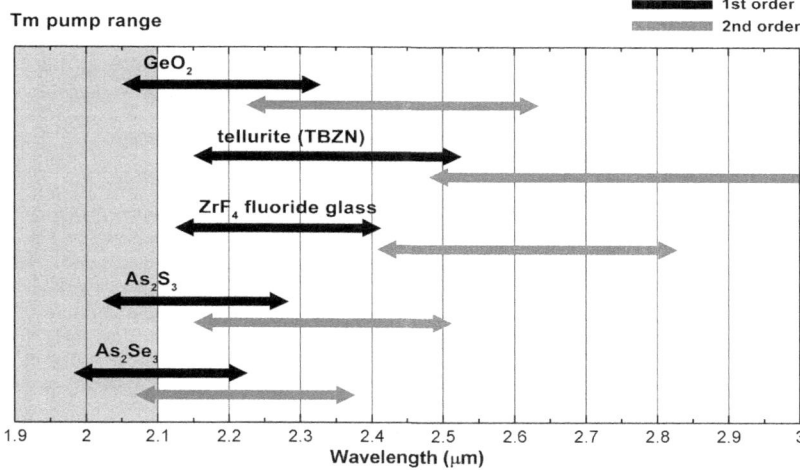

Fig. 3.10 Potential spectral coverage of Raman cavities pumped around 2 μm with a Tm^{3+}:SiO_2 fiber laser

commercial laser diodes emitting around 975 nm. Although this transition is self-terminating, the energy transfer upconversion process from level $^4I_{13/2}$ toward levels $^4I_{15/2}$ and $^4I_{9/2}$ [see ETU1 in Fig. 3.11a] helps to repopulate the upper laser level [118]. A high doping level of Er^{3+} ions enhances this process and can lead to performances exceeding the quantum efficiency. Efficient lasing is also possible with lightly doped Er^{3+} fibers if it is co-doped with Pr^{3+} ions to depopulate the lower laser level [119] or if co-lasing with the 1.55 μm transition is allowed [19, 117].

In FG fibers, the broad emission cross section of Er^{3+} ions allows tuning of the laser emission from 2.71 [120] to 3.005 μm [29]. Among the highest-power demonstrations in the CW regime, two 30 W class passively cooled all-fiber lasers were reported at 2.825 μm (27 W) [21] and 2.94 μm (30 W) [23], as shown in Fig. 3.11b. Furthermore, a lower power version of this laser is now commercially available [78].

There is also a significant interest in developing high-energy and high-peak power pulsed sources from this laser transition. Currently, the best performances from a Q-switched Er^{3+} cavity rely on an active control element (e.g., an acousto-optic modulator). In one demonstration, 90 ns pulses having an energy of 100 μJ (0.9 kW peak power) were generated [121]. The 120 kHz pulsetrain reached a maximum average power of 12 W. Pulse generation in the mode-locked (ultrafast) regime has also made significant progress in the last few years. The picosecond pulse duration regime was first reached with passive mode-locking cavities based on saturable absorbers [122]. More recently, femtosecond mode-locked fiber lasers were demonstrated near 2.8 μm based on nonlinear polarization rotation [123–125].

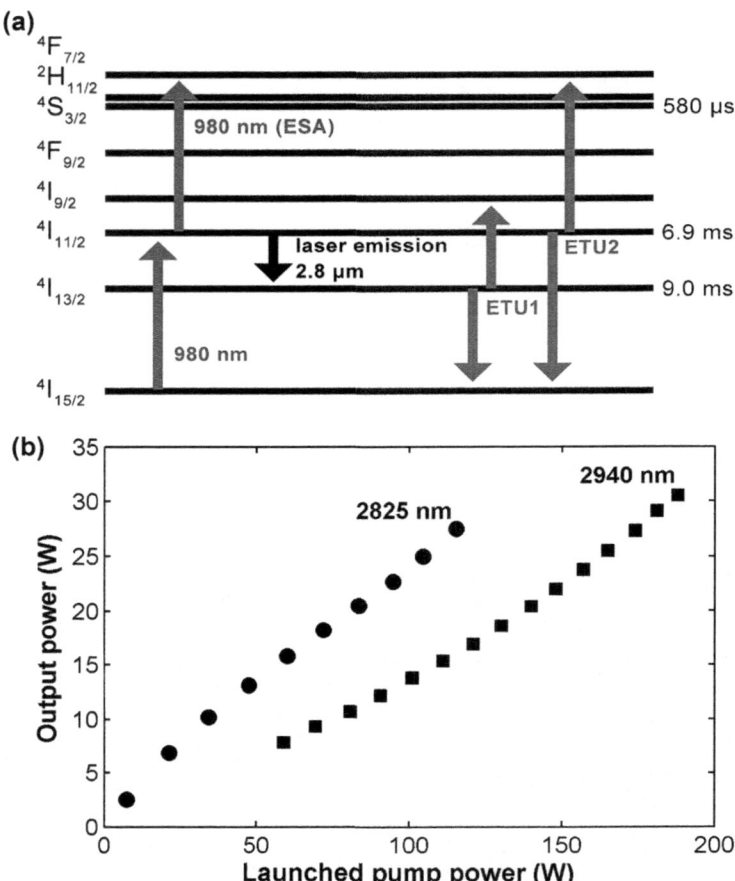

Fig. 3.11 (a) Simplified energy diagram for Er^{3+} ions in fluoride glasses (From [117]). (b) Output power as a function of the launched pump power for 2825 nm [21] and 2940 nm [23] Er^{3+}:FG fiber lasers

A record of ultrafast pulse duration of 207 fs was demonstrated by Duval et al. with an estimated peak power of 3.5 kW [123].

With their broad gain spectrum and high-power capability, Er^{3+} fiber lasers are generally regarded as the best choice for pumping Raman cavities between 3 and 4 μm. In fact, with no more than two Stokes shifts, any given wavelength can be reached in principle within this spectral range if a suitable Raman fiber is chosen (As_2S_3, As_2Se_3 fluoride, or tellurite glasses; see Fig. 3.12).

Erbium ions also have a second transition in the MIR ($^4F_{9/2} \rightarrow {}^4I_{9/2}$) that can produce 3.5 μm laser radiation. The first lasing demonstration was performed in the early 1990s by directly pumping the upper laser level at 650 nm with a dye laser [126]. It was not until recently that such lasers were proved to be much more

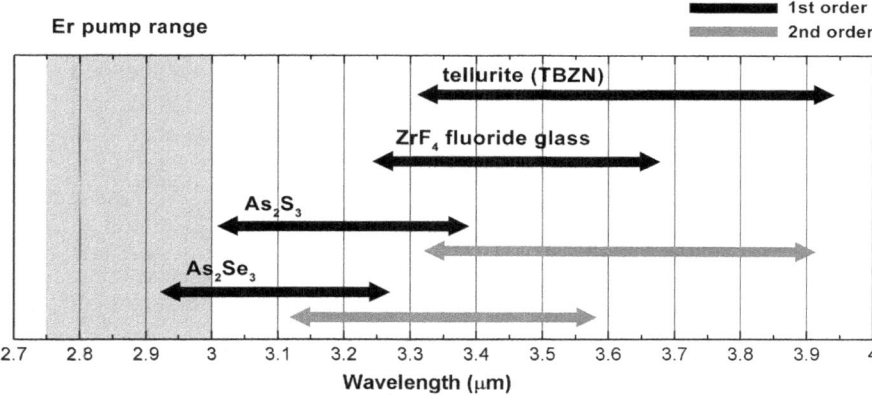

Fig. 3.12 Potential spectral coverage of Raman cavities pumped around 2.8 μm with an Er^{3+}:FG fiber laser

efficient under dual-wavelength pumping, i.e., by using a first pump at 974 nm pump to generate a virtual ground state on the longer-lived $^4I_{11/2}$ level and a second pump at 1976 nm to promote the ions to the upper laser level [127]. Recent demonstrations based on this new approach have reached up to 1.5 W of continuous output power at 3.5 μm [26, 27]. This laser transition can also operate on a broad range of wavelengths, from 3.33 to about 3.78 μm [27]. As a result, it could be a potential pump source for Raman cavities operating at longer wavelengths, especially to reach the spectral range beyond 3.9 μm with a single Stokes shift. Further development is however required to reach multi-watt power levels and meet the requirements of a Raman cavity pump.

3.2.3.3 Pump Coupling in the Raman Cavity

Launching an intense pump laser beam into a high-confinement single-mode fiber always requires great care to maximize efficiency and stability of the system. The situation becomes even more challenging when fluoride and chalcogenide fibers are involved because of their weak mechanical resistance and low laser damage threshold.

When fiber splicing is not an option, the pump coupling into the Raman cavity must be performed by manual alignment. If the mode field diameter (MFD) of the fibers is similar, the simplest approach is to proceed with a butt-coupled alignment. Alignment of the fiber cores is then carried out with precision mounts and often results in good launch efficiency exceeding 70%. To ensure long-term stability of the junction point, the fibers must be slightly pressed against each other. In general, the high-power stability can be improved by adding Peltier effect modules or a water-cooled heat sink to cool the fiber joint.

In most cases, however, the MFD of the pump does not match that of the Raman fiber. In fact, a large-diameter core is generally preferred for the doped (pump) fiber to inhibit nonlinear effects, whereas the fibers designed for Raman cavities often have a strong modal confinement (i.e., small core) to maximize Raman conversion. Given these conditions, one must usually turn to a two-lens system to adapt the modes and achieve sufficient launching efficiency. An aspherical lens made of MIR materials such as sapphire, silicon, germanium, chalcogenides (ZnSe, ZnS, GeSbSe, Black Diamond, etc.), or fluoride compounds (CaF_2, MgF_2, BaF_2, etc.) is usually selected to that purpose. As some of these materials have a high refractive index, they require antireflection coatings deposited on both sides.

A fusion splice is obviously the ideal junction between pump and Raman fibers. These low-maintenance junctions typically lead to lower losses and offer much better stability. Splicer systems based on resistive filaments, such as the ones manufactured by Thorlabs (e.g., models FFS-2000 and GPX-3000), provide sufficient temperature control to achieve a high-quality splice between soft-glass fibers. While it is possible to develop splice recipes for soft-glass fibers with similar compositions, their repeatability is usually limited (i.e., requiring several attempts) due to varying cleave angles and core/cladding eccentricity along the fiber. Nevertheless, doped and undoped fluoride fiber splices were reported in several high-power demonstrations [21, 23]. This type of splice is generally quite robust and can have very low losses (less than 10%) if an active alignment is performed. When the two fibers are made of different materials (i.e., silica/soft glass, FG/ChG, etc.), achieving a robust junction is not always possible. The challenge stems primarily from the differences in glass transition temperatures (T_g) and in thermal expansion coefficients. To deal with these differences, the filament can be offset toward the fiber with the higher T_g so as to generate a temperature gradient [128]. In some cases, it is also possible to strengthen the bond between dissimilar fibers by applying a coating on one of the fibers prior to the splicing process [129]. In the last few years, significant progress was made in splicing silica to different soft glasses, namely, to fluoride [23, 129, 130] and chalcogenide [131] glasses. To date, however, the splicing of FG fibers to ChG fibers has not been reported. This component would be a key technological advance for developing Raman all-fiber lasers between 3 and 4 μm.

3.2.4 Special Considerations Related to MIR Raman Fiber Lasers

3.2.4.1 Hot Spots in Fiber Polymer Jacket

UV curable polyacrylate polymers are the most standard coatings used on optical fibers [132–134]. The structure of these materials is complex, but it is mostly composed of hydrogen, carbon, oxygen, and nitrogen atoms. Since the mid-infrared region is where most fundamental molecular absorption bands are found, these

Fig. 3.13 Absorption coefficient of a standard polyacrylate fiber coating between 2.2 and 3.2 µm

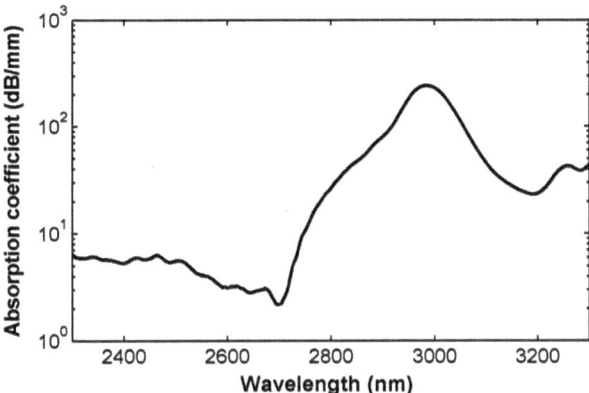

materials typically suffer from a limited transmission in this spectral range. This is especially true around the common molecular absorption peaks, namely, near 2.75–2.9 µm (O–H stretch), 2.85–3.2 µm (N–H stretch), and 3.3–3.45 µm (C–H stretch) [32]. Figure 3.13 presents the transmission spectrum of a typical polyacrylate fiber coating in the 2.2–3.3 µm spectral range. The attenuation reaches a maximum value of 240 dB/mm at 2990 nm due to a resonance with the N–H bond. Because of such strong absorption, any optical component that introduces radiation leakage in the MIR laser cavity will in turn induce a local temperature increase in the surrounding polymer coating. Extra care should thus be taken to minimize the losses of splices and fiber Bragg gratings. Thermal management of these components with metallic heat sinks is usually required to avoid catastrophic failure of the fiber. Fiber segments containing core defects (e.g., due to local crystallization, impurities, etc.) should also be removed since they can scatter light toward the polymer coating. These defects can be localized with an IR imaging camera and a laser source [135].

3.2.4.2 Fiber Tip Deterioration

Catastrophic fiber tip photodegradation was reported to occur in fluoride glass fibers at high power in the neighborhood of 2.8 µm [136]. This power-dependent avalanche-type effect was found to be related to the diffusion of OH impurities in the fluoride glass fiber, which can be modeled according to Fick's laws. Since OH bonds strongly absorb light near 2.8 µm, the fiber tip's temperature is raised until catastrophic damage occurs. Figure 3.14 shows the typical evolution of a fluoride fiber tip for several watts of output power at 2.8 µm. Note that both the pump fiber (e.g., Er^{3+}:FG) and the Raman fiber can experience this type of damage. To date, this problem was mainly observed in FG fibers [23, 136]; it is not clear if chalcogenides will suffer from a similar effect. Possible solutions include the fabrication of endcaps to reduce the laser intensity on the tip (and in some cases, to

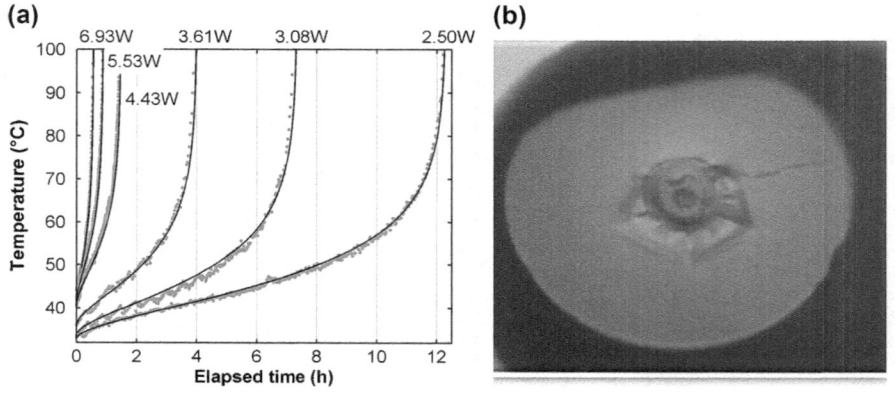

Fig. 3.14 (a) Evolution of the temperature of a fluoride fiber tip for different 2.8 μm output powers, ultimately leading to fiber damage [136]. (b) Image of a thermally damaged fluoride fiber tip

introduce a material with high chemical resistance), the addition of a nitrogen flux to lower the presence of moisture in the atmosphere, and a good thermal management with proper heat sinks.

3.3 Oxide-Based Raman Fiber Sources

Silica-based fibers have been used for more than 40 years to build near-infrared Raman fiber lasers [137]. With their high mechanical robustness and exceptionally low attenuation between 1 and 2 μm, these fibers were undoubtedly the best choice in this spectral range. As we move on to longer wavelengths, however, their attenuation raises significantly and the design of efficient RFLs becomes challenging. As discussed in Sect. 3.2.1, one successful approach for reaching longer wavelengths is based on germanosilicate fibers having extremely high germania (GeO_2) content, often more than 50 mol%. High concentration of GeO_2 not only raises the Raman gain coefficient, but it can also widen the glass transparency window toward MIR wavelengths. In addition, it also improves the photosensitivity of the glass which eases the fiber Bragg gratings inscription process. In the present section, we review the best results achieved with such fibers. Potential future directions for oxide-based RFLs emitting at long wavelengths are also discussed.

3.3.1   CW Raman Laser Cavities Based on Heavily GeO$_2$-Doped Silicate Fibers

The first investigations of CW GeO$_2$-based RFLs emitting beyond 2 μm were carried on by Dianov et al. in 2004 [76, 138]. Their Raman oscillator relied on a 75 mol.% GeO$_2$-doped silica fiber having a 2 μm diameter core and a single-mode cutoff wavelength near 1.42 μm. While the attenuation of this fiber was below 40 dB/km for wavelengths around 2 μm, it increased to a value of 150 dB/km at 2.2 μm. Cascade Raman conversion using a Yb/Er co-doped fiber laser pump at 1608 nm was first investigated [76]. Two different experimental Raman cavities were reported: a 13-m three-cascade cavity at 2027 nm and an 8-m four-cascade cavity reaching a maximum wavelength of 2193 nm. In the latter demonstration, a maximum output power of 210 mW at 2193 nm was achieved for a pump power of 4.6 W [138]. However, the slope efficiency was rather low at 6%. Figure 3.15 illustrates both the setup used in this experiment and the output spectrum

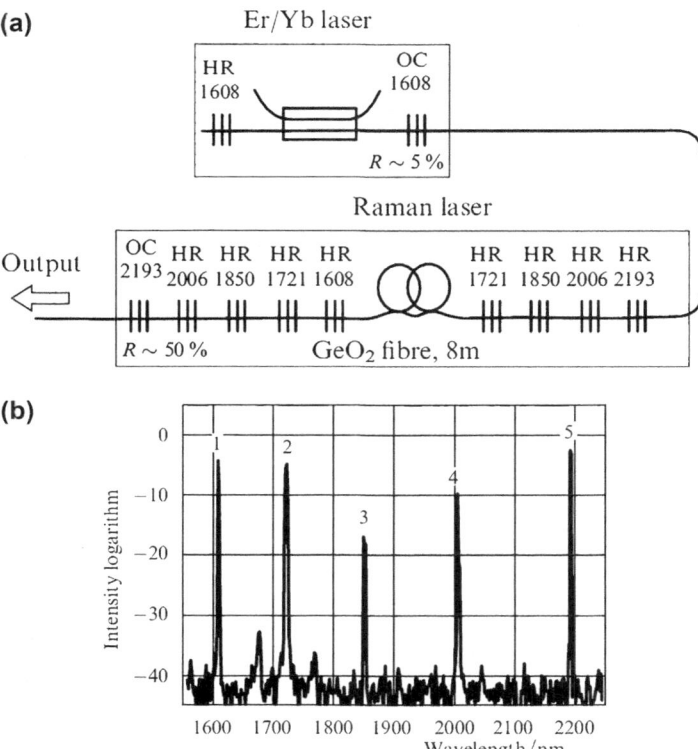

Fig. 3.15 (**a**) Schematic of a cascaded GeO$_2$ Raman fiber laser emitting at 2193 nm [76]. (**b**) Output spectrum showing the different Stokes orders generated by the nested Raman cavities [138]

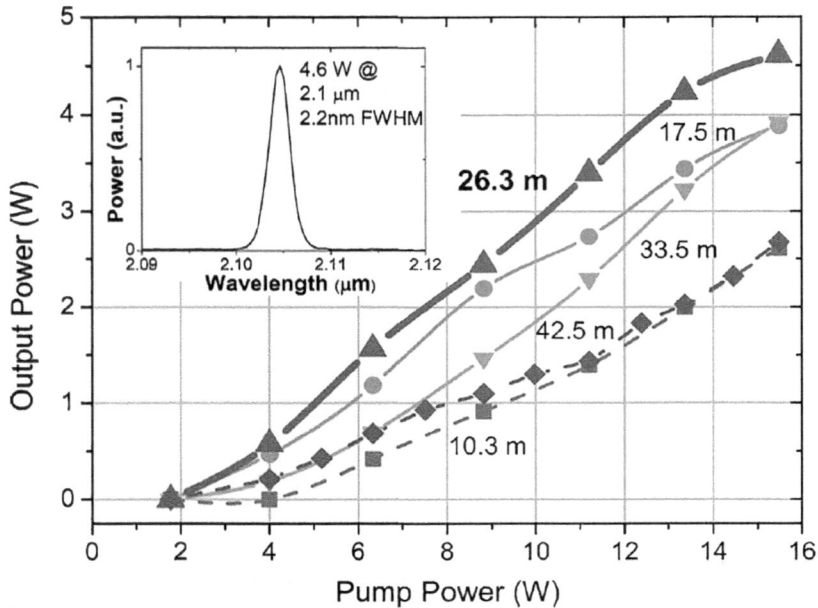

Fig. 3.16 Output power curves of a high-power GeO$_2$ Raman laser cavity made of different fiber lengths (10.3, 17.5, 26.3, 33.5, 42.5 m) [139]. The inset shows the spectrum at maximum power

composed of the different Stokes orders emitted by the cascaded cavities. The other demonstration, at 2027 nm, proved much more efficient (39% slope efficiency) since the fiber's attenuation was much lower at this Stokes wavelength.

Note that in all these demonstrations, a significant amount of power leaked out for each of the intermediate Stokes orders even though highly reflective output gratings were used. This observation suggests that intracavity spectral broadening was present.

In a later experiment, the power-scaling capability of these lasers was demonstrated by replacing complex pump sources by a far more convenient Tm-doped fiber laser operated at 1938 nm [139]. Such pump laser allows the generation of Stokes wavelengths beyond 2 μm with a single Raman shift. Raman conversion was investigated with GeO$_2$ fiber lengths from 42.5 to 10.3 m by performing a cutback between each test. The optimal cavity length was found to be 26.3 m and led to the generation of 4.6 W at 2.105 μm for a pump power of 22 W (Fig. 3.16). A lasing efficiency of more than 30% was also measured. In addition, the authors detected spontaneous emission near 2.3 μm (second Stokes order) and discussed the possibility of reaching longer wavelengths with the same fiber.

More recently, an alternative approach was reported for generating high power beyond 2 μm through Raman amplification in germanosilicate fibers [140]. It relies on a MOPA configuration, in which two seeds [one at the Raman pump wavelength (1963 nm) and a second one at the first-order Stokes wavelength (2147 nm)] are

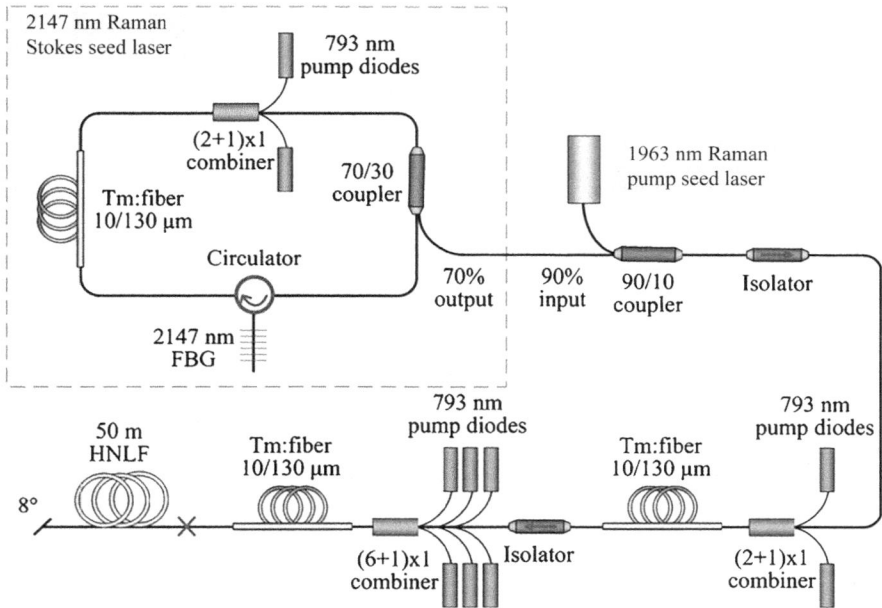

Fig. 3.17 Schematic of the high-power Tm-Raman MOPA source [140]

generated by low-power Tm-doped fiber cavities. To generate such a long Stokes seed wavelength with a Tm-doped cavity, one has to use a very long fiber to increase reabsorption of the signal at shorter wavelengths and inhibit parasitic lasing. Both seeds are then launched into a two-stage Tm fiber amplifier. Naturally, the 1963 nm signal (i.e., the Raman pump) is preferentially amplified since it is subjected to a higher gain. Then, a 50 m germanosilicate fiber acting as a Raman amplifier is spliced on the Tm power amplifier to convert the 1963 nm light to the first-order Stokes wavelength at 2147 nm. A schematic of this innovative laser setup is shown in Fig. 3.17. This laser system exhibited a 38.5% Raman amplifier efficiency. In addition, it delivered a maximum output power of 14.3 W (CW) at 2147 nm, thus surpassing all previous Raman conversion experiments in this spectral range. A good stability was also observed, with 0.5% power fluctuations over a time period of 30 min. This MOPA architecture has significant potential for further power scaling. In fact, a similar source was previously demonstrated around 1 μm with Yb-doped fibers and has reached more than 1.5 kW at the first Stokes wavelength (1120 nm) [141].

3.3.2 Pulsed Raman Sources Based on Heavily GeO₂-Doped Silicate Fibers

A number of pulsed Raman sources based on germanosilicate fibers are also found in the literature. Generally, these sources simply consist of a long heavily doped GeO_2 fiber segment without reflectors. As the peak powers are often very high, an efficient Stokes conversion can arise from spontaneous Raman emission. One of the first demonstrations of these devices was reported in 2008 [142]. By sending 2 ns pulses emitted from an erbium-doped pulsed amplifier (1.53 μm wavelength, 170 W peak power) in a 50 m long commercial fiber (Nufern UHNA7), a five-order Stokes cascade was generated up to 2.41 μm. Moreover, up to 37% of the power was converted to the fourth-order Stokes (at 2.14 μm) and about 16% to the fifth-order Stokes (at 2.41 μm). To reach wavelengths longer than 2 μm, it is however far more convenient to pump the Raman fiber using a Q-switched Tm fiber laser, which can deliver several kilowatts of peak powers in the spectral range of interest. In this regard, a first-order Raman shift was reported from 75 ns pulses pumping at 1960 nm with a Tm fiber laser [143]. The Raman converter was made from a 63% GeO_2-doped silicate fiber, which had low losses of 8.7 dB/km at the pump wavelength and 44.6 dB/km at the Stokes wavelength (2.14 μm). For a pump average power of 3 W (750 W peak power), the first-order signal reached an average power of 0.6 W. No higher Stokes orders were observed in this experiment.

In recent years, new studies were conducted with the aim of increasing the output power and reaching longer wavelengths. Jiang et al. demonstrated second-order shifts at 2.43 μm and 2.48 μm under pumping at 2.008 μm and 2.04 μm, respectively [77]. The pump pulses were delivered by a Q-switched Tm fiber laser and had a 100 ns duration and a   2.3 kW peak power (230–240 mW average power). This pump was converted to the second-order Stokes with an efficiency of 16.5% at 2.43 μm but only of 7.9% at 2.48 μm, due to the sharp increase of the fiber attenuation in this spectral range (from 0.7 to 4.5 dB/m). In a more recent experiment, the authors re-optimized the fiber length to suppress the second Stokes order and maximize the power at the first Stokes order [144]. The best lasing efficiency was estimated at 36%. This was the first report of a watt-level (average power) silicate Raman source having an emission wavelength at or above 2.2 μm. Figure 3.18 shows the output power curves for the two Stokes orders and the residual pump as well as the temporal profile of the pulses generated at 2.2 μm.

3.3.3 Outlook: Generation of Longer Wavelengths

From the demonstrations described above, it is clear that heavily doped GeO_2 silicate RFLs are limited by the glass attenuation at long wavelengths. While CW sources were restricted to ≤2.2 μm, pulsed pumping was able to push their long wavelength edge to 2.48 μm, near the glass transparency limit. GeO_2 glass

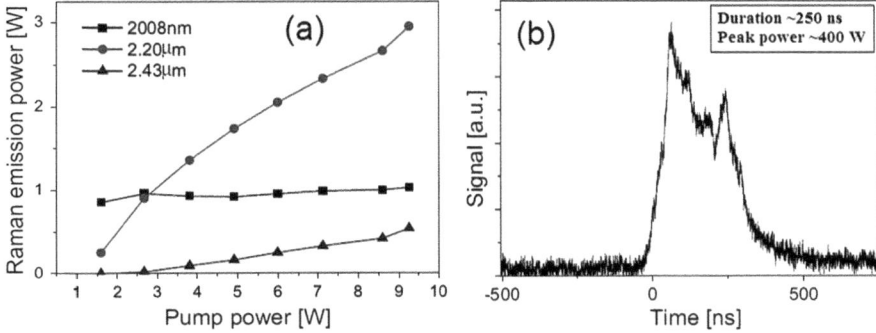

Fig. 3.18 (**a**) Output laser curve of the first watt-level silicate Raman converter emitting at 2.2 μm or above. (**b**) Typical output pulse characteristics at 2.2 μm [144]

compositions that do not contain silica, such as germanate-lead glasses (a heavy metal oxide) [145], could help to reach longer wavelengths.

Tellurite glasses also have a significant potential for MIR Raman fiber lasers. While no such demonstrations currently exist, a numerical study was recently published on a tellurite-based (TBZN) RFL pumped with an Er-doped FG fiber laser at 2.8 μm [46]. At the first Stokes order, it suggested that a 10 W output power could be obtained at 3.53 μm with a 20 W pump. On the other hand, since the attenuation of such fibers is still relatively high (0.25 dB/m at 3.5 μm), it was also suggested that short-fiber lengths (< 3 m) and high-reflectivity output couplers (>80%) were preferable to raise the lasing efficiency to acceptable values. The authors also performed calculations for a second-order Stokes located between 4 and 5 μm. A maximum output power of 1.5 W was predicted at 4.36 μm. We note, however, that first-order Stokes spectral broadening was not considered in these calculations and might reduce the first- to second-order Raman conversion efficiency [146]. Simulations based on 2 μs pulsed pumping with a Q-switched Er-doped fiber laser were even more promising for generating >4 μm light. The second Stokes efficiency was in fact 2.5 times higher in these pumping conditions (compared to CW). Overall, this study provided useful guidelines for future development of tellurite-based Raman fiber lasers.

3.4 Fluoride-Based Raman Fiber Sources

Fluoride glass fibers are good candidates to generate wavelengths outside the transmission window of silicate glasses. As seen in Sect. 3.2.1, their minimum attenuation is located around 2.5 μm, making them perfectly suited for Raman experiments in the 2.1–2.6 μm spectral range. Although they are not as robust as silicates, FGs can still sustain quite high average powers as required in most core-pumped Raman applications. The following section reviews the main results based on these materials and briefly discusses possible future directions.

3.4.1 Early Raman Amplification Experiments

The first Raman experiments with FG fibers investigated amplification in pump-probe configurations [65, 147]. The aim of these studies was to evaluate the Raman characteristics (i.e., gain magnitude and spectrum) rather than to demonstrate high-power Raman amplifiers. The first report by Durteste et al. [65] studied Raman amplification by launching two signals near 1064 nm in a 10 m long fluorozirconate fiber. The pump laser was a Q-switched Nd:YAG providing 250 ns pulses, while the probe consisted of another Q-switched Nd:YAG that was broadened in a silica fiber and filtered around the Stokes wavelength at 1.12 μm. With pump peak powers of a few kWs, they were able to identify the main Raman bands and obtain the first estimation of the Raman gain coefficient.

A few years later, a similar experiment was carried out in an 80-m fluoride fiber using two tunable dye lasers near 580 nm as pump and probe [147]. The purpose of this study was to determine the material Raman gain coefficient with a better accuracy and compare it to the value obtained in a silica fiber. A coefficient about six times higher than silica was found. Figure 3.19 shows the experimental setup used for this measurement as well as the gain recorded as a function of the input Stokes intensity, from which the Raman coefficient was evaluated.

In the last few years, there has been a renewed interest to study Raman characteristics of FG fibers, a consequence of their intensive use in supercontinuum generation. In the early demonstrations described above, a gain coefficient several times higher than silica was found for FGs. More recently, it was shown that these demonstrations may have overestimated the Raman gain, and a value comparable to that of silica was more realistic [148]. This result was also in agreement with the gain coefficient inferred from Raman laser experiments [68, 85].

3.4.2 CW Raman Laser Cavities Between 2 and 2.5 μm

3.4.2.1 First Demonstration at 2185 nm

Fluoride fiber technology has made significant progress in the last decade, leading to the availability of low-loss fibers (1 dB/km minimum attenuation around 2.5 μm) [16]). During the same period, fiber Bragg grating inscription was achieved in FGs using high-peak power femtosecond pulses [14]. These advances were of key importance for the onset of FG-based Raman laser cavities.

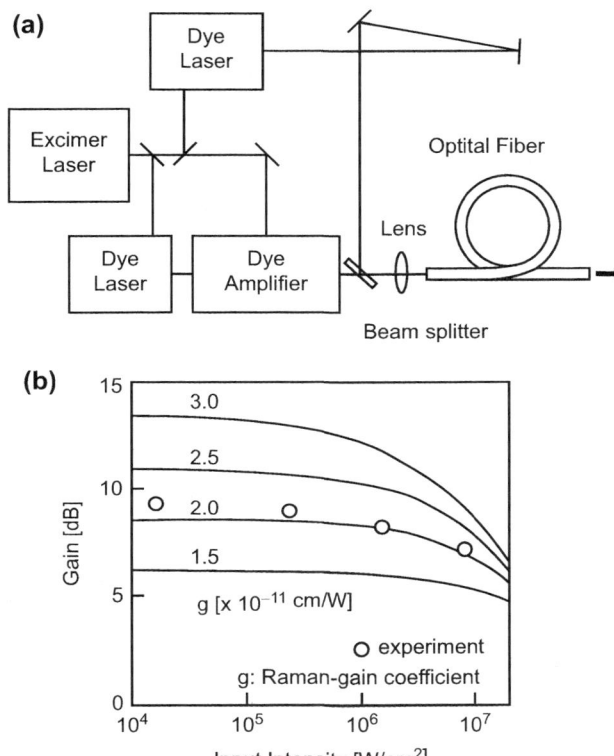

Fig. 3.19 (**a**) Schematic of a Raman amplification experiment using an FG fiber and tunable dye lasers [147]. Amplifier gain as a function of Stokes intensity for different Raman gain coefficients. The experimental data points (*circles*) suggest a Raman gain coefficient of about 2×10^{-13} m/W at 580 nm

The first demonstration of a fluoride fiber-based Raman cavity was reported in 2011. It was made by butt-coupling a Tm fiber laser with a 29 m long FG (cf. Fig. 3.20) [68]. The 1940 nm pump wavelength resulted in a peak Stokes wavelength near 2185 nm. The oscillator at the Stokes wavelength was composed of two FBGs directly written in the same fluoride fiber [14]. Very high reflectivities (respectively 99% and 95%) were chosen to reduce the laser threshold to an acceptable level. The output grating was deliberately shifted (by 3 nm) by applying tension during the writing process to allow a precise measurement of its transmission spectrum. As a result, a similar tension had to be applied during laser operation to restore the overlap between the two gratings. To enhance the long-term stability of the gratings, they were thermally annealed at 100 °C after inscription.

A maximum output power of 580 mW was achieved when the launched pump power was about 7 W. At low pump powers, the efficiency was rather high, at 29%, but it decreased to 14% near the maximum pump power. This significant rollover

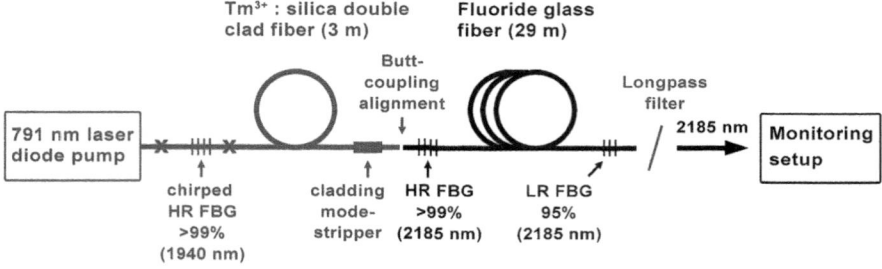

Fig. 3.20 Experimental setup of the first Raman laser based on an FG fiber [68]

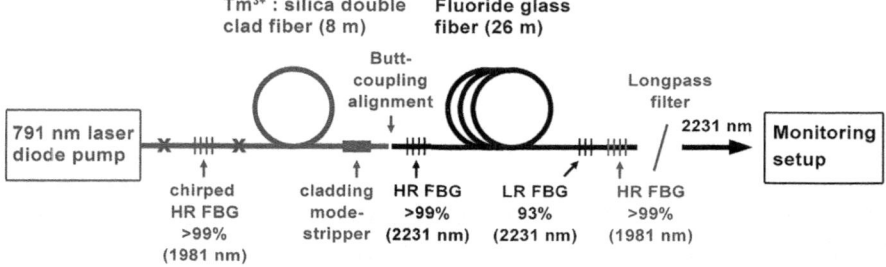

Fig. 3.21 Experimental setup of a nested cavity fluoride Raman fiber laser delivering 3.7 W at 2231 nm [85]

was thought to be caused by a thermal (spectral) shift of the output FBG as the power was increased. In addition, there was also some evidence that intracavity spectral broadening reduced the effective reflectivity of the output grating away from its optimal value. Since the fiber and cavity parameters were known, it was also possible to evaluate the absolute Raman gain coefficient using the threshold of the laser curve. A value of 3.5×10^{-14} m/W was obtained at 1940 nm.

3.4.2.2 Nested Raman Laser Cavity at 2231 nm

An improved watt-level version of the fluoride glass RFL was reported by Fortin et al. [85]. By nesting the Raman cavity inside the Tm pump cavity, the Stokes signal was exposed to a much higher local pump, resulting in a better Raman conversion efficiency as well as a lower threshold. This was simply achieved by writing the output grating of the pump cavity at the end of the fluoride fiber, as shown in Fig. 3.21. The intracavity losses experienced by the pump were kept low because the mode mismatch between the two fibers (Tm:silica and fluoride) was negligible.

This demonstration also implemented highly reflective gratings to form the Raman cavity (respectively 99% and 93%). These gratings did not require any post-writing spectral tuning as they were written to overlap at rest. A highly reflective

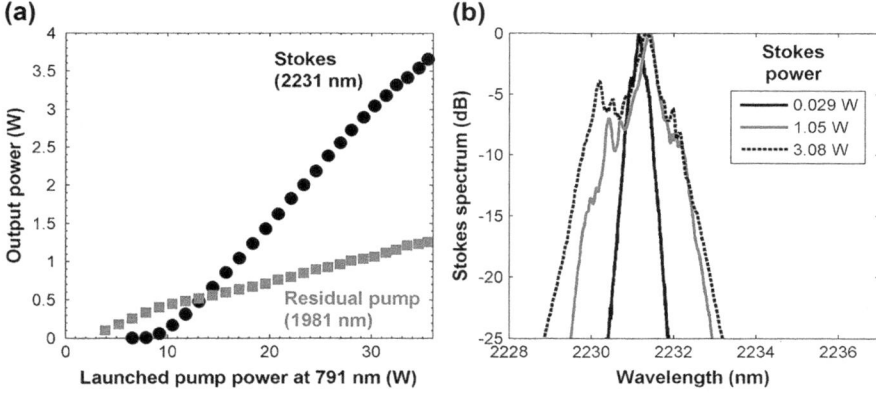

Fig. 3.22 (a) Laser curve and residual pump power of the nested cavity fluoride Raman fiber laser. (b) Output spectrum for different Stokes powers showing strong spectral broadening [85]

(>99%) FBG at 1981 nm was also used to recycle the residual pump that was not converted to Stokes wavelength in the first pass.

The nested Raman cavities allowed the generation of 3.7 W at 2231 nm for a 791 nm diode pump power of about 35 W [Fig. 3.22a], leading to a 15% overall slope efficiency. Note that due to the nested cavity nature of the RFL, it was impossible to have direct access to the 1981 → 2231 nm Raman conversion efficiency. The reported 15% efficiency was in fact the result of the two wavelength conversion processes, i.e., from 791 to 1981 nm and from 1980 to 2231 nm. The output spectrum is displayed in Fig. 3.22b, showing the growing effect of spectral broadening as the power is increased. However, the authors anticipated the decrease in effective reflectivity (caused by spectral broadening) by selecting an output FBG's reflectivity higher than its optimal value. Note that spectral broadening was also observed for the pump wavelength and is the primary explanation for the large residual pump power measured at the end of the FG fiber, as seen in see Fig. 3.22a. A model was also developed to describe the nested cavity Raman laser, and a very good agreement was achieved with the experimental data [85].

3.4.3 Outlook

3.4.3.1 Reaching Longer Wavelengths

Very few demonstrations of FG-based Raman fiber lasers have been reported to date. A natural follow-up to the first-order cavities near 2.2 μm would be a second-order shift to reach the 2.4–2.6 μm spectral range. High efficiencies are expected even for a second-order cavity since the newly generated wavelengths would fall precisely on the minimum attenuation of zirconium fluoride glasses. The lack of specific applications in this spectral range might be one of the reasons why no attempts were made until now.

The mid-infrared spectrum beyond 3 μm presents a much wider interest due to its possible applications in spectroscopy, material processing, and defense. However, the relatively low Raman gain of fluoride fibers hinders their potential for CW lasing in this spectral range. The raising attenuation of fluorozirconate glasses beyond 2.5 μm [16] also contributes to this problem. Nevertheless, modeling efforts have shown that it is possible to reduce the threshold of these CW sources provided highly reflective gratings with virtually no losses are written in the fluoride fiber [149]. However, this is generally not a reasonable expectation as intracavity spectral broadening will lead to power leaks and thus raise the effective losses of the cavity. A better alternative to generate wavelengths beyond 3 μm would be to pump the FG cavity using high-peak power pulses (ns to μs pulse duration) to enhance the conversion efficiency. Such pump sources are currently being developed based on Er-doped FG fiber lasers [121, 150], as discussed in Sect. 3.2.3. Other fluoride glass compositions such as fluoroindates (InF$_3$-based) could also prove useful as their minimum attenuation is shifted toward longer wavelengths (i.e., near 3.8 μm [57]).

3.4.3.2   Alternative Approach Based on Soliton Self-Frequency Shift

One alternative Raman process-based strategy has been proposed recently based on soliton self-frequency shift (SSFS) to efficiently and continuously tune pulsed laser sources to longer wavelengths. FGs are well suited for SSFS between 2 and 4 μm since they exhibit anomalous dispersion and low losses in this spectral range. In a recent report, it was shown that 2 μm soliton pulses could be shifted up to 4.3 μm by an appropriate choice of fiber design and length [151]. MIR SSFS can also benefit from the recent advent of femtosecond fiber lasers near 2.8 μm [123–125]. As a matter of fact, a 2.8–3.6 μm tunable SSFS source was recently demonstrated based on an Er:FG fiber amplifier seeded at 2.8 μm [152]. Accordingly, the generation of 160 fs pulses at 3.4 μm with a peak power above 200 kW and an average output power surpassing 2 W was demonstrated.

3.5   Chalcogenide-Based Raman Fiber Sources

Chalcogenide glass fibers benefit from a broad MIR transmission window as well as a high Raman gain coefficient. Although they suffer from the lowest damage threshold of all the glasses described in this chapter, they remain the best option for generating >3 μm light through Raman conversion. Most demonstrations reviewed in this section have been reported in the last few years spurred by the recent availability of high-quality glass fibers. Chalcogenide Raman conversion in both CW and pulsed regimes is covered here.

3.5.1 Raman Laser Cavities

3.5.1.1 Early Demonstrations Near 2 μm

The first report of a ChG-based Raman laser cavity was published in 2006 by
Jackson et al. [153]. It was based on a commercial As_2Se_3 step-index fiber (IRT-
SE-06-01-0105, CorActive) having a 6 μm diameter core. The free-running Raman
resonator was composed of a 1 m long chalcogenide fiber segment bounded by
a broadband mirror and Fresnel reflection on the fiber tip (22%) as the cavity
reflectors. By using a 10 W (CW) Tm fiber laser pump centered at 2051 nm, several
Raman bands were observed, namely, at 2062 nm (0.64 W), 2102 nm (0.2 W),
and 2166 nm (0.016 W). They attributed the first band to vibrations between the
"polymerlike" layers in the two-dimensional As_2Se_3 network, while the second and
third bands were associated respectively with bond bending and bond stretching
within the layers. Note that the latter peak is the usual maximum in the Raman gain
spectrum (as seen in Fig. 3.2, for instance). Figure 3.23 shows the output power
measured in each band as a function of the launched pump power.

3.5.1.2 All-Fiber Laser Cavities Operating Beyond 3 μm

It took nearly a decade before further progress was made in ChG-based Raman laser
cavities. A renewed interest was seen once it was demonstrated that fiber Bragg
gratings could be written directly through the polymer jacket of the chalcogenide

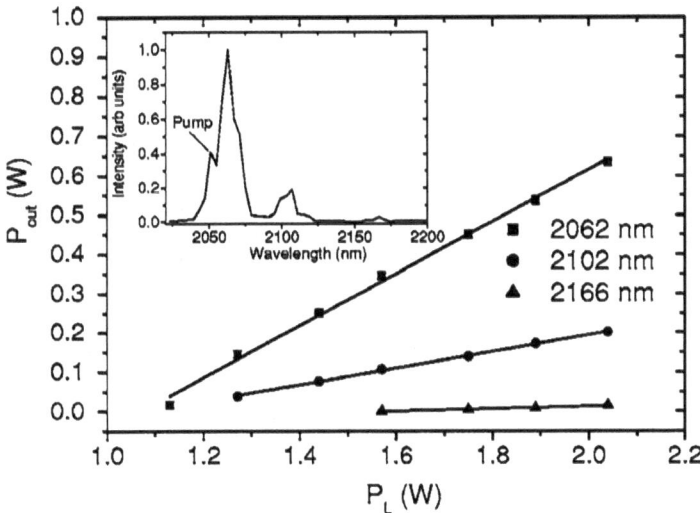

Fig. 3.23 Measured output power for three different first-order Stokes bands generated in an
As_2Se_3 Raman cavity [153]. The inset plots the output spectrum at maximum power

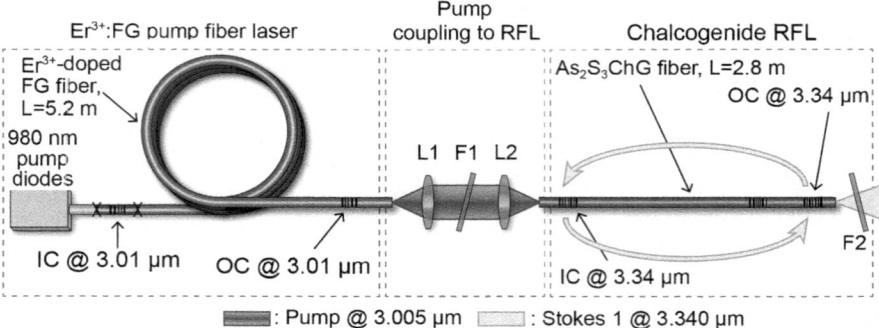

Fig. 3.24 Experimental setup of a first-order Raman cavity based on an As_2S_3 fiber [29]. *F1–F2* long-pass filters, *L1–L2* aspheric lenses, *IC* input coupler at 3.34 μm, *OC* output coupler at 3.34 μm

fiber [103]. In addition, newly available high-power pumps near 3 μm (i.e., Er:FG glass fiber lasers) made it possible to target MIR wavelengths.

In the first demonstration of a MIR chalcogenide Raman laser, a 4 μm diameter core As_2S_3 fiber was employed as the gain medium [29]. As shown in Fig. 3.24, FBGs with respective reflectivities of >99% and 63% were written in the 3 m chalcogenide fiber to form a first-order Stokes oscillator at 3.34 μm. The Er^{3+}:FG pump fiber laser was set to emit at 3.005 μm (i.e., significantly away from the typical 2.8 μm emission wavelength) to ensure the first-order Stokes wavelength was located far from the OH absorption band of the ChG fiber. As a result, the attenuation coefficient at the Stokes wavelength was lower than 0.1 dB/m. The pump source was operated in a quasi-CW regime, in which pulses having 5 ms duration and 10 W peak power were emitted at a repetition rate of 20 Hz. These operating conditions prevented thermomechanical instabilities of the free-space coupling setup.

This MIR laser system delivered up to 0.6 W of peak power (47 mW average power) at 3.34 μm. Moreover, the laser slope (Fig. 3.25) showed no sign of saturation and had an impressive efficiency of 39% with respect to the launched pump power. It is important to note, however, that a large portion of the pump power was wasted due to the free-space coupling setup (i.e., up to 74%).

A similar experimental setup was also used to investigate second-order Stokes emission around 3.77 μm [146]. The laser source relied on nested Fabry–Perot cavities formed by two pairs of FBGs, respectively, centered at 3.34 μm and 3.77 μm. The 3.34 μm first-order Stokes gratings had a very broad bandwidth (≥10 nm) and were strongly reflective (>99%) to limit the loss of power resulting from spectral broadening. As for second-order Stokes gratings, they had initial reflectivities of >99% and 98%. The output coupler was in fact thermally annealed in three iterative steps to optimize the laser performances. The best performances were achieved once the output coupler was reduced to 80%, where a maximum output peak power of 112 mW (9 mW average power) was generated at 3.77 μm for a peak launched pump power of 3.9 W. Under these conditions, the laser slope efficiency

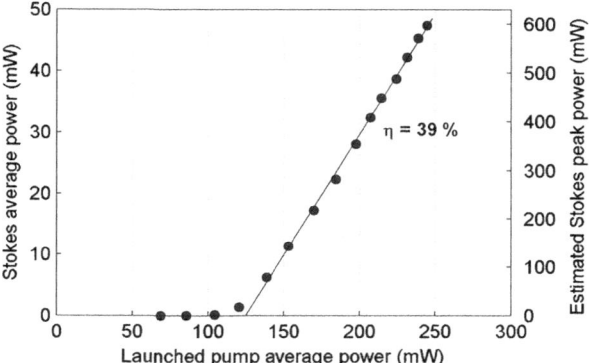

Fig. 3.25 Output power (at 3.34 μm) of a first-order Raman cavity based on an As_2S_3 fiber [29]

was estimated at 8.3%. Figure 3.26a shows the output power at 3.77 μm for the three different output coupler reflectivities (98%, 92%, and 80%). The relatively low efficiency is mostly due to power losses of the first-order Stokes through the FBGs, a result of intracavity spectral broadening. The spectra of the Stokes signals as well as those of the FBGs are displayed in Fig. 3.26b. This demonstration constitutes the longest wavelength RFL cavity reported to date.

3.5.1.3   Alternative Raman Cavities

Other chalcogenide cavity designs seem promising to generate new wavelengths through Raman conversion. Tapers (or microwires) can enhance the effective Raman gain coefficient by a large factor and reduce the Raman threshold accordingly. Based on this approach, Ahmad et al. reported a Fabry-Perot cavity made of a 0.95 μm diameter As_2Se_3 microwire, which displayed a low peak pump power threshold of 470 mW but a limited conversion efficiency of 0.25% [154]. The cavity feedback was provided by Fresnel reflections on both fiber ends. In a later demonstration, a distributed Bragg reflector (DBR) was written inside the ChG microwire to increase the efficiency to >2% [155]. While these studies were limited to the near-infrared spectral range, the approach seems to be applicable to longer MIR wavelengths.

ChG microresonators are another promising strategy to generate Stokes emission with a low threshold and a very compact package. These devices can be easily fabricated using a CO_2 laser or heated filament to melt the fiber tip into a sphere. High-quality factor microresonators made of As_2S_3 glass were reported in the last few years [156, 157]. Experiments carried out with a 1.88 μm Tm fiber laser pump showed cascaded Stokes emission up the third order at 2.35 μm [156].

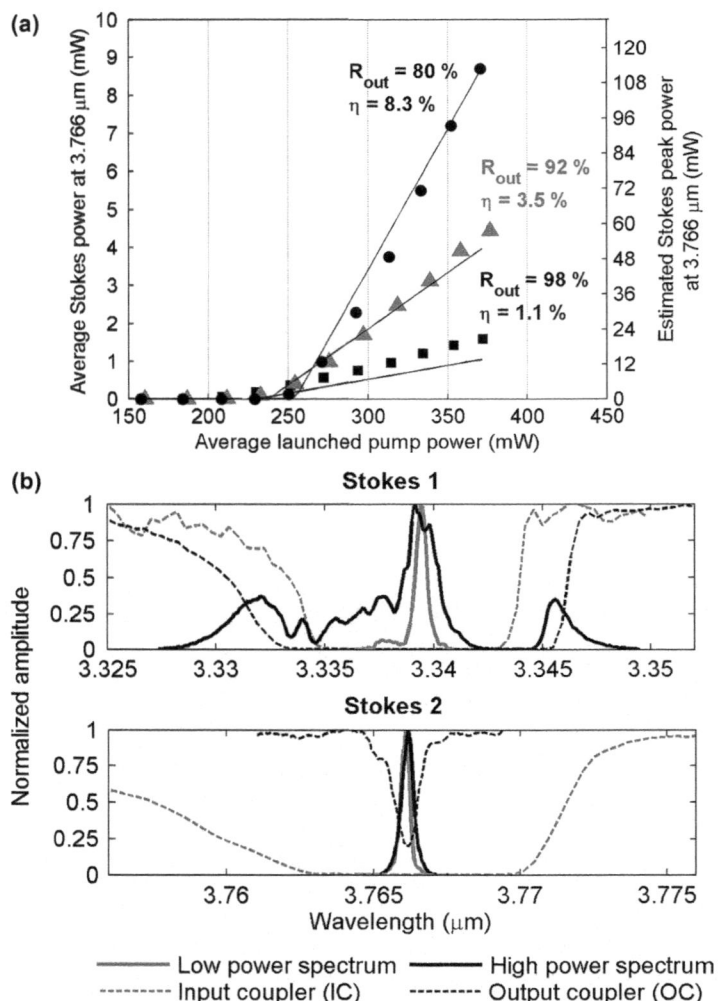

Fig. 3.26 (a) Output power curves for a second-order As_2S_3 Raman cavity with different output couplers (98%, 92%, and 80%). (b) Spectra of the first and second Stokes orders for low and high pump levels

3.5.2 Single-Pass Cascaded Raman Shifts with Pulsed Pumping

High Raman conversion efficiencies can be achieved in ChG fibers under pulsed pumping. With sufficient peak powers, stimulated Raman scattering can take place without the need for a resonant cavity. Pulsed cascaded Raman shifts were first observed in chalcogenide fibers by Kulkarni et al. near 1.55 μm [158], where three

Stokes orders were generated using a 1.55 μm nanosecond source. A few years later, the first demonstration at a wavelength beyond 2 μm was reported. While different fiber geometries were tested, the best results were obtained with an AsSe microstructured fiber (MOF) having a suspended core geometry (diameter ≈ 3 μm) [159]. By pumping the core using a gain-switched nanosecond Tm fiber laser at 1.995 μm, a third-order Stokes cascade was generated up to 2.33 μm. The pump peak power required to reach the (first-order) Raman threshold was approximately 0.8 W, while damage to the fiber structures occurred at 10 W. In a later study based on a similar setup, an additional Stokes order (fourth) centered on 2.45 μm was measured.

Meanwhile, cascaded Raman shift was also studied in larger core ChG fibers [71]. Such fibers usually possess a better mechanical robustness (compared to MOFs) and can withstand significantly higher pulse energies and peak powers. Experiments were carried out with both As_2S_3- and As_2Se_3-based fibers, which had a core diameter of 65 μm. The fiber lengths were, respectively, 4.3 m and 4.9 m. To "emulate" Tm fiber laser pumping, a 1.9 μm wavelength was selected from a Nd:YAG/OPO Q-switched laser source. The system produced 2 ns pulses with mJ energies at a repetition rate of 10 Hz. The output spectrum showed up to the third Stokes order (2.38 μm) in the As_2S_3 fiber, while the fourth order (2.33 μm) was observed in the As_2Se_3 fiber. The total output pulse energy (including the residual pump and all Stokes orders) reached 110 μJ and 78 μJ, respectively. Figure 3.27 shows the output spectrum for both fibers, which is composed of multiple Stokes bands.

More recently, cascaded Raman shifts were investigated in a novel all-solid core (i.e., multi-glass) microstructured ChG fiber [160]. While such fiber benefits from a strong mode confinement, it does not have the typical problems of standard air-clad MOFs, namely, a low damage threshold and glass aging due to environmental contamination [161]. The core and clad were respectively composed of $AsSe_2$ and As_2S_5 glasses. The best Raman conversion efficiency (-15 dB at first order) was achieved under picosecond pulse pumping at 1958 nm. For low pump powers, the Stokes intensity grew steadily; however, saturation was observed at high power levels as the spectrum was broadened through self-phase modulation.

3.5.3 Outlook

Not surprisingly, the reliability of ChG-based Raman sources is still an issue. At this time, the lack of robust (and low-loss) fusion splices between the pump and ChG fiber limits not only the robustness of the system but also its power scaling. This is especially true for Raman sources beyond 3 μm, where the pump lasers are essentially made of FG fibers. In chalcogenide Raman cavities, the long-term stability of the FBGs is also problematic (see Sect. 3.2.2). An in-depth study of the FBG's aging under different writing conditions could lead to improved performances.

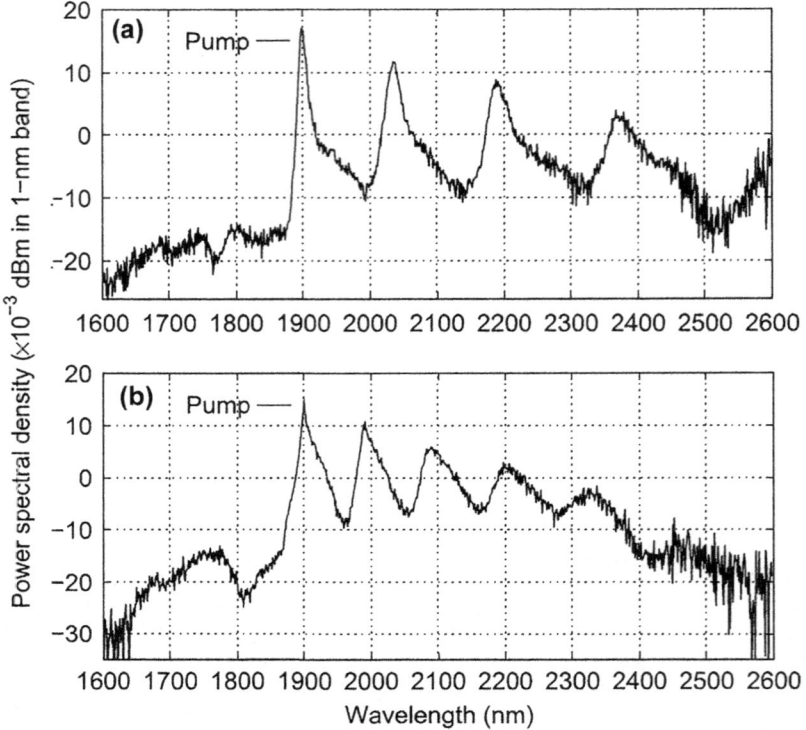

Fig. 3.27 Output spectrum of a single-pass Raman cascade in (**a**) As$_2$S$_3$ and (**b**) As$_2$Se$_3$ fibers [71]. The fibers were pumped at 1.9 μm with 2 ns pulses

Chalcogenide Raman cavities such as those presented in Sect. 3.5.1 have not yet revealed their full potential. In fact, numerical simulations have shown that they could be designed to operate on virtually any wavelengths between 3 and 4 μm [73] based on currently available Er:FG fiber laser pumps near 2.8 μm. In addition, they also have excellent prospects for reaching even longer wavelengths. Erbium-doped fiber lasers operating near 3.45 μm, which have made recent progress [26, 27], would be convenient pump sources for generating wavelengths between 3.5 and 4.5 μm.

3.6 Conclusion

Mid-infrared Raman fiber lasers have attracted considerable interest in recent years. While Raman-based sources are not a replacement for rare-earth-doped lasers, they represent a convenient complement to access the hard-to-reach wavelengths where no electronic emission bands exist. Currently, this is especially true for the spectral range around 2.4 μm (2.2–2.7 μm) and above 3.7 μm. A complete coverage of the MIR could open up a number of applications in the biomedical, spectroscopy, and defense and security fields. Evidently, such niche applications could become technology drivers for the development of specific RFLs in the following years.

Early studies have focused primarily on extending the spectral coverage of silicate-based fibers, essentially through high doping levels of GeO_2. This approach has led to several notable demonstrations in the 2–2.5 μm spectral range: a multi-watt all-fiber cavity near 2.1 μm in CW operation [139] and a single-pass Stokes cascade shifted above 2.4 μm under pulsed pumping [77]. However, a transition from silicate glasses to low phonon energy soft glasses was required to improve the conversion efficiency beyond 2.2 μm. The very first attempts with soft glass Raman fiber lasers were made with FGs, resulting in more than 3 watts around 2.23 μm [85]. Although fluorides displayed a smaller gain coefficient compared to the former GeO_2 silicate glasses, they benefited from a higher FOM (see Sect. 3.2.1) because of their lower propagation losses, thus allowing for a higher conversion efficiency. Nonetheless, ChG fibers have made an even stronger statement by enabling the first Raman fiber laser demonstrations above 3 μm (i.e., at 3.34 μm [29] and 3.77 μm [146]). The cavities were also all-fibered through the use of fiber Bragg gratings. In parallel, alternative strategies for ChG fibers were proposed based on microwires and microresonators and could potentially result in MIR demonstrations in the near future.

Clearly, the progress in MIR Raman fiber sources is closely linked to a number of key technological advances, such as low-loss fiber manufacturing, FBG inscription capabilities, and high-power pump source availability. Although these technologies have reached a sufficient maturity for Raman fiber lasers, much work remains to improve their long-term reliability and high-power handling.

References

1. Endo, M., Walter, R.F.: Gas Lasers. CRC Press, Boca Raton (2006)
2. Myrabo, L.N., Knowles, T.R., Bagford, J.O., Seibert Ii, D.B., Harris, H.M.: Laser-boosted light sail experiments with the 150-kW LHMEL II CO_2 laser. Proc. SPIE. **4760**, 774–798 (2002)
3. Farson, D.F., Ready, J.F., Feeley, T.: LIA Handbook of Laser Materials Processing. Springer, New York (2001)
4. Kaplan, I., Giler, S.: CO_2 Laser Surgery. Springer Berlin Heidelberg (2012)
5. Ionin, A.A., Sinitsyn, D.V., Suchkov, A.F.: High-power N_2O laser as alternative to CO_2 laser. Proc. SPIE. **2206**, 287–292 (1994)
6. Arnold, S.J., Foster, K.D.: A purely chemical HBr laser. Appl. Phys. Lett. **33**, 716–717 (1978)
7. Faist, J., Capasso, F., Sivco, D.L., Sirtori, C., Hutchinson, A.L., Cho, A.Y.: Quantum cascade laser. Science. **264**, 553–556 (1994)
8. Yao, Y., Hoffman, A.J., Gmachl, C.F.: Mid-infrared quantum cascade lasers. Nat. Photonics. **6**, 432–439 (2012)

9. DeLoach, L.D., Page, R.H., Wilke, G.D., Payne, S.A., Krupke, W.F.: Transition metal-doped zinc chalcogenides: spectroscopy and laser demonstration of a new class of gain media. IEEE J. Quant. Electron. **32**, 885–895 (1996)
10. Moskalev, I., Mirov, S., Mirov, M., Vasilyev, S., Smolski, V., Zakrevskiy, A., Gapontsev, V.: 140 W Cr:ZnSe laser system. Opt. Express. **24**, 21090–21104 (2016)
11. IPG Photonics. www.ipgphotonics.com
12. Maidment L., Schunemann P.G., Reid D.T.: Molecular fingerprint-region spectroscopy from 5–12 μm using an orientation-patterned gallium phosphide optical parametric oscillator. arXiv:1606.09613 (2016)
13. Lockheed Martin Corporation. www.lockheedmartin.com
14. Bernier, M., Faucher, D., Vallée, R., Saliminia, A., Androz, G., Sheng, Y., Chin, S.L.: Bragg gratings photoinduced in ZBLAN fibers by femtosecond pulses at 800 nm. Opt. Lett. **32**, 454–456 (2007)
15. Berrou, A., Kieleck, C., Eichhorn, M.: Mid-infrared lasing from H_o^{3+} in bulk InF_3 glass. Opt. Lett. **40**, 1699–1701 (2015)
16. Le Verre Fluoré. www.leverrefluore.com
17. Jackson, S.D.: Towards high-power mid-infrared emission from a fibre laser. Nat. Photonics. **6**, 423–431 (2012)
18. Schneider, J., Carbonnier, C., Unrau, U.B.: Characterization of a H_o^{3+} −doped fluoride fiber laser with a 3.9-μm emission wavelength. Appl. Opt. **36**, 8595–8600 (1997)
19. Aydın, Y.O., Fortin, V., Maes, F., Jobin, F., Jackson, S.D., Vallée, R., Bernier, M.: Diode-pumped mid-infrared fiber laser with 50% slope efficiency. Optica. **4**, 235–238 (2017)
20. El-Agmy, R.M., Al-Hosiny, N.M.: 2.31 μm laser under up-conversion pumping at 1.064 μm in Tm^{3+}:ZBLAN fibre lasers. Electron. Lett. **46**, 936–937 (2010)
21. Fortin, V., Bernier, M., Caron, N., Faucher, D., El Amraoui, M., Messaddeq, Y., Vallée, R.: Towards the development of fiber lasers for the 2 to 4 μm spectral region. Opt. Eng. **52**, 054202 (2013)
22. Jackson, S.D.: High-power and highly efficient diode-cladding-pumped holmium-doped fluoride fiber laser operating at 2.94 μm. Opt. Lett. **34**, 2327–2329 (2009)
23. Fortin, V., Bernier, M., Bah, S.T., Vallée, R.: 30 W fluoride glass all-fiber laser at 2.94 μm. Opt. Lett. **40**, 2882–2885 (2015)
24. Carbonnier, C., Tobben, H., Unrau, U.B.: Room temperature CW fibre laser at 3.22 μm. Electron. Lett. **34**, 893–894 (1998)
25. Majewski, M.R., Jackson, S.D.: Highly efficient mid-infrared dysprosium fiber laser. Opt. Lett. **41**, 2173–2176 (2016)
26. Fortin, V., Maes, F., Bernier, M., Bah, S.T., D'Auteuil, M., Vallée, R.: Watt-level erbium-doped all-fiber laser at 3.44 μm. Opt. Lett. **41**, 559–562 (2016)
27. Henderson-Sapir, O., Jackson, S.D., Ottaway, D.J.: Versatile and widely tunable mid-infrared erbium doped ZBLAN fiber laser. Opt. Lett. **41**, 1676–1679 (2016)
28. Xiushan, Z., Jain, R.: Watt-level 100 nm tunable 3 μm fiber laser. IEEE Photon. Technol. Lett. **20**, 156–158 (2008)
29. Bernier, M., Fortin, V., Caron, N., El-Amraoui, M., Messaddeq, Y., Vallée, R.: Mid-infrared chalcogenide glass Raman fiber laser. Opt. Lett. **38**, 127–129 (2013)
30. Tittel, F.K., Richter, D., Fried, A.: Mid-infrared laser applications in spectroscopy. In: Sorokina, I., Vodopyanov, K. (eds.) Solid-State Mid-infrared Laser Sources, pp. 458–529. Springer, Germany (2003)
31. Stuart, B.H.: Infrared Spectroscopy: Fundamentals and Applications. Wiley, Chichester (2004)
32. Lin-Vien, D., Colthup, N.B., Fateley, W.G., Grasselli, J.G.: The Handbook of Infrared and Raman Characteristic Frequencies of Organic Molecules. Elsevier Science, San Diego (1991)
33. Klingbeil, A.E., Jeffries, J.B., Hanson, R.K.: Temperature- and pressure-dependent absorption cross sections of gaseous hydrocarbons at 3.39 μm. Meas. Sci. Technol. **17**, 1950 (2006)

34. Lundqvist, S., Kluczynski, P., Weih, R., von Edlinger, M., Nähle, L., Fischer, M., Bauer, A., Höfling, S., Koeth, J.: Sensing of formaldehyde using a distributed feedback interband cascade laser emitting around 3493 nm. Appl. Opt. **51**, 6009–6013 (2012)
35. Sigrist, M.W., Bartlome, R., Marinov, D., Rey, J.M., Vogler, D.E., Wächter, H.: Trace gas monitoring with infrared laser-based detection schemes. Appl. Phys. B Lasers Opt. **90**, 289–300 (2008)
36. Fotia, M.L., Sell, B.C., Hoke, J., Wakefield, S., Schauer, F.: 1 kHz mid-IR absorption spectroscopy for CO and CO_2 concentration and temperature measurement. Combust. Sci. Technol. **187**, 1922–1936 (2015)
37. Serebryakov, V.A., Boiko, É.V., Petrishchev, N.N., Yan, A.V.: Medical applications of mid-IR lasers. Problems and prospects. J. Opt. Technol. **77**, 6–17 (2010)
38. Apitz, I., Vogel, A.: Material ejection in nanosecond Er:YAG laser ablation of water, liver, and skin. Appl. Phys. A Mater. Sci. Process. **81**, 329–338 (2005)
39. Frenz, M., Pratisto, H., Konz, F., Jansen, E.D., Welch, A.J., Weber, H.P.: Comparison of the effects of absorption coefficient and pulse duration of 2.12-μm and 2.79-μm radiation on laser ablation of tissue. IEEE J. Quantum Electron. **32**, 2025–2036 (1996)
40. Johnson, D.E., Cromeens, D.M., Price, R.E.: Use of the holmium:YAG laser in urology. Laser Surg. Med. **12**, 353–363 (1992)
41. Schwartz, L.W., Moster, M.R., Spaeth, G.L., Wilson, R.P., Poryzees, E.: Neodymium-YAG laser Iridectomies in glaucoma associated with closed or Occludable angles. Am J. Ophthalmol. **102**, 41–44 (1986)
42. Ostertag, M., Walker, R., Weber, H., van der Meer, L., McKinley, J.T., Tolk, N.H., Jean, B.J.: Ablation in teeth with the free-electron laser around the absorption peak of hydroxyapatite (9.5 μm) and between 6.0 and 7.5 μm. Proc. SPIE. **2672**, 181–192 (1996)
43. Beck, O.J.: The use of the Nd-YAG and the CO_2 laser in neurosurgery. Neurosurg. Rev. **3**, 261–266 (1980)
44. Lowe, N.J., Lask, G., Griffin, M.E.: Laser skin resurfacing. Dermatol. Surg. **21**, 1017–1019 (1995)
45. Bekman, H.H.P.T., van den Heuvel, J.C., van Putten, F.J.M., Schleijpen, R.: Development of a mid-infrared laser for study of infrared countermeasures techniques. Proc. SPIE. **5615**, 27–38 (2004)
46. Zhu, G., Geng, L., Zhu, X., Li, L., Chen, Q., Norwood, R.A., Manzur, T., Peyghambarian, N.: Towards ten-watt-level 3-5 μm Raman lasers using tellurite fiber. Opt. Express. **23**, 7559–7573 (2015)
47. Ebendorff-Heidepriem, H., Kuan, K., Oermann, M.R., Knight, K., Monro, T.M.: Extruded tellurite glass and fibers with low OH content for mid-infrared applications. Opt. Mater. Express. **2**, 432–442 (2012)
48. Zhu, X., Peyghambarian, N.: High-power ZBLAN glass fiber lasers: review and prospect. Adv. Optoelectron. **2010**, 501956 (2010)
49. Dianov, E.M., Mashinsky, V.M.: Germania-based core optical fibers. J. Lightwave Technol. **23**, 3500–3508 (2005)
50. Sidharthan, R., Yoo, S., Ho, D., Zhang, L., Qi, W., Yue, M.S., Zhu, L., Dong, X., Tjin, S.C.: Fabrication of 74 mol% GeO_2-doped fibers and mid-IR supercontinuum generation. Paper presented at conference on lasers and electro-optics, San Jose (2016)
51. Mossadegh, R., Sanghera, J.S., Schaafsma, D., Cole, B.J., Nguyen, V.Q., Miklos, R.E., Aggarwal, I.D.: Fabrication of single-mode chalcogenide optical fiber. J. Lightwave Technol. **16**, 214–217 (1998)
52. Savage, J.A.: Materials for infrared fibre optics. Mater. Sci. Rep. **2**, 99–137 (1987)
53. Digonnet, M.J.F.: Rare-Earth-Doped Fiber Lasers and Amplifiers. Marcel Dekker, New York (2001)
54. Sanghera, J.S., Brandon Shaw, L., Aggarwal, I.D.: Chalcogenide glass-fiber-based mid-IR sources and applications. IEEE J. Sel. Topics Quantum Electron. **15**, 114–119 (2009)
55. FiberLabs Inc. www.fiberlabs-inc.com
56. Thorlabs. www.thorlabs.com

57. Gauthier, J.-C., Fortin, V., Carrée, J.-Y., Poulain, S., Poulain, M., Vallée, R., Bernier, M.: Mid-IR supercontinuum from 2.4 to 5.4 μm in a low-loss fluoroindate fiber. Opt. Lett. **41**, 1756–1759 (2016)

58. Handbook of Minerals Raman Spectra, Laboratoire de géologie de Lyon. www.ens-lyon.fr/LST/Raman

59. Micoulaut, M., Cormier, L., Henderson, G.S.: The structure of amorphous, crystalline and liquid GeO_2. J. Phys. Condens. Matter. **18**, R753 (2006)

60. Henderson, G.S., Neuville, D.R., Cochain, B., Cormier, L.: The structure of GeO_2–SiO_2 glasses and melts: a Raman spectroscopy study. J. Non-Cryst. Solids. **355**, 468–474 (2009)

61. Galeener, F.L., Mikkelsen, J.C., Geils, R.H., Mosby, W.J.: The relative Raman cross sections of vitreous SiO_2, GeO_2, B_2O_3, and P_2O_5. Appl. Phys. Lett. **32**, 34–36 (1978)

62. Bromage, J., Rottwitt, K., Lines, M.E.: A method to predict the Raman gain spectra of germanosilicate fibers with arbitrary index profiles. IEEE Photon. Technol. Lett. **14**, 24–26 (2002)

63. Bürger, H., Kneipp, K., Hobert, H., Vogel, W., Kozhukharov, V., Neov, S.: Glass formation, properties and structure of glasses in the TeO_2-ZnO system. J. Non-Cryst. Solids. **151**, 134–142 (1992)

64. Mori, A., Masuda, H., Shikano, K., Shimizu, M.: Ultra-wide-band tellurite-based fiber Raman amplifier. J. Lightwave Technol. **21**, 1300–1306 (2003)

65. Durteste, Y., Monerie, M., Lamouler, P.: Raman amplification in fluoride glass fibres. Electron. Lett. **21**, 723–724 (1985)

66. Saïssy, A., Botineau, J., Macon, L., Maze, G.: Diffusion Raman dans une fibre optique en verre fluoré. J. Phys. Lett. **46**, 289–294 (1985)

67. Almeida, R.M., Mackenzie, J.D.: Vibrational spectra and structure of fluorozirconate glasses. J. Chem. Phys. **74**, 5954–5961 (1981)

68. Fortin, V., Bernier, M., Carrier, J., Vallée, R.: Fluoride glass Raman fiber laser at 2185 nm. Opt. Lett. **36**, 4152–4154 (2011)

69. Malherbe, C., Gilbert, B.: Direct determination of the NaF/AlF_3 molar ratio by Raman spectroscopy in NaF–AlF_3–CaF_2 melts at 1000 °C. Anal. Chem. **85**, 8669–8675 (2013)

70. Almeida, R.M., Pereira, J.C., Messaddeq, Y., Aegerter, M.A.: Vibrational spectra and structure of fluoroindate glasses. J. Non-Cryst. Solids. **161**, 105–108 (1993)

71. White, R.T., Monro, T.M.: Cascaded Raman shifting of high-peak-power nanosecond pulses in As_2S_3 and As_2Se_3 optical fibers. Opt. Lett. **36**, 2351–2353 (2011)

72. Lucovsky, G., Martin, R.M.: A molecular model for the vibrational modes in chalcogenide glasses. J. Non-Cryst. Solids. **8-10**, 185–190 (1972)

73. Fortin, V., Bernier, M., El-Amraoui, M., Messaddeq, Y., Vallée, R.: Modeling of As_2S_3 Raman fiber lasers operating in the mid-infrared. IEEE Photon. J. **5**, 1502309 (2013)

74. Buck, J.A.: Fundamentals of Optical Fibers. Wiley, Hoboken (2004)

75. Agrawal, G.: Nonlinear Fiber Optics. Academic Press, San Diego (2001)

76. Dianov, E.M., Bufetov, I.A., Mashinskii, V.M., Neustruev, V.B., Medvedkov, O.I., Shubin, A.V., Mel'kumov, M.A., Gur'yanov, A.N., Khopin, V.F., Yashkov, M.V.: Raman fibre lasers emitting at a wavelength above 2 μm. Quant. Electron. **34**, 695 (2004)

77. Jiang, H., Zhang, L., Feng, Y.: Silica-based fiber Raman laser at > 2.4 μm. Opt. Lett. **40**, 3249–3252 (2015)

78. CorActive High-Tech Inc. www.coractive.com

79. Pask, H.M., Carman, R.J., Hanna, D.C., Tropper, A.C., Mackechnie, C.J., Barber, P.R., Dawes, J.M.: Ytterbium-doped silica fiber lasers: versatile sources for the 1-1.2 μm region. IEEE J. Sel. Topics Quantum Electron. **1**, 2–13 (1995)

80. Fermann, M.E., Hartl, I.: Ultrafast Fiber Laser Technology. IEEE J. Sel. Topics Quantum Electron. **15**, 191–206 (2009)

81. Richardson, D.J., Nilsson, J., Clarkson, W.A.: High power fiber lasers: current status and future perspectives [Invited]. J. Opt. Soc. Am. B. **27**, B63–B92 (2010)

82. Erdogan, T.: Fiber grating spectra. J. Lightwave Technol. **15**, 1277–1294 (1997)

83. Krause, M., Cierullies, S., Renner, H.: Stabilizing effect of line broadening in Raman fiber lasers. Opt. Commun. **227**, 355–361 (2003)
84. Vallée, R., Bélanger, E., Déry, B., Bernier, M., Faucher, D.: Highly efficient and high-power Raman fiber laser based on broadband chirped fiber bragg gratings. J. Lightwave Technol. **24**, 5039–5043 (2006)
85. Fortin, V., Bernier, M., Faucher, D., Carrier, J., Vallée, R.: 3.7 W fluoride glass Raman fiber laser operating at 2231 nm. Opt. Express. **20**, 19412–19419 (2012)
86. Kitcher, D.J., Nand, A., Wade, S.A., Jones, R., Baxter, G.W., Collins, S.F.: Directional dependence of spectra of fiber Bragg gratings due to excess loss. J. Opt. Soc. Am. A. **23**, 2906–2911 (2006)
87. Zeller, M., Lasser, T., Limberger, H.G., Maze, G.: UV-induced index changes in undoped fluoride glass. J. Lightwave Technol. **23**, 624–627 (2005)
88. Williams, G.M., Tsung-Ein, T., Merzbacher, C.I., Friebele, E.J.: Photosensitivity of rare-earth-doped ZBLAN fluoride glasses. J. Lightwave Technol. **15**, 1357–1362 (1997)
89. Taunay, T., Poignant, H., Boj, S., Niay, P., Bernage, P., Delevaque, E., Monerie, M., Xie, E.X.: Ultraviolet-induced permanent Bragg gratings in cerium-doped ZBLAN glasses or optical fibers. Opt. Lett. **19**, 1269–1271 (1994)
90. Poignant, H., Boj, S., Delevaque, E., Monerie, M., Taunay, T., Niay, P., Bernage, P., Xie, W.X.: Ultraviolet-induced permanent Bragg gratings in Ce-doped fluorozirconate glasses or optical fibres. J. Non-Cryst. Solids. **184**, 282–285 (1995)
91. Saad, M., Chen, L.R., Gu, X.: Highly reflective fiber Bragg gratings inscribed in Ce/Tm co-doped ZBLAN fibers. IEEE Photon. Technol. Lett. **25**, 1066–1068 (2013)
92. Snopatin, G.E., Shiryaev, V.S., Plotnichenko, V.G., Dianov, E.M., Churbanov, M.F.: High-purity chalcogenide glasses for fiber optics. Inorg. Mater. **45**, 1439–1460 (2009)
93. Asobe, M., Ohara, T., Yokohama, I., Kaino, T.: Fabrication of Bragg grating in chalcogenide glass fibre using the transverse holographic method. Electron. Lett. **32**, 1611–1613 (1996)
94. Florea, C., Sanghera, J.S., Shaw, B., Aggarwal, I.D.: Fiber Bragg gratings in As$_2$S$_3$ fibers obtained using a 0/−1 phase mask. Opt. Mater. **31**, 942–944 (2009)
95. Bernier, M., Asatryan, K.E., Vallée, R., Galstian, T.M., Sergei, A.V.e., Medvedkov, O.I., Plotnichenko, V.G., Gnusin, P.I., Evgenii, M.D.: Second-order Bragg gratings in single-mode chalcogenide fibres. Quant. Electron. **41**, 465–468 (2011)
96. Brawley, G.A., Ta'eed, V.G., Bolger, J.A., Sanghera, J.S., Aggarwal, I., Eggleton, B.J.: Strong photoinduced Bragg gratings in arsenic selenide optical fibre using transverse holographic method. Electron. Lett. **44**, 846–847 (2008)
97. Zou, L.E., Kabakova, I.V., Mägi, E.C., Li, E., Florea, C., Aggarwal, I.D., Shaw, B., Sanghera, J.S., Eggleton, B.J.: Efficient inscription of Bragg gratings in As$_2$S$_3$ fibers using near bandgap light. Opt. Lett. **38**, 3850–3853 (2013)
98. Ahmad, R., Rochette, M., Baker, C.: Fabrication of Bragg gratings in subwavelength diameter As$_2$Se$_3$ chalcogenide wires. Opt. Lett. **36**, 2886–2888 (2011)
99. Gattass, R.R., Mazur, E.: Femtosecond laser micromachining in transparent materials. Nat. Photonics. **2**, 219–225 (2008)
100. Davis, K.M., Miura, K., Sugimoto, N., Hirao, K.: Writing waveguides in glass with a femtosecond laser. Opt. Lett. **21**, 1729–1731 (1996)
101. Mihailov, S.J., Smelser, C.W., Lu, P., Walker, R.B., Grobnic, D., Ding, H., Henderson, G., Unruh, J.: Fiber Bragg gratings made with a phase mask and 800-nm femtosecond radiation. Opt. Lett. **28**, 995–997 (2003)
102. Mihailov, S.J., Grobnic, D., Smelser, C.W.: Efficient grating writing through fibre coating with femtosecond IR radiation and phase mask. Electron. Lett. **43**, 442–443 (2007)
103. Bernier, M., El-Amraoui, M., Couillard, J.F., Messaddeq, Y., Vallée, R.: Writing of Bragg gratings through the polymer jacket of low-loss As$_2$S$_3$ fibers using femtosecond pulses at 800 nm. Opt. Lett. **37**, 3900–3902 (2012)
104. Vallee, R., Bernier, M., Faucher, D.: System and method for permanently writing a diffraction grating in a low phonon energy glass medium. US Patent US8078023

105. Bernier, M., El-Amraoui, M., Messaddeq, Y., Vallee, R.: Mid-infrared Bragg grating in chalcogenide fiber. Paper presented at Advanced Photonics Congress, Colorado Springs, Colorado (2012)
106. Jackson, S.D., King, T.A.: Theoretical modeling of Tm-doped silica fiber lasers. J. Lightwave Technol. **17**, 948–956 (1999)
107. Moulton, P.F., Rines, G.A., Slobodtchikov, E.V., Wall, K.F., Frith, G., Samson, B., Carter, A.L.G.: Tm-doped fiber lasers: fundamentals and power scaling. IEEE J. Sel. Top. Quant. Elect. **15**, 85–92 (2009)
108. Jackson, S.D.: Cross relaxation and energy transfer upconversion processes relevant to the functioning of 2 μm Tm^{3+} -doped silica fibre lasers. Opt. Commun. **230**, 197–203 (2004)
109. Qpeak. www.qpeak.com
110. Ehrenreich, T., Leveille, R., Majid, I., Tankala, K., Rines, G., Moulton, P.: 1 kW, all-glass Tm:fiber laser. Paper presented at SPIE Photonics West 2010: LASE, Fibre Lasers VII: Technology, Systems and Applications, San Francisco (2010)
111. Yin, K., Zhu, R., Zhang, B., Liu, G., Zhou, P., Hou, J.: 300 W-level, wavelength-widely-tunable, all-fiber integrated thulium-doped fiber laser. Opt. Express. **24**, 11085–11090 (2016)
112. Keopsys Inc. www.keopsys.com
113. Nufern. www.nufern.com
114. Eichhorn, M., Jackson, S.D.: High-pulse-energy actively Q-switched Tm^{3+}-doped silica 2 μm fiber laser pumped at 792 nm. Opt. Lett. **32**, 2780–2782 (2007)
115. Jackson, S.D.: Passively Q-switched Tm^{3+}-doped silica fiber lasers. Appl. Opt. **46**, 3311–3317 (2007)
116. Ouyang, D.Q., Zhao, J.Q., Zheng, Z.J., Ruan, S.C., Guo, C.Y., Yan, P.G., Xie, W.X.: 110 W all fiber actively Q-switched thulium-doped fiber laser. IEEE Photon. J. **7**, 1–6 (2015)
117. Jackson, S.D., Pollnau, M., Jianfeng, L.: Diode pumped erbium cascade fiber lasers. IEEE J. of Quant. Electron. **47**, 471–478 (2011)
118. Pollnau, M., Jackson, S.D.: Energy recycling versus lifetime quenching in erbium-doped 3-μm fiber lasers. IEEE J. Quant. Electron. **38**, 162–169 (2002)
119. Jackson, S.D., King, T.A., Pollnau, M.: Diode-pumped 1.7-W erbium 3-μm fiber laser. Opt. Lett. **24**, 1133–1135 (1999)
120. Tokita, S., Hirokane, M., Murakami, M., Shimizu, S., Hashida, M., Sakabe, S.: Stable 10 W Er:ZBLAN fiber laser operating at 2.71–2.88 μm. Opt. Lett. **35**, 3943–3945 (2010)
121. Tokita, S., Murakami, M., Shimizu, S., Hashida, M., Sakabe, S.: 12 W Q-switched Er:ZBLAN fiber laser at 2.8 μm. Opt. Lett. **36**, 2812–2814 (2011)
122. Wei, C., Zhu, X., Norwood, R.A., Peyghambarian, N.: Passively continuous-wave mode-locked Er^{3+} doped ZBLAN fiber laser at 2.8 μm. Opt. Lett. **37**, 3849–3851 (2012)
123. Duval, S., Bernier, M., Fortin, V., Genest, J., Piché, M., Vallée, R.: Femtosecond fiber lasers reach the mid-infrared. Optica. **2**, 623–626 (2015)
124. Hu, T., Jackson, S.D., Hudson, D.D.: Ultrafast pulses from a mid-infrared fiber laser. Opt. Lett. **40**, 4226–4228 (2015)
125. Duval, S., Olivier, M., Fortin, V., Bernier, M., Piché, M., Vallée, R.: 23-kW peak power femtosecond pulses from a mode-locked fiber ring laser at 2.8 μm. Proc. SPIE. **9728**, 972802–972806 (2016)
126. Tobben, H.: Room temperature CW fibre laser at 3.5 μm in Er^{3+} doped ZBLAN glass. Electron. Lett. **28**, 1361–1362 (1992)
127. Henderson-Sapir, O., Munch, J., Ottaway, D.J.: Mid-infrared fiber lasers at and beyond 3.5 μm using dual-wavelength pumping. Opt. Lett. **39**, 493–496 (2014)
128. Yablon, A.D.: Optical Fiber Fusion Splicing. Springer, Berlin/Heidelberg (2005)
129. Okamoto, H., Kasuga, K., Kubota, Y.: Efficient 521 nm all-fiber laser: splicing Pr^{3+}-doped ZBLAN fiber to end-coated silica fiber. Opt. Lett. **36**, 1470–1472 (2011)
130. Yin, K., Zhang, B., Yao, J., Yang, L., Chen, S., Hou, J.: Highly stable, monolithic, single-mode mid-infrared supercontinuum source based on low-loss fusion spliced silica and fluoride fibers. Opt. Lett. **41**, 946–949 (2016)

131. Thapa, R., Gattass, R.R., Nguyen, V., Chin, G., Gibson, D., Kim, W., Shaw, L.B., Sanghera, J.S.: Low-loss, robust fusion splicing of silica to chalcogenide fiber for integrated mid-infrared laser technology development. Opt. Lett. **40**, 5074–5077 (2015)
132. Shustack, P.J.: Ultraviolet radiation-curable coatings for optical fibers. US Patent 5352712 A, 4 Oct 1994
133. Coady, C.J., Krajewski, J.J., Bishop, T.E.: Polyacrylated oligomers in ultraviolet curable optical fiber coatings. EP Patent 0204160 A2, 10 Dec 1986
134. Chien, C.K., Fewkes, E.J., Urruti, E.H., Winningham, M.J.: Coating composition for optical fibers. WO Patent 2002072498 A1, 19 Sep 2002
135. Gagnon, M.-A., Fortin, V., Vallée, R., Farley, V., Lagueux, P., Guyot, É., Marcotte, F.: Non-destructive testing of mid-IR optical fiber using infrared imaging. Proc. SPIE. **9861**, 986110–986116 (2016)
136. Caron, N., Bernier, M., Faucher, D., Vallée, R.: Understanding the fiber tip thermal runaway present in 3 μm fluoride glass fiber lasers. Opt. Express. **20**, 22188–22194 (2012)
137. Stolen, R.H., Ippen, E.P., Tynes, A.R.: Raman oscillation in glass optical waveguide. Appl. Phys. Lett. **20**, 62–64 (1972)
138. Dianov, E.M., Bufetov, I.A., Mashinskii, V.M., Shubin, A.V., Medvedkov, O.I., Rakitin, A.E., Mel'kumov, M.A., Khopin, V.F., Gur'yanovb, A.N.: Raman fibre lasers based on heavily GeO$_2$-doped fibres. Quant. Electron. **35**, 435–441 (2005)
139. Cumberland, B.A., Popov, S.V., Taylor, J.R., Medvedkov, O.I., Vasiliev, S.A., Dianov, E.M.: 2.1 μm continuous-wave Raman laser in GeO$_2$ fiber. Opt. Lett. **32**, 1848–1850 (2007)
140. Liu, J., Tan, F., Shi, H., Wang, P.: High-power operation of silica-based Raman fiber amplifier at 2147 nm. Opt. Express. **22**, 28383–28389 (2014)
141. Zhang, H., Tao, R., Zhou, P., Wang, X., Xu, X.: 1.5-kW Yb-Raman combined nonlinear fiber amplifier at 1120 nm. IEEE Photon. Technol. Lett. **27**, 628–630 (2015)
142. Rakich, P.T., Fink, Y., Soljačić, M.: Efficient mid-IR spectral generation via spontaneous fifth-order cascaded-Raman amplification in silica fibers. Opt. Lett. **33**, 1690–1692 (2008)
143. Gruppi, D., Eichhorn, M., Hirth, A., Pfeiffer, P.: Numerical modeling of pulsed Raman fiber converters at 2 μm. IEEE J. Quantum Electron. **45**, 446–453 (2009)
144. Jiang, H., Zhang, L., Yang, X., Yu, T., Feng, Y.: Pulsed amplified spontaneous Raman emission at 2.2 μm in silica-based fiber. Appl. Phys. B Lasers Opt. **122**, 1–4 (2016)
145. Munasinghe, H.T., Winterstein-Beckmann, A., Schiele, C., Manzani, D., Wondraczek, L., Afshar, S., Monro, T.M., Ebendorff-Heidepriem, H.: Lead-germanate glasses and fibers: a practical alternative to tellurite for nonlinear fiber applications. Opt. Mater. Express. **3**, 1488–1503 (2013)
146. Bernier, M., Fortin, V., El-Amraoui, M., Messaddeq, Y., Vallée, R.: 3.77 μm fiber laser based on cascaded Raman gain in a chalcogenide glass fiber. Opt. Lett. **39**, 2052–2055 (2014)
147. Mizunami, T., Iwashita, H., Takagi, K.: Gain saturation characteristics of Raman amplification in silica and fluoride glass optical fibers. Opt. Commun. **97**, 74–78 (1993)
148. Petersen, C., Dupont, S., Agger, C., Thøgersen, J., Bang, O., Rud Keiding, S.: Stimulated Raman scattering in soft glass fluoride fibers. J. Opt. Soc. Am. B. **28**, 2310–2313 (2011)
149. Luo, H., Li, J., Li, J., He, Y., Liu, Y.: Numerical modeling and optimization of mid-infrared fluoride glass Raman fiber lasers pumped by Tm^{3+} doped fiber laser. IEEE Photon. J. **5**, 2700211 (2013)
150. Gongwen, Z., Xiushan, Z., Norwood, R.A., Peyghambarian, N.: Experimental and numerical investigations on Q-switched laser-seeded fiber MOPA at 2.8 μm. J. Lightwave Technol. **32**, 3951–3955 (2014)
151. Tang, Y., Wright, L.G., Charan, K., Wang, T., Xu, C., Wise, F.W.: Generation of intense 100 fs solitons tunable from 2 to 4.3 μm in fluoride fiber. Optica. **3**, 948–951 (2016)
152. Duval, S., Gauthier, J.-C., Robichaud, L.-R., Paradis, P., Olivier, M., Fortin, V., Bernier, M., Piché, M., Vallée, R.: Watt-level fiber-based femtosecond laser source tunable from 2.8 to 3.6 μm. Opt. Lett. **41**, 5294–5297 (2016)
153. Jackson, S.D., Anzueto-Sanchez, G.: Chalcogenide glass Raman fiber laser. Appl. Phys. Lett. **88**, 221106 (2006)

154. Ahmad, R., Rochette, M.: Raman lasing in a chalcogenide microwire-based Fabry-Perot cavity. Opt. Lett. **37**, 4549–4551 (2012)
155. Ahmad, R., Rochette, M.: All-Chalcogenide Raman-parametric laser, wavelength converter, and amplifier in a single microwire. IEEE J. Sel. Top. Quant. Elect. **20**, 299–304 (2014)
156. Vanier, F., Peter, Y.-A., Rochette, M.: Cascaded Raman lasing in packaged high quality As_2S_3 microspheres. Opt. Express. **22**, 28731–28739 (2014)
157. Vanier, F., Rochette, M., Godbout, N., Peter, Y.-A.: Raman lasing in As_2S_3 high-Q whispering gallery mode resonators. Opt. Lett. **38**, 4966–4969 (2013)
158. Kulkarni, O.P., Xia, C., Joon Lee, D., Kumar, M., Kuditcher, A., Islam, M.N., Terry, F.L., Freeman, M.J., Aitken, B.G., Currie, S.C., McCarthy, J.E., Powley, M.L., Nolan, D.A.: Third order cascaded Raman wavelength shifting in chalcogenide fibers and determination of Raman gain coefficient. Opt. Express. **14**, 7924–7930 (2006)
159. Troles, J., Coulombier, Q., Canat, G., Duhant, M., Renard, W., Toupin, P., Calvez, L., Renversez, G., Smektala, F., El Amraoui, M., Adam, J.L., Chartier, T., Mechin, D., Brilland, L.: Low loss microstructured chalcogenide fibers for large non linear effects at 1995 nm. Opt. Express. **18**, 26647–26654 (2010)
160. Gao, W., Cheng, T., Xue, X., Liu, L., Zhang, L., Liao, M., Suzuki, T., Ohishi, Y.: Stimulated Raman scattering in $AsSe_2$-As_2S_5 chalcogenide microstructured optical fiber with all-solid core. Opt. Express. **24**, 3278–3293 (2016)
161. Toupin, P., Brilland, L., Mechin, D., Adam, J.L., Troles, J.: Optical aging of chalcogenide microstructured optical fibers. J. Lightwave Technol. **32**, 2428–2432 (2014)

Chapter 4
Infrared Super-continuum Light Sources and Their Applications

Mohammed N. Islam

4.1 Introduction

Super-continuum (SC) lasers are broadband light sources that share desirable traits from lamps and lasers. The bandwidth of the SC laser is broad like a lamp, while the spatial coherence and high intensity or brightness of the output is like a laser. Hence, some have even dubbed SC laser as "the ultimate white light" [1, 2]. However, up to now SC lasers have been used primarily in laboratory settings, because large, tabletop, mode-locked lasers have been often used to pump nonlinear media, such as optical fibers. We now replace those large pump lasers with laser diodes and fiber amplifiers used in the mature telecommunications and fiber-optic industry, thereby enabling practical applications in defense, homeland security, metrology, spectroscopy, and healthcare.

By exploiting the nature physics in a fiber, our group has made three major breakthroughs in SC lasers. First, we have extended the wavelength range for SC sources into the mid-infrared, covering most of the near- and mid-infrared wavelengths simultaneously. For example, our lasers provide continuous spectrum ranging from ~0.47 microns in the visible through the near-infrared from ~1 to 2 microns and reaching through most of the mid-infrared from ~2 to 4.5 microns. Research is also ongoing to extend the wavelength range out to the long-wave infrared, reaching to wavelengths of 12 microns or more. Second, by initiating the SC generation

M.N. Islam (✉)
Department of Electrical Engineering and Computer Science, University of Michigan, Ann Arbor, MI, USA

Department of Internal Medicine, Division of Cardiovascular Medicine, University of Michigan Medical School, Ann Arbor, MI, USA

Omni Sciences, Inc., Ann Arbor, MI, USA
e-mail: mni@eecs.umich.edu

© Springer International Publishing AG 2017
Y. Feng (ed.), *Raman Fiber Lasers*, Springer Series in Optical Sciences 207,
DOI 10.1007/978-3-319-65277-1_4

through naturally occurring phenomena in fibers, we eliminate the need for the mode-locked laser, replacing what in many cases is a large, table-top laser with commercial off-the-shelf (COTS) parts from the mature telecommunications and fiber-optic industry. Finally, we can scale the power to 64 W or higher by simply increasing the repetition rate and using a high-power fiber amplifier. Of course, along with the power increase, engineering challenges arise from packaging and thermal management.

SC generation describes the process by which narrowband optical pulses undergo substantial spectral broadening through the interplay of a number of nonlinear optical interactions in the medium, to yield a broadband spectrally continuous output. Since its first observation in bulk media by Alfano et al. [1], SC generation has been studied extensively, and numerous applications using SC have been proposed and demonstrated. Broadband SC generation in optical fibers has been of particular interest due to the unique advantages offered by their long optical interaction lengths, high nonlinearity, and potential applications in optical telecommunications. In addition, fiber-based SC lasers are potentially compact, reliable, and robust, which make them attractive candidates over conventional bulk laser sources for practical applications. With the development of mature gain fibers, high-power pump diodes, optical fibers of various materials, geometries, and dispersion profiles, it is now possible to construct a broadband SC fiber laser platform for almost any wavelength of interest from the UV to the mid-wave infrared and even long-wave infrared.

The use of picosecond and nanosecond pump pulse regimes with modulational instability-initiated SC generation has enabled the development of high average power, broadband SC sources. In addition, this pump regime provides easier access to a range of attractive SC properties such as a high degree of spectral flatness and relative simplicity in implementation compared to many SC systems that use mode-locked lasers. Figure 4.1 illustrates our architecture for modulational instability-initiated SC generation and provides a platform for generating SC in multiple wavelength regions. We utilize this framework to demonstrate SC systems in the visible, near-infrared, and mid-wave infrared wavelengths by selecting the appropriate gain fiber and SC generation fiber. In addition to the simplicity of implementation, we demonstrate that this architecture for SC generation also allows for the scalability of the SC time-averaged power by simply increasing the repetition rates and the pump power in the amplifier stages.

4.1.1 Modulational Instability-Initiated SC Generation

Our SC lasers are an all-fiber-integrated, high-powered light source that are elegant not because of its complexity, but rather because of their simplicity. For example, a block diagram of a SC laser is illustrated in Fig. 4.1. We start with a distributed

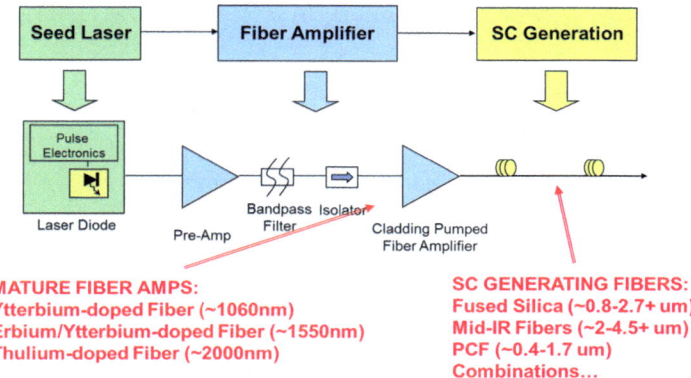

Fig. 4.1 Top-level SC generation design using all-COTS components

feedback laser diode, whose ~0.5–2 ns pulsed output is then amplified in a multiple-stage fiber amplifier. The first-stage preamplifier can be a standard erbium-, ytterbium-, or erbium/ytterbium-doped fiber amplifier, and its configuration is designed for optimal noise performance. Between amplifier stages band-pass filters can be used to block amplified spontaneous emission, and isolators may be used to prevent spurious reflections. The power amplifier stage is often made in cladding-pumped fiber amplifier, and its performance is optimized to minimize nonlinear distortion. This pump laser configuration is a pretty common and standard telecom design.

The novel SC generation occurs in the relatively short lengths of fiber that follow the pump laser. As shown in Fig. 4.1, we use exemplary just one or two meters of standard single-mode fiber SMF after the power amplifier stage, and this fiber is then followed by several meters of SC generation fiber. In the SMF, the peak power may be several kilowatts, and the pump light falls in the anomalous group velocity dispersion regime, which is often called the soliton regime. For these high peak powers in this dispersion regime, the nanosecond pulses are unstable due to a phenomenon known as modulational instability, which is basically parametric amplification in which the fiber nonlinearity helps to phase-match [3]. As a consequence, the nanosecond pump pulses are broken into many shorter pulses, because modulational instability is trying to form soliton pulses from the quasi-CW background. Although the DFB laser diode and amplification starts with nanosecond-long pulses, through modulational instability in the short length of SMF fiber, about 0.5 ps to several picosecond-long pulses are formed with high intensity. Thus, the few meters of SMF fiber results in an output similar to that produced by mode-locked lasers, except in a much simpler and cost-effective manner.

4.1.2   Raman Effect Dominates Long-Wavelength Expansion

The short pulses created through modulational instability are coupled into a nonlinear fiber for generation of SC. The nonlinear mechanisms leading to broadband SC include four-wave mixing or self-phase modulation along with the optical Raman effect [4]. Since the Raman effect is self-phase matched and shifts light to longer wavelength by emission of optical phonons, the SC spreads to longer wavelengths very efficiently. Therefore, for much of the long-wavelength expansion, the Raman effect is the dominant nonlinear mechanism. The short-wavelength edge arises from four-wave mixing, and oftentimes the short-wavelength edge is limited by increasing group velocity dispersion in the fiber. In our experience, for sufficient peak power and SC fiber length, the SC generation process will fill the long-wavelength edge up to the transmission window of the particular fiber used.

The configuration of Fig. 4.1 is actually a platform for generating SC over multiple wavelength ranges. For example, some of the mature fiber amplifiers for the power amplifier stage include ytterbium-doped fibers ($\sim$1060 nm), erbium-/ytterbium-doped fibers ($\sim$1550 nm), or thulium-doped fibers ($\sim$2000 nm). In addition, some of the candidates for the SC fiber include fused silica fibers (for generating SC between $\sim$0.67 and 2.5 microns); mid-infrared fibers such as fluorides, chalcogenides, or tellurites (for generating SC out to 4.5 microns or longer); photonic crystal fibers (for generating SC between $\sim$0.4 and 1.7 microns); or combinations of these fibers. Therefore, by selecting the appropriate fiber amplifier doping and nonlinear fiber, SC can be generated in the visible, near-infrared, or mid-infrared wavelength region.

4.1.3   Summary of Book Chapter

In this book chapter, we will describe in more detail the physics behind modulational instability (MI)-initiated SC generation, along with examples of SC lasers operating in the visible, short-wave infrared, mid-wave infrared, and even extensions into the long-wave infrared wavelength ranges. To be specific, in this chapter, we will use the following rough definitions of wavelength ranges: visible (VIS) covers $\sim$400 to $\sim$750 nm, near-infrared (NIR) is roughly 800 to 1100 nm, short-wave infrared (SWIR) will be approximately 1000 to 2500 nm, mid-wave infrared (MWIR) will be about 2000 nm (2 microns) to 5000 nm (5 microns), and long-wave infrared (LWIR) will be roughly 7500 nm (7.5 microns) to 12,000 nm (12 microns).

This chapter is organized as follows. The next section will provide a literature overview of different SC mechanisms and lasers. Then, more details of the physics of MI-initiated SC generation will be provided. SC lasers using fused silica fiber will be described, which operate in the visible, near-infrared, and short-wave infrared wavelengths. Next, mid-infrared SC lasers based on fluoride fibers will be described that extend the wavelength range out to approximately 4.5 microns. Prospects for extending to yet longer wavelengths will then be described, including using fluoride, tellurite, chalcogenide fibers, or combinations of these fibers.

The book chapter turns next to exemplary applications of SC lasers. Active remote sensing and hyper-spectral imaging experiments conducted with 5 W and 64 W short-wave infrared SC lasers will be detailed. Then, SC laser experiments for stand-off detection of solid targets are illustrated. A unique application of SC lasers in additive manufacturing or 3D printing of plastics or photopolymers will also be described. A number of medical applications of SC lasers will also be illustrated, such as detection of atherosclerotic plaque and noninvasive glucose monitoring. Finally, the chapter will conclude with a summary of SC lasers and their applications.

4.2   Literature Overview

In this section, we briefly review some of the recent SC setups and results reported in literature focusing more on the MWIR SC regime [5]. We also briefly describe the SC generation process relevant to the systems discussed in this chapter. Table 4.1 provides a summary of the pump configuration, SC wavelength range, and average power of the SC systems discussed in this section. The works presented here serve only to provide a background of some of the SC laser systems and is not an exhaustive list of reported SC sources. Detailed reviews of the works published with respect to continuum generation are provided by Dudley et al. [29], Genty et al. [30], and Taylor [31].

SC generation in optical fibers was first observed by Lin and Stolen for pumping in the normal group velocity dispersion regime in standard silica fiber, where the spectral broadening was attributed to a combination of Raman scattering, self- and cross-phase modulation, and four-wave mixing [32]. Since then, various groups have studied SC generation in silica fibers by pumping around 1310 nm, near the zero-dispersion wavelengths, or around 1550 nm, in the anomalous dispersion region, where the spectral broadening was attributed mainly to soliton propagation dynamics. However, since most of the SC investigations were carried out in silica fibers, the long-wavelength edge of the SC is generally <3 microns, which is then limited by the soaring absorption of silica glass [33]. In the past decade or so, fibers made of various materials, geometries, and dispersion profiles have been investigated for SC generation, and broadband SC has been reported in fibers covering the spectrum from the ultraviolet to the MWIR.

SC generation in the visible wavelengths has been widely studied using photonic crystal fibers and various pump sources. For example, mode-locked picosecond pulses from an ytterbium (Yb) fiber lasers at 1060 nm have been used to generate >1 W in the visible region [12, 13]. SC generation extending down to 400 nm was demonstrated using a microchip laser at 1064 nm to pump a photonic crystal fiber (PCF) with modified group index in the infrared to effectively phase-match with deeper blue wavelengths [7]. Another method, as demonstrated by Kudlinski

Table 4.1 Examples for SC sources with various pump and SC fiber configurations

Pump configuration	SC fiber	SC wavelength	SC average power	
~0.6 ns, 1.064 μm, Q-switched Nd:Yag microchip laser	PCF	~0.4–0.7 μm	–	[6]
600 ps, ~1.064 μm, microchip laser	PCF	~0.4–2.45 μm	~12.4 mW	[7]
600 ps, ~1.064 μm, Q-switched Nd:Yag laser	PCF	~0.5 to >1.75 μm	<30 mW	[8]
~0.6 ns, 1.064 μm, Q-switched Nd:Yag microchip laser	PCF	~0.5 to >1.5 μm	<40 mW	[9]
1 ps, ~1.5 μm, chirped-pulse amplification system	Photonic bandgap fiber	~0.43–1.45 μm	~0.16 W	[10]
~2 ns, 1.5 μm, DFB laser diode	PCF	~0.45–1.2 μm	~0.74 W	[11]
ps, 1.06 μm, mode-locked Yb fiber laser	PCF	~0.525–1.8 μm	>1.3 W	[12]
ps, 1.06 μm, mode-locked Nd-glass oscillator	PCF	~0.5–1.8 μm	~5 W	[13]
~3–4 ps, 1.064 μm, mode-locked Yb pump laser	PCF	~0.37–1.75 μm	~3.5 W	[14]
30 ps, 1.55 μm, gain-switched DFB laser	Dispersion-shifted fiber	~0.9 to >1.8 μm	<50 mW	[15]
34 fs, 1.55 μm, mode-locked Er-doped fiber laser	HiNL	0.85 to >2.6 μm	~0.4 W	[16]
CW, ~1.3 μm, Raman Laser	HiNL	~1.2 to >1.7 μm	< ~5 W	[17]
~2 ns, ~1.55 μm, DBR laser	HiNL	~0.8–2.8 μm	~5.3 W	[18]
CW, ~1.07 μm, Yb fiber laser	PCF	~1.06–1.67 μm	~29 W	[19]
~21 ps, ~1.06 μm, YDFA MOPA	PCF	~0.4–2.25 μm	~39 W	[20]
900 fs, 1.55 μm, mode-locked Er-doped fiber laser	ZBLAN	~1.8–3.4 μm	~5 mW	[21]
100 fs, ~2.5 μm, OPA	Chalcogenide	~2.1–3.2 μm	–	[22]
110 fs, 1.55 μm, OPO	Soft-glass PCF	0.35–3 μm	~70 mW	[23]
100 fs, 1.55 μm OPO	Tellurite PCF	~0.79–4.87 μm	~70 mW	[24]
180 fs, 1.45 μm, laser tunable OPA	ZBLAN	UV–6.28 μm	<20 mW	[25]
~2 ns, 1.55 μm, DFB laser diode	ZBLAN	~0.8–4.5 μm	~23 mW	[26]
~2 ns, 1.55 μm, DBR laser diode	ZBLAN	~0.8–4 μm	~1.3 W	[27]
~2 ns, 1.55 μm, DFB laser diode	ZBLAN	~1.9–4.5 μm	~5.2 W	[28]
0.4–2 ns, ~1.54 μm, DFB laser diode	ZBLAN	~0.8–4 μm	~10.5 W	[4]

et al., uses tapers with continuously decreasing dispersion to generate SC with a wavelength edge at 400 nm and >2 mW/nm spectral density [14]. Multiple wavelength pumping schemes involving a pump and its second harmonic [6] or four-wave mixing pump conversion [9] have also been demonstrated to increase spectral coverage in the visible region. Most of the techniques, however, use a mode-locked laser and suffer from one primary drawback, which is the lack of average power scalability due to a fixed repetition rate. SC systems based on microchip laser [8, 9] and master oscillator power amplifier (MOPA)-type pumps with the ability to vary repetition rates have been demonstrated. For example, Matos et al. demonstrated a repetition rate tunable chirped-pulse amplification erbium system which was frequency doubled to generate SC in PCF with ∼160 mW average power [10]. Kumar et al. demonstrated an SC extending from 0.45 to 1.2 microns with up to 0.74 W of time-averaged power, by pumping a 1.5 m PCF with a frequency-doubled, amplified, gain-switched, telecom laser diode [11]. An advantage of the setup shown by Kumar et al. is that the average power can be scaled up by increasing the repetition rate of the laser diode while keeping the peak power (and hence the spectral shape) constant. A detailed review of the ways to generate visible SC and to extend the short-wavelength side of the SC further into the UV is given by Travers [34].

SC lasers in the IR have also been demonstrated in various pump and fiber configurations. Two approaches of generating SC are widely used: pumping a short length of nonlinear fiber using femtosecond pulses with high peak power or using continuous wave (CW)/quasi-CW with lower peak power to pump longer lengths of fiber. For example, a mode-locked femtosecond erbium fiber laser was used to generate SC ranging from 0.8 to 2.7 microns in just 12 cm of highly nonlinear silica fiber [16]. Moon et al. reported an SC extending from ∼0.8 to 1.7 microns in dispersion-shifted fiber using amplified laser diode pulses with 30 ps pulse width [15]. In contrast, Xia et al. reported a ∼5.3 W SC ranging from ∼0.8 to 3 microns in highly nonlinear silica fiber using nanosecond diode pulses amplified with a multistage amplifier [18]. In the CW regime, Abeeluck et al. reported an SC generation extending beyond 1.75 um and ∼5 W average output power in 500 m of highly nonlinear fibers pumped using a CW Raman fiber laser [17]. Ytterbium-doped fiber amplifier (YDFA)-based pumps have also been used to demonstrate high-power SC generation in silica-based PCFs. For example, time-averaged powers of ∼29 W in an SC extending from 1.06 to 1.67 microns using a CW ytterbium fiber laser have been demonstrated [19]. Time-averaged powers of ∼39 W in an SC extending from 0.4 to 2.25 microns using a picosecond YDFA MOPA pump source have also been reported [20]. Since the pump wavelength is limited to ∼1–1.2 microns for YDFA-based pumps, the SC in this case extends out only to ∼2.25 microns on the long-wavelength side. In addition, since the pump wavelength for ytterbium-based laser sources lies in the normal dispersion regimes for standard silica fibers, specialty dispersion-engineered PCFs will have to be used to utilize the advantages of pumping in the anomalous dispersion regime to obtain broader SC generation.

In order to generate SC in the MWIR, optical fibers with low loss in the MWIR windows, such as fluoride, chalcogenide and tellurite are required. For example, Sanghera et al. used a Ti-sapphire laser to generate 100-pJ, 100 fs duration pulses near 2.5 microns using nonlinear frequency conversion in bulk crystals; these pulses were coupled into various chalcogenide fibers to generate SC spanning $\sim$2.1–3.2 microns (10 dB width) [35]. Kulkarni et al. reported spectral shifting from 1.55 microns to $\sim$1.9 microns in chalcogenide fibers, where low optical damage threshold of $\sim$1 GW/cm^2 and high normal GVD limit further redshifting of the spectrum [36]. Sulfide and selenide fibers have also been used to demonstrate SC ranging from 2 to 3 microns using a 2.5 micron OPA pump laser [22]. Omenetto et al. reported a broadband SC extending from 0.35 to beyond 3 microns in a short piece of high-nonlinearity soft-glass PCF pumped using an OPO laser source that provides tunable femtosecond pulses around 1550 nm [23]. Domachuk et al. reported a broadband SC extending from $\sim$0.79 to 4.87 microns, with $\sim$90 mW of average output power, has been demonstrated by coupling 100 fs long, 1.55 micron pulses into an 8 mm length of highly nonlinear tellurite microstructured PCF fiber [24].

Recently, ZBLAN fibers have been of great interest for SC generation in the MWIR. Although ZBLAN has a lower nonlinearity than that of silica, ZBLAN glasses are superior to other soft-glass fibers for high-power super-continuum generation in the mid-IR due to their much lower background loss, relatively higher strength, and a wide transparency window in the mid-IR region [37]. The first SC generation in ZBLAN fibers extending from $\sim$1.8 to 3.4 microns, with a total average power of 5 mW, was demonstrated by Hagen et al., where a commercial mode-locked, 1550 nm, Er3+-doped fiber laser with a 900 fs pulse duration and a repetition rate of 200 KHz was used to pump a 21 cm long standard single-mode silica fiber. The output was then coupled to a $\sim$91 cm long ZBLAN fiber with a core diameter of 8.5 microns and an NA of 0.21. SC generation in this case was based almost entirely on the cascaded Raman soliton self-shifting process [21]. Qin et al. demonstrated an ultra-broadband SC generation expanding from the UV to $\sim$6.28 microns where a 1450 nm femtosecond laser (20 mw average power, 50 MW peak power) with a pulse width of $\sim$180 fs and repetition rate of $\sim$1 KHz from a tunable OPA pumped by a Ti-sapphire femtosecond laser was launched into a 2 cm ZBLAN fiber. Since the pump wavelength is in the normal dispersion region, initial spectral broadening is caused mainly by SPM, and further spectral broadening occurs due to SPM, Raman scattering, and four-wave mixing [25]. Thus, femtosecond lasers have been used to generate SC beyond 3 microns using both fluoride and tellurite fibers, but with modest average powers (<0.1 W).

In contrast to ZBLAN SCs generated with femtosecond pump pulses, nanosecond pump pulses can be readily amplified using multiple fiber amplifier stages before the peak power-related damage becomes of concern. This allows for the scalability in the total average power of the SC, enabled by the recent advances in high-power fiber amplifiers and fiber-coupled laser diode pumps. The possible disadvantage, may be that the confinement loss is higher due to the longer lengths of ZBLAN required and the spectrum does not extend out as far as in the case of femtosecond pulses. A novel technique to eliminate the need for a mode-locked setup for high peak power pulse generation has been previously demonstrated by Xia et al. [18]. In this technique, modulational instability in standard SMF is used to initiate pulse breakup of quasi-CW laser diode pulses to form shorter pulses with high peak powers. Xia et al. have demonstrated SC generation in ZBLAN fibers, extending from $\sim$0.8 to 4.5 microns with an average power of $\sim$23 mW [26]. They have also demonstrated the linear scaling up of the time-averaged power to 5.3 W [27] and then up to 10.5 W in a continuum extending from $\sim$0.8 to 4 microns in ZBLAN fibers, using more powerful amplifier stages and higher repetition rates [4]. Recently, Kulkarni et al. demonstrated a MWIR SC laser, using a thulium-doped fiber amplifier (TDFA)-based system, extending from $\sim$1.9 to 4.5 microns and total power of $\sim$2.6 W in the SC, with $\sim$0.7 W of time-averaged power in wavelengths beyond 3.8 microns [28]. The SC in this case is also modulated at 50% duty cycle. Geng et al. also recently demonstrated a TDFA-based MWIR SC with a lower average power, extending from $\sim$1.7 to 4.2 microns using ZBLAN fibers and $\sim$1.8–2.9 microns using chalcogenide fibers [38].

SC generation mechanisms have also been extensively studied, and a number of reviews have been written, devoted to understanding the physical processes involved [29, 30, 39–41]. Figure 4.2 provides a summary of the dominant mechanisms for the different pulse and dispersion regimes as described in Genty et al. [30]. Here, we will only briefly cover the physical process relevant to the SC systems presented in this chapter, which is picosecond and larger pulses pumping in the anomalous dispersion regime. For this pulse regime, the initial soliton dynamics is different from the sub-picosecond case. For pulses of long durations with high peak power, the soliton order N becomes very large ($N = (L_\mathrm{D}/L_\mathrm{NL})^{1/2} \gg 10$), where L_D and L_NL are the dispersion and nonlinear lengths, respectively. In this case, the soliton fission mechanism becomes less prominent. This is expected, since the characteristic length scale over which fission occurs, $L_\mathrm{fiss} = L_\mathrm{D}/N \propto \Delta\tau$, and increases with pulse duration $\Delta\tau$ [29, 42]. Instead, MI occurs and begins to dominate the initial soliton propagation mechanism. MI is the equivalent of parametric sidebands generated by

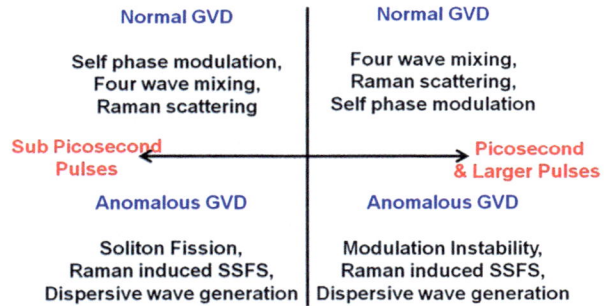

Fig. 4.2 Summary of dominant mechanisms for SC generation in different pulse and dispersion regimes

four-wave mixing and occurs on the same characteristic length scale, $L_{MI} = L_{NL}/2$ [39], independent of the pulse duration. In the time domain, this is characterized by the splitting of the pulse into shorter temporal sub-pulses.

The frequency shift of the first-order MI gain peaks from the pump is given by $\Delta\omega_{MI} = \sqrt{2\gamma P/|\beta_2|}$, where γ is the nonlinearity coefficient, P is the pump peak power, and β_2 is the group velocity dispersion parameter of the fiber at the pump wavelength [39]. The MI time period is then given by $T_{MI} = 2\pi/(\Delta\omega_{MI})$ and is the deciding factor for whether soliton fission or MI dominates the initial SC generation process. MI is expected to occur if the MI time period is sufficiently smaller than the pump pulse duration. Since the time period is a function of peak power, a specific pump pulse duration, below which fission occurs, cannot be defined as such. However, due to the similarity in scaling, it can reasonably be defined in terms of the soliton order [34]. Travers [34] calculated $N \sim 22$ for the upper limit on N for soliton fission to occur and is close to value of $N \sim 15$ calculated by Genty et al. [30]. Dudley et al. calculated N to be ~ 10, above which soliton fission will be overcome by MI [29]. The $N \sim 10$–22 range is not an issue for ultrashort-pulse sources such as the mode-locked lasers, but for picoseconds and longer-pulse sources with high peak powers, enough to generate SC, N would be much larger. Thus, we would not expect any soliton fission to occur in this case. It has further been pointed out by Travers [34], that the products of MI do not themselves undergo any soliton fission, since only fundamental solitons are created from MI. Once the MI dynamics breaks up the pump pulses into shorter-duration sub-pulses, further spectral broadening follows the same mechanism as the short pump pulses, whereby the short-wavelength side is due to the solitons that excite dispersive waves in the normal dispersion region through four-wave mixing and dispersive wave generation and the long-wavelength generation is due to the Raman-induced soliton self-frequency shift (SSFS). Pumping too far in the anomalous dispersion region could, however, result in reduced spectral broadening since the initial MI dynamics does not generate sufficient bandwidth to effectively seed the dispersive wave transfer into the normal dispersion regime [30]. Thus, in order to achieve the broadest SC in the picosecond and larger pulse regime, we should use a pump wavelength in the anomalous dispersion regime but closer to the zero-dispersion wavelength of the fiber.

Coherence properties of SCs have also been widely studied in the literature [30, 43]. For the SC systems presented in this chapter, which utilize picosecond and larger pulses, SC generation is initiated with a fast modulation of the pump envelope. In the anomalous dispersion case, this modulation can arise from MI and/or Raman scattering. In both cases, the modulation arises spontaneously from noise at frequencies that do not overlap with the pump bandwidth [30]. Thus, MI-initiated SC, which is a noise-seeded process, destroys any temporal coherence the pump pulse may have possessed and is therefore incoherent [34].

4.3   MI-Initiated SC Generation and Raman Nonlinearity

Our SC light sources are simple and can use standard telecom or fiber-optic components because the SC generation is initiated by MI. To illustrate the concept, a cartoon can help elucidate the pulse generation mechanism. We start with a quasi-continuous wave input, which can be pulses in the range of 30 ps to several nanoseconds (left side of Fig. 4.3). When we launch the quasi-CW input into a fiber operating in the anomalous group velocity dispersion regime at high powers (e.g., in the kilowatts range), then the quasi-CW input is unstable and begins to experience undulations that grow along the fiber length. Through a combination of group velocity dispersion and nonlinearity, the quasi-CW wave wants to break up into what are known as "soliton" pulses, which are the stable solution for the fiber in this regime. The middle of Fig. 4.3 illustrates the formation of soliton pulses through the MI process. As the solitons continue to propagate down the fiber, they may start to overlap and collide and exchange energy through the Raman effect in the fiber (e.g., shift energy from shorter wavelengths to longer wavelengths through emission of optical phonons in the fiber). This interaction is shown on the right side of Fig. 4.3. This cartoon shows the real benefit of MI-initiated SC generation: we do not need mode-locked lasers to generate pulses, but rather we use the nature physics in the fiber to generate the pulses in relatively short lengths of fiber. Since mode-locked lasers can be quite complicated, using MI introduces a tremendous simplification in the SC generation process. Note that there are two requirements to take advantage of MI: the quasi-CW input be in the anomalous dispersion regime and relatively high powers where nonlinear effects are significant.

We generate a broad continuum by implementing the SC generation as a two-step process (Fig. 4.4) [18]. In the first stage, laser diode pulses are launched into single-mode fiber, where the interaction between nonlinearity and anomalous dispersion breaks the quasi-CW input into a train of solitons through MI [15, 26, 39] and

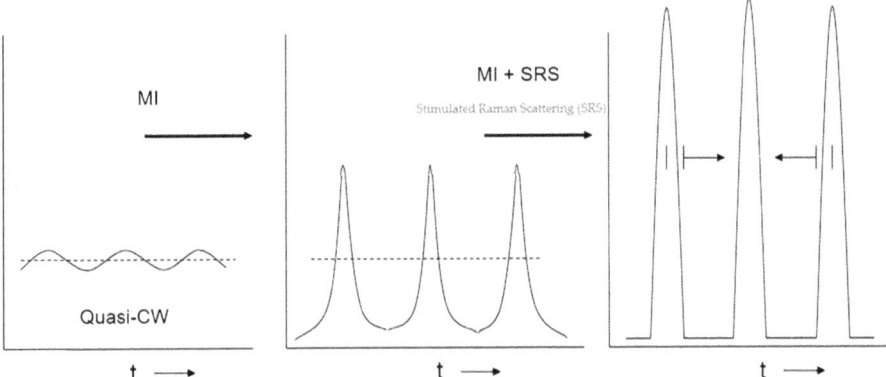

Fig. 4.3 Cartoon illustrating the MI-initiated process leading to SC generation

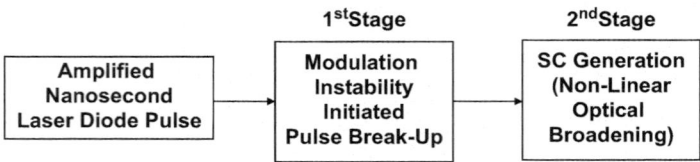

Fig. 4.4 Two-step SC generation process – pulse breakup in single-mode fiber followed by spectral broadening in nonlinear fiber

significantly increases the peak power. The generated solitons will then undergo frequency downshift through soliton self-frequency shifting (SSFS) [44]. In the second-stage fiber, the SC spectrum is further broadened primarily through the combination of self-phase modulation and the Raman effect of the ultrashort solitons [26]. By choosing a fiber with a zero-dispersion wavelength close to the pump wavelength, the SC generation also benefits from parametric four-wave mixing effects due to the reduced dispersion of the nonlinear fiber [39]. The ensemble average of these solitons, which spread over the entire nanosecond wide pulse, gives rise to the broad, smooth SC spectrum [44]. The nanosecond laser diode pumping also simplifies the scaling up of the time-averaged power. Since the nonlinear phenomena responsible for SC generation are related to the peak power of the pump pulses, the average power in the continuum can be increased simply by increasing the pulse duty cycle while keeping the peak power approximately constant.

To further investigate and understand the SC generation mechanism under different pump systems, we numerically solve the generalized nonlinear Schrödinger equation (NLSE). The complex envelope $A(z, \tau)$ of a pulse, under the slowly varying approximation, satisfies the generalized NLSE given by [39]

$$\frac{\partial A}{\partial z} = \left(\hat{D} + \hat{N}\right)A$$

$$\hat{D} = -\frac{i}{2}\beta_2\frac{\partial^2 A}{\partial \tau^2} + \frac{1}{6}\beta_3\frac{\partial^3 A}{\partial \tau^3} + \frac{i}{24}\beta_4\frac{\partial^4 A}{\partial \tau^4} - \frac{\alpha}{2}$$

$$\hat{N} = i\gamma\left(1 + \frac{i}{\omega_0}\frac{\partial}{\partial t}\right)\int_{-\infty}^{+\infty}[(1 - f_R)\,\delta(t) + f_R h_R(t)]$$
$$*|A\,(z, t - t')|^2 dt',$$

where the pulse moves along z in the retarded time frame $\tau = t - z/v_g$ with the center angular frequency of ω_0. The linear terms in the differential operator $\hat{D}$ account for the second- (β_2), third- (β_3), and fourth-order (β_4) dispersion as well as the loss (α) of the fiber. The terms in the operator $\hat{N}$ result from nonlinear interactions, which describe self-phase modulation, self-steepening, and stimulated Raman scattering effects. In particular, the effective nonlinearity is defined as $\gamma = n_2\omega_0/cA_{\mathrm{eff}}$, where

n_2 is the nonlinear refractive index and A_{eff} is the effective mode area of the fiber, respectively. In addition, $h_R(t)$ represents the Raman response function, and f_R is the fractional contribution of the Raman response to the nonlinear polarization.

As a specific example that helps to elucidate the MI process and SC generation, we will study the SC generation in a 2.5 m length of SMF (first stage in Fig. 4.4) followed by a 15 cm length of high-nonlinearity (HiNL) fused silica fiber (second stage in Fig. 4.4) [18]. The NLSE described above has been solved by an adaptive split-step Fourier method with the initial pulse shape as the known boundary value [18]. To reduce the computation time, we assume a 100 ps super-Gaussian pulse at 1553 nm as the input to our simulator. Simulations with pulse widths ranging 30–200 ps have also been conducted and show the same results. Furthermore, the step size is determined and dynamically adjusted by the nonlinear gain in each section. Because of the large bandwidth of the SC compared to the Raman gain spectrum of the silica glass, approximation of the Raman gain as a linear function of frequency is not valid any more. Therefore, we take into account the actual Raman gain spectrum of the specific fiber in the simulator. For the single-mode fiber, the effective nonlinearity γ is 1.6 W^{-1} km^{-1}, and Raman gain peak g_R is 6.4×10^{-14} m/W [18]. The HiNL fiber used in the experiment has a zero-dispersion wavelength at 1544 nm with the effective nonlinearity coefficient assumed to be 9.6 W^{-1} km^{-1}.

The simulation is performed in three distinct stages in the low average power setup. The first stage simulates the output after the amplifier, i.e., before the single-mode fiber, by including the nonlinearity in the gain fiber. Due to large peak powers developed in the last-stage amplifier and same core size of the gain fiber, we observe some spectral broadening at this stage itself, which is primarily attributed to the modulational instability seeded by the 1530 nm ASE from the amplifier.

The output from the amplifier stage is then coupled into a 2.5 m length of single-mode fiber. The spectrum is redshifted to ∼2.5 μm in the simulation, as shown in Fig. 4.5. In the corresponding time domain, we observe the 100 ps super-Gaussian pulse breaks up into a series of short pulses. To further clarify the evolution of the pulse breakup process, Fig. 4.6 illustrates both the pulse profile and spectrum of the 100 ps input pulse. The quasi-CW input pulse from the amplifier first breaks up into a train of solitons through the MI in the fiber. In the spectrum domain, multiple side bands can be observed with the band separation inversely proportional to the temporal separation of the soliton pulses. The generated solitons then experience frequency downshifting due to the SSFS, with the wavelength redshifting in the corresponding spectrum.

The SC spectrum is further broadened in the following HiNL fiber with the simulated spectrum and pulse profile shown in Fig. 4.7. By using a short piece of highly nonlinear fiber, i.e., 15 cm, the spectrum is primarily broadened through SPM, combined with the contribution from other nonlinear effects, e.g., the generation of non-solitonic dispersive wave and cross-phase modulation. The SC spectrum is smooth and flat from ∼1.8 to 2.8 μm, whose power density is ∼20 dB higher than the short-wavelength side and is consistent with experimental results on the HiNL fiber [18].

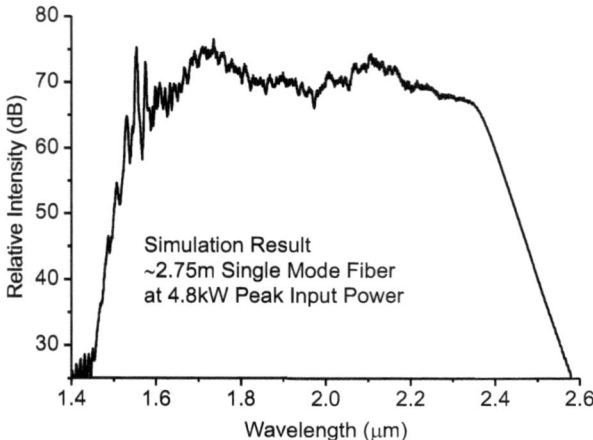

Fig. 4.5 Simulation results of low average power continuum after 2.5 m single-mode fiber at 4.8 kW peak power

Compared to the SC generation by using femtosecond laser pumping, we generate the SC spectrum through the ensemble average of a series of independent solitons spread in the nanosecond scale [18]. Figure 4.8 shows spectra of three sub-pulses sliced from the different parts of the 100 ps simulated pulse. Sub-pulses in the middle and trailing edge of the pulse are composed of multiple solitons and contain the entire SC spectrum. The spectrum of the sub-pulse from the leading edge is slightly shorter in the long-wavelength side. Because the redshifted solitons travel slower than the blueshifted spectral components in the anomalous dispersion environment, fewer redshifted solitons remain in the leading edge of the pulse. Nevertheless, all three sub-pulses contain most of the SC spectra with comparable spectral power intensity. Therefore, the entire spectrum is a superposition of the spectra from these sub-pulses. Although each soliton may have a different nonlinear wavelength shift, the ensemble average of these solitons gives rise to a stable and smooth SC spectrum.

To further confirm our hypothesis, Fig. 4.9 simulates the SC spectrum and the corresponding pulse profile generated in 1 m length of HiNL fiber by using a femtosecond laser with pulse width of ~150 fs at 15 kW peak power. The femtosecond pulse breaks up into multiple solitons and dispersive waves in the time domain and generates a similar SC spectrum as using nanosecond pulses [18]. Hence, by using MI to break up the nanosecond pulses into femtosecond soliton trains, laser diode pulses can be used to generate SC in a similar way as femtosecond lasers. Furthermore, compared to the relatively high fluctuation and instability associated with the SC spectrum generated by using femtosecond pump laser, we estimate the amplitude fluctuation over the entire spectrum from pulse to pulse to be less than 1% in our system.

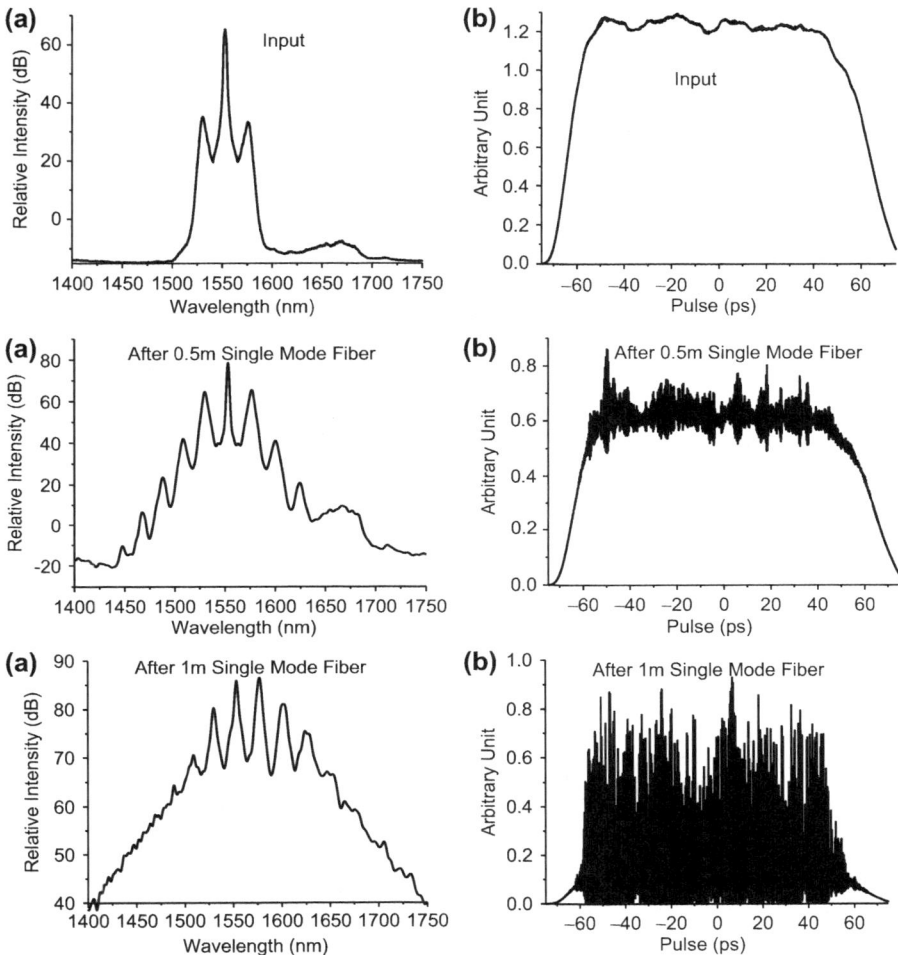

Fig. 4.6 (**a**) Spectrum and (**b**) pulse evolution in 1 m single-mode fiber as a function of propagation distance at 4.8 kW peak power

4.4 Visible, NIR, and SWIR Experiments in Fused Silica Fiber

In this section we will review some of the experiments conducted in fused silica fiber (e.g., the type of fiber material that telecommunications systems operate on) that cover parts of the visible, near-infrared (NIR), and short-wave infrared (SWIR). First, we will review experiments in high-nonlinearity (HiNL) fiber. We will follow this discussion with a simple prototype that was used in field trials described later in the chapter, followed by a broadband system that covers the VIS, NIR, and

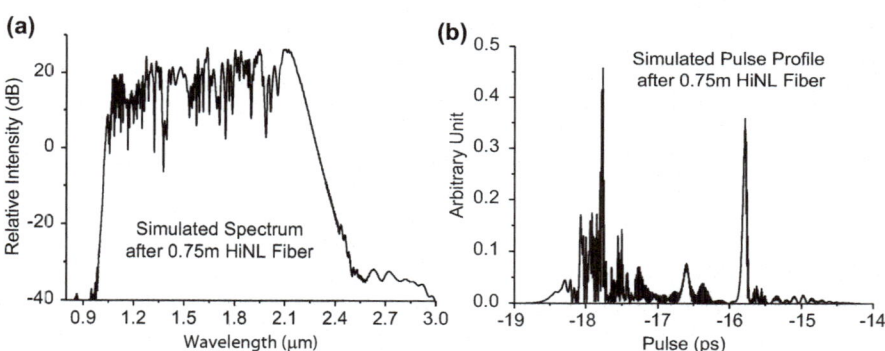

Fig. 4.7 Simulation results after 2.75 m single-mode fiber plus 15 cm HiNL fiber

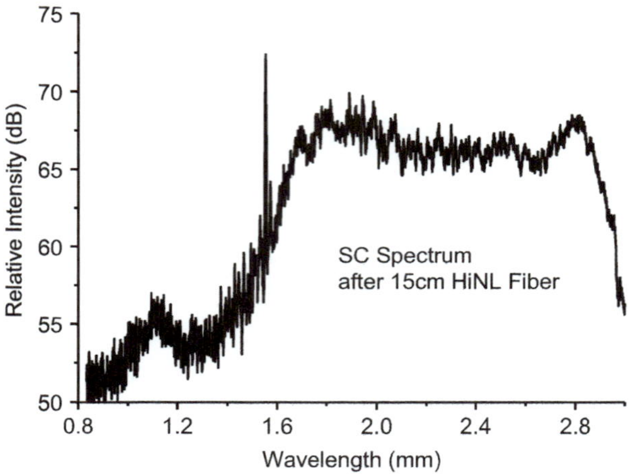

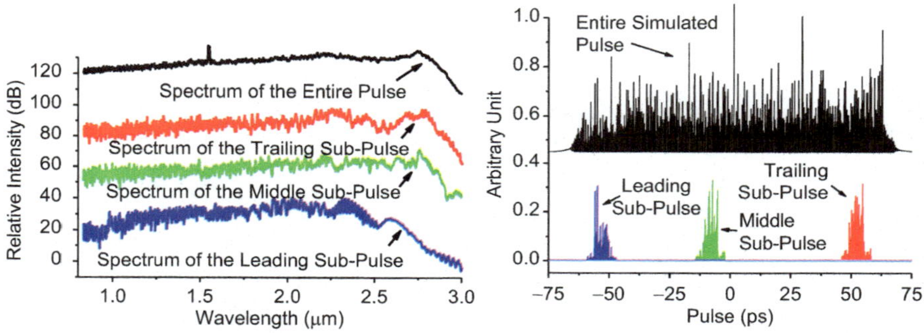

SC Spectrum after 15cm HiNL Fiber

Spectrum of the Entire Pulse

Spectrum of the Trailing Sub-Pulse

Spectrum of the Middle Sub-Pulse

Spectrum of the Leading Sub-Pulse

Entire Simulated Pulse

Leading Sub-Pulse

Trailing Sub-Pulse

Middle Sub-Pulse

Fig. 4.8 Simulation results of the SC spectrum from different locations of the 100 ps pulse. The spectra of the sub-pulses have the similar spectral intensity and are plotted with vertical shifts for the clarification purpose

(a)

Simulated Spectrum after 0.75m HiNL Fiber

(b)

Simulated Pulse Profile after 0.75m HiNL Fiber

Fig. 4.9 Simulated SC (**a**) spectrum and (**b**) pulse profile from 1 m HiNL fiber pumped by 150 fs pulse with 15 kW peak power

SWIR. We next discuss a high-power, 64 W prototype that as used in hyper-spectral imaging applications. Fused silica-based systems have key advantages in that all the fiber parts can be spliced together, fused silica has among the highest damage thresholds, and the power can be scaled to 100 W or more. On the other hand, fused silica fiber cannot extend beyond the SWIR due to the inherent absorption in the fiber. Therefore, in the following sections, we show how to extend to MWIR and LWIR by using softer glasses, such as fluorides, tellurites, and chalcogenides.

4.4.1   SC Generation in HiNL Fused Silica Fiber

We demonstrate a fused silica fiber-based SC light source extending to ~3 μm with time-averaged power scaled up to 5.3 W. The pump consists of 2 ns laser diode (LD) pulses, which are amplified to ~4 kW peak power in a multistage fiber amplifier. The pulses are launched into 2–3 m of standard single-mode fiber to initiate MI-induced pulse breakup and then spectrally broadened in <1 m length of highly nonlinear fiber to obtain a continuum ranging from ~0.8 to 3 μm. The average power of the SC is scalable from ~27 mW to 5.3 W by varying the pulse repetition rate and using a cladding-pumped fiber amplifier.

SC generation experiments are performed at two different repetition rates – 5 kHz and 1 MHz. The experimental setup for the low average power system (5 kHz) is illustrated in Fig. 4.10. A distributed feedback (DFB) LD at 1553 nm is driven by a pulse generator to provide 2 ns signal pulses at a 5 kHz repetition rate, corresponding to a duty cycle of 100,000:1. The pulses are amplified to ~4.7 kW peak power in three stages of single-mode erbium-doped fiber amplifiers (EDFAs). We use an electro-optic modulator to suppress the in-band ASE, while a 200 GHz optical add/drop multiplexer (OADM) removes the out of band ASE. The setup is made entirely of telecommunications components and fusion spliced together with no free space elements. To generate a wide SC, the output of the power-amp is spliced to the first-stage single-mode fiber to break up the amplified 2 ns pulses into ultrashort soliton pulses. The single-mode fiber output is then spliced to the second-stage HiNL fiber to achieve the desired spectral broadening. The HiNL fiber used

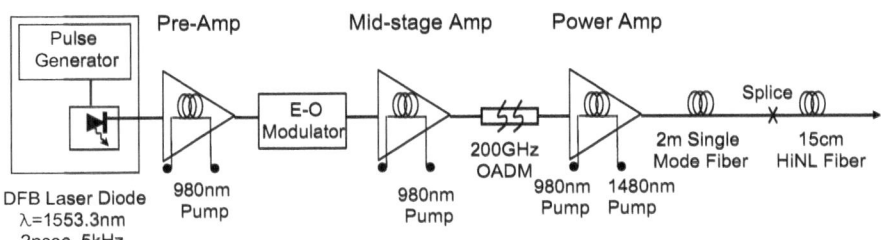

Fig. 4.10 Low average power experimental setup – LD pulses amplified by single-mode EDFAs

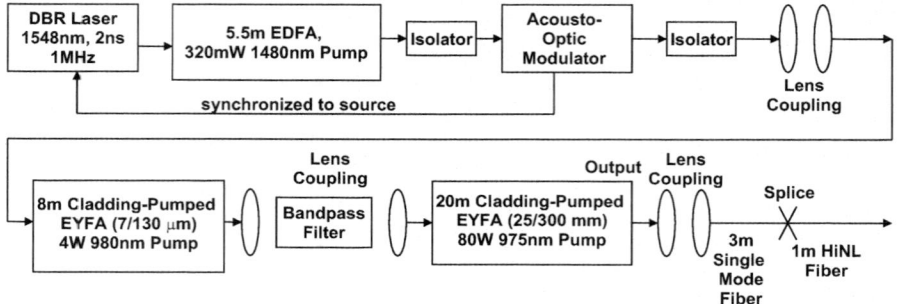

Fig. 4.11 High average power experimental setup – LD pulses amplified by cladding-pumped erbium/ytterbium co-doped fiber amplifiers

in the experiment has a zero-dispersion wavelength of 1544 nm and a dispersion slope of 0.044 ps/(km·nm²) at the pump wavelength. The HiNL fiber is extra water-dried to minimize the water composition in the fiber. With an effective mode area of $\sim$10 μm², the HiNL fiber has a nonlinearity coefficient of $\sim$9.6 W^{-1} km^{-1}. An optical spectrum analyzer is used to measure the lower end of the SC spectrum (800–1750 nm), while the long-wavelength spectrum is acquired by a nitrogen-purged grating spectrometer. A low-noise thermoelectric-cooled InAs detector connected to a lock-in amplifier is used to reliably measure the long-wavelength spectrum with more than 40 dB of dynamic range.

To scale up the average power in the SC, the pulse repetition rate of the pump is increased from 5 kHz to 1 MHz, while keeping the peak pump power almost constant. The high-power system (Fig. 4.11) at 1 MHz (duty cycle of 500:1) has a similar experimental setup as the 5 kHz system, with a distributed Bragg reflector (DBR) LD producing 2 ns pulses that are amplified through three stages of fiber amplifiers. To suppress the stimulated Brillouin scattering, the signal spectrum of the laser is broadened to $\sim$1 nm bandwidth by chirping the phase segment of the diode. The system consists of three main amplification stages: a standard single-mode core preamplifier, a 7 μm core double-clad fiber amplifier, and a 25 μm core double-clad fiber amplifier. The preamplifier stage is a 5.5 m long EDFA pumped by a 320 mW 1480 nm pump diode. The second stage is an 8 m long single-mode erbium/ytterbium co-doped fiber amplifier (EYFA), which has a 7 μm $NA = 0.17$ core and a 130 μm $NA = 0.46$ inner pump cladding. The amplified spontaneous emission of the first- and second-stage fiber amplifiers is suppressed by using an acousto-optic modulator and a band-pass filter. The last stage is a 20 m long large-mode-area EYFA, which has a 25 μm $NA = 0.1$ core and a 300 μm $NA = 0.46$ cladding. Large-mode-area fiber is used to minimize nonlinear effects in the fiber core, and single-mode output can be achieved by properly matching the mode of the input beam with the fundamental mode of the gain fiber. The power-amp of this system provides a free space output with 8.8 kW peak (17.6 W average) power with $\sim$80 W pump power, of which 3.8 kW (7.6 W) is coupled into the first-stage single-mode fiber.

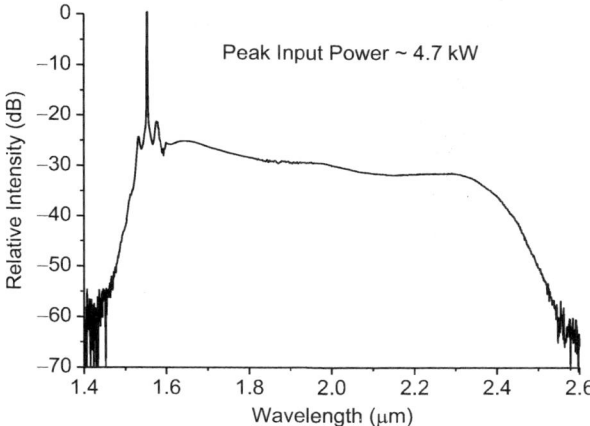

Fig. 4.12 Spectrum after 2 m single-mode fiber for 4.7 kW peak input power in the low power setup

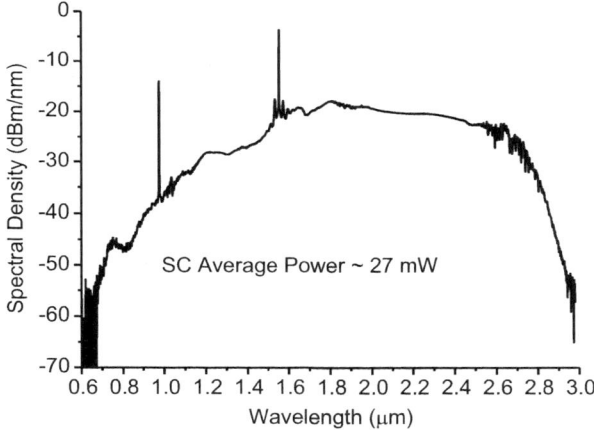

Fig. 4.13 27 mW average power SC spectrum after 2 m single-mode fiber + 15 cm HiNL fiber in the low power setup

A broad continuum extending from ∼0.8 to ∼3 μm is obtained with the 5 kHz system for 2 m single-mode fiber followed by 15 cm length of HiNL fiber. The resulting spectrum after the single-mode fiber for 4.7 kW peak (47 mW average) input power is shown in Fig. 4.12. The redshifted spectrum after the single-mode fiber extends from 1.4 to 2.5 μm and is primarily attributed to the SSFS effect. After the HiNL fiber stage, there is SPM-induced spectral broadening, which further pushes out the spectrum to ∼3 μm on the long-wavelength side and ∼800 nm on the short-wavelength side. The resulting super-continuum is shown in Fig. 4.13. Both spectra have been calibrated by including the wavelength response of the grating

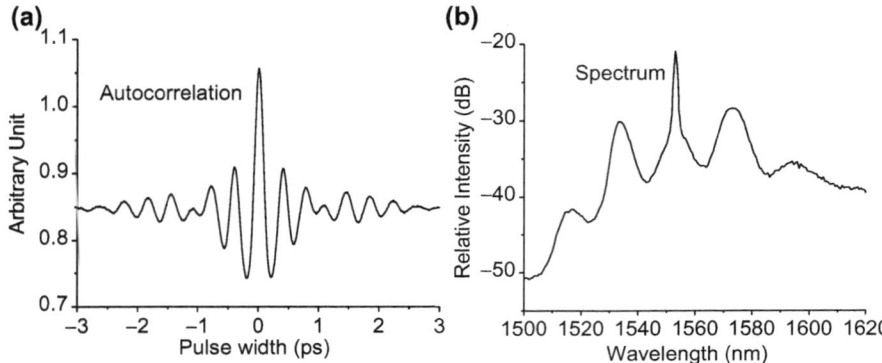

Fig. 4.14 MI-induced pulse breakup in single-mode fiber (**a**) autocorrelation and (**b**) spectrum. The results are shown for 3 m of single-mode fiber at 1 kW peak input power

and InAs detector. We measure the average power in the continuum to be ∼27 mW, thus achieving a ∼60% power conversion efficiency from pump to continuum. The spectral peak around 1550 nm is due to the LD pump, while the peak at 980 nm is due to the undepleted counter-propagating pump of the power-amp. We also observe multiple absorption lines between 2.5 and 2.8 μm and attribute these to the water absorption in the spectrometer.

To confirm that the SC is initiated by MI, we measure the temporal auto-correlation and spectrum after the first-stage single-mode fiber. The results after propagation through 3 m single-mode fiber at 1 kW peak input power are shown in Fig. 4.14. The autocorrelation (Fig. 4.14a) shows the formation of short pulses with pulse envelop of ∼500 fs, and the background level is due to the lower-intensity pulse wings. The spectrum shows two sets of characteristic MI sidebands, with the first-order sidebands separated by ∼18 nm from the pump at 1553 nm. The frequency shift of the first-order MI gain peaks from the pump is given by $\Delta\omega = \sqrt{2\gamma P / |\beta_2|}$, where γ is the nonlinearity coefficient, P_0 is the peak power of the pump, and β_2 is the group velocity dispersion parameter of the fiber at the pump wavelength. For single-mode fiber pumped by 1 kW pulses at 1.55 μm, $\gamma = 1.6\,\mathrm{W}^{-1}\,\mathrm{km}^{-1}$, $\beta_2 = -18\,\mathrm{ps}^2/\mathrm{km}$, and the calculated frequency shift $\Delta\nu = 2.12$ THz. The corresponding wavelength separation is ∼17 nm, which is close to the experimentally observed value in Fig. 4.14b. We observe that the 1530 nm MI gain peak coincides with the ASE peak from the EDFA, and thus the pulse breakup process is seeded by the ASE.

To optimize the second stage, we studied the spectral evolution of the SC as a function of HiNL fiber length for a fixed length of single-mode fiber. Figure 4.15 shows the long-wavelength spectrum for a 2 m length of single-mode fiber followed by varying lengths of HiNL fiber, with each plot being on the same scale relative to the others. As the HiNL fiber length is cut back from 20 m down to 15 cm, the long-wavelength edge of the SC increases from 2.6 μm to ∼3 μm along with an increase

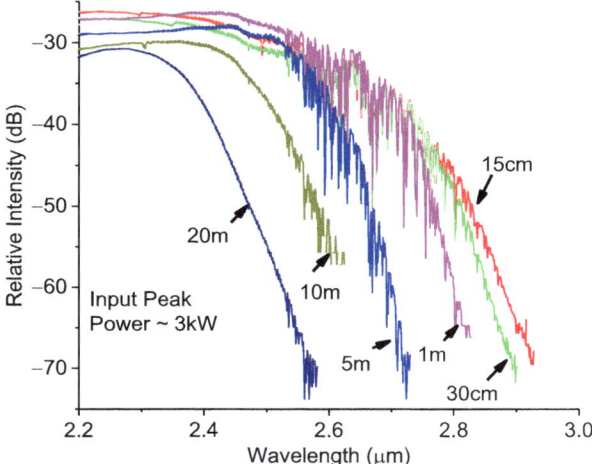

Fig. 4.15 Optimization of HiNL fiber length to achieve the longest SC edge – spectrum after 2 m single-mode fiber followed by varying lengths of HiNL fiber

Fig. 4.16 Transmission loss in HiNL fiber. Dotted curve shows experimental results while solid curve shows previously published data

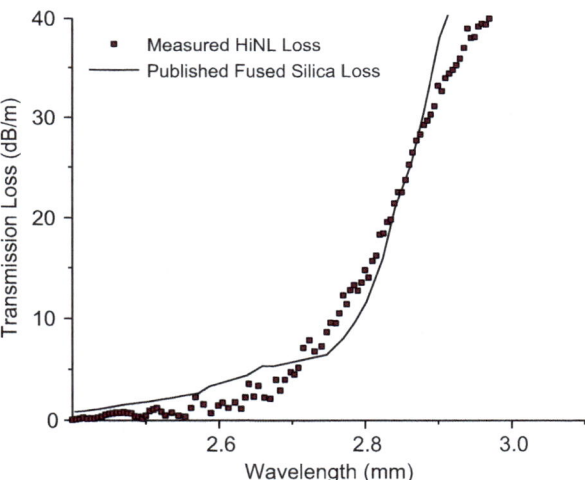

in the overall spectral density. These results indicate that the nonlinear processes responsible for the wide-spectrum generation occur within the first few centimeters of the HiNL fiber, and propagation through longer lengths merely attenuates the spectrum due to the high loss associated with the silica glass absorption around 3 μm. Thus, obtaining the broadest spectrum requires the shortest fiber length where nonlinearity dominates over loss.

Furthermore, we measure the transmission loss of the HiNL fiber as a function of wavelength. The loss curve is shown in Fig. 4.16. It is found that the loss increases rapidly beyond 2.7 μm and is as high as 40 dB/m at ~2.9–3 μm. The loss can be

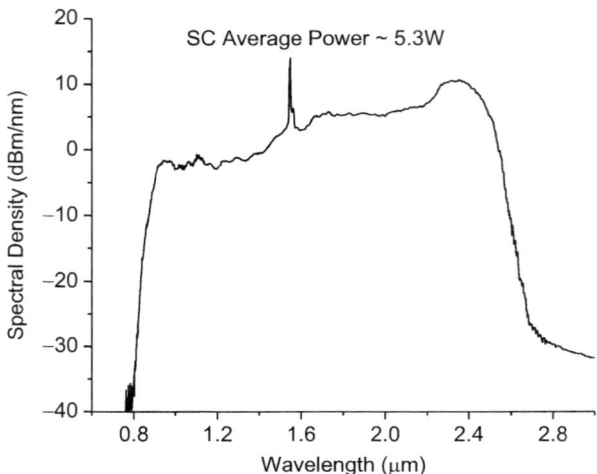

Fig. 4.17 5.3 W average power SC – spectrum after 3 m of single-mode fiber + 1 m HiNL fiber for ~3.8 kW peak input power

possibly attributed to the water absorption, bend-induced loss, and intrinsic silica glass absorption. By comparing the SC from non-dried fibers with dried fibers of 3–5 m lengths, the edge only extends ~100 nm and still does not go beyond 2.8 μm, which suggests that water absorption plays a role but is not the dominant limitation. Fibers with different core sizes and a similar index profile also give comparable edges to the SC spectrum, which downplays the significance of the bend-induced loss. Finally, the measured loss curve is in good agreement with previously published results and, thus, shows that the intrinsic silica loss due to vibrational absorption limits the long-wavelength edge of the SC [18].

The time-averaged power in the continuum is scaled up to 5.3 W by increasing the repetition rate of the system from 5 kHz to 1 MHz and the use of cladding-pumped fiber amplifiers. The spectrum after 3 m single-mode fiber followed by 1 m HiNL fiber for 3.8 kW peak (7.6 W average) input power is shown in Fig. 4.17 and extends from 0.8 to 2.8 μm. For the 5 kHz system, the power in the continuum is ~27 mW, while for the 1 MHz system, it is ~5.3 W. Thus, the power is scaled up by a factor of ~195 and is consistent with the 200× increase in the repetition rate.

The difference in the single-mode fiber lengths is attributed to the different fiber amplifiers used in the two systems. We notice that the broadest SC requires 3 m single-mode fiber at 1 MHz but only 2 m at 5 kHz. For the 5 kHz system, the pulse breakup is enhanced by the EDFA ASE, and the nonlinear effects start within the power-amp itself. But the large-mode-area EYFA in the 1 MHz system is designed for minimal nonlinear broadening, and thus longer lengths of single-mode fiber are required to achieve a similar pulse breakup. In addition, the high average power system has about 20% lower peak power (3.8 kW) compared to the low average power system (4.7 kW). Furthermore, the difference in spectral shape of the SC in Fig. 4.17 compared to Fig. 4.13 may be attributed to the lower peak power and longer HiNL fiber length in the 1 MHz system.

4.4.2   All-Fiber SWIR SC Laser Prototype

We also demonstrate a 5 W all-fiber SC laser spanning the SWIR wavelength band from ∼1.55 to 2.35 microns [45]. The optical layout of the all-fiber 5 W SWIR SC laser is shown in Fig. 4.18a and consists of an amplified 1.54 micron laser source followed by a spectrum broadening fused silica fiber. The amplified 1.54 micron laser source consists of a 1542 nm seed laser diode that is driven by electronic circuits to provide a 0.5 ns pulse at variable repetition rates from ∼20 MHz down to a few KHz. These pulses are amplified by two erbium/ytterbium fiber amplifier (EYFA) stages designated as the preamplifier and the power amplifier, respectively. The preamplifier consists of a ∼2 m length of 12/130 microns (core/cladding diameter) EYFA pumped by a 940 nm diode laser, and the power amplifier consists of ∼7 m length of 12/130 microns (core/cladding diameter) EYFA pumped by a ∼25 W 940 nm diode. A 100 Ghz band-pass filter is used after the preamplifier to filter out the amplified spontaneous emission. An in-line polarizer at the filter

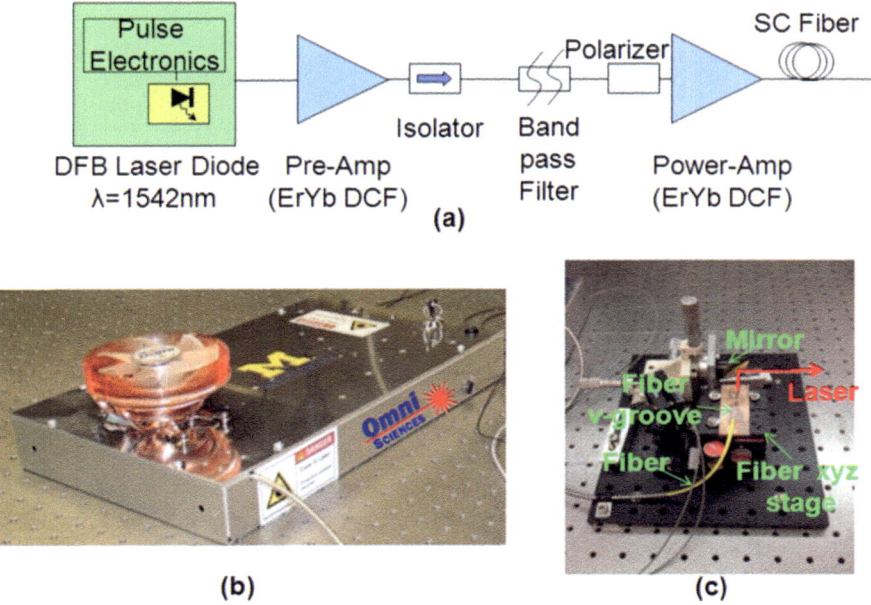

Fig. 4.18 (**a**) Optical layout of the all-fiber-integrated 5 W SWIR SC laser. (**b**) Packaged 5 W SC laser prototype. (**c**) Collimation setup for the SC final output

Fig. 4.19 (a) SC output
spectrum spanning from
~1.55 to 2.35 μm with an
average power of ~5 W
across the continuum. (b) SC
output power scaling with
940 nm pump power in the
power amplifier

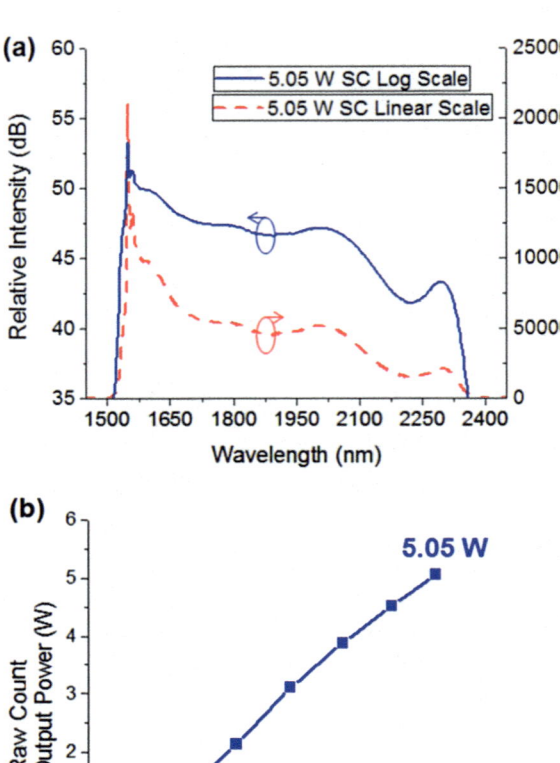

output ensures that the input to the power amplifier is in the optimum polarization. The amplified 1.54 μm light is then spliced onto ~10 m length of 8/125 microns (core/cladding diameter), 0.125 NA, PM1550 fiber, where the interaction between the nonlinearity and the anomalous dispersion breaks up the quasi-CW input pulses into a train of solitons through modulational instability and significantly increases the peak power.

The SC laser is packaged in a box with dimensions of 10″ × 17″ × 2″ and is shown in Fig. 4.18b. The SC fiber output is also collimated and mounted on to a breadboard to allow for easy integration into various field trials and measurements, as shown in Fig. 4.18c. The SC laser box and the collimation setup weigh ~7 pounds each. The SC collimation is achieved using a 90 degree off-axis parabolic gold-coated mirror with a 25.4 mm focal length. The collimated beam diameter ($1/e^2$) at 1 m is measured to be ~6.5 mm.

Figure 4.19a shows the spectral output from the SC laser prototype corrected for the detector and grating response. The SC spectrum extends from ~1.55 to

Table 4.2 5 W SWIR amplitude fluctuations at different wavelengths in laboratory

Wavelength (nm)	% Fluctuation (integration time 100 ms)
1600	0.22%
1800	0.23%
2000	0.22%
2200	0.57%

$\sim$2.35 μm with a time-averaged power of $\sim$5.05 W in the entire continuum. The input comprises of 1.54 μm laser diode input pulses of $\sim$0.5 ns duration at 8.3 MHz repetition rate. By pumping the power amplifier with $\sim$25 W of 940 nm pump power in the counter propagation configuration, we are able to generate $\sim$7 W of average power output around 1.54 μm. The pump (940 nm) to signal (1540 nm) efficiency at the output of the power amplifier is observed to be $\sim$28% and is typical of EYFA-based SC systems. The SC output power scaling with the power amplifier pump power is also shown in Fig. 4.19b and shows that $\sim$25 W of 940 nm pump in the power amplifier gives rise to the 5.05 W of final SC output in a $\sim$10 m length of the PM1550 fiber.

The SC amplitude fluctuations are also measured in-lab at various wavelengths tuned using a spectrometer and lock-in amplifier. The SC relative variability is shown in Table 4.2. The measurements are performed at an integration times of 100 ms (2 Hz sampling) using an InGaAs detector. The percent fluctuation at each wavelength is calculated as the ratio of the RMS sample to sample deviation and the mean of the measured amplitudes. The fluctuations are measured at 1.6, 1.8, 2, and 2.2 μm. The measured amplitude fluctuations show that the SC laser is stable with <0.6% fluctuation across the spectrum.

The output power stability over long periods is another factor to consider in a practical light source for long-term measurements. We have performed power stability measurements for a continuous >43 h time period on the final system. The final power is measured at the fiber output without a collimating mirror. The results are shown in Fig. 4.20 and show an average power of $\sim$5.09 W with a standard deviation of 0.012 W, corresponding to a fluctuation (standard deviation/average) of $\sim$0.23%. The small variation is most likely due to the temperature fluctuations in the room temperature. Since our pump diodes are passively cooled, it is possible for the pump center wavelength to fluctuate with temperature, which could affect the final SC power level.

4.4.3   Simple VIS-NIR-SWIR SC Laser

A simple SC laser that can span from the visible, through the NIR into the SWIR has also been demonstrated using commercial off-the-shelf components. The optical block diagram of the laser is shown in Fig. 4.21, where all of the fibers are solid-core fibers and the SC fiber is a NuFern SM-1950 fiber. The optical configuration

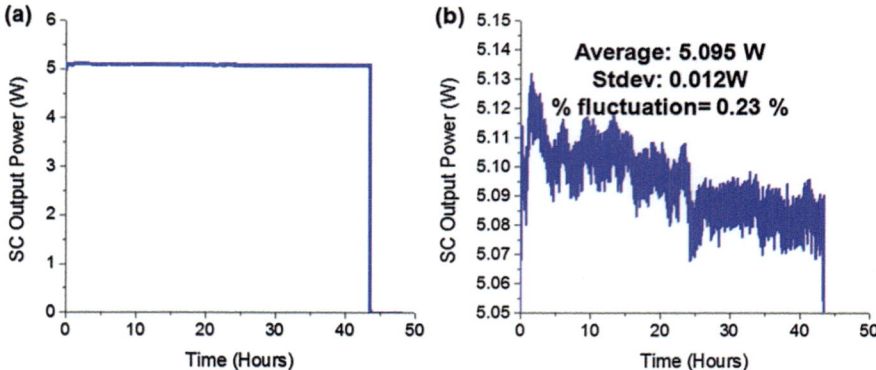

Fig. 4.20 (**a**) Long-term power stability measurements of the 5 W SC prototype before the collimating mirror. (**b**) Zoomed in view of the stability measurements

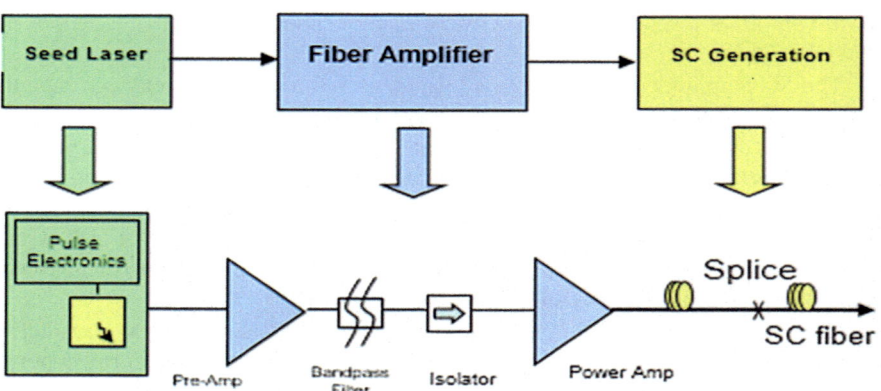

Fig. 4.21 Block diagram of SC laser spanning visible to SWIR wavelength range

of the all-fiber 5 W NIR-SWIR SC laser starts with a 1063 nm laser diode as the seed laser. The seed laser is driven by pulsed electronics to provide the seed laser signal at 1 ns with variable repetition rate up to 5 MHz. The seed pulses are then amplified by two ytterbium fiber amplifier stages designated as the preamplifier and the power amplifier, respectively. The preamplifier consists of a ~4 m length of 10/130 microns YDFA pumped by a 915 nm diode laser, and the power amplifier consists of ~10 m length of 10/130 microns YDFA pumped by a ~20 W 915 nm diode. Between the power amplifier and the preamplifier, a 2 nm bandwidth bandpass filter is used to suppress the amplified spontaneous emission. The amplified laser output is then spliced onto different lengths of 7/125 microns, SM1950 fiber, generating the final SC output through a mixture of different nonlinear effects such as modulational instability and stimulated Raman scattering.

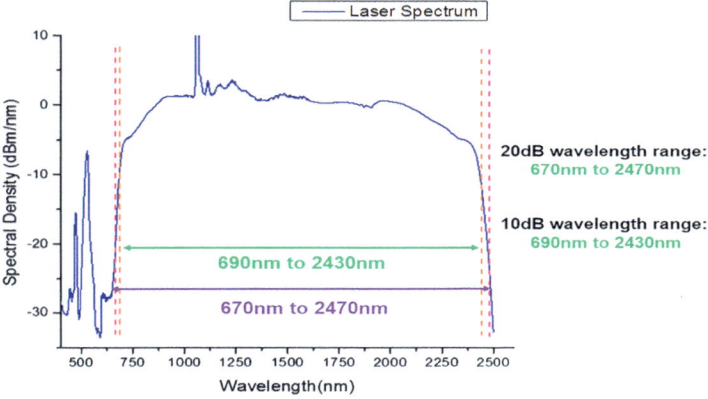

Fig. 4.22 Spectral output for configuration of Fig. 4.21

Fig. 4.23 Comparison of spectra for different SC fiber lengths

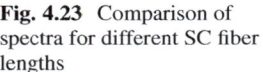

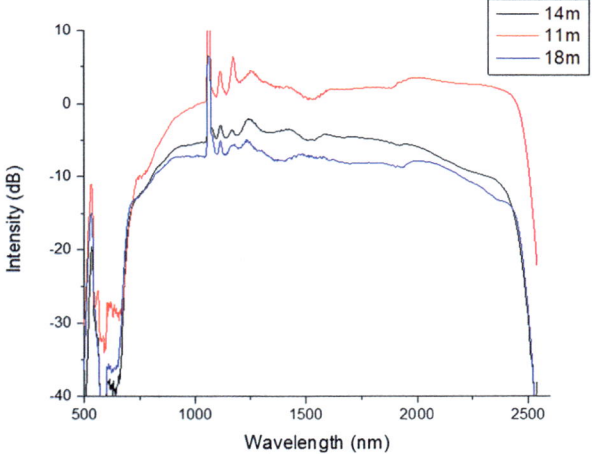

The output spectrum from the SC laser is shown in Fig. 4.22 with a time-averaged output power level of 4.7 W using an 18 m length of SC fiber. The −20 dB bandwidth stretches from ∼670 to ∼2470 nm, while the −10 dB bandwidth encompasses ∼690 to ∼2430 nm. The length of SC fiber can be varied, and the spectrum can be tailored with the fiber length. For example, Fig. 4.23 shows that using 11 m of SC fiber enhances the long-wavelength edge, while 18 m of SC fiber enhances the short-wavelength edge. Also, for the 18 m length, the evolution of the spectrum as a function of pump power is shown in Fig. 4.24 (pump power into the power amplifier). Starting from the pump wavelength of ∼1064 nm, the spectrum spreads from the nonlinear activity as the pump power is increased.

The wall-plug efficiency of the SC laser has also been measured. For the tabletop system where the SC fiber is directly spliced to the power amplifier pump combiner, the measured wall-plug efficiency is ∼7.1%, where we define wall-plug efficiency

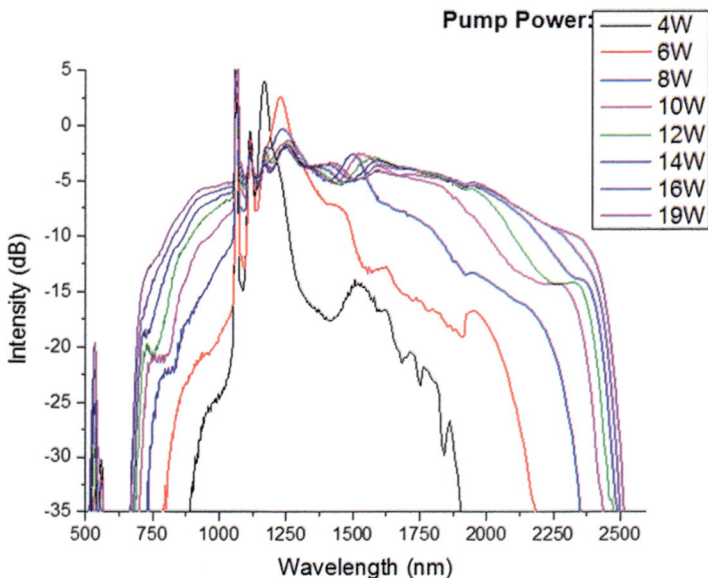

Fig. 4.24 Spectral evolution as a function of pump power to last-stage amplifier

Table 4.3 Wall-plug efficiency for SC laser in Fig. 4.21

Parts	Efficiency
Electrical to electrical	80%
Electrical to 915 nm pump power	40%
Pump power to 106 nm signal	55%
Power-amp to SC splice	60%
SC conversion efficiency	68%
Wall-plug efficiency[a] (measured)	7.1%
Wall-plug efficiency with MFA[b]	10.65%

[a]Wall plug efficiency = SC power output/ electrical power drawn
[b]MFA = mode field adapter used to transition between fibers

as SC power output measured at the end of the fiber divided by total electrical power drawn by everything in the unit. A breakdown of the various factors that contribute to the wall-plug efficiency is included in Table 4.3. It should be noted that the power amplifier to SC fiber splice has a transmission of about 60%. If we used a mode field adapter to transition between the fibers (e.g., use a component that tapers the fibers and adiabatically couples the light), then we could nearly eliminate this loss, which would increase the wall-plug efficiency closer to 10.65%.

The spectrum from this SC laser can also be extended further into the visible by adding a length of photonic crystal fiber (PCF) to the end of the SC fiber. In solid-core fibers, the zero-dispersion wavelength is ~1310 nm or longer in

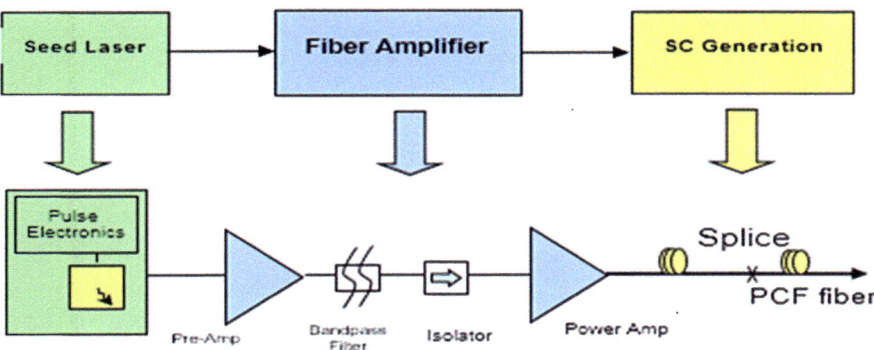

Fig. 4.25   Addition of PCF at SC fiber end to extend further into the visible

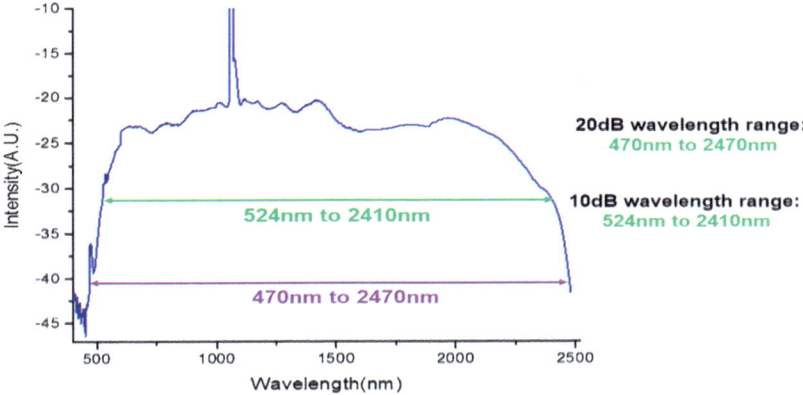

Fig. 4.26   Spectral output of SC laser from Fig. 4.25

wavelength. This is problematic for extending the spectrum further into the visible because the group velocity dispersion becomes excessive. By using a PCF fiber, the zero-dispersion wavelength can be shifted down in wavelength, which reduces the dispersion in the visible and enables further bandwidth expansion into the short-wavelength side. For example, Fig. 4.25 shows the block diagram, which is the same as in Fig. 4.21 with a PCF attached at the end. In particular, the PCF is a SC-5.0-1040 fiber from NKT Photonics with a zero-dispersion wavelength at 1040 nm and 4.8 micron core size. The length of fiber used in the experiment is 5 m. For the purpose of this experiment, we merely mechanically spliced the PCF to the end of the SC fiber. However, it should be noted that the two fused silica fiber can be fusion spliced to reduce the loss at the joint.

The output spectrum for this VIS-NIR-SWIR SC source is illustrated in Fig. 4.26. The bandwidth into the visible extends down to the blue wavelength range, at the expense of reducing the long-wavelength edge somewhat. In particular, the −20 dB wavelength range is ∼470 to ∼2470 nm, while the −10 dB wavelength range is

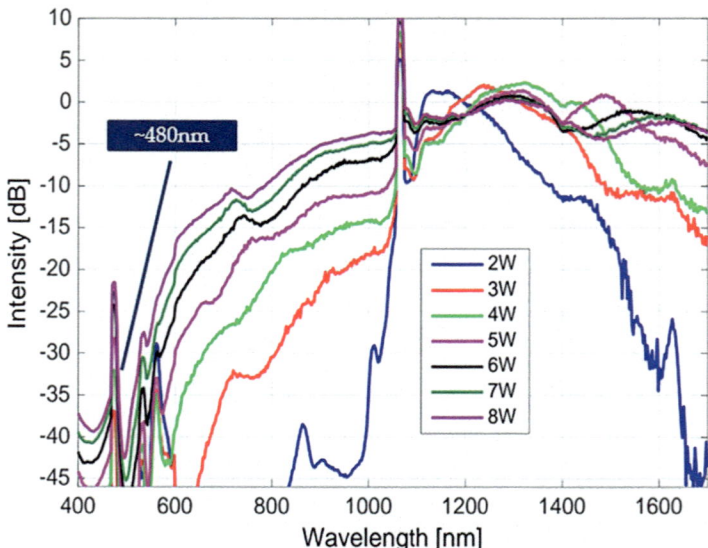

Fig. 4.27 Spectral evolution in SC fiber including PCF at end

~524 to ~2410 nm. Moreover, the spectral evolution as a function of pump power is illustrated in Fig. 4.27. We believe by improving the coupling to the PCF and increasing the power into the PCF, we should be able to shift the short-wavelength edge down to ~400 nm. In comparison, commercial units from NKT Photonics achieve this wavelength range using PCF and a mode-locked laser pumping scheme, which turns out to be a much more complicated and expensive optical system. Therefore, with the simple configuration of Fig. 4.25 and initiating the SC through MI mechanism, we can achieve comparable spectral extent as mode-locked laser systems with actually a smoother spectrum.

4.4.4   64 w High-Power SWIR SC Laser Prototype

A block diagram of a high-powered 64 W SWIR SC laser is shown in Fig. 4.18a. Also, a photograph of the SWIR-SCL prototype is presented in Fig. 4.28b [46]. The configuration is a master oscillator power amplifier with a 0.45 ns pulsed DFB laser diode at 1064 nm as the seed oscillator. Amplification of the 1064 nm radiation is achieved with three ytterbium-doped fiber amplifier stages each with their own high-power 915 nm pump laser diodes that provide fiber cladding pumping. The first-stage preamplifier is designed for optimal noise performance. Between amplifier stages, band-pass filters and polarizers block amplified spontaneous emission

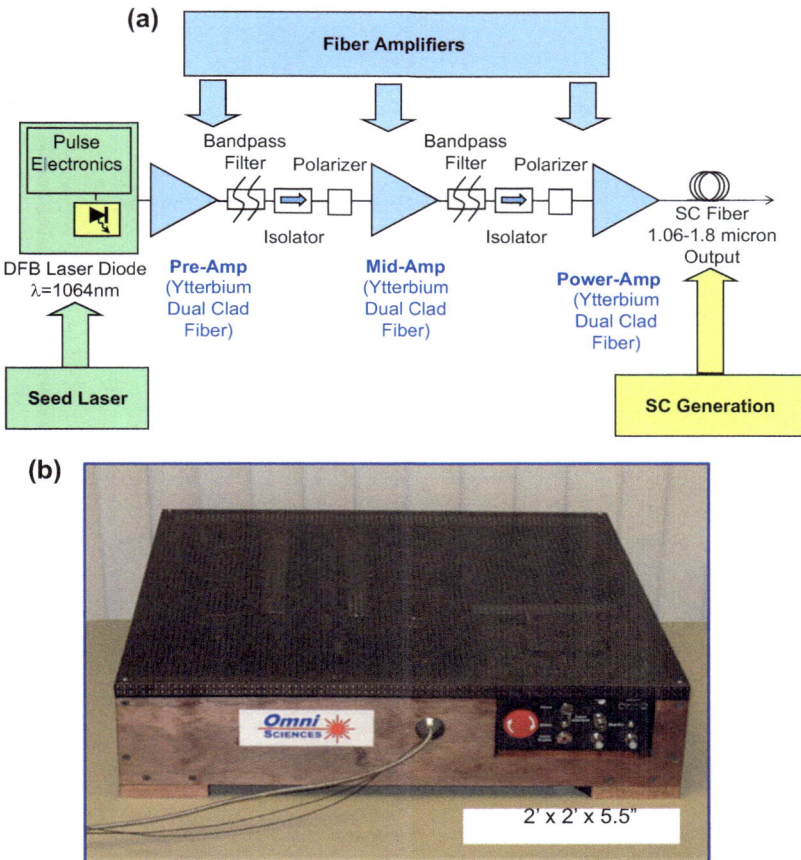

Fig. 4.28 (**a**) Laser configuration for the 64 W SWIR super-continuum laser. (**b**) Photograph of the SWIR-SCL prototype used in the field test experiments

(ASE), and isolators suppress spurious reflections and backward propagating ASE. The power amplifier stage is optimized to minimize nonlinear distortion.

The high average power of our SC laser was achieved with three separate amplifier stages, as shown in Fig. 4.28. Each amplifier is double-clad (dual-core) polarization-maintaining fiber with 1064 nm radiation propagating in the Yb-doped core and CW 915 nm pump radiation propagating in the cladding. The preamp stage delivers ∼28 dB of gain, the mid-stage ∼28 dB of gain, and the power-amp stage ∼23 dB of gain, while the between-stage components exhibit ∼4 dB of loss. The output from the power amplifier is a train of ∼0.45 ns pulses with peak power of about 24 KW. These intense pulses are coupled to the SC generation fiber.

The spectral broadening for the SC laser occurs in the un-doped fiber that follows the power amplifier. The SC generation fiber is approximately 10 m long, has a

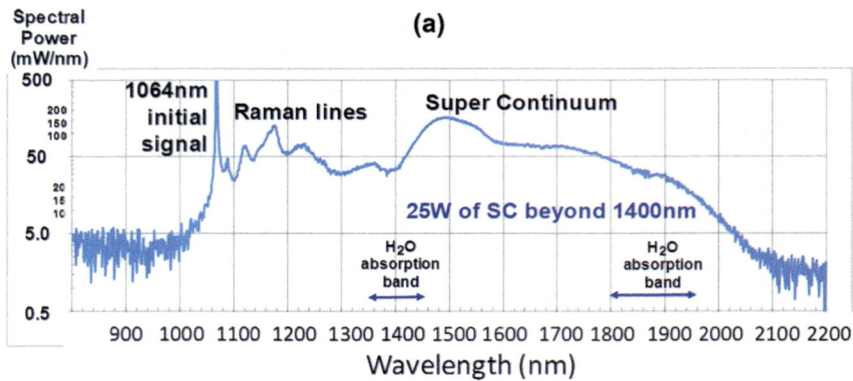

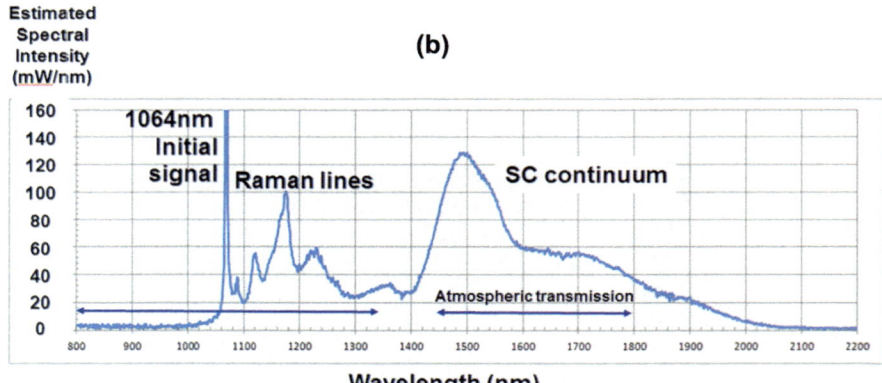

Fig. 4.29 64 Watt average SWIR-SCL spectral output at 6.25 MHz repetition rate plotted on (**a**) logarithmic and (**b**) linear scale

core diameter of 25 μm, and has a 0.065 numerical aperture. Figure 4.29 shows the SWIR SC laser output spectral power from 800 to 2200 nm measured with a monochromator and a TE-cooled HgCdTe (5.2 μm cutoff wavelength) detector. The raw spectrum was corrected for the approximately linear spectral response of the detector. Figure 4.29a shows the spectral output on a logarithmic scale, while Fig. 4.29b shows the spectral output on a linear scale.

The features exhibited in the SWIR-SCL spectrum may be explained as follows. Since the ∼1064 nm pump wavelength is well below the zero-dispersion wavelength of the fiber (near 1300 nm), the dominant nonlinear mechanism for the ∼0.45 ns long pulses in the fiber is stimulated Raman scattering at the pump wavelength [3, 36]. For fused silica fiber as used in the SC generation here, the peak Raman gain is about 440 cm^{-1} or ∼13.2 THz lower energy (or longer wavelength) compared to the pump. Therefore, the peaks observed in the spectrum between roughly 1064 and 1300 nm correspond to different cascaded Raman frequency-shifted

Fig. 4.30 Output SWIR-SCL
power versus repetition rate
of the seed laser

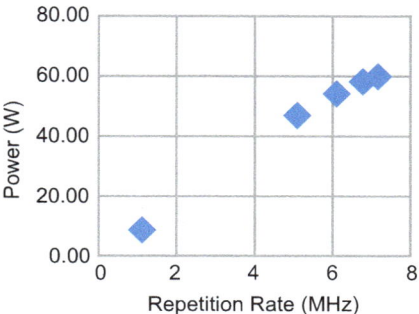

orders [36]. Once the cascaded Raman shifting process downshifts significant energy to wavelengths longer than ~1300 nm, the Raman-shifted light is in the anomalous group velocity dispersion regime, and the MI-initiated super-continuum generation process generates the longer-wavelength spectrum. Hence, the spectrum from ~1300 to ~2200 nm in Fig. 4.29 corresponds to the smoother, continuous SC-generated light.

The time-averaged power for the SWIR SC can be varied by varying the repetition rate of the seed laser pulses [47]. As the repetition rate is increased, the output power increases approximately linearly, and a nearly constant spectrum can be obtained if the parameters are adjusted to maintain a nearly constant peak power (e.g., peak power determines the spectrum, while repetition rate determines the time-averaged power). For example, Fig. 4.30 plots the output SC power versus repetition rate, as the repetition rate is varied between 1 and 7 MHz. For each of these data points, the power amplifier pump level was adjusted to obtain a peak power of approximately 24 kW, which is the peak power from the power amplifier going into the SC fiber.

We also characterized the amplitude stability of the output from the SWIR SC light source. The quasi-continuous power output of our 64 W SWIR SC laser was measured over a period of 15 h and was found to have excellent amplitude stability. Fig. 4.31 shows average laser output power as a function of time over the long measurement. After the SWIR-SCL has warmed up for a time on the order of 20 min (e.g., all the optics in the box has reached a steady-state temperature), the output power remains quite stable. The box in the figure shows an expanded power and time scale to show details of the amplitude fluctuations. In particular, the power fluctuation is measured to be ±0.2% peak-to-peak or 0.04% RMS.

The wall-plug efficiency of the 64 W SWIR-SCL was also measured in the laboratory. The power drawn from the standard wall-plug was measured using a "Kill-A-Watt" electricity monitor. We define the wall-plug efficiency as the output power from the SWIR-SCL as measured by a time-averaging power meter divided by the power drawn from the standard wall-plug. It is important to note that this metric includes the electrical power supply, all of the electronics, the optical-to-electrical efficiency of the pump lasers, as well as the pump light to SC generation

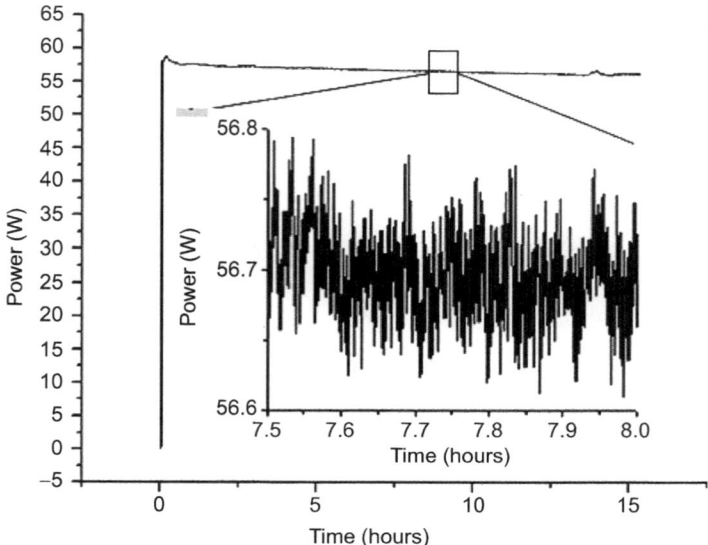

Fig. 4.31 SWIR-SCL output power stability vs. time. The insert shows a blowup of the power in the time window indicated by the box

conversion efficiency. At full-power operation with 64 W measured from the SC output, we measured a wall-plug efficiency of 15.7%. Such a high wall-plug efficiency is important for any application where the power has to be provided from an onboard system (e.g., aircraft or drone), and the high efficiency also means that the heat generated from power dissipation will be minimized.

4.5 MWIR SC Sources Using Fluoride Fibers

We next move to describing experiments in the MWIR. Since fused silica fibers do not transmit beyond ∼2.7–3 microns, it is necessary to use softer glasses, such as fluorides, tellurites, or chalcogenides. Here we focus on experiments using ZBLAN fluoride fiber. All setups follow the framework described in Fig. 4.1, consisting of amplified picosecond/nanosecond laser diode pulses, MI-induced pulse breakup, and, finally, the SC generation fiber.

MWIR SC sources have a wide variety of applications, such as spectroscopy [48], IR countermeasures [49], free space communications [49], and optical tissue ablation [50] to name a few. Conventional mid-IR laser sources including optical parametric amplifiers [51], quantum cascaded lasers [49], synchrotron lasers [52], and free-electron lasers [53] have been used to demonstrate some of these applications. In comparison, an all-fiber, mid-IR SC laser has no-moving-parts output in a single spatial mode and operation at room temperature. SC lasers also generate a broad spectrum covering the entire near-IR and the mid-IR simultaneously, which can improve the selectivity of remote chemical sensing [54] and real-time optical metrology [55]. In addition, direct signal modulation functionality is also desirable for SC laser sources to eliminate the need for external signal modulation or chopping equipment, which is difficult to implement in the mode-locked laser-based systems. By initiating the SC generation through modulational instability, we eliminate the need for a mode-locked laser in our SC systems and are able to use commercial off-the-shelf parts from the mature telecommunications and fiber-optic industry.

For the SC systems presented in this section, there are basically three adjustable parameters, which are the pulse width, the repetition rate, and the pump power [5]. To obtain the maximum SC power and to generate the broadest spectrum, all three parameters must be optimized. The pulse width is set based on the following two criteria. First, the pulse width should be short enough to mitigate any transient thermal effects in the fiber. On the other hand, a nanosecond pulse scale is also preferable for the intra-pulse nonlinear interaction and spectrum broadening. Therefore, we drive our distributed feedback (DFB) LD with pulse width in the range of 0.4–2 ns. Furthermore, the pulse repetition rate couples with the pulse width to determine the duty cycle of the laser system. Since the DFB LD pulses remove the energy provided by the pump lasers of the power amplifier of the SC system in time between DFB pulses, the peak output power varies inversely with the pulse duty cycle for a fixed pump power supply. In other words, the peak output power increases with the reduction of the pulse repetition rate, i.e., pulse duty cycle and vice versa. Thus, by increasing the repetition rate and the pump power accordingly, we can maintain the peak power for the required spectral extent while scaling up the total average power in the SC.

We begin with a relatively high-power setup, where the average power is scaled up to $\sim$10.5 W in an SC extending from $\sim$0.8 to 4 microns. Next, we push the spectrum out further to $\sim$4.3 microns by increasing the power-amp output peak power to $\sim$20 kW. Modulation capability is also demonstrated, where the entire system and the SC are modulated with a 500 Hz square wave at 50% duty cycle, without the need for any external modulation or chopping mechanism. Finally, we present the SC laser specifically for higher efficiency in generating mid-IR wavelengths (> than $\sim$3.8 microns) by incorporating a thulium-doped fiber amplifier (TDFA) as the power-amp stage. As we will see below, the TDFA system is an extension of the three-stage approach, where we start by generating a first-stage SC extending out to $\sim$2 microns to serve as the seed for the TDFA power-amp. The amplified 2 micron light from the TDFA output is then coupled into the ZBLAN fiber to generate the second-stage mid-IR SC out to $\sim$4.5 microns.

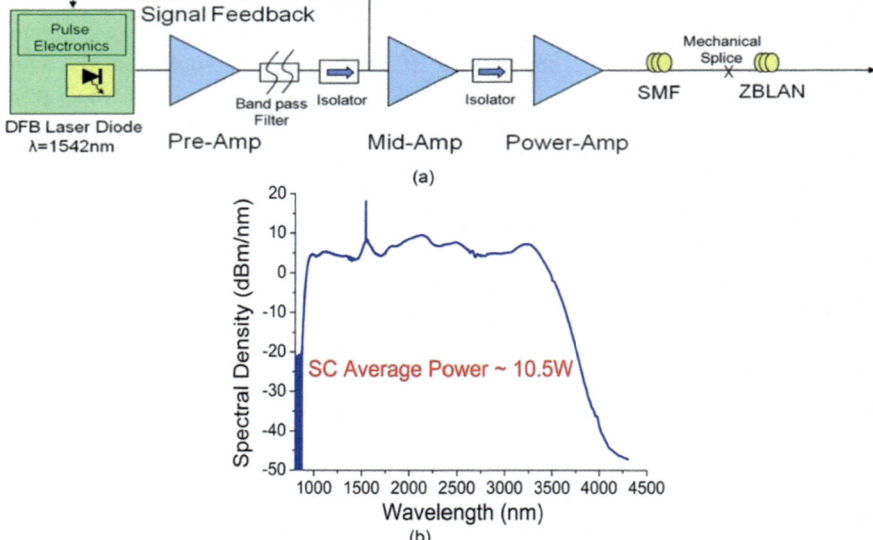

Fig. 4.32 (**a**) High-power (10.5 W SC) all-fiber-integrated SC laser setup. (**b**) SC spectrum for 2 m SMF followed by ~7 m ZBLAN fiber

4.5.1 Increasing Average Power in MWIR SC System

The experimental setup used for increasing the output power from mid-IR SC generation is shown in Fig. 4.32a [4]. A 10 mW DFB laser diode at 1542 nm is driven by electronic circuits to provide 400 ps to 2 ns pulses at variable repetition rates and serves as the seed. The optical pulses are amplified by three stages of fiber amplifiers: an EDFA preamplifier followed by EYFA mid-stage amplifier and power amplifier stages. The preamplifier uses a 1 m length of 4/125 micron (core/cladding diameter) single-mode EDFA, and the mid-amplifier stage uses a 1.5 m length of 7/125 (core/cladding diameter) cladding-pumped gain fiber. To lower the ASE, the amplifier is separated into one preamplifier and one mid-amplifier, and the ASE after the first stage is filtered by using a 100 GHz band-pass filter. Optical isolators are also placed between the stages to protect the system from any back reflection damage as well as to reduce the noise figure and improve the efficiency of the combined amplifier system. Under typical operating conditions, we obtain ~20 dB of gain in both the pre- and mid-amplifier stages for the optical signal, while the ASE-to-signal ratio is measured to be less than 1%. The power from the mid-amplifier is then boosted in an all-fiber-spliced, cladding-pumped EYFA before coupling into the SC fiber. A cladding-pumped fiber amplifier increases the gain volume and enables the coupling of multiple pump diodes. In addition, to minimize the nonlinearity in the amplifier, we use 5 m of EYFA with 15/200 micron (core/cladding diameter) and a core NA of 0.15. Ten 8 W 976 nm diodes and two 8 W 940 nm uncooled multimode pump diodes are coupled into the gain fiber through an 18 × 1 combiner. Single spatial mode operation is maintained in the EYFA by carefully splicing the gain fiber to the signal-input SMF fiber and the pump combiner. By pumping the system with ~75 W average powers in the counter propagation configuration, the EYFA can provide ~20.2 W average power at the output of ~2 m SMF fiber that is spliced on to the power-amp output. This corresponds to a peak power of ~6.1 kW (~15 dB signal gain) for 1 ns pulses at a 3.33 MHz repetition rate and ~27% pump-to-signal efficiency, which is slightly lower than the 30–35% value because of the high peak power-induced nonlinear wavelength generation in both the fiber amplifier and the SMF. For the 10.5 W SC generation, the ZBLAN fiber used is 7 m long and has a core/cladding diameter of 8.9/125 microns and an NA of 0.21. The ends of the SMF and the ZBLAN fibers are both angle cleaved and clamped on to aluminum v-grooves before mechanically coupling them. The fiber cladding is also covered with high refractive index optical glue to strip out any residual cladding modes.

The output from the high-power SC setup is shown in Fig. 4.32b. The SC spectrum extends from ~0.8 microns to beyond 4 microns, with a time-averaged power of ~10.5 W in the continuum. The generated SC is smooth and relatively flat across majority of the spectrum with a spectral power density >0 dBm/nm (1 mW/nm). The SC average power beyond 1600, 2500, and 3000 nm is measured to be ~7.3 W, ~3.0 W, and ~1.1 W respectively, by using long-pass filters with the corresponding cutoff wavelengths. In the first-stage SMF, we utilize MI to break up the nanosecond pulses into femtosecond pulses to enhance the nonlinear optical effects and redshift the spectrum to beyond 2 microns. The SC spectrum is further broadened in the ZBLAN fiber through the interplay of SPM, Raman scattering and FWM. The long-wavelength edge of the SC is limited by the input peak power and the length of the ZBLAN fiber in conjunction with other optical effects, including bend-induced loss and material absorption.

The repetition rate should be optimized for maximum MWIR power. For example, Fig. 4.33a shows the average power of the SC spectral components beyond 3 microns for varying pulse repetition rate with four 8 W 976 nm pump diodes. We see that the total SC spectral power beyond 3 microns increases from ~0.19 to ~0.48 W, by reducing the pulse repetition rate from 1.33 to 0.67 MHz. This increase of spectral power is attributed to the boost of the peak power coupled into the SMF and ZBLAN fibers to enhance the nonlinear SC generation process. However, we also see that the SC power drops down to ~0.26 W with further decrease in the repetition rate to 0.25 MHz. This drop in spectral power can be caused by the increased power loss in the SC long-wavelength edge and reduced amplifier conversion efficiency. Since the SC long-wavelength edge is limited by the loss of the fiber, the additional spectrum generated by the additional peak power is highly attenuated, which, in turn, reduces the total SC power in the MWIR regime.

The time-averaged power of the SC is seen to be linearly scalable with respect to the input pump power without changing the spectral shape. This is possible, since the entire SC spectrum is generated in each amplified nanosecond laser pulse, which does not interact with adjacent pulses, and the SC average power can be boosted by

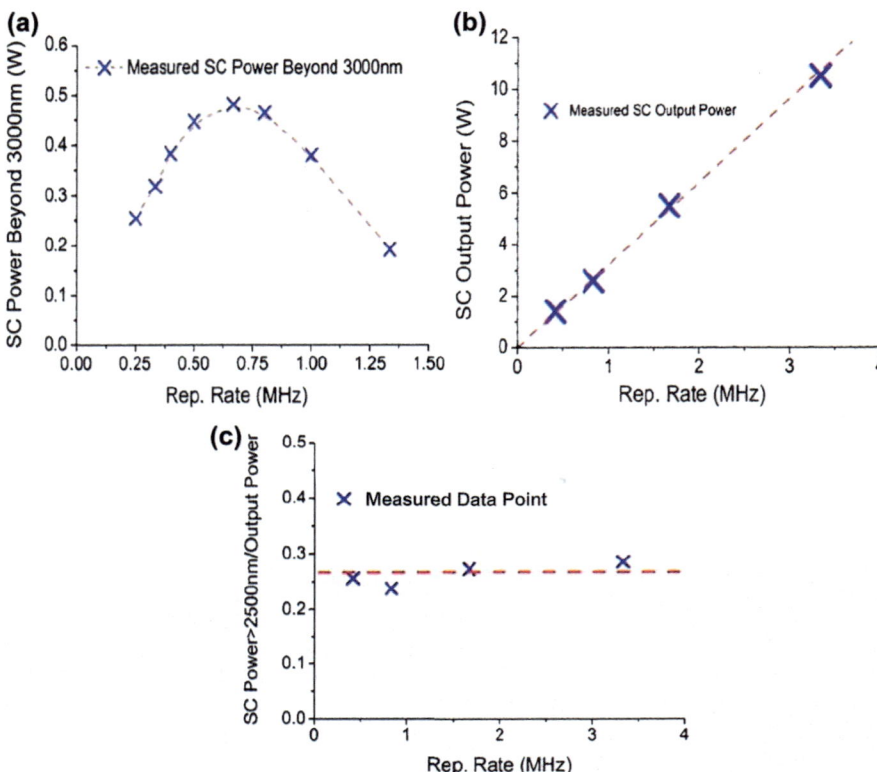

Fig. 4.33 (a) Average power of SC spectral components beyond 3 microns for varying pulse repetition rate with four 976 nm diodes. (b) Average SC power scaling by varying the pulse repetition rate and pump power (c) Ratio of the SC power of the spectral components beyond 2.5 microns with respect to the total SC power under different operating repetition rates

simply increasing the number of optical pulses in a given time period, i.e., increase the repetition rate. To ensure that the SC spectrum, which is a function of the peak power, remains unchanged as the average power increases, it is necessary to boost the pump power as the repetition rate is increased, thus, keeping the peak power the same. As illustrated in Fig. 4.33b, we scale the SC average power from 1.4 to 10.5 W by varying the pulse repetition rate proportionally from 0.42 to 3.33 MHz and increasing the pump power accordingly. To further confirm that the SC spectrum remains unchanged as the total SC power is increased, the ratio of the SC spectral power beyond 2.5 microns over the total SC power is measured and shown in Fig. 4.33c. We observe that the ratio says approximately constant across the entire power range, which indicates that the power generated in the MWIR scales up proportionally to the increasing total SC power. Therefore, the output power of our SC light source can be linearly and continuously varied with respect to the input pump power while keeping the same spectral shape.

4.5.2   Extending the SC Wavelength Edge

As seen in Fig. 4.32b, the ZBLAN SC output spectrum is seen to extend only to
~4 microns. In an effort to push the SC spectrum out further, we increase the
peak power of the power-amp output from ~6 to 8 kW in the previous setup, to
~20 kW, and as a result see the wavelength edge of the SC extend further to ~4.3
microns. Additionally, we simplify the previous setup by using a single cladding-
pumped EYFA (12/130, core/cladding diameter) stage to combine the preamp and
the mid-amp into a single preamp stage capable of providing sufficient signal gain.
This setup is shown in Fig. 4.34a. Modulation capability is also demonstrated in
this setup, where the entire system is modulated at 500 Hz with a 50% duty cycle
without any external chopping mechanism. The new preamp is pumped by a 10 W
976 nm diode. The power-amp uses the same EYFA as the previous setup. Four
25 W 976 nm multimode pump diodes are coupled into the gain fiber through a 6 × 1
combiner. At the highest 976 nm pump power of 52 W (104 W during ON phase),
the 1543 nm signal output from ~1 m combiner fiber of the power-amp is ~9.6 W
(19.2 W during the ON phase). This corresponds to a peak power of ~19.2 kW
for 0.5 ns pulse at a 2 MHz repetition rate and ~19% pump-to-signal efficiency.

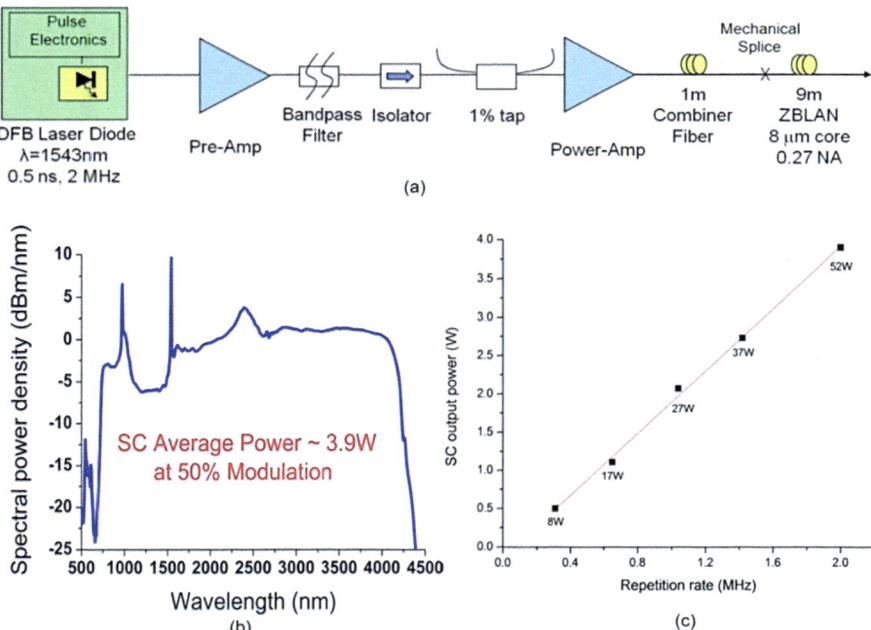

Fig. 4.34 (**a**) Setup for MWIR SC generation in ZBLAN fiber with ~0.51 W (at 50% modulation)
in wavelengths beyond 3.8 microns. (**b**) 3.9 W (at 50% modulation) SC output spectrum from the
ZBLAN fiber (**c**) SC output power scaling with repetition rate

The lower efficiency is again due to the higher peak power, which is responsible for spectral generation outside the gain band in the fiber amplifier. The output of the power-amp is then coupled into ∼9 m ZBLAN with a core/cladding diameter of 8/125 microns and an NA of 0.27. Both the combiner fiber and the ZBLAN are angle cleaved and coupled using a mechanical fiber alignment stage.

The ZBLAN SC output corrected for the grating and detector response is shown in Fig. 4.34b and extends from ∼0.75 to 4.3 microns with a time-averaged power of ∼3.9 W in the entire continuum, of which ∼0.51 W lies in wavelengths beyond 3.8 microns. Since the system is modulated at 50% duty cycle, this corresponds to ∼7.8 W in the continuum and ∼1 W in wavelengths >3.8 microns during the ON phase of the modulation cycle. The modulated SC average power beyond 1800, 2500, and 3000 nm is measured to be ∼2.7 W, ∼2 W, and ∼1.3 W respectively, by using long-pass filters with the corresponding cutoff wavelengths.

Power scalability is also shown in this system by increasing the repetition rate along with a proportional increase in the 976 nm pump power, while maintain the peak power going into the ZBLAN fiber. Figure 4.34c shows the linear increase in the SC average output power with an increase in the seed laser repetition rate, with the corresponding 976 nm power-amp pump power labeled next to the data point. Further increase in output power is limited by the available pump power, and we expect the results to scale up with higher pump power and thermal management of the critical splice points and power-amp gain fiber. The wavelength edge in this setup is now limited mainly by the inherent material absorption loss of the ZBLAN glass.

4.5.3   Improving MWIR SC Generation Efficiency

By using a thulium-doped fiber amplifier stage in a MWIR SC laser, a higher pump-to-signal conversion efficiency is demonstrated in TDFAs compared to EYFAs. For example, slope efficiency as high as ∼56% is demonstrated in Tm-doped gain fibers [56] compared to EYFAs, which have demonstrated slope efficiencies of ∼38% in comparable fiber geometries [57]. Since the TDFA gain band is in the 1.9–2.1 micron region, a signal input at ∼2 microns is required. Various techniques have been reported in the literature to generate ∼2 microns pulses as input to the TDFA. For example, by using a mode-locked Er fiber laser with polarization-maintaining components operating at 1557 nm followed by an EYFA, a Raman shifting soliton pulse to ∼2 microns in a 12 m length of high NA fiber has been demonstrated by Imeshev et al. [58]. A subsequent dispersion-managed large-mode-area TDFA stage generates pulses with up to 230 kW peak power at ∼2 microns. Another approach generates ∼750 fs long ∼1.95 micron pulses using a saturable absorber based carbon nanotubes in a Tm-doped fiber laser cavity [59]. Other approaches using 1.55 micron laser diode pulses, instead of mode-locked lasers have also been demonstrated. For instance, one approach involves three successive orders of cascaded Raman wavelength shifting from 1.53 micron to 1.94 microns in

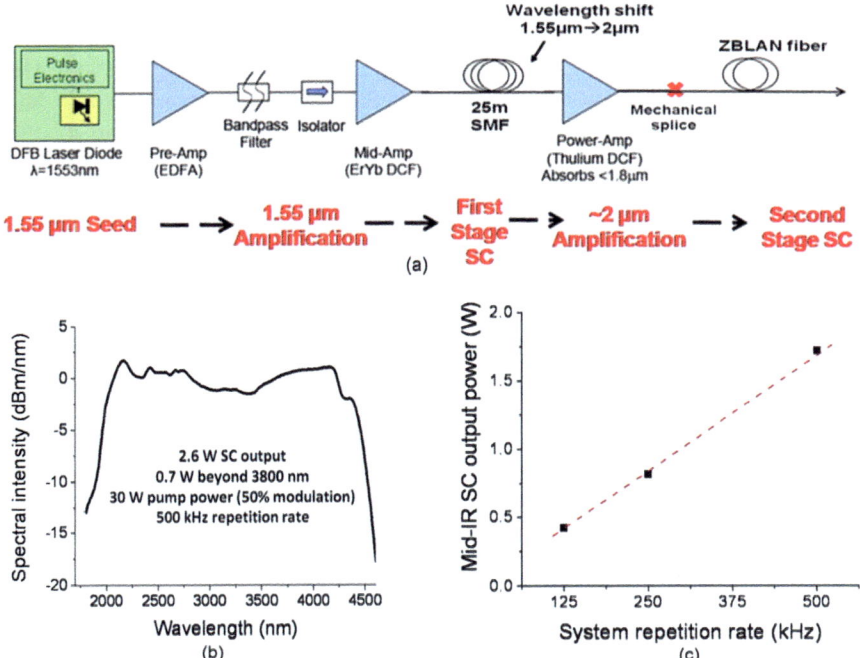

Fig. 4.35 (**a**) Experimental setup depicting two-stage MWIR SC generation using thulium-doped dual-cladding fiber. (**b**) MWIR SC spectrum, extending from ∼1.9 to 4.5 microns from ∼8.5 m ZBLAN fiber pumped with TDFA. (**c**) SC output power scaling with repetition rate

germanium-doped fused silica fibers with an NA of ∼0.41 [60]. Another approach involves gain switching a Tm-doped fiber cavity with a modulated 1.55 micron pump to generate 10 ns long, 2 micron pulses [61]. In our approach, we first generate an SC extending across the thulium amplifier gain band, which is then further amplified in the TDFA power-amp stage to act as a pump for further spectral broadening in the MWIR.

For SC generation in wavelengths beyond ∼2 microns into the MWIR, a TDFA-based SC laser is capable of providing a higher MWIR (wavelengths >3.8 microns) SC generation efficiency compared to EYFA-based systems [28]. In our approach, as shown in Fig. 4.35a, the seed laser consists of a 1553 nm DFB laser diode with ∼1 ns pulse output and a repetition rate adjustable from a few MHz down to ∼10 KHz. The diode pulses are amplified to a peak power of a few kilowatts through the use of two EYFA stages pumped by ∼980 nm diodes, in the preamp (2 m, 7/130 microns, core/cladding diameter, 400 mW pump) and the mid-amp stages (5 m, 12/130 microns, core/cladding diameter, 10 W pump), respectively. The output of the mid-amp is then spliced on to ∼25 m SMF. This is the first SC generation stage and is done in order to generate light in the Tm gain band, which extends from ∼1.9

to 2.1 microns with a peak at 1.98 microns, for the power-amp stage. This first-stage SC then serves as the seed for the power-amp that generates the high peak power input for the second-stage SC output in the MWIR. Although the first-stage SC could be replaced with a seed laser near 2 microns in principle, we prefer to use the 1.55 micron seed stages, as this permits the use of standard off-the-shelf telecommunications technology.

The output of the 25 m SMF-based 2 micron light source is then spliced on to ∼4.5 m of large-mode area TDFA, which serves as the power-amp. The TDFA fiber has a core/cladding diameter of 25/250 microns and a low core NA of ∼0.1 and is chosen to reduce the nonlinear effects in the gain fiber. The mode conversion from ∼10/125 to 25/250 microns, in this case, is done using a pump combiner/mode adapter on both sides of the TDFA. Two 35 W 793 nm pump diodes are coupled into the gain fiber through a 6×1 combiner in the counter propagation configuration. The pumps are also modulated by a square wave with 50% duty cycle at ∼250 Hz. The output of the TDFA is then mechanically spliced to the input end of ∼8.5 m of ZBLAN fiber with a core/cladding diameter of 8/125 microns and an NA of ∼0.27. The zero-dispersion wavelength is expected to be ∼1.65 microns based on the material dispersion parameter of ZBLAN. The output of the ZBLAN fiber is measured with a thermal power meter, and a power distribution in various spectral bands is measured using suitable cutoff long-pass filters.

The final SC output spectrum is shown in Fig. 4.35b and extends from ∼1.9 to 4.5 microns, with a time-averaged power of ∼2.6 W in the continuum, of which ∼0.7 W lies in wavelengths beyond 3.8 microns. The TDFA is pumped with ∼30 W of 790 nm pump power modulated with a 50% duty cycle, 250 Hz square wave, and outputs ∼8 W of average power around 2 microns. The input to the TDFA comprises the output from a ∼ 25 m SMF with ∼2.5 KW peak power 1.55 micron input pulses of ∼1 ns duration at 500 kHz repetition rate. Once again, since the system is modulated, the observed SC output power of 2.6 W corresponds to ∼5.2 W of SC output during the ON phase of the modulation signal, and the 0.7 W of average power measured beyond 3.8 microns indicates ∼1.4 W of output during the ON phase. The dip in the spectrum at ∼4.25 microns is attributed to the CO_2 absorption in the atmosphere while traversing the ∼0.75 m path length inside the monochromator.

Figure 4.35c shows the power scalability of the TDFA-based MWIR SC laser as function of the pulse repetition rate and the 790 nm pump power. As can be seen, by varying the pulse repetition rate electronically, the thulium system shows versatility in its ability to adjust the output power with similar spectral extent. The spectrum out of the ZBLAN fiber is observed to have similar spectral extent and shape in each case; however, the average power is seen to scale linearly with the repetition rate. Thus, it is possible to increase the average power in the continuum by increasing the repetition rate and the pump power, while maintaining the pulse peak power. It is also seen that the amount of 790 nm pump power required to double the output average power with repetition rate does not scale linearly. For example, at 125 kHz, the SC outputs ∼0.4 W average power at 12 W of modulated pump power. However, to double the output average power to ∼0.8 W at 250 kHz, only ∼3 W additional

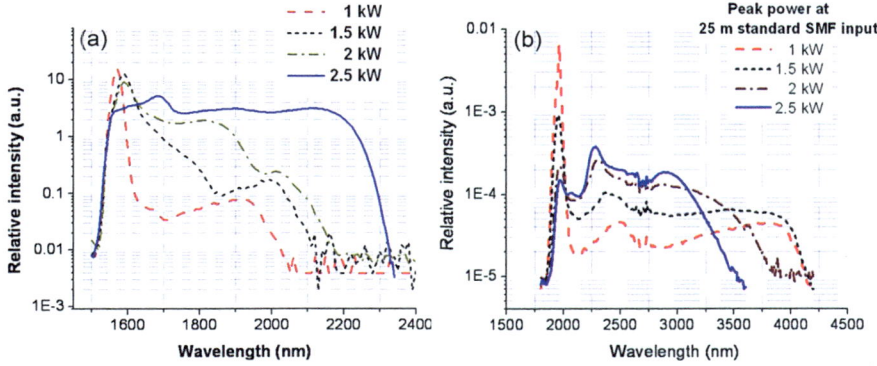

Fig. 4.36 Peak power optimization for stage 1 of the TDFA-based SC source. (**a**) Output spectrum from ∼25 m SMF for various peak input powers of 1.55 microns pulses. (**b**) Corresponding output spectrum from 12 m ZBLAN at ∼8 W CW TDFA pump power for various 1.55 micron peak powers used in stage 1

modulated pump power is required. The increase in efficiency with repetition rate is attributed to the threshold of the TDFA (∼6 W) and the increase in the amplifier efficiency at higher pulse duty cycles, due to lower losses to spontaneous emission.

The first-stage SC (output from the ∼25 m SMF) should be optimized to obtain the broadest MWIR SC at the output of the ZBLAN fiber. Figure 4.36a shows the first-stage SC spectra for various 1.55 micron pulse peak powers at the input of the 25 m of SMF. Figure 4.36b shows the corresponding second-stage SC measured at the output of ∼12 m of ZBLAN for a fixed CW TDFA pump power of ∼8 W. We observe that the lower input peak powers used for the 2 micron light generation gives the broadest ZBLAN output spectrum. However, the ASE peak from the TDFA in the ZBLAN output is higher for the lower peak power case due to insufficient strength of the input signal in the TDFA gain band. Therefore, even though the ZBLAN output spectrum extends further for the low peak power case, the efficiency of SC generation is poor compared to the high peak power case. We see that the higher spectral shifting efficiency (defined as the ratio of power in wavelengths >2.5 microns to the total SC power) is achieved at ∼2.5 kW peak power compared to the lower peak powers, with saturation observed beyond 2 kW peak power. Above 2.5 kW, the standard SMF generates additional spectral components beyond 2.1 microns, which are not efficiently amplified in the TDFA.

Figure 4.37 shows the modulated power measurements for varying 790 nm pump powers. The TDFA configuration shows an overall efficiency of ∼27% with ∼8 W output power at ∼30 W of 790 nm pump power. The observed SC output power of ∼2.6 W corresponds to an overall mid-IR efficiency of ∼9% with respect to the 790 nm pumps. ∼0.7 W of the ZBLAN output lies in wavelengths beyond 3.8 microns, leading to an observed efficiency of ∼2.35% in generating wavelengths >3.8 microns with respect to the TDFA pump power. We currently observe a 790 nm pump to 2 micron signal efficiency of ∼27% for the TDFA system. However, this

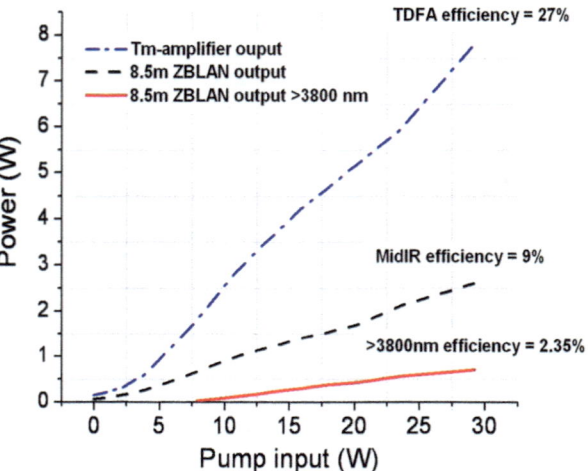

Fig. 4.37 Power measurements of the modulated TDFA output, ZBLAN SC output, and output average power > 3.8 microns versus the 790 nm pump power

is lower than the absolute efficiency of ∼53% demonstrated in a CW lasing cavity configuration. This discrepancy for our system is attributed to three main losses. First, the signal loss in the combiner-gain fiber splice and the insertion loss of the output combiner amount up to ∼1 dB of loss and account for ∼7% reduction in the efficiency. In addition, the generation of high peak powers and the phonon losses associated with the spectral shifting process in the ∼0.5 m length of the 10 microns core diameter fiber at the combined output, ∼11% reduction is observed in the power-amp efficiency. Finally, the remainder of the gap is attributed to the low duty cycle of the input signal to the amplifier which results in larger losses to the spontaneous emission in the amplifier and reduces the efficiency of the system.

4.6 Extending Further into the MWIR and LWIR

As the last section illustrated, ZBLAN-based fluoride fibers limit the long-wavelength SC edge to 4.5 microns or shorter due to the inherent transmission of these fibers. However, there are fibers made from other types of fluorides, tellurites, and chalcogenides that have shown expansion of SC all the way to ∼13.3 microns. In this section we give a brief review of some of the state-of-the-art experimental demonstrations of the long-wavelength edge expansion. For expanding to ∼5.5 microns, three fiber options that have been demonstrated are tellurite, indium fluoride (InF$_3$), and sulfides (one type of chalcogenide). For expanding to ∼13.3 microns, most of the demonstrations to date have been in chalcogenides based on selenides in the fiber core region. Particularly as we move to the LWIR, some of limitations in the experiments include a complicated pumping system and very limited time-averaged output power. However, the field is rapidly growing, and in the near future, these limitations may be solved by borrowing techniques from SC generation in the SWIR and MWIR, as described earlier in this chapter.

4.6.1   Expanding the Wavelength Edge out to ~5.5 Microns

Tellurite fibers are one fiber type that have been used to expand the wavelength edge through the entire MWIR. Tellurite glasses have relatively high nonlinearities, approximately 40–60 times fused silica, glass transition temperatures above 350 Celsius, and the potential to be more mechanically robust than fluorides or chalcogenides [62]. Using a specially fabricated W-type tellurite fiber, Ref. [62] reports SC generation over the wavelength range of ~1.3 microns to ~4.4 microns. The experimental setup consists of a passively mode-locked thulium-doped fiber oscillator based on carbon nanotube saturable absorber in a ring configuration, a fiber amplifier system, and the specialty tellurite fiber for SC generation. Using ~2.6 W of pump power centered at 1917 nm, they generate SC output powers of ~1.4 W.

Photonic crystal fiber made from tellurites has also been used to obtain wider wavelength ranges, although with lower output powers. For example, Ref. [24] reported broad-bandwidth, MWIR SC generation using an 8 mm length of highly nonlinear tellurite microstructured photonic crystal fiber. By pumping with 100 fs pulses at 1550 nm with an energy of 1.9 nJ, they obtain SC extending from 789 to 4870 nm. The combination of microstructure and highly nonlinear glass provides enhanced optical nonlinearity responsible for SC output after only 8 mm of pulse propagation. Also, the use of the sub-centimeter fiber also mitigates the relatively high loss of the tellurite glass and the fiber design. Moreover, the fiber design with a zero-dispersion wavelength of 1380 nm allows pumping at 1550 nm, which is a common telecommunications wavelength.

The experimental setup for the [24] results is as follows. Pulses are generated by an optical parametric oscillator pumped by a Ti-sapphire laser centered at 810 nm. This parametric oscillator generates pulses with a center wavelength of 1550 nm, pulse width of 110 fs, and average power of 250 mW at a repetition rate of 80 MHz. The pulses generated are focused using a 40 × 0.5 NA lens into the tellurite fiber. The SC output power is measured to be 70 mW. The long-wavelength edge at 4870 nm is probably limited by the loss window of the tellurite fibers.

Fluoride glasses other than ZBLAN have also been used to expand the wavelength coverage in the MWIR. In particular, solid-core, single-mode InF$_3$ fibers have been fabricated and used in SC generation experiments. In one experimental report, coherent MWIR SC is generated in a dispersion-engineered, step-index, indium fluoride fiber pumped near 2 microns. The SC spectrum spans from 1.25 to 4.6 microns. Optical fibers made of InF$_3$ glass are supposed to have robust fabrication processes, environmentally stability, and broad transmission window from the visible wavelengths to ~5.5 microns [63]. The material dispersion for

InF$_3$ crosses zero at ∼1.7 microns and monotonically increases with wavelength. To generate their MWIR SC, Ref. [63] engineer their fiber to move the zero-dispersion wavelength and dispersion flatness. By controlling the core size and numerical aperture of a step-index fiber, they moved their zero-dispersion wavelength into the gain spectrum of thulium, while maintaining relatively flat dispersion from 2 to ∼4.5 microns.

The experimental setup for SC generation by [63] is as follows. A femtosecond, mode-locked fiber laser at 1560 nm with a 50 MHz repetition rate is amplified in an erbium-doped fiber amplifier. The amplified pulses are sent into a highly nonlinear fiber, which shifts more than half of the pulse energy to 1960 nm through Raman soliton self-frequency shifting. A cladding-pumped, thulium-doped fiber amplifier pumped in the forward direction using a 793 nm multimode pump diode is used for boosting the signal power in the 2 micron region to above 500 mW. By pumping with 570 mW of pump power in a 97.5 fs pulse, they generate 258 mW of SC from a 30 cm length of InF$_3$ fiber from 1.25 to 4.6 microns. Beyond this wavelength, the group dispersion climbs rapidly, and the mode field diameter expands rapidly; either or both may be responsible for the shorter-wavelength expansion compared to the InF$_3$ transparency band.

However, other experimental demonstrations have been able to obtain spectra reaching out to the transparency edge of InF$_3$ fiber. For example, Ref. [64] demonstrate a MWIR SC source by cascading an erbium-doped fluoride fiber amplifier and a low-loss InF$_3$ fiber. They obtain spectral broadening from 2.4 to 5.4 microns, which is up to the long-wavelength edge of the InF$_3$ glass transparency window. In this SC source, the amplifier is directly spliced to a low-loss fluoroindate fiber seeded by a 2.75 micron optical parametric generation source. The output has a maximum average power of ∼10 mW with up to 82% of the power located in the mid-IR beyond 3 microns.

The SC generation system employed by [64] consists of a forward-pumped erbium-doped fluoride fiber amplifier spliced to a low-loss fluoroindate fiber. The amplifier is pumped by a 970 nm multimode diode and seeded with a 2.75 micron optical parametric generator source, which has a diode-pumped passively Q-switched microchip laser followed by a periodically poled lithium niobate nonlinear crystal. The amplifier is made of a 1.25 m long erbium-doped ZrF$_4$ double-clad fiber. The doped fiber is then spliced using a filament-based splicer to a piece of InF$_3$ fiber, which has a 100 micron diameter cladding and an average core diameter size of 13.5 microns (increasing from 12.5 to 14.5 microns along the 31 m fiber). With a numerical aperture of 0.3, it is slightly multimode for most of the wavelength range covered by the super-continuum. The InF$_3$ fiber has an attenuation coefficient below 30 dB/km for all wavelengths between 2.2 and 4.2 microns, with a minimum value of 12 dB/km at 3.8 microns and a low O-H absorption peak of 28 dB/km. The broadest spectrum is obtained with a 15 m length of InF$_3$ fiber, and at a pump power of 978 mW, the SC spans from 2.4 to 5.4 microns.

Another alternative to generate spectra out to 5.5 microns is to use chalcogenide fibers, namely, sulfide fibers (As$_2$S$_3$). For example, a tunable femtosecond SC source is demonstrated in [65] with a maximum output power of 550 mW and a spectral range from 2.3 microns to 4.9 microns by tuning the pump wavelength. Sulfide chalcogenide fibers with core diameters of 7 and 9 microns are pumped at different wavelengths from 2.5 to 4.1 microns by means of a post-amplified optical parametric oscillator pumped by a Yb:KGW laser.

The SC is generated by Ref. [65] in a fairly complicated setup. A Yb:KGW oscillator (450 fs, 42 MHz, 1.03 microns, 8 W output) is used to pump a fiber-feedback optical parametric oscillator. The 300 fs signal pulses are further amplified with an optical parametric amplifier, thereby delivering $\sim$1 W output pulses with pulse duration between 300–450 fs and tunable from 2.5 to 4.1 microns. The As_2S_3 fiber has a step-index design with a core diameter of 7 or 9 microns and a 0.3 numerical aperture. Fiber lengths of 6, 13, and 23 cm were used in the experiments. For 23 cm length of the 9 micron fiber pumped at 3.83 microns, SC was obtained with 550 mW between 2.95 and 4.9 microns.

Another example of SC generation through the MWIR using sulfide fiber is reported by [66]. An all-fiber sulfide-based SC source covering 1.9–4.8 microns is demonstrated based on a combination of silica commercial off-the-shelf components and an As_2S_3 step-index fiber. The SC spectrum has a 10 dB spectral flatness from 2 to 4.6 microns and −20 dB points from 1.9 to 4.8 microns with a total output power of 565 mW. They also show that the long-wavelength limit of the system arises from an extrinsic absorption in the fiber, meaning that further broadening is still possible in the sulfide glass system. It turns out that the multi-photon absorption edge for sulfide fibers is about 7.4 microns.

The all-fiber SC system developed by Ref. [66] uses a multistage master oscillator power amplifier geometry. A 40 ps, 10 MHz erbium fiber-based laser with 200 mW average output power serves as the seed for the system. The seed is amplified in multiple stages, and the power is soliton shifted into the thulium amplification band. The thulium power amplifier increases the power, and the output is coupled to a sulfide fiber. The chalcogenide fiber used in the last stage is a step-index core-clad As_2S_3 fiber with a 10 micron core diameter and numerical aperture of $\sim$0.3 and 2 m length.

4.6.2   Further Wavelength Expansion into the LWIR

To further extend the long-wavelength edge of the SC into the LWIR requires the use of yet softer glasses, such as selenide-based chalcogenides. For example, the infrared long-wavelength transparency limit (as defined by a loss of 1 dB/m) for silicate glasses is $\sim$2.4 microns, for zirconium fluoride glasses is $\sim$4.3 microns, for indium fluoride glasses is $\sim$5.4 microns, for chalcogenide sulfide glasses is $\sim$7 microns, and for chalcogenide selenide glasses is $\sim$9 microns or longer. Moreover, the nonlinear refractive index of the selenide fibers (As_2Se_3) is about 600 times higher than that of silica-based glasses [67]. One reason the LWIR is interesting for spectroscopy is that the fundamental molecular vibration absorption bands are two to three orders of magnitude stronger than the overtones and combinational bands in the NIR and SWIR.

One of the widest spectral range for SC generation is reported by [108]. By launching intense ultrashort pulses with a central wavelength of either 4.5 microns or 6.3 microns into short pieces of ultrahigh numerical aperture, step-index, chalcogenide glass optical fibers, they demonstrate SC spanning 1.5 to 11.7 microns (for 4.5 micron pumping) and 1.4 to 13.3 microns (for 6.3 micron pumping). The chalcogenide fiber used is specifically designed for both ultrahigh numerical aperture and thermal compatibility of the core and cladding glasses. The fiber has a $\sim$16 micron diameter with a $As_{40}Se_{60}$ core surrounded by a $Ge_{10}As_{23.4}Se_{66.6}$ cladding. Because of the large core and numerical aperture, the fiber is effectively multimode over much of the wavelength range. Also, the zero-dispersion wavelength is calculated to be 5.83 microns.

The pump was generated from a tunable optical parametric amplifier and a noncollinear difference frequency generation module with central wavelength tunable from 2.5 to 11 microns. The amplifier was pumped by a millijoule pulse energy Ti-sapphire laser operating with $\sim$60 fs pulses and a repetition rate of 1 KHz. SC generation is achieved by launching $\sim$100 fs pulses into 85 mm of the selenide fiber. The average pump power was increased without observing fiber damage up to $\sim$350 (760) microwatts in the 4.5 (6.3) micron pump case, and the highest output power is $\sim$150 microwatts. When pumping at 6.3 microns, just above the zero-dispersion wavelength of 5.83 microns, the pump pulse transforms into a higher-order soliton that rapidly breaks up into multiple fundamental solitons through soliton fission and radiates dispersive waves at a wavelength that is phase matched to the solitons in the normal dispersion regime [68].

In another experiment, Ref. [67] generate SC spanning from 3 to 8 microns using a low-loss As_2Se_3 commercial step-index fiber. A maximum average output power of 1.5 mW is obtained at a low repetition rate of 2 kHz (the incident power on the SC fiber for this case is 0.82 W). The pump source consists of an erbium-doped ZrF_4-based in-amplifier SC source spanning from 3 to 4.2 microns. The seed source is an optical parametric generator, whose output is coupled to the amplifier fiber. The amplifier output is filtered using a long-pass filter, and the filtered spectrum is launched into the selenide step-index fiber. It is noted that only 1 m of SC fiber is sufficient for the SC's edge to reach 6 microns. Since the fiber has all normal dispersion up to $\sim$8.9 microns, the spectrum broadens mainly through self-phase modulation. The best SC performance is obtained with a fiber length of 3.5 m. For longer fiber lengths, the background losses become more important and reduce the overall SC power density.

In yet another experiment, an SC source is demonstrated covering 3–10 microns in a 4 cm long As_2Se_3 core and As_2S_5 cladding step-index chalcogenide fiber pumped with the MWIR pulse generated by difference frequency generation [69]. The output wavelengths of the MWIR pulse source, which could deliver $\sim$170 fs pulses with a tunable wavelength range from 2.5 to 10 microns, are tuned to pump the fabricated chalcogenide fiber in the anomalous dispersion regime. The core size of the fiber is $\sim$27 microns, and the refractive index difference between the core and the cladding is as large as $\sim$0.57 at a 6 micron wavelength with an ultrahigh numerical aperture of 1.69. The simulated chromatic dispersion predicts a zero-dispersion wavelength at $\sim$6.35 microns and a flattened dispersion profile over a wide range from 3 to 12 microns.

The details of the experimental setup are as follows [69]. The MWIR pulse source starts with a Ti-sapphire mode-locked seed laser, whose output is sent to a pulse picker regenerative amplifier for boosting the pulse energy to about 1 mJ at a low repetition rate of 1 kHz. The amplified pulse passes through a traveling-wave optical amplifier to generate a signal and idler. The signal and idler beams are collinear combined together and passed through a difference frequency generator to generate the MWIR pulses from 2.5 to 10 microns. Despite this elaborate setup, Ref. [69] does not report the output SC power that is generated.

Finally, in another experiment, by pumping an 11 cm long step-index chalcogenide fiber with $\sim$330 fs pulses at 4 microns from an optical parametric amplifier, SC spanning from $\sim$1.8 to $\sim$10 microns has been demonstrated [70]. In particular, they fabricate a small-core step-index chalcogenide fiber with a $Ge_{12}As_{24}Se_{64}$ core and a $Ge_{10}As_{24}S_{66}$ cladding. These two chalcogenide glasses show broad transparency and low losses out to 9 microns with good properties for fiber fabrication. The pump source was a custom femtosecond optical parametric amplifier and allowed an SC containing a few milliwatts of average power to generated, similar to that of a typical MWIR or LWIR beam line of a synchrotron. They report that the fiber core could damage at a peak intensity of $\sim$30 GW/cm^2 and an average power density of $\sim$100 kW/cm^2 at the input facet. Reference [69] concludes that damage remains the limitation of chalcogenide-based devices and according to their measurements depends significantly on the composition of the glass. The SC spectrum broadens progressively until the damage threshold of the chalcogenide material is reached.

4.7   Applications of Super-Continuum Sources

In this section we survey some of the applications demonstrated for SC light sources, both in the SWIR as well as the MWIR. Although by no means an exhaustive list of applications, the examples presented feature how the attributes of SC lasers can benefit various applications. One application for SWIR SC light sources is in active remote sensing and hyper-spectral imaging (HSI). The high average power, the broad spectrum, and the convenience of an all-fiber-integrated laser make the SC laser an attractive light source for active illumination in the SWIR wavelengths for long-distance remote sensing and hyper-spectral imaging applications. Field trials using SWIR SC sources were performed at Wright-Patterson Air Force Base (WPAFB). Using a 5 W SWIR source, targets a mile away were identified when the detectors were placed near the target [45]. When the power is increased to 64 W, the diffuse reflected signal could be detected by a sensor or camera adjacent to the SC laser [46].

Another application for MWIR and SWIR SC sources is stand-off detection of solid targets [71–75]. For example, in the defense and security industry, the MWIR SC spectrum can be used to emulate the black body radiation of hot objects. The MWIR SC also overlaps the vibrational and rotational resonances of many solids, and this feature can be exploited for stand-off detection of solid targets such as explosives and other chemicals [48, 76–78].

The SWIR and MWIR SC sources can also conveniently target hydrocarbon bonds, either the fundamental resonance (MWIR) or the overtone and combinational bands (SWIR). Taking advantage of this feature, in another application, hydrocarbon-based thermoplastics used in additive manufacturing or 3D printing can be sintered using SC sources operating between 2 and 2.5 microns. For example, we sinter 11 different materials and fabricate rods with strengths up to 5× that from carbon dioxide lasers [79].

In the healthcare industry, the MWIR and SWIR SC sources can be a useful tool to perform diagnostics and therapeutics [80–110]. For example, we demonstrate the use of a MWIR SC source to perform atherosclerotic plaque detection and to cause preferential damage to lipids, which are a main component of such plaques [5, 111]. The plaque detection and preferential damage is possible since the absorption of hydrocarbon bonds that are the main building blocks of lipids falls within this wavelength range. Whereas the fundamental absorption of hydrocarbons occur in the MWIR around $\sim$3.4 to $\sim$3.6 microns, the first and second overtones fall in the SWIR around $\sim$1.72 microns and $\sim$1.21 microns and the combinational bands between $\sim$2 and $\sim$2.4 microns. Another example of medical diagnostics is the use of SWIR SC sources to perform noninvasive glucose monitoring. For instance, we measure glucose solutions down to 1 mg/dL based on the C-H combinational bands, and we estimate that the SWIR SC increased the measurement system signal-to-noise ratio by >50× compared with lamps [112]. Details of these exemplary applications of SWIR and MWIR SC sources are provided in the following.

4.7.1 Active Remote Sensing Using A SWIR SC Laser

Passive hyper-spectral imaging systems that are illuminated by sunlight have a number of limitations that we aim to overcome using the SC laser technology. Useful HSI data requires sun angles available for no more than 7 h each clear day, depending on the location and time of year. Cloud cover also reduces illumination and restricts the utility of the HSI data. Moving beyond these limitations requires active illumination, which can extend the operational envelope of the technology to 24 h. Additional advantages of active illumination include the elimination of shadows (when the source and receiver are co-located) that alter the measured spectral signature and increase false alarms, improved change detection by providing a consistent illumination geometry, direct retrieval of surface reflectance at short ranges, and with modulated sources the possibility of ranging and forest canopy penetration.

There are two possibilities for illuminating a surface for spectral interrogation: lamps and lasers. Lamps can produce smooth, broadband illumination that can be customized by optical filtering, but to maintain brightness at multi-kilometer ranges, the beam needs to be highly collimated and spatially coherent, neither of which have been efficiently achieved with lamps thus far. Lasers, on the other hand, have the correct spatial characteristics to deliver high brightness at a distance but have been limited in wavelength range.

We first present the field trial results of a 5 W all-fiber broadband SC laser covering the SWIR wavelength bands from ~1.55 to 2.35 microns [45]. The SC laser is kept on a 12 story tower at WPAFB and propagated through the atmosphere to a target 1.6 km away. Beam quality measurements of the SC laser after propagating through 1.6 km are studied using a SWIR camera and show a near-diffraction-limited beam, with an M^2 value of <1.3. The SC laser is used as the illumination source to perform spectral reflectance measurements of various samples at 1.6 km, and the results are compared to measurements using a conventional lamp source.

The field trial layout at WPAFB is shown in Fig. 4.38, and the SC laser used was the packaged prototype from Fig. 4.18. The SC laser is placed on a 76 m tall tower, and the collimated beam is propagated thorough the atmosphere to an $8' \times 8'$ target panel kept 1.6 km away from the tower. The target panel consists of a plywood stand for placement of the materials of interest used in our measurement and is slightly

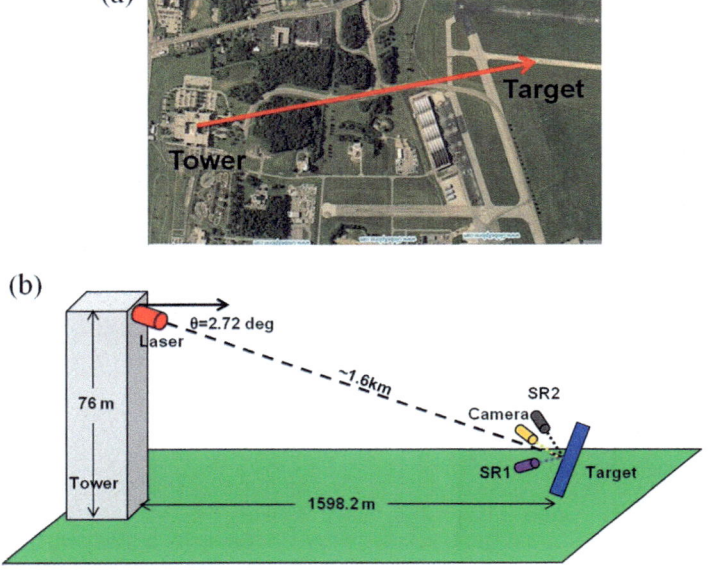

Fig. 4.38 (a) Map view of 1.6 km tower to ground path at Wright-Patterson Air Force Base. (b) Diagram of tower-target test layout

(a) **Camera Image of Beam**

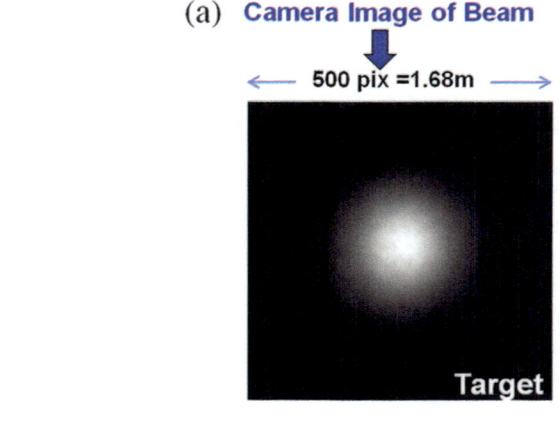

(b) **3D Image of Beam**   (c) **Gaussian Fit of Beam**

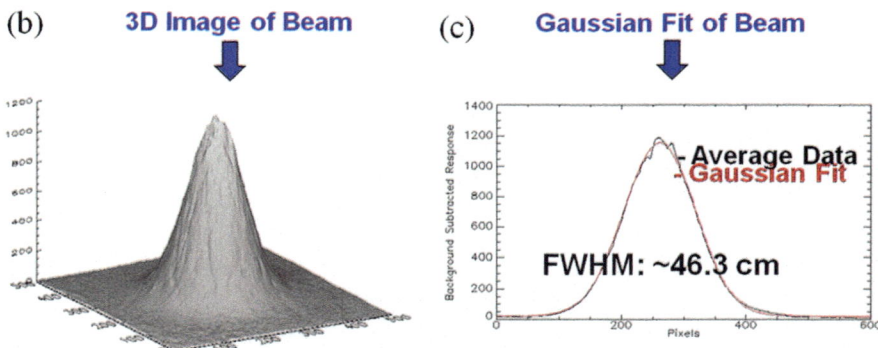

Fig. 4.39 SC laser (from Fig. 4.18) beam profile measurements at ∼1.6 km; 1000 frame average. (**a**) Camera image of the beam. (**b**) 3D image of the beam. (**c**) Gaussian fit and beam width measurement

tilted to create a normal surface with respect to the laser beam propagation angle of ∼2.72°. The SWIR camera and the spectro-radiometers (SR1 and SR2) for the various measurements are placed on the field close to the target. The tower-target beam alignment is verified by observing the reflected beam at the tower from a retro-reflecting mirror placed at the target site. All field measurements are performed after sun set to minimize the effects of solar illumination.

Figure 4.39a shows a camera image of the collimated SC laser beam profile as projected on a target after 1.6 km of atmospheric propagation. Tyvek material is used as the target for these measurements, since this material possesses high reflectance in this wavelength region. The profile shown is an average of 1000 frames measured using a 4 ms integration time. The beam diameter is measured as follows. First, the beam profile obtained using the camera is fit to a Gaussian profile. Then, the beam full width at half maxima (FWHM) is measured from the Gaussian fit. Figure 4.39a, b shows a picture of the beam as projected on the target and the 3D image of the beam corresponding to the camera image. Figure 4.39c also shows the beam profile through the beam cross section and the corresponding Gaussian fit.

The M^2 factor is a common measure of the laser beam quality and is used to quantify the ratio of divergence of the actual laser beam to an ideal Gaussian beam. An ideal Gaussian beam has an M^2 value of 1. Thus, the M^2 value indicates how close in divergence and, hence, diffraction limited the laser beam is, compared to an ideal Gaussian beam. The SC M^2 value is calculated using the ratio of the SC beam divergence and the ideal Gaussian beam divergence and is seen to be $\theta_{SC\,Beam}/\theta_{Ideal} = \sim 0.487/0.392 = \sim 1.24$. The calculated SC M^2 value of ~ 1.24 from the measurements at 1.6 km indicates that the SC beam is nearly diffraction limited.

Figure 4.40 shows the retrieved spectral reflectance for the various samples measured using the SC laser averaged over at least five measurements for each sample and the corresponding in-lab reflectance measurements. The measurements are performed in wavelength steps of 1 nm. The in-lab measurements are collected in a more controlled setting using a quartz-halogen filament lamp as the light source and a fiber-optic contact probe. The SC data in the wavelengths from ~ 1.8 to 1.95 microns is absorbed by the water in the atmosphere and has been removed from the figure. As can be seen in Fig. 4.40, the spectral reflectance curves for the various samples agree closely with the lab measurements, especially with respect to the spectral shape of the curves. For many of the materials, a reflectance offset exists between the lab measurements and the SC laser measurements. This is most likely the result of non-lambertian target surfaces and differences that exist between the illumination/viewing geometry for the lab and field measurements.

4.7.2 Field Tests for Round-Trip Imaging at a 1.4 km Distance

In addition, we have conducted field trials showing the feasibility of using a high-power SWIR SC laser (SWIR-SCL) as an active illuminator for HSI. The SWIR-SCL with light intensity comparable to the sun was used in field trials at WPAFB where the laser was placed in a 12-story tower and the targets were placed on the ground at a distance of 1.4 km or more. The SWIR-SCL (from Figs. 4.28 and 4.29) delivered 64 W time-averaged output power over the continuous wavelength range from 1064 to >1800 nm with a wall-plug efficiency of 15.7% and a beam quality that is nearly diffraction limited with $M^2 < 1.3$ over the wavelength range. Also, the SWIR-SCL produces ~ 0.45 ns pulses with a repetition rate near 6.25 MHz and an average relative variability or fluctuation of $\pm 0.2\%$. This SWIR-SCL along with various diagnostics including a hyper-spectral imaging camera was placed in the 12-story tower, while targets on an $8' \times 8'$ plywood board were placed on the ground, normal to the beam, at a distance of 1.4 km or more.

The field tests demonstrated three significant outcomes. First, because of the high brightness from the SWIR-SCL, the diffuse reflection from targets at a distance of 1.4 km could be detected by the hyper-spectral imaging camera placed adjacent

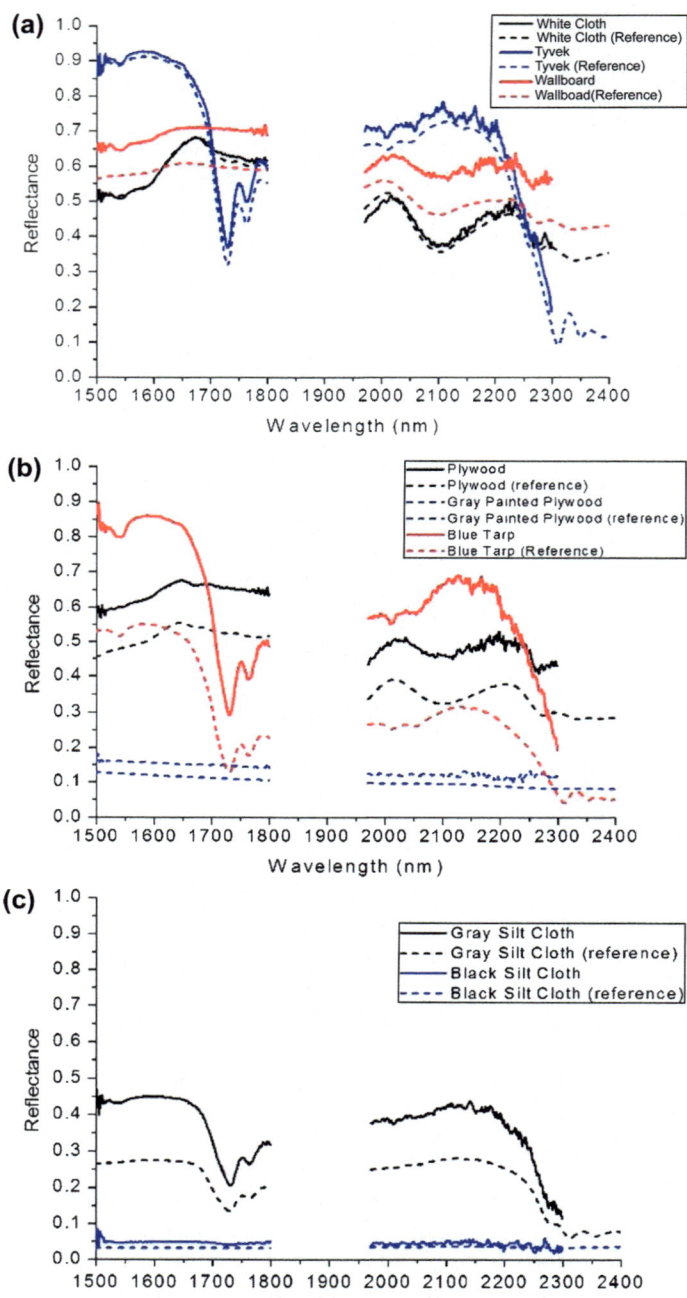

Fig. 4.40 Spectral reflectance measurements at ~1.6 km using the SC laser (*solid lines*) and their comparison to in-lab measurements performed using a quartz-halogen lamp (*dashed lines*). (**a**) Retrieved reflectance of white cloth, Tyvek, and wallboard. (**b**) Retrieved reflectance of *plywood*, *gray* painted *plywood*, and *blue tarp*. (**c**) Retrieved reflectance of *gray* and *black* silt cloth

to the laser. In particular, we demonstrated end-to-end stand-off sensing with a SWIR hyper-spectral camera at a distance of 1.4 km on a slant path near the ground. Second, the SWIR-SCL was successfully used for a broadband change detection demonstration experiment, where the laser was on-off modulated at 15 Hz synchronized to the 30 Hz frame rate of an observing SWIR camera and receiver telescope. In the change detection process, camera frames with laser off are subtracted from the frames with laser on, thereby canceling or minimizing artifacts associated with the sun illumination, the change in angle of the sun, and shadows. Third, similar to LIDAR techniques, ranging was used to measure the distance of the target with an absolute accuracy of ∼1.5 cm even at distances of 1.4 km or more. The ranging measured the ∼0.45 ns pulses from the SWIR-SCL using a time-of-flight and modulated laser technique. These field test results confirm the feasibility of using the SWIR-SCL as an active illuminator for hyper-spectral imaging, thereby permitting 24/7 operation and in shadowed conditions, rather than being restricted to sunlight operation. Moreover, beyond simply providing sun-equivalent illumination, the SWIR-SCL also permits broadband change detection and ranging, which are not possible with passive hyper-spectral imaging using the sun.

Initial calculations showed that to match sunlight, it would be necessary to generate at least 50–100 Watts per micron per square meter, so we set a goal of SC output power density of 100 W/μm as the initial benchmark [46]. Figure 4.41 plots the output power density spectra of the SWIR SC of Fig. 4.28 (from 1.06 to 1.8 microns) as compared to solar irradiance at a 50° elevation (40° zenith angle). We also built and tested a thulium-based SC laboratory prototype in the 2–2.5 μm region with 25 W of SC output or 50 W/μm [47]. In the wavelength range of 2–2.5 microns, Fig. 4.41 shows the power density spectrum produced by the 25 W breadboard. Both power spectral densities were measured against a Spectralon™ target at approximately 1 m from the source. Thus, over a significant fraction of the wavelength range covered by the 64 W SWIR SC of Fig. 4.28, the output power density is comparable to or exceeds the solar spectrum when projected over a 1 m² area.

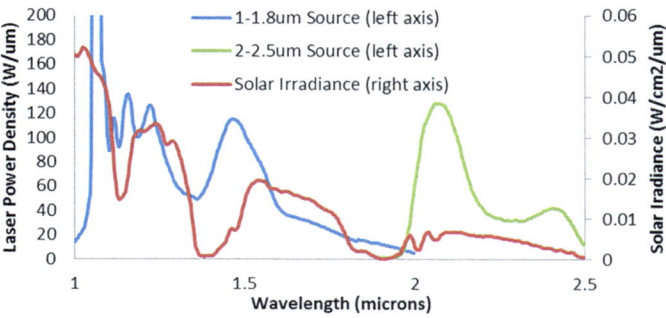

Fig. 4.41 Super-continuum laser power density spectra from 1 to 1.8 micron source in this paper (*blue curve*) and 2–2.5 micron source from 25 W laboratory SC unit (*green curve*) [47]. The *red curve* corresponds to the solar irradiance spectrum over the SWIR wavelength range

The performance of the 64 W SWIR-SCL as a source for a bistatic active HSI sensor was evaluated at WPAFB, and we assessed the beam propagation characteristics of the laser and demonstrated reflectance retrieval. In the first tests, the majority of testing occurred after sunset to remove any solar contribution to the measured signal (next section will discuss change detection results, which were collected throughout the day). The SWIR-SCL was placed in a tower laboratory and directed along a slant path to a target panel on the ground, similar to Fig. 4.38. The materials used on the target panel include Tyvek, black silt cloth, gray silt cloth, a blue tarp, and gray painted plywood. A broadband SWIR camera made by FLIR with spectral response from 1 to 5 microns was utilized to provide beam shape and turbulence analysis. Additionally, a slit-based SWIR hyper-spectral imaging camera made by Headwall Photonics was used to measure the characteristics of the laser return signal as a function of wavelength. Both the broadband and HSI cameras were equipped with 100 mm lenses to provide multiple pixels across the 0.9 m laser spot diameter at a 1.4 km slant range. Note that due to losses in the mirrors used to direct the laser beam from the tower, the effective laser power is approximately 55 W rather than the lab measured 64 W [46].

We started by first performing short-range characterization of the SC laser (e.g., in the tower only). The ratio of the measured beam divergence and the theoretical diffraction-limited divergence gives us an $M^2 = 1.07$ at 1400 nm and 1.28 at 1064 nm. For the spread of wavelengths of our SC laser, we conclude a conservative $M^2 < 1.3$ over the full wavelength range. In addition, the hyper-spectral camera was used to collect spectral radiance data along a vertical line down the center portion of the beam. Figure 4.42 shows the results, including the average radiance spectrum compared to the temporal variability and the sensor's dark noise and the beam profile for a subset of spectral bands across the spectral range of the laser. The beam profile is not perfectly Gaussian, which may be due to imperfect optics used to direct the beam as well as the angle cleave and the fiber end cap.

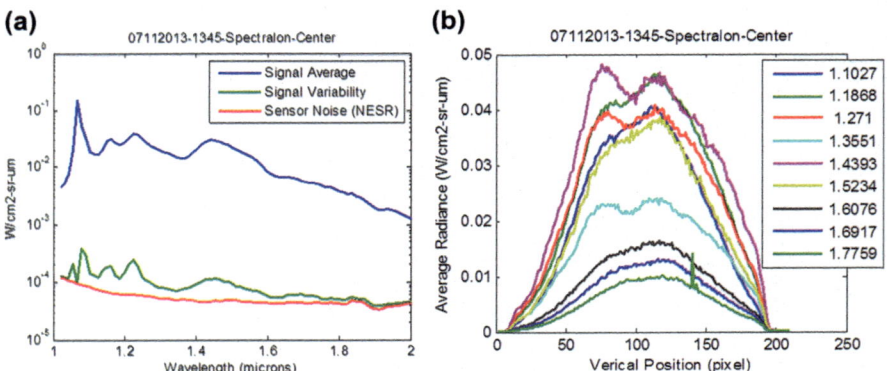

Fig. 4.42 Spatial-spectral characterization of the SWIR SC beam profile in the lab. Graph (**a**) shows the average radiance spectrum compared to the temporal standard deviation and the sensor dark noise; (**b**) contains the vertical beam profiles at multiple wavelengths

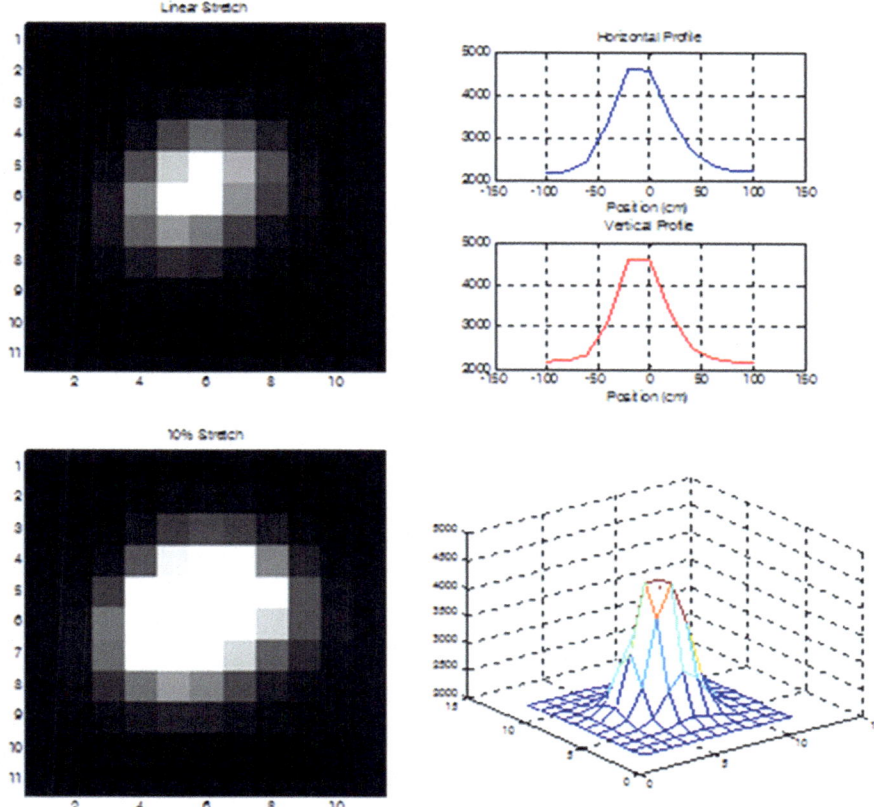

Fig. 4.43 Images taken with FLIR camera placed next to SWIR-SCL in tower. The image is of the illumination of the target 1.4 km, as seen from the diffuse reflected light

We next performed long-range characterization of the SC laser of Fig. 4.28. The SWIR-SCL light was directed from the tower laboratory through seven relay mirrors to the 8′ × 8′ Tyvek target. Because the beam is less than a meter across at the 1.4 km distance, and the camera is looking at the image from 1.4 km away, only a few pixels show the beam spot. Figure 4.43 shows the spatial characterization. The FLIR camera was placed next to the SWIR-SCL source in the tower, and the image was taken using a 100 mm lens. The beam directing optics were cleaned and adjusted to minimize astigmatism due to heating of the mirror surfaces, although some distortion is clearly visible in the images. The images of Fig. 4.43 represent an average of 5000 samples taken at a 100 Hz frame rate. As shown on the right side of Fig. 4.43, the disk-equivalent diameters measured are 61 cm at 50% light intensity and 97 cm at $1/e^2$ intensity.

Figure 4.44 shows the spatial-spectral characterization analogous to Fig. 4.42 but collected after two-way propagation of the collimated beam over 1.4 km. The

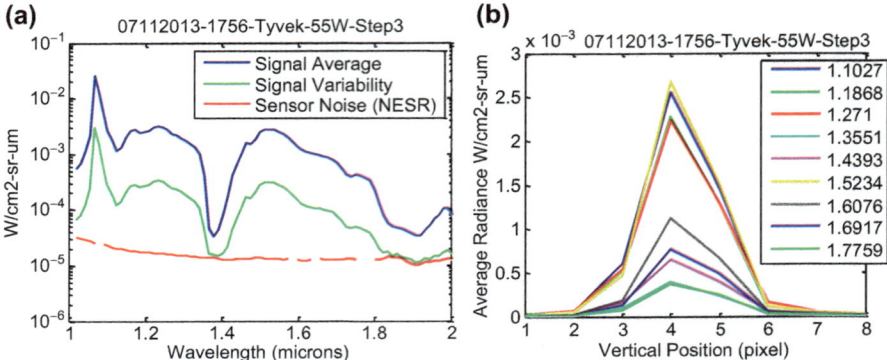

Fig. 4.44 Spatial-spectral characterization of the SWIR SC beam profile for 1.4 km round trip. Graph (**a**) shows the average radiance spectrum compared to the temporal standard deviation and the sensor dark noise; (**b**) contains the vertical beam profiles at multiple wavelengths

atmospheric spectral effects are clearly visible in the spectral radiance measurements (particularly the water absorption band at 1.4 microns). The measured signal variability is much greater than in the lab due to the strong effects of turbulence in the test geometry.

Six materials were used as targets: Tyvek, gray painted wood, black silt cloth, grass (by pointing the source off to the side of the target), gray silt cloth over Tyvek, and a blue plastic tarp. The Tyvek, black cloth, and blue tarp are all made primarily of polyethylene. The gray silt cloth is made of polypropylene which is spectrally similar to the polyethylene in the spectral region of the measurement. These materials were chosen to span the range of reflectance from ~4% (black cloth) to ~90% (Tyvek).

We compare the estimated apparent reflectance spectra of these materials against reference measurements. Since we did not have a reflectance standard such as Spectralon™ to provide a reference, we used the empirical line method (ELM) to estimate the illumination and atmospheric parameters needed to convert the measured radiance to apparent reflectance. The brightest radiance spectrum was extracted from each image. The ELM parameters were computed by linear regression using all except the Tyvek spectrum, because inclusion of this measurement caused the results to be distorted and the spectral features were significantly diminished [46]. This is likely due to the large specular component of the material, which is not captured by the reference measurements. The specular component is significant in the monostatic configuration that was used. Figure 4.45 shows the results for the grass and the blue tarp compared to the reference spectra of the materials. The spectral shapes match very well, with some noise caused by the sharp radiance spike from the pump residual at 1.06 μm.

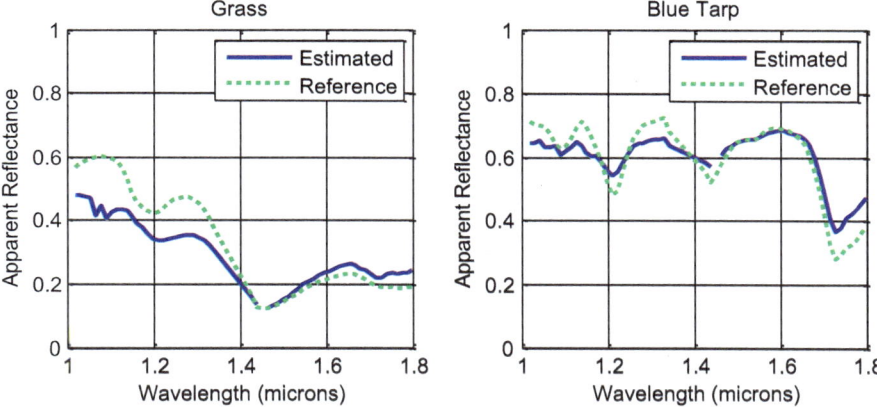

Fig. 4.45 Estimated apparent reflectance of the grass and *blue* tarp targets compared to the reference spectra. The target materials were placed 1.4 km away from the SWIR-SCL

4.7.2.1   Change Detection by Syncing SWIR-SCL and Camera

In the change detection experiment, the camera frames with laser off are subtracted from the frames with laser on, thereby canceling out artifacts associated with the sun illumination, the change in angle of the sun, and shadows. To perform the change detection experiments, the output of the SWIR-SCL was modified so that it could be modulated and synchronized with the frame rate of the SWIR camera (a Sensors Unlimited, Inc., SU640CS InGaAs Camera). Before going into the experimental details, we provide examples of these images at various times of the day at the WPAFB field tests in Fig. 4.46. During the morning and mid-afternoon, the laser spot was not visible to the eye in the SWIR camera as the solar irradiance dominated the signal, but the difference imagery did show the laser spot. However, in the early evening, the solar irradiance was reduced such that the laser and solar irradiance became similar in magnitude and the laser spot became visible. Another observation is the captured imagery had a larger dynamic range in the evening as the max and min pixel values indicate. This is because the integration time and gain of the camera were varied throughout the day to avoid saturation of the scenery. The laser spot was also noticeably smaller in the early evening collect due to refocusing that took place at 1600. Fitting of the spot size using a 2D Gaussian fit indicates that the SWIR band peak irradiance increased from approximately 180 W/m^2 to around 500 W/m^2 after refocusing.

For the change detection, the 30 Hz camera trigger output was used to control the laser modulation. Since alternate frames are to be "on" and then "off," a programmable microcontroller was used to trigger the laser using every other frame (15 Hz). Figure 4.47 presents a schematic diagram of the experiment, along with a time diagram showing the laser pulse train during the laser "on" time. As drawn, the camera begins integrating after the laser pulse intensity has stabilized and stops integrating before the pulse train ends.

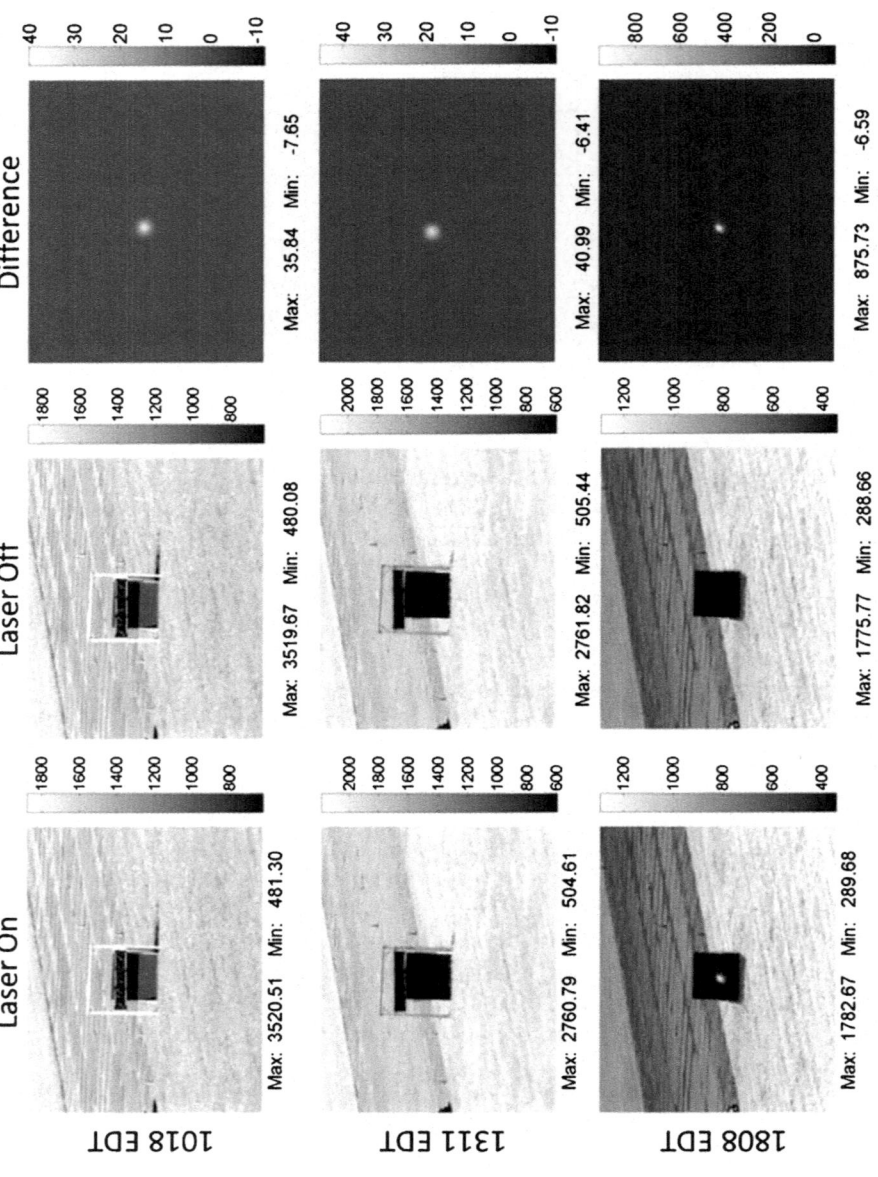

Fig. 4.46 Examples of change detection images throughout the day at WPAFB

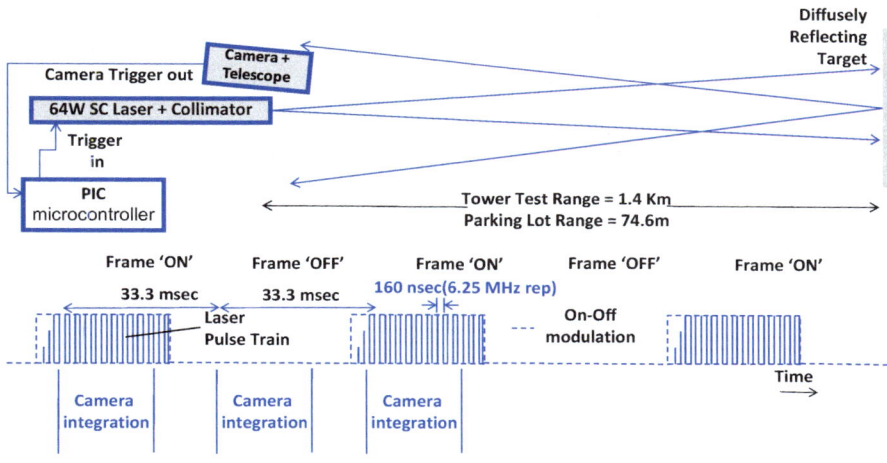

Fig. 4.47 Schematic and timing diagram for the change detection measurements

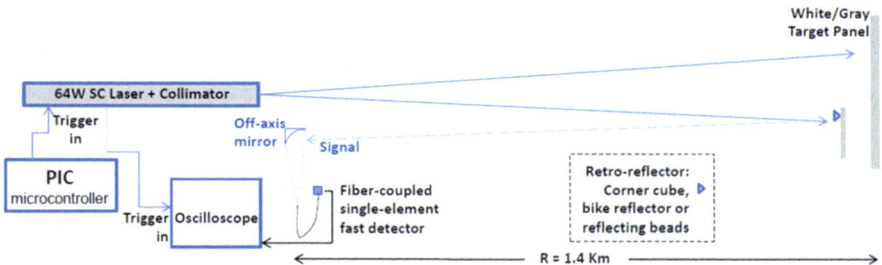

Fig. 4.48 Basic experimental configuration for laser ranging

4.7.2.2   Ranging at Distances of ~1.4 km in Field Tests

Next, we take advantage of the pulsed nature of the SWIR-SCL output to measure the distance to the target. Since the SWIR-SCL is a pulsed laser, it has the potential for making time-of-flight absolute range measurements, and, since the pulses are sub-nanosecond, the range measurements can be quite accurate. Figure 4.48 shows schematically the measurement concept. The pulse train generated by the 64 W SWIR-SCL is expanded and sent through a collimator to the target scene. The receiver is comprised of a telescope adjacent to the SWIR-SCL, a telescope mount, and a SWIR camera. A REIGL Lasertape FG21 laser range finder (resolution of ~1 m) was also integrated, to act as a range truth measurement, onto the same tripod mounting plate as the SWIR camera and telescope.

We designed a range receiver to measure the last pulse received in a pulse train. A 3 inch diameter 9 inch focal length gold-coated off-axis parabolic mirror gathers return radiation from the target and focuses it onto a PD-LD Inc. fiber-coupled

(9 μm core) fast 75 μm diameter InGaAs PIN photodiode. A 160 mW 1490 nm laser diode with fiber pigtail was spliced to a 50:50 beam combiner along with the receiver fiber to serve as a back-propagating bore-sighted alignment laser.

The time of flight is measured by accurately determining the time between the last pulse leaving the laser and its arrival at the range receiver's off-axis parabolic mirror. The laser provides a trigger output signal which precisely indicates when the last pulse is generated. To allow the last pulse in an on-frame pulse train to be distinguished from the others arriving at the receiver, a delay is added between the last full-power pulse and the first mini-pulse. This delay is 120 ns longer than the typical time between pulses while the laser is on (i.e., the pulse repetition time), providing a clear signature indicating the last pulse. This signature allows for automated as well as manual detection of the last pulse in an on-frame.

The data acquisition and measurements result from comparing the transmitted and received optical signals as shown in Fig. 4.49. This figure shows the general method used to measure the time of flight. The camera integration times (shaded regions) repeat at the 30 Hz frame rate. The top row shows the waveform representing the transmitted SCL pulse train which turns on during every other frame's integration time. The middle waveform represents the received optical signal (Chap. 1) that is delayed by approximately twice the time of flight (TOF) to the target (t_{Return}). The bottom waveform represents the output from the laser's trigger output (TO) (Chap. 2). The measured time of flight was determined experimentally by comparing the beginning of the pulse time count (red pulses in Chap. 2) to the

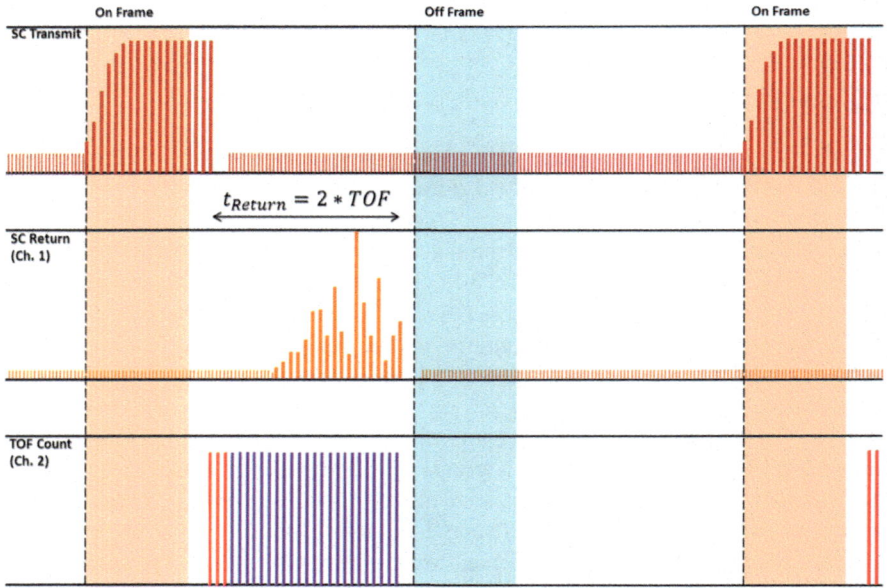

Fig. 4.49 Schematic of timing diagram to compare the measured oscilloscope signals

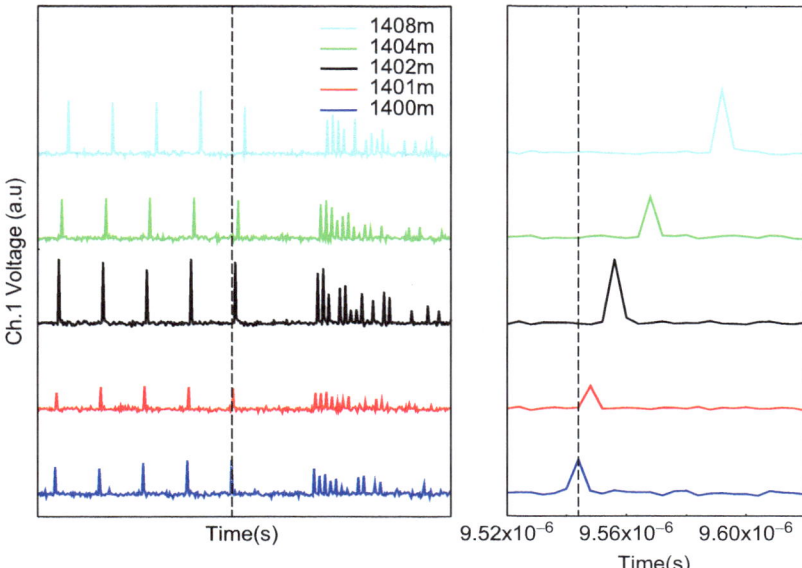

Fig. 4.50 Range return pulse comparisons

final pulse of the received signal (Chap. 1). This could be offset from the true time of flight due to electronic delays and, therefore, must first be calibrated with a known standard (this study used the commercially available range finder with 1 m resolution).

Figure 4.50 shows comparisons between the pulse waveforms measured using the corner cube at different ranges. This figure shows a zoomed out (left) and zoomed in (right) waveform. The end of the laser pulse train was identified by a gap in the pulse train followed by an exponentially decaying higher frequency pulse train. By comparing the shifts in positions of the last pulse of the received laser signal at different ranges, the resolution of the ranging sensor was calculated. It is worth noting that the pulse-to-pulse variability shown in Figs. 4.50 and 4.51 is probably due to sampling artifacts for the very narrow laser pulses.

The measured time of flight versus truth range distance are shown in Fig. 4.51. By comparing these results to the theoretical fit and using the formula

$$\text{TOF} = \frac{2*\text{Range}}{c} + t_{\text{Offset}}$$

gives a $t_{\text{Offset}} = 201.4$ ns. The error bars in the range direction reflect the certainty of the truth range finder that was used to calibrate the slant range distances. The sensitivity of the range measurements was tested experimentally by taking several waveforms using the oscilloscope and measuring the time of the last pulses

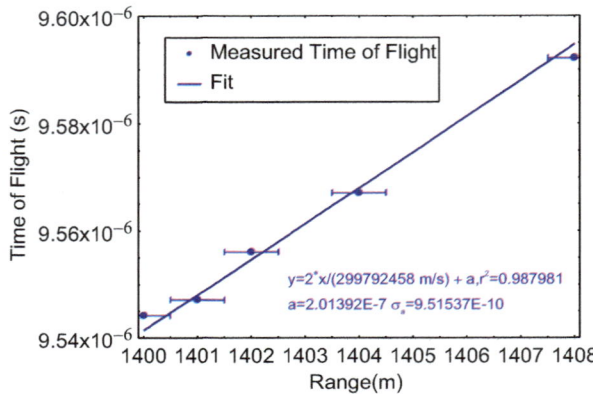

Fig. 4.51 Measured time of flight with fit

maximum value. These results suggested that the sensitivity of the manual method was accurate to within 0.1 ns (∼1.5 cm range resolution), much less than the 1 m resolution of the range finder.

In summary, we have used a SWIR-SCL in field tests at WPAFB to investigate the use of SCL technology to provide active illumination for SWIR hyper-spectral sensors in order to extend their operational window into the night and heavy overcast periods. The SWIR-SCL prototype output 64 W over the continuous wavelength ranges from 1064 to >1800 nm with a wall-plug efficiency of 15.7%, a beam quality that is nearly diffraction limited with $M^2 < 1.3$ over the wavelength range, and output variability or fluctuations of ±0.2%. The SWIR-SCL is an all-fiber-integrated, no-moving-parts laser that leverages mature fiber-optic and telecommunications technologies. The MOPA configuration exploiting the modulational instability mechanism is a simple, cost-effective design and a platform for SC generation over the visible, SWIR, and even mid-wave infrared.

In the field tests at WPAFB, the SWIR-SCL was placed in a 12-story tower, and the beam was directed to targets that were placed on an 8′ × 8′ board on the ground at a distance of 1.4 km or more. The beam quality was measured with a SWIR camera, and the laser beam propagated a distance of 1.4 km with the end diameter of ∼0.9 m. Also, radiometric analysis was performed of different target materials including Tyvek, plywood, blue tarp, and black and gray cloth, and the SNR was estimated for the different materials over the spectral range. In addition, a SWIR camera was placed adjacent to the laser in the tower to measure the diffused reflectance signal from targets placed 1.4 km away. To our knowledge, this was the first ever experiment of two-way propagation of active hyper-spectral imaging illumination over a long distance, and it confirmed the feasibility of the overall concept of using a SWIR-SCL as the broadband illuminator. The 64 W SWIR-SCL has the capability of providing near-sunlight-equivalent illumination over multiple square meters.

Using the SWIR-SCL as an active illuminator for HSI can lead to additional benefits beyond extending the operational hours. Since we can control the light source, modulate the output, and synchronize the light source to a SWIR camera, the SWIR-SCL also enables change detection. In change detection camera frames with the laser off are subtracted from the frames with the laser on, thereby permitting cancelation of at least some of the artifacts associated with the sun illumination, the change in angle of the sun, and shadows. Change detection was performed in the field tests with measurements at a distance of 1.4 km using a board and reflector placed in front of the target at different times throughout the day (from early morning dawn through nighttime). The experiments demonstrate that using the SWIR-SCL during daytime and performing image differencing holds potential for reducing some of the uncertainties involved in HSI data reduction due to unknown environmental irradiance. Our tests verify that it is possible to compete with direct solar irradiance and retrieve the reflected signal using difference imaging.

Moreover, by taking advantage of the pulsed nature of the SWIR-SCL, time-of-flight measurements were used to calculate the distance to the target. In our experiments, due to limitations from the collection optics and sensitivity of the receiver, the field tests used the return signal from a 5 inch corner cube placed on the ground at distances of 1.4 km or more. The experimental results suggest that the resolution of our method was on the order of ~1.5 cm even over the 1.4 km distances. Thus, the SWIR-SCL has the potential to serve as an active illuminator for hyper-spectral imaging, with brightness comparable to the sun in the SWIR, and an HSI system using this laser could include enhancements not possible simply using sun illumination, such as change detection and laser ranging.

4.7.3   Stand-Off Detection of Solid Targets

Diffusion reflection spectroscopy is a widely used technique for both qualitative and quantitative analysis of IR active samples. We demonstrate the use a MWIR SC laser to detect solid targets at a stand-off distance of 5 m using diffuse reflection spectroscopy [48]. The SC source is used to obtain the reflection spectra of a wide range of samples including explosives, fertilizers, and paint coatings. We observe unique spectral fingerprints in the NIR and MWIR wavelength regions of the reflection spectrum. Thus, the SC light source is well suited for spectroscopy applications in a wide range of fields such as defense, homeland security, remote sensing, and geology.

The SC source used for this experiment is described in Fig. 4.34 and the accompanying text. Figure 4.52 shows the experimental setup used in the diffusion reflectance spectroscopy setup. The SC output is first collimated using a parabolic mirror and is incident on a sample kept at ~5 m from the collimating mirror.

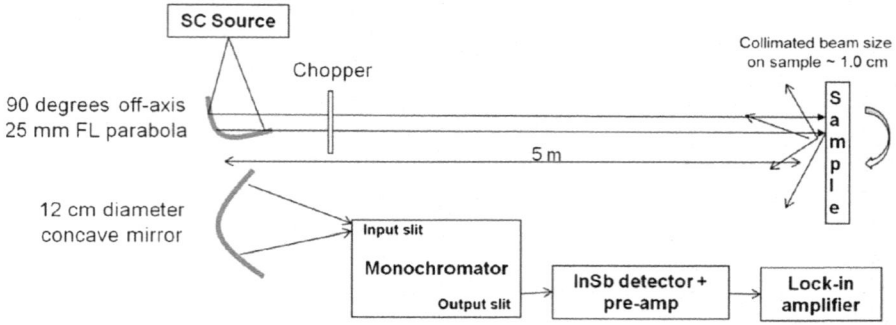

Fig. 4.52 Experimental setup for SC-based stand-off diffuse reflection spectroscopy

The beam is incident normally on the sample but detected at a reflection angle of ~2.3 degrees. A 12 cm diameter concave mirror with a 75 cm focal length is used to collect a fraction of the reflected light from the sample and focus it to the input slit of the monochromator. The output slit is set to 2 mm corresponding to a spectral resolution of 10.8 nm, and the light is detected using a liquid-nitrogen-cooled indium antimonide detector. The chopper frequency is 400 Hz, and the lock-in time constant is set to 100 ms, corresponding to a noise bandwidth of ~1 Hz. The scan is performed across the wavelength region from 1.2 to 4.2 microns. Appropriate wavelength filters were used to avoid contribution from the higher wavelength orders from the grating. The reflection spectrum of a given sample is normalized to the reflectance form Infragold, an electrochemically plated diffuse gold metallic coating with a nearly ideal lambertian scattering profile that is kept at the same distance as the sample. A 45 degree fold mirror was used for samples such as powders that could not be oriented in the vertical plane normal to the collimated SC beam.

For the spectra measurements, our sample consists of 4–8% of explosives, such as TNT, RDX, PETN, and potassium nitrate, on a fused silica substrate. The absorbance spectra measured using the SC laser are shown in Fig. 4.53. In each case, the spectral features are characteristic of the stretching vibrations specific to the molecular makeup of the samples. The strongest absorption features in TNT, RDX, and PETN are in the 3200–2500 nm band and arise due to the fundamental aromatic and aliphatic C-H stretch. The spectral features around 3600 nm in potassium nitrate are due to the first overtone of the N-O asymmetric stretch. The common broad feature at 2720 nm in all four samples is due to the O-H stretch from absorbed water in the fused silica host. In order to verify the validity of our measurement system, we also compare the spectral features with reported measurements in literature and see them to be in good agreement.

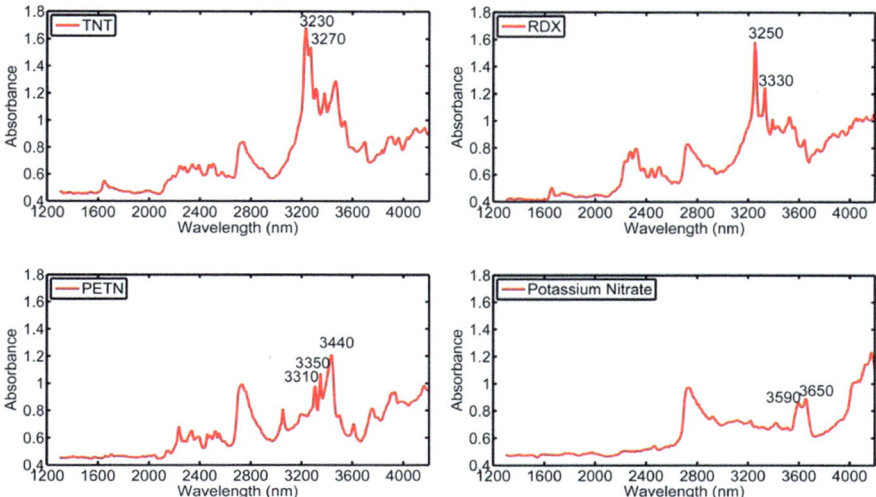

Fig. 4.53 Mid-IR SC absorbance spectra of explosives – TNT, RDX, PETN, and potassium nitrate

4.7.4   3D Printing of Thermoplastics Using SWIR SC Laser

We demonstrate that a SWIR SC laser sinters a wide variety of thermoplastics used in additive manufacturing or 3D printing since the 2–2.5 micron wavelength range corresponds to the combinational bands in most hydrocarbon-based materials. Furthermore, we show that rods created using our SWIR SC laser have a 1.5–5× larger flexural modulus than those made using a CO_2 laser. SWIR SC lasers can increase the number of viable materials for laser sintering and improve the strength of existing materials by creating a more uniform heat profile on the surface of a powder bed, thus serving as a viable light source for 3D printing of thermoplastics [79].

Current challenges in additive manufacturing of plastics include that different light sources are required for different plastics, and the 3D printed structures lack laminar strength due to the layer-by-layer manufacturing. For example, CO_2 lasers are used to sinter nylon materials, while UV light is used for polyethylene. Also inferior mechanical properties of sintered polyether ether ketone (PEEK), high-density polyethylene (HDPE), and polypropylene can be attributed to porous structures resulting from poor layer adhesion [79].

Common to almost all of the thermoplastics used in 3D printing is that they contain hydrocarbons, so they have SWIR absorption bands from their C-H combinational bands. Figure 4.54 shows this range of absorption peaks from 2.15 to 2.5 μm in Nylon 12, PEEK, and HDPE, which is exemplary of the C-H combinational band in thermoplastics. The range of penetration depths associated

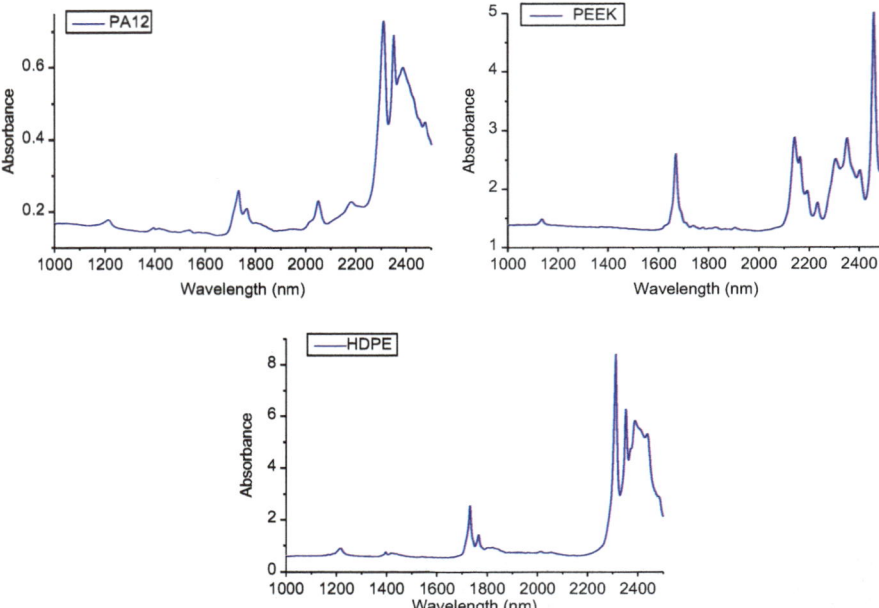

Fig. 4.54 Representative absorption spectra depicting C-H Combinational bands in the 2.15–2.5 μm range and first overtone near 1.7 μm

with these absorptions allows a laser to more uniformly heat sintering zones, thus creating better adhesion between layers and leading to increased laminar strength in sintered structures.

The setup for testing different photopolymers uses a variety of lasers whose light are collimated and then focused onto a powder bed, as shown in Fig. 4.56. To benchmark the performance for different cases, four light sources are used for plastic machining. First, we use an all-fiber-based SWIR SC laser utilizing a seed laser, two-stage amplification, and a fiber for nonlinear spectral broadening to output 25 W in the 2–2.5 μm wavelength range. Further details are provided in [47]. Next, a band-pass filter is placed in front of the SWIR output to give the filtered SC source with a 2320–2380 nm wavelength range. Third, we use a 30 W, 10.6 μm CO_2 laser (Diamond C-30, Coherent) to benchmark our results. Finally, we use a 12 W, 1685 nm diode laser (BriteLase 6017, QPC Lasers) to inspect the first overtone absorption peak near 1700 nm (Fig. 4.54). A wide array of thermoplastics are sintered with the SWIR SC laser. We target Nylon 11, Nylon 12, polystyrene, polycaprolactone (PCL), poly(glycerol-dodecanoate), polyvinyl alcohol (PVA), PEEK, HDPE, low-density polyethylene, polylactic acid, and poly-l-lactic acid.

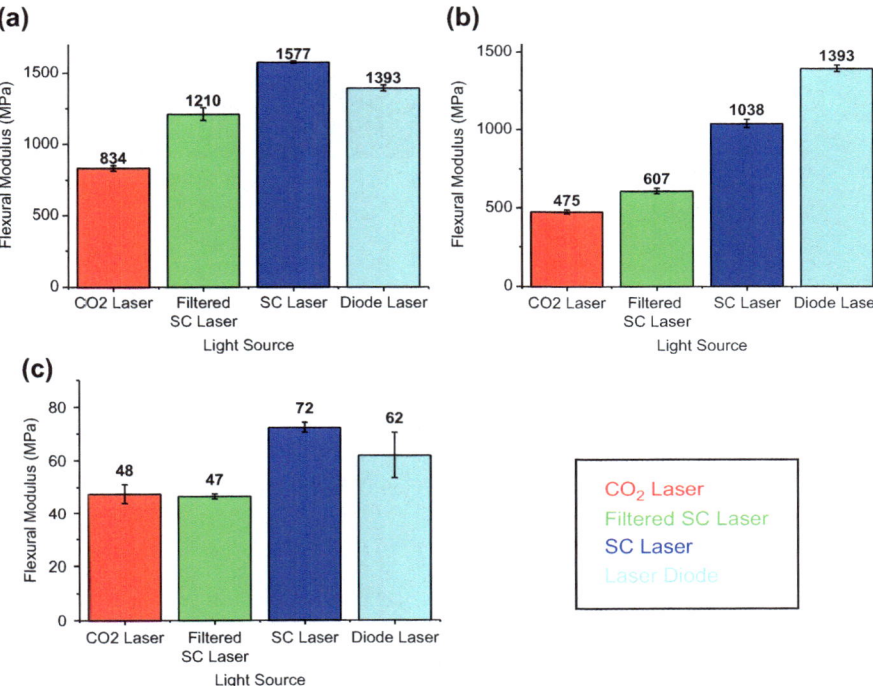

Fig. 4.55 Flexural moduli of rods created from (**a**) Nylon 12, (**b**) Nylon 11, and (**c**) PCL powders

Rods are created from Nylon11, Nylon 12, PEEK, PVA, and PCL powder using all four light sources and tested for flexural properties. The rods are sintered by placing each powder bed on a computer controlled X-Y translational stage (Fig. 4.56). To achieve a maximum rod density without thermal degradation, the laser power for each source is tuned to just below the damage threshold at a translational speed of 1064 μm/s. Three sets of ten rods are made from each powder/light source combination, and five rods from each set are chosen randomly for testing. The flexural modulus is calculated from the results of 3 pt. bend tests run on an MTS machine.

We compare the flexural moduli of rods created from the five powders. Figure 4.55 depicts our results from (A) Nylon 12, (B) Nylon 11, and (C) PCL rods. From left to right, the columns in each graph represent the average flexural moduli of rods created with the CO_2 laser, the filtered SC laser, the SWIR SC laser, and, finally, the diode laser. In all samples we see that rods created with the SWIR SC laser have a 1.5–5× larger flexural modulus than rods created with the CO_2 laser. Furthermore, filtered SC sintered rods consistently exhibit flexural moduli between CO_2 and SC sintered flexural moduli.

To elucidate the trade-offs between power and strength, Fig. 4.57 plots the flexural modulus versus the power used for sintering in Nylon 11 and Nylon 12.

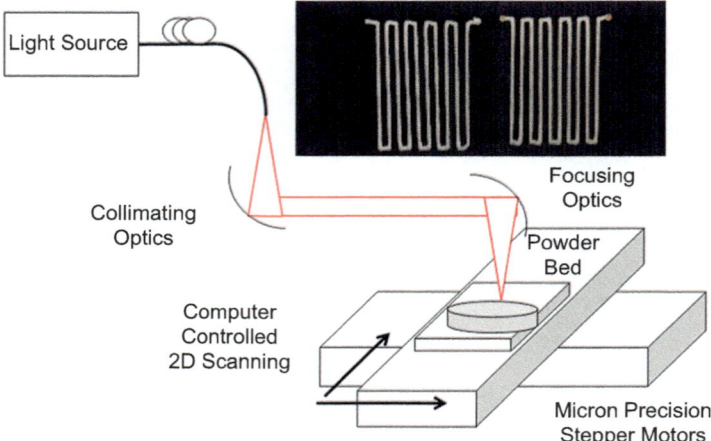

Fig. 4.56 Experimental setup and rods made in Nylon 11 and Nylon 12

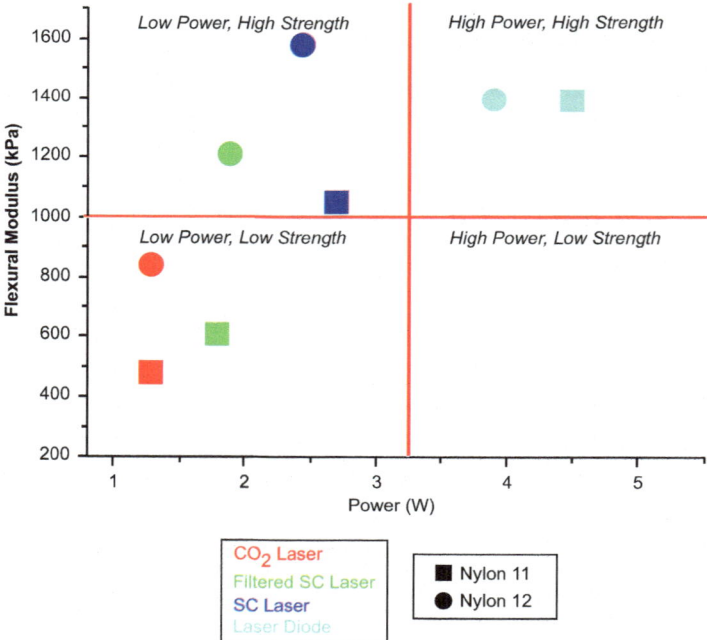

Fig. 4.57 Flexural modulus vs. sintering laser power in nylons

The powers are chosen to be just below the power at which thermal degradation would begin to occur given our sintering parameters. In Fig. 4.57 we see that the CO_2 laser requires the lowest power but produces the weakest rods, while the diode laser produces strong structures with a much higher power requirement. SWIR SC rods require moderate power but at the same time have a high flexural modulus. Depending on your system requirements, you can use a different laser for different tasks. If minimal incident power is preferable, choose a CO_2 laser. If power is less of a concern and high strength is needed, it is worthwhile to choose a SWIR or diode laser for sintering.

Our results show that the SWIR SC is a viable source for thermoplastic 3D printing because it can sinter thermoplastics and create stronger materials by combining the characteristics seen in the CO_2 and 1685 nm diode lasers. All 11 materials tested can be sintered via their absorptions in the combinational band (c.f. Fig. 4.54). We believe that the increased strength in SC samples is due to a range of penetration depths within this band. The CO_2 laser's light at 10.6 μm is calculated to penetrate $\sim$100 μm in nylons, whereas SC light penetrates 2−6× deeper. High absorption wavelengths, such as 10.6 μm, promote efficient sintering. On the other hand, overtone band wavelengths near 1685 nm provide deep penetration and uniform heating. The SWIR SC laser combines both these features to produce parts with high strength at a low required power as shown in Fig. 4.57.

In summary, we have sintered 11 different thermoplastics and found that up to a fivefold increase in flexural modulus can be achieved using a SWIR SC laser over a CO_2 laser. Thus, our data supports that the SWIR SC laser can be a viable light source for the laser sintering of thermoplastics. Furthermore, because the SWIR SC laser is similar to fiber lasers used in metal powder sintering, the SWIR SC fiber laser should share similar reliability traits.

4.7.5   Medical: Atherosclerotic Plaque Detection and Preferential Damage of Lipids

The SC source used for this experiment is shown in Fig. 4.32. Figure 4.58 shows the experimental setup used and consists of three main parts: the SC laser, the MWIR spectroscopy, and the laser ablation [111]. It should be noted that although the SC laser has been demonstrated with up to 10.5 W average output power, the spectroscopy measurements and laser damage are conducted with only a few of the pump laser diodes on and $\sim$1.5 W SC output power. For the absorption spectroscopy measurements, the SC is first collimated using a parabolic mirror and then split into one signal arm that passes through the sample under test and one reference arm. The two arms are modulated by choppers at different frequencies, and the recombined light is coupled into a grating-based monochromator, where the signal is collected by a liquid-nitrogen-cooled indium antimonide detector. For the preferential damage setup, shown in Fig. 4.58b, the SC light for targeting lipids is chosen by filtering out light below 3.2 microns using a long-pass filter. The light is then focused by a 25.4 mm calcium fluoride lens to a spot size of $\sim$3 mm on to the sample. The spectroscopy and the selective ablation are performed in vitro on the tissue and cultured cells, respectively.

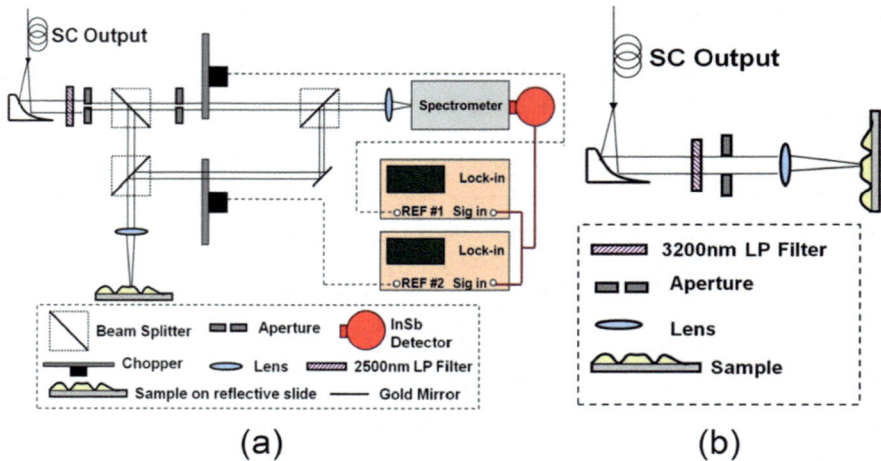

Fig. 4.58 (a) Reflection-absorption spectra measurement setup. (b) Preferential laser ablation setup

MWIR SC absorption spectra of the components of normal artery, which includes endothelial cells and smooth muscle cells, are illustrated in Fig. 4.59a, b. The broad absorption feature from 2.8 to 3.2 microns with a peak at $\sim$3.05 microns is observed and can be attributed to the vibrational bands of O-H stretching in the hydroxyl group and N-H stretching present in the protein amino acid. The spectra for the constituents of atherosclerotic plaque, including macrophages, adipose tissue, and foam cells, are also illustrated in Fig. 4.59. In the lipid-rich samples such as the adipose tissue and foam cells, we can distinguish the absorption lines in the 3.2–3.6 microns window, for example, the C-H stretching vibration at $\sim$3.33 microns, CH_3 stretching vibration at $\sim$3.39 microns, and CH_2 stretching vibration at $\sim$3.42 microns and $\sim$3.510 microns. It is worth noting that while the macrophages exhibit a similar absorption spectrum as compared to the normal artery cells, prominent spectral characters between $\sim$3.2 and 3.6 microns with two absorption peaks at $\sim$3.42 and 3.51 microns are observed in the macrophage-transformed foam cells and adipose tissue absorption spectra, revealing the pathological relationship between these two cell types, i.e., active macrophages engulfing lipid-rich substance to become foam cells.

Preferential damage takes advantage of the differential absorption of different tissues where by the targeted tissue is preferentially heated to a temperature lethal to cells by laser radiation within the corresponding signature absorption band. We hypothesize that by using laser radiation targeting the lipid absorption band from 3.2 to 3.6 microns, we can heat adipose tissue to a damage threshold faster than the artery tissue. Preferential heating (or thermal modification) of adipose tissue without damaging artery tissue is observed under light microscopy (Fig. 4.60) with an SC laser fluence starting as low as $\sim$15 mJ/mm^2, while no damage was observed to the

Fig. 4.59 Mid-IR SC laser-based reflection-absorption spectra of normal artery compositions, (**a**) endothelial cells and (**b**) smooth muscle cells, and atherosclerotic plaque constituents: (**c**) macrophages, (**d**) adipose tissue, and (**e**) foam cells

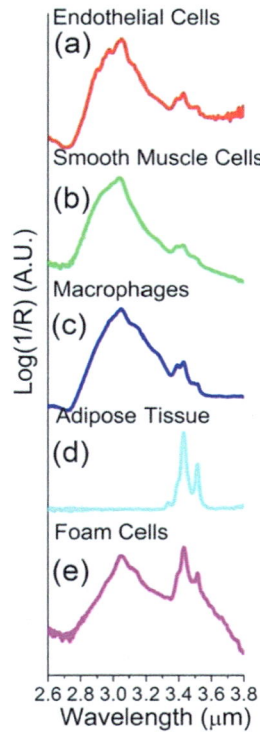

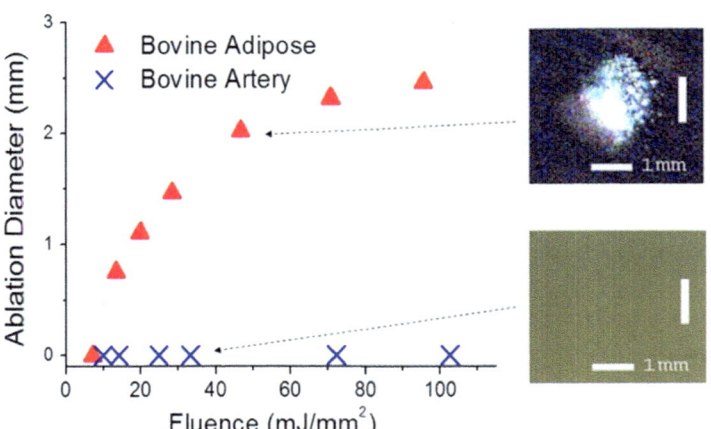

Fig. 4.60 Power dependence of the damage diameter in adipose tissue and bovine artery. Right, images showing the difference before and after laser damage at ∼40 J/mm²

artery up to ∼100 mJ/mm². The beam diameter used is ∼3 mm and the exposure time used is 5 s. Figure 4.60 also shows the image subtraction of before and after laser treatments to identify damage of the adipose and artery tissues. For the adipose tissue, the damaged region is in bright color, while the unaffected region is dark color. Histochemistry study was also performed to better understand the biological integrity and damage to the artery, and we observed that while laser-induced damage can be observed at a laser fluence of ∼102 mJ/mm², no such effects are present at ∼33 mJ/mm², and very subtle wounds exist at ∼72 mJ/mm².

Therefore, with the dual capability of locating atherosclerotic plaque and differential damage, the MWIR SC laser could potentially be used to diagnose atherosclerosis disease. Treatment might be accomplished in a catheter-based, minimally invasive procedure that could be an add-on to the angiogram (i.e., use the guide wire already in place to insert the MWIR catheter. Obviously, considerable work remains to explore the possibility of using this technology in clinical settings.

4.7.6 Medical: High SNR Glucose Monitoring Using SWIR SC Laser

We demonstrate that low-concentration glucose solutions down to 1 mg/dL can be measured using an all-fiber-integrated SWIR SC laser as the light source (e.g., Fig. 4.18) in a balanced arm spectroscopy system [112]. Measuring glucose concentrations in aqueous solutions down to 1 mg/dL is proposed as the first step in a "test of technologies" toward a viable noninvasive glucose monitoring approach in Ref [113]. To our knowledge, no other noninvasive technique has been able to detect glucose in aqueous solution at this concentration, and almost all infrared spectroscopic techniques to date have relied on tungsten or halogen lamps. We estimate that our SWIR SC laser improves the system SNR by at least 50 times compared to lamps due to the high brightness reaching the detector without saturation, even after passing through a highly absorbed background such as water. In addition, the glucose combinational bands from ∼2000 to ∼2400 nm are used, since they fall in water absorption minimum valleys without interference from hemoglobin features (e.g., FTIR spectra in Fig. 4.61). Therefore, a SWIR SC could be a key enabling technology for noninvasive glucose monitoring.

The spectroscopy system used for the measurements is a balanced arm spectroscopy system with a SC laser as the light source shown (Fig. 4.62). The SWIR SC is an all-fiber-integrated SC laser covering the 1500–2400 nm wavelength range with an output up to 5 W (Fig. 4.18).

The system in Fig. 4.62 is designed to achieve the high repeatability and SNR required for detecting 1 mg/dL glucose solution. The light from the SWIR SC enters a balanced arm subsystem followed by a detection subsystem. The wavelength of light is selected by sending the SWIR SC output to a monochromator with ∼5 nm output bandwidth. Only ∼250 mW optical power is sent into the spectrometer and

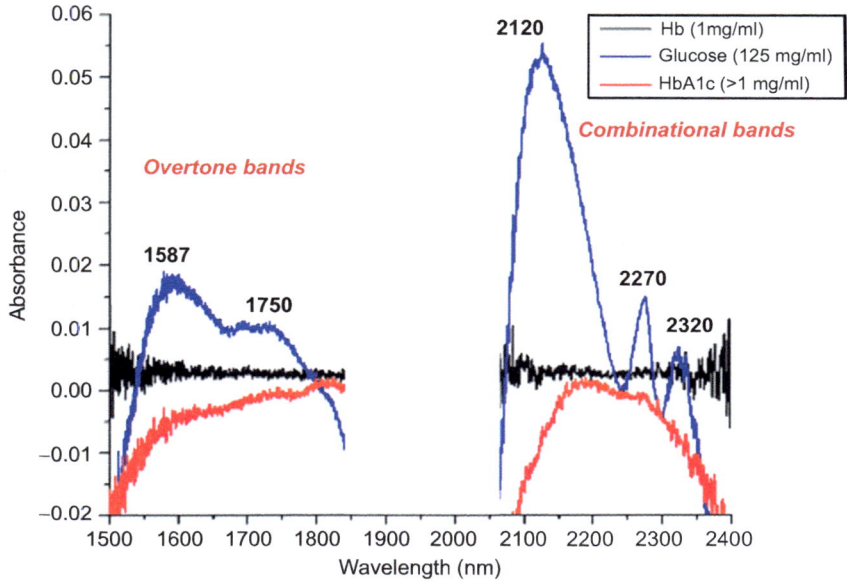

Fig. 4.61 Glucose absorption in first overtone and combinational bands

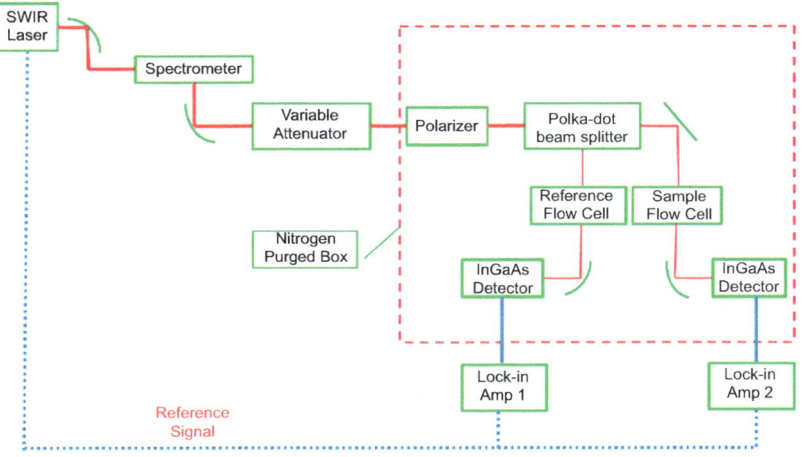

Fig. 4.62 Schematic diagram of SC spectroscopy system

∼50 mW after the attenuator in the experiments to prevent detector saturation. The balanced arm subsystem comprises a broadband beam splitter and two flow cells used for sample and reference solutions, respectively. A matched pair of InGaAs detectors is used for light detection, and data is collected through two lock-in amplifiers synchronized to the laser output. The SNR at highest intensity

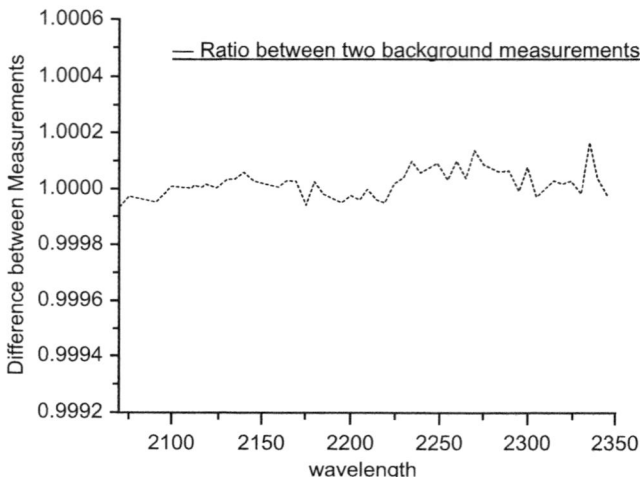

Fig. 4.63 Difference between two background measurements

is measured to be ∼47 dB, and the repeatability of the system is within −40 dB, as tested by comparing the spectral difference between two consecutive background measurements (shown in Fig. 4.63).

A series of measurements are done on glucose aqueous solutions of different concentrations. The concentrations used in this test are 100 mg/dL, 50 mg/dL, 25 mg/dL, 10 mg/dL, 5 mg/dL, 2.5 mg/dL, and 1 mg/dL. The SWIR spectrums are taken from 2000 to 2400 nm with a 5 nm step. Five spectrums are recorded and averaged at each concentration. At the beginning of the experiment, background spectrums are taken by injecting deionized (DI) water into both sample and reference cells. During the experiment, glucose solutions, starting with the 1 mg/dL solution, are injected into the sample cell, while DI water is injected into the reference cell. The time taken for one measurement is ∼10 min due to the slow mechanical scanning rate of the monochromator. The acquired data is processed by standard spectral processing techniques, smoothing, and baseline correction. Amplitude drift between measurements is corrected by aligning spectrum to a known absorption valley at 2240 nm.

Exemplary glucose spectrums taken at different concentrations are shown in Fig. 4.64. Not shown here is the 2.5 and 1 mg/dL spectrum, since their amplitude is too low to be shown in this scale. The first (2120 nm) and second (2280 nm) absorption peaks are very distinct, although the spectrum features after 2300 nm show background fluctuations caused by temperature dependence of water absorption. Also, we believe the etalon effects in the spectrum may be due to reflections in the detector cover glass.

To validate that our system is properly measuring glucose, we verify that the absorption at the first two absorption peaks are linearly proportional to glucose concentration, as predicted by Beer's Law. A linear fitting analysis was conducted

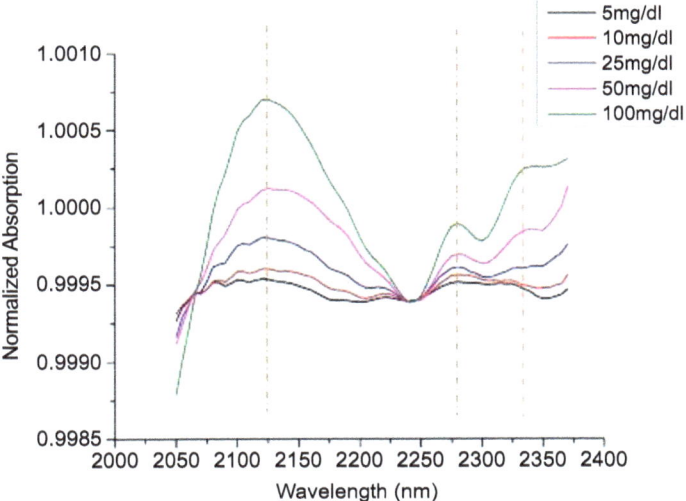

Fig. 4.64 Measured glucose spectra of different concentrations

on the peak value of the first and second glucose absorption peaks of lower concentration solutions (1–50 mg/dL), and the highly linear relationship (Fig. 4.65) between concentration and absorption confirms that we are measuring glucose. Also note that at the lower concentrations, the linearity is maintained although the absorption becomes negative because of the temperature dependence of the large water absorption background.

We show that infrared spectroscopy of glucose can be done in the combinational bands, where water absorption has a local minimum and hemoglobin is featureless. With the high brightness provided by the SWIR SC, it is possible to detect glucose solution absorption down to 1 mg/dL, which corresponds to absorption changes in the fourth to fifth decimal place. A properly balanced detection setup is necessary, since at such low signal level, any environmental fluctuation such as temperature or humidity could distort the measured signal (c.f. Fig. 4.64). The SNR upper limit is set by the maximum SNR of the detection electronics and detector saturation power, not by the input optical power. Our estimate of better than 50× improvement (>47 dB) by using the SWIR SC compared with a halogen or tungsten lamp is based on considering not only the laser power but also the detector saturation.

In summary, we measure glucose concentrations down to 1 mg/dL in aqueous solution in a balanced arm spectroscopy system using a SWIR SC, which improves the SNR of the setup by >50× compared to lamps. We show linearity of the absorption strength of the C-H combinational band features with glucose concentration,

Fig. 4.65 (a) Linear
relationship between
absorptivity and glucose
solution concentration at
2120 nm. (b) Linear
relationship between
absorptivity and glucose
solution concentration at
2280 nm

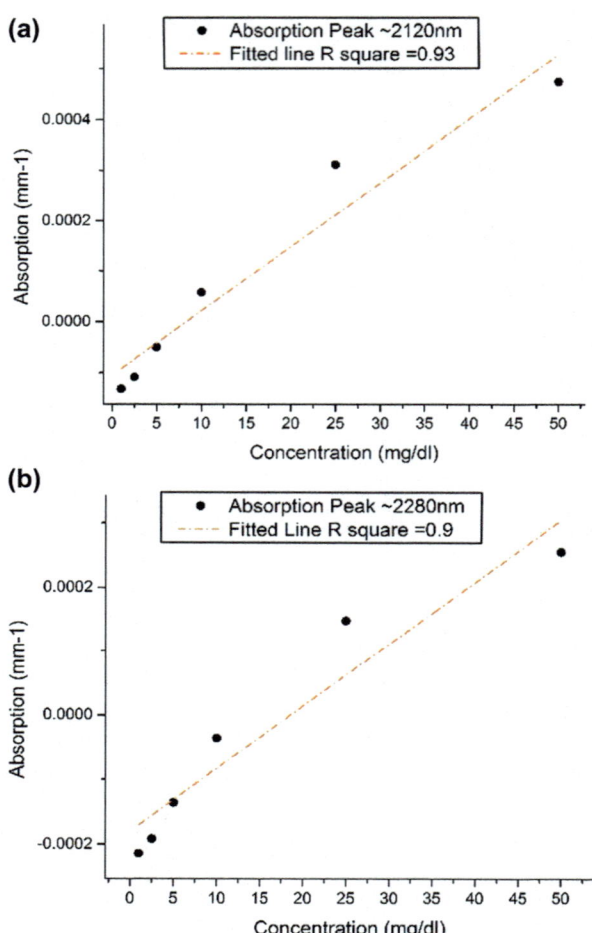

4.8 Summary

In this chapter we have focused on MI-initiated SC generation. Whereas many of the other approaches on SC generation have used mode-locked lasers or optical parametric oscillator or amplifier systems, MI-initiated SC light sources use the natural physics of the fiber to obtain the necessary pulse generation. We describe a simple SC light source architecture (c.f., Fig. 4.1) that starts from a seed laser diode, amplifies the output using one or more stages of fiber-optic amplifiers, and then adds a short length of anomalous dispersion fiber for MI initiation. Finally, the light is coupled to a length of nonlinear fiber for SC generation. The power amplifier may be an ytterbium-doped fiber, erbium/ytterbium-doped fiber, or thulium fiber, while the SC generation fiber may be a fused silica fiber, a fluoride fiber (e.g., ZBLAN or indium fluoride), a tellurite fiber, a chalcogenide fiber (e.g., sulfide or selenide), a photonic crystal fiber, or some combination of these fibers. Some of the advantages of the MI-initiated SC architecture include:

- All-fiber-integrated, no moving parts. This means that the laser has the potential to be compact and robust, with high reliability and low maintenance cost.
- Leverage mature fiber-optic and telecommunications technologies. All of the parts used in the laser are COTS – commercial off-the-shelf components. This means the price will continue to decrease and the performance will continue to increase due to a large market requiring the components.
- Simple, cost-effective design. This design does not use mode-locked lasers and can therefore have relatively low parts costs. Also, the cost of assembly can also be potentially low, since all the parts are fiber based and can be spliced together with commercially available splicers.
- Platform for SC over the SWIR, mid-wave infrared, and visible. In fact, for properly selected parameters, the long-wavelength edge of SC can fill the transparency band of the particular fiber

In this chapter, after a review of the literature in Sect. 4.2, we investigated details of the MI-initiated SC generation process in Sect. 4.3. MI is a parametric amplification in a fiber where the nonlinear index of refraction enters into the phase matching process. In particular, when we launch a quasi-CW (typically $\sim$0.5–2 ns wide, although it can be anything over roughly 30 ps) into a fiber operating in the anomalous group velocity dispersion regime at high powers (e.g., typically in the kilowatts range), through a combination of group velocity dispersion and nonlinearity, the quasi-CW wave breaks up into a number of soliton pulses. The pulses generated by this mechanism lead to SC generation, obviating the need for mode-locked or other short-pulse lasers. The system is further simplified by using a two-step SC generation process: pulse breakup in a single-mode fiber through the MI process followed by spectral broadening in a nonlinear fiber or waveguide. Beyond the simplicity of the resulting system, we also greatly benefit from expansion into the longer wavelengths through the Raman effect. In general, for sufficient fiber length and pulse intensity, we find that SC generation can extend through the low-loss, transparency window of the fiber.

Section 4.4 describes visible, NIR, and SWIR SC experiments in fused silica fiber. The beauty of the all-fused silica fiber configuration is that everything can be spliced together and the fibers have a very high damage threshold while also being environmentally stable. Using a short length of HiNL fiber, SC light extending out to $\sim$3 microns is demonstrated with a time-averaged output power of 5.3 W. Also, a 5 W, all-fiber-integrated SC laser prototype spanning the SWIR wavelength band from $\sim$1.55 to 2.35 microns is demonstrated and used in field tests at WPAFB. In addition, a simple, VIS-NIR-SWIR SC laser is described using commercial off-the-shelf components. Using solid-core fiber, the wavelength spans from $\sim$0.67 to 2.47 microns. By adding a length of PCF to the end of the solid-core fiber, the short-wavelength edge can be moved down to $\sim$0.47 microns, while the long-wavelength edge goes out to $\sim$2.47 microns. Finally, a high-powered SC prototype is described that is used in two-way active remote sensing experiments. The SWIR SC laser delivers 64 W time-averaged output power over the continuous wavelength range from 1064 to >1800 nm with a wall-plug efficiency of 15.7%, an average fluctuation of ±0.2%, and a beam quality that is nearly diffraction limited with $M^2 < 1.3$ over the wavelength range.

MWIR SC sources using fluoride fibers are described in Sect. 4.5. Since fused silica fibers do not transmit beyond ∼2.7–3 microns, it is necessary to use softer glasses, such as fluorides, tellurites, or chalcogenides. This section focuses on experiments using ZBLAN fluoride fibers. First, in a relatively high-power setup, the average power is scaled up to ∼10.5 W in an SC extending from ∼0.8 to 4 microns. Next, the spectrum is pushed out further to ∼4.3 microns by increasing the power-amp output peak power to ∼20 kW. Modulation capability is also demonstrated, where the entire system and the SC are modulated with a 500 Hz square wave at 50% duty cycle, without the need for any external modulation or chopping mechanism. Finally, an SC laser is presented specifically for higher efficiency in generating MWIR wavelengths (> than ∼3.8 microns) by incorporating a thulium-doped fiber amplifier as the power-amp stage, which produces a MWIR SC out to ∼4.5 microns.

Extending the long-wavelength edge further into the MWIR and LWIR is explored in Sect. 4.6. Whereas ZBLAN-based fluoride fibers limit the edge to ∼4.5 microns, other types of fluorides, tellurites, and chalcogenides have shown expansion of SC all the way to ∼13.3 microns. For expanding to ∼5.5 microns, three fiber options that have been demonstrated are tellurites, indium fluoride, and sulfides. As an example, a PCF made from tellurite has generated SC from 0.79 to 4.87 microns. Using indium fluoride fibers, SC has been demonstrated from 2.4 to 5.4 microns with a maximum average power of ∼10 mW. Moreover, using sulfide chalcogenide fibers, two demonstrations have been reported: one with a range of 2.3–4.9 microns by tuning the pump wavelength and with a maximum output power of 550 mW and a second all-fiber SC source covering 1.9–4.8 microns with total output power of 565 mW. To further extend the long-wavelength edge into the LWIR requires the use of yet softer glasses, such as selenide-based chalcogenides. Using high numerical aperture selenide fibers, a wavelength range of 1.4–13.3 microns has been reported, although all of the selenide-reported results produce relatively low output power and require complicated pumping schemes. However, the field is rapidly growing, and in the near future, these limitations may be solved by borrowing techniques from SC generation in the SWIR and MWIR discussed herein.

Finally, Sect. 4.7 surveys some of the applications demonstrated for SC light sources both in the SWIR and the MWIR. Several key attributes of the SC light sources are a continuous spectrum covering several octaves of light, spatially coherent beam that can propagate nearly transform limited, and the ability of the pulsed laser to be modulated and synchronized. One application for SWIR SC light sources is in active remote sensing and hyper-spectral imaging. Field trials using SWIR SC sources were performed at Wright-Patterson Air Force Base. Using a 5 W SWIR source, targets a mile away were identified when the detectors were placed near the target. When the power is increased to 64 W, the diffuse reflected signal could be detected by a sensor or camera adjacent to the SC laser. The MWIR

SC also overlaps the vibrational and rotational resonances of many solids, and this feature can be exploited for stand-off detection of solid targets such as explosives and other chemicals.

The SWIR and MWIR SC sources can also conveniently target hydrocarbon bonds, either the fundamental resonance (MWIR) or the overtone and combinational bands (SWIR). Taking advantage of this feature, in another application, hydrocarbon-based thermoplastics used in additive manufacturing or 3D printing can be sintered using SC sources operating between 2 and 2.5 microns. In the healthcare industry, the MWIR and SWIR SC sources can be a useful tool to perform diagnostics and therapeutics. For example, a MWIR SC source has been used to perform atherosclerotic plaque detection and to cause preferential damage to lipids, which are a main component of such plaques. Another example of medical diagnostics is the use of SWIR SC sources to perform noninvasive glucose monitoring. For instance, glucose solutions are measured down to 1 mg/dL based on the C-H combinational bands, and it is estimated that the SWIR SC increased the measurement system signal-to-noise ratio by >50× compared with lamps.

Although SC generation is a relatively old field dating back to the 1960s [1], in the past decade, SC sources have become more practical and may enable important commercial applications. In this chapter we have described a very simple SC light source architecture that uses the nature physics in the fiber (MI initiation) to enable the use of all commercial off-the-shelf components. Moreover, the MI-initiated SC can be constructed into an all-fiber-integrated system that is cost-effective and reliable due to no moving parts. The applications described in this chapter demonstrate the potential, but they probably represent only the tip of the iceberg. For example, SWIR and MWIR sources can be used for spectroscopy and to identify targets or objects based on their chemical composition. Applications for these sources range from defense/homeland security to product inspection and in-line process control for many arenas in agriculture, pharmaceuticals, petroleum and natural gas, etc., all the way to healthcare and biomedical diagnostics and therapeutics. So, with the promising developments described in this chapter, the readers should stay tuned for new demonstrations and applications. We are optimistic that the best is yet to come.

References

1. Alfano, R.R.: The Supercontinuum Laser Source: Fundamentals with Updated References. Springer, New York (2006)
2. Alfano, R.R.: The ultimate white light. Sci. Am. 86–93 (2006)
3. Islam, M.N.: Ultrafast Fiber Switching Devices and Systems. Cambridge Studies in Modern Optics 12, Cambridge university press, NewYork (1992)
4. Xia, C., Xu, Z., Islam, M.N., Terry, J.F.L., Freeman, M.J., Zakel, A., Mauricio, J.: 10.5 W time-averaged power mid-IR supercontinuum generation extending beyond 4 um with direct pulse pattern modulation. IEEE J. Sel. Top. Quantum. Electron. 15, 422–434 (2009)
5. Vinay, V.A., Kulkarni, O.P., Kumar, M., Xia, C., Islam, M.N., Terry Jr., F.L., Welsh, M.J., Ke, K., Freeman, M.J., Neelakandan, M., Chan, A.: Modulational instability initiated high power all-fiber supercontinuum lasers and their applications. Optical Fiber Technology. 18, 349–374 (2012)

6. Champert, P.A., Couderc, V., Leproux, P., Tombelaine, F.S.V., Labonte, L., Roy, P., Froehly, C., Nerin, P.: White-light supercontinuum generation in normally dispersive optical fiber using original multiwavelength pumping system. Opt. Express. **12**, 4366–4371 (2004)

7. Stone, J.M., Knight, J.C.: Visibly "white" light generation in uniform photonic crystal fiber using a microchip laser. Opt. Express. **16**, 2670–2675 (2008)

8. Wadsworth, W.J., Joly, N., Knight, J.C., Birks, T.A., Biancalana, F., Russel, P.: Supercontinuum and four-wave mixing with Q-switched pulses in endlessly single-mode photonic crsytal fibres. Opt. Express. **12**, 299–309 (2004)

9. Xiong, C., Leon-Saval, W.A.S.G., Birks, T.A., Wadsworth, W.J.: Enhanced visible continuum generation from a microchip 1064 nm laser. Opt. Express. **14**, 6188–6193 (2006)

10. de Matos, C.J.S., Kennedy, R.E., Popov, S., Taylor, J.R.: 20-kW peak power all-fiber 1.57-um source based on compression in air-core photonic bandgap fiber, its frequency doubling, and broadband generation from 430 to 1450 nm. Opt. Lett. **30**, 436–438 (2005)

11. Kumar, M., Xia, C., Ma, X., Alexander, V.V., Islam, M.N., Terry, J.F.L., Aleksoff, C.A., Klooster, A., Davidson, D.: Power adjustable visible supercontinuum generation using amplified nanosecond gain-switched laser diode. Opt. Express. **16**, 6194–6201 (2008)

12. Rulkov, A., Vyatkin, M., Popov, S., Taylor, J., Gapontsev, V.: High brightness picosecond all-fiber generation in 525-1800nm range with picosecond Yb pumping. Opt. Express. **13**, 377–381 (2005)

13. Schreiber, T., Limpert, J., Zellmer, H., Tunnermann, A., Hansen, K.: High average power supercontinuum generation in photonic crystal fibers. Opt. Comm. **228**, 71–78 (2003)

14. Kudlinski, A., George, A.K., Knight, J.C., Travers, J.C., Rulkov, A., Popov, S., Taylor, J.: Zero dispersion wavelength decreasing photonic crystal fibers for ultraviolet-extended supercontinuum. Opt. Express. **14**, 5715–5722 (2006)

15. Moon, S., Kim, D.Y.: Generation of octave-spanning supercontinuum with 1550 nm amplified diode-laser pulses and a dispersion-shifted fiber. Opt. Express. **14**, 270–278 (2006)

16. Nicholson, J.W., Yablon, A.D., WEstbrook, P.S., Feder, K.S., Yan, M.F.: High power, single mode, all-fiber source of femtosecond pulses at 1550 nm and its use in supercontinuum generation. Opt. Express. **12**, 3025–3044 (2004)

17. Abeeluck, A.K., Headley, C.: Continuous-wave pumping in the anomalous- and normal dispersion regimes of nonlinear fibers for supercontinuum generation. Opt. Lett. **30**, 61–63 (2005)

18. Xia, C., Kumar, M., Cheng, M.-Y., Kulkarni, O.P., Islam, M.N., Galvanauskas, A., Terry, J.F.L., Freeman, M.J., Nolan, D.A., Wood, W.A.: Supercontinuum generation in silica fibers by amplified nanosecond laser diode pulses. IEEE J. Sel. Top. Quant. Elec. **13**, 865–871 (2007)

19. Cumberland, B.A., Travers, J.C., Popov, S.V., Taylor, J.R.: 29 W High power CW supercontinuum source. Opt. Express. **16**, 5954–5962 (2008)

20. Chen, K.K., Alam, S.-u., Price, J.H.V., Hayes, J.R., Lin, D., Malinowski, A., Codemard, C., Ghosh, D., Pal, M., Bhadra, S.K., David, J.: Richardson, Picosecond fiber MOPA pumped supercontinuum source with 39 W output power. Opt. Express. **18**, 5426–5432 (2010)

21. Hagen, C.L., Walewski, J.W., Sanders, S.T.: Generation of a continuum extending to the midinfrared by pumping ZBLAN fiber with an ultrafast 1550-nm source. IEEE Photonics Tech Lett. **18**, 91–93 (2006)

22. Sanghera, J.S., Shaw, L.B., Florea, C.M., Pureza, P.C., Nguyen, V.Q., Gibson, D., Kung, F.H., Aggarwal, I.D.: Non-Linearity in Chalcogenide Glasses and Fibers and their Applications, Quantum Electron. Laser Sci. Conf. (QELS), San Jose (2008.) Paper QTuL5

23. Omenetto, F.G., Wolchover, N.A., Wehner, M.R., Ross, M., Efimov, A., Taylor, A.J., Kumar, V.V.R.K., George, A.K., Knight, J.C., Joly, N.Y., Russell, P.S.J.: Spectrally smooth supercontinuum from 350 nm to 3 μm in sub-centimeter lengths of soft-glass photonic crystal fibers. Opt. Express. **14**, 4928–4934 (2006)

24. Domachuk, P., Wolchover, N.A., Cronin-Golomb, M., Wang, A., George, A.K., Cordeiro, C.M.B., Knight, J.C., Omenetto, F.G.: Over 4000 nm bandwidth of Mid-IR supercontinuum generation in sub-centimeter segments of highly nonlinear tellurite PCFs. Opt. Express. **16**, 7161–7168 (2008)
25. Qin, G., Yan, X., Kito, C., Liao, M., Chaudhari, C., Suzuki, T., Ohishi, Y.: Ultrabroadband supercontinuum generation from ultraviolet to 6.28 μm in a fluoride fiber. Appl. Phys. Lett. **95**, 161103 (2009)
26. Xia, C., Kumar, M., Kulkarni, O.P., Islam, M.N., Terry, J.F.L., Freeman, M.J., Poulain, M., Maze, G.: Mid-infrared supercontinuum generation to 4.5 um in ZBLAN fluoride fibers by nanosecond diode pumping. Opt. Lett. **31**, 2553–2555 (2006)
27. Xia, C., Kumar, M., Cheng, M.-Y., Hegde, R.S., Islam, M.N., Galvanauskas, A., Terry, J.F.L., Poulain, M., Maze, G.: Power scalable mid-infrared supercontinuum generation in ZBLAN fluoride fibers with up to 1.3 W time-averaged power. Opt. Express. **15**, 865–871 (2007)
28. Kulkarni, O.P., Alexander, V.V., Kumar, M., Freeman, M.J., Islam, M.N., Terry, J.F.L., Neelakandan, M., Chan, A.: Supercontinuum generation from ~1.9 to 4:5 μm in ZBLAN fiber with high average power generation beyond 3:8 μm using a thulium-doped fiber amplifier. J. Opt. Soc. Am. B. **28**, 2486–2498 (2011)
29. Dudley, J.M., Genty, G., Coen, S.: Supercontinuum generation in photonic crystal fiber. Rev. Mod. Phys. **78**, 1135–1184 (2006)
30. Genty, G., Coen, S., Dudley, J.M.: Fiber supercontinuum sources. J. Opt. Soc. Am. B. **24**, 1771–1785 (2007)
31. Taylor, J.R.: Supercontinuum Generation in Optical Fibers. Cambridge University Press, Cambridge (2010)
32. Lin, C., Stolen, R.H.: New nanosecond continuum for excited state spectroscopy. Appl. Phys. Lett. **28**, 216–218 (1976)
33. Izawa, T., Shibata, N., Takeda, A.: Optical attenuation in pure and doped fused silica in their wavelength region. Appl. Phys. Lett. **31**, 33–35 (1977)
34. Travers, J.C.: Blue extension of optical fibre supercontinuum generation. J. Opt. **12**, 113001 (2010)
35. Sanghera, J.S., Aggarwal, I.D., Busse, L.E., Pureza, P.C., Nguyen, V.Q., Kung, F.H., Shaw, L.B., Chenard, F.: Chalcogenide optical fibers target midIR applications. Laser Focus World. **41**, 83–87 (2005)
36. Kulkarni, O.P., Xia, C., Lee, D.J., Kumar, M., Kuditcher, A., Islam, M.N., Terry, J.F.L., Freeman, M.J., Aitken, B.G., Currie, S.C., McCarthy, J.E., Powley, M.L., Nolan, D.A.: Third order cascaded Raman wavelength shifting in chalcogenide fibers and determination of Raman gain coefficient. Opt. Express. **14**, 7924–7930 (2006)
37. Zhu, X., Peyghambarian, N.: High-power ZBLAN glass fiber lasers: Review and prospect. Adv. Opto. Electron. **2010**, 1–23 (2010)
38. Geng, J., Wang, Q., Jiang, S.: High-spectral-flatness mid-infrared supercontinuum generated from a Tm-doped fiber amplifier. Appl. Optics. **51**, 834–840 (2011)
39. Agrawal, G.P.: Nonlinear Fiber Optics, 3rd edn. Academic, San Diego (2001)
40. Taylor, J.R.: Optical Solitons: Theory and Experiment. Cambridge University Press, Cambridge (1992)
41. Skryabin, D.V., Gorbach, A.V.: Colloquium: Looking at a soliton through the prism of optical supercontinuum. Rev. Mod. Phys. **82**, 1287–1299 (2010)
42. Dianov, E.M., Nikanova, Z.S., Prokhorov, A.M., Serkin, V.N.: Optimal compression of multi-soliton pulses in optical fibers. Sov. Tech. Phys. Lett. **12**, 311–313 (1986)
43. Comanescu, G., Manka, C.K., Grun, J., Nikitin, S., Zabetakis, D.: Identification of explosives with two-dimensional ultraviolet resonance raman spectroscopy. Appl. Spectrosc. **62**, 833–839 (2008)
44. Islam, M.N., Sucha, G., Bar-Joseph, I., Wegener, M., Gordon, J.P., Chemla, D.S.: Femtosecond distributed soliton spectrum in fibers. J. Opt. Soc. Amer. B. **6**, 1149–1158 (1989)

45. Vinay, V.A., Shi, Z., Islam, M.N., Ke, K., Kalinchenko, G., Freeman, M.J., Ifarraguerri, A., Meola, J., Absi, A., Leonard, J., Zadnik, J.A., Szalkowski, A.S., Boer, G.J.: Field Trial of Active Remote Sensing using a high-power short-wave infrared supercontinuum laser. Appl. Optics. **52**(27), 6813–6823 (2013)
46. Mohammed, N.I., Freeman, M.J., Peterson, L.M., Ke, K., Ifarraguerri, A., Bailey, C., Baxley, F., Wager, M., Absi, A., Leonard, J., Baker, H., Rucci, M.: Field tests for round-trip imaging at 1.4km distance with change detection and ranging using a short-wave infrared supercontinuum laser. Appl. Optics. **55**(7), 1584–1602 (2016)
47. Vinay, V.A., Shi, Z., Islam, M.N., Ke, K., Freeman, M.J., Ifarraguerri, A., Meola, J., Absi, A., Leonard, J., Zadnik, J., Szalkowski, A.S., Boer, G.J.: Power Scalable >25W Supercontinuum Laser from 2 to 2.5 microns with near diffraction limited beam and low output variability. Opt. Lett. **38**(13), 2292–2294 (2013)
48. Kumar, M., Islam, M.N., Terry, J.F.L., Freeman, M.J., Chan, A., Neelakandan, M., Manzur, T.: Stand-off detection of solid targets with diffuse reflection spectroscopy using a high power mid-infrared supercontinuum source. Appl. Opt in publication. **51**, 2794–2807 (2012)
49. Razeghi, M., Slivken, S., Bai, Y., Darvish, S.R.: The quantum cascade laser: A versatile and powerful tool. Opt. Photon. News. **19**, 42–47 (2008)
50. Anderson, R.R., Farinelli, W., Laubach, H., Manstein, D., Yaroslavsky, A.N., Gubeli, J., Jordan, K., Neil, G.R., Shinn, M., Chandler, W., Williams, G.P., Benson, S.V., Douglas, D.R., Dylla, H.F.: Selective photothermolysis of lipid-rich tissues: A free electron laser study. Lasers Surg. Med. **38**, 913–919 (2006)
51. Sorokina, I.T., Vodopyanov, K.L.: Solid-State Mid-Infrared Laser Sources. Springer, Berlin/Germany (2003)
52. Wille, K.: Synchrotron radiation sources. Rep. Prog. Phys. **54**, 268 (1991)
53. Edwards, G.S., Austin, R.H., Carroll, F.E., Copeland, M.L., Couprie, M.E., Gabella, W.E., Haglund, R.F., Hooper, B.A., Hutson, M.S., Jansen, E.D., Joos, K.M., Kiehart, D.P., Lindau, I., Miao, J., Pratisto, H.S., Shen, J.H., Tokutake, Y., Van Der Meer, A.F.G., Xie, A.: Free-electron-laser based biophysical and biomedical instrumentation. Rev. Sci. Instrum. **74**, 3207–3245 (2003)
54. Mandon, J., Sorokin, E., Sorokina, I.T., Guelachvili, G., Picqu, N.: Supercontinua for high-resolution absorption multiplex infrared spectroscopy. Opt. Lett. **33**, 285–287 (2008)
55. Diddams, S.A., Jones, D.J., Ye, J., Cundiff, S.T., Hall, J.L., Ranka, J.K., Windeler, R.S., Holzwarth, R., Udem, T., Hansch, T.W.: Direct link between microwave and optical frequencies with a 300 THz femtosecond laser comb. Phys. Rev. Lett. **84**, 5102–5104 (2000)
56. Frith, G.P., Lancaster, D.G.: Power scalable and efficient 790nm pumped Tm3þ-doped fiber lasers. Proc. SPIE. **6102**, 610208 (2006)
57. Carter, A., Farroni, J., Tankala, K., Samson, B., Machewirth, D., Jacobson, N., Torruellas, W., Chen, Y., Cheng, M., Galvanauskas, A., Sanchez, A.: Robustly single-mode polarization maintaining Er/Yb co-doped LMA fiber for high power applications,, CLEO/QELS, 2007, pp. paper CTuS6
58. Imeshev, G., Fermann, M.: 230kW peak power femtosecond pulses from a high power tunable source based on amplification in Tm-doped fiber. Opt. Express. **13**, 7424–7431 (2005)
59. Kieu, K., Wise, F.W.: Soliton thulium-doped fiber laser with carbon nanotube saturable absorber. IEEE Photon. Technol. **21**, 128–130 (2009)
60. Rakich, P.T., Fink, Y., Soljačić, M.: Efficient mid-IR spectral generation via spontaneous fifth-order cascaded-Raman amplification in silica fibers. Opt. Lett. **33**, 1690–1692 (2008)
61. Jiang, M., Tayebati, P.: Stable 10 ns, kilowatt peak-power pulse generation from a gain-switched Tm-doped fiber laser. Opt. Lett. **32**, 1797–1799 (2007)
62. Thapa, R., Ronehouse, D., Nguyen, D., Wiersma, K., Smith, C., Zong, J., Chavez-Pirson, A.: Mid-IR supercontinuum generation in ultra-low loss, dispersion-zero shifted tellurite glass fiber with extended coverage beyond 4.5 microns. In: Titteton, D.H., Richarson, M.A., Grasso, R.J., Ackermann, H., Bohn, W.L. (eds.) Technologies for Optical Countermeasures X; High-Power Lasers 2013, Technology and Systems, vol. 8898, p. 889808. Proc. of SPIE, Bellingham/Washington, DC (2013)

63. Salem, R., Jiang, Z., Liu, D., Pafchek, R., Gardner, D., Foy, P., Saad, M., Jenkins, D., Cable, A., Fendel, P.: Mid-infrared supercontinuum generation spanning 1.8 octaves using step-index indium fluoride fiber pumped by a femtosecond fiber laser near 2 microns. Opt. Express. **23**(24), 30592–30602 (2015)
64. Gauthier, J.-C., Fortin, V., Carree, J.-Y., Poulain, S., Poulain, M., Vallee, R., Bernier, M.: Mid-IR supercontinuum from 2.4-5.4 microns in a low-loss fluoroindate fiber. Opt. Lett. **41**(8), 1756–1759 (2016)
65. Kedenburg, S., Steinle, T., Morz, F., Steinmann, A., Giessen, H.: High-power mid-infrared high repetition-rate supercontinuum source based on chalcogenide step-index fiber. Opt. Lett. **40**(11), 2668–2671 (2015)
66. Rafael, R.G., Brandon Shaw, L., Nguyen, V.Q., Pureza, P.C., Ishwar, D.A., Sanghera, J.S.: All-fiber chalcogenide-based mid-infrared supercontinuum source. Optical Fiber Technology. **18**, 345–348 (2012)
67. Robichaud, L.-R., Fortin, V., Gauthier, J.-C., Chatigny, S., Couillard, J.-F., Delarosbil, J.-L., Vallee, R., Bernier, M.: Compact 3–8 micron supercontinuum generation in a low-loss, As2Se3 step-index fiber. Opt. Lett. **41**(20), 4605–4608 (2016)
68. Petersen, C.R., Moller, U., Kubat, I., Zhou, B., Dupont, S., Ramsay, J., Benson, T., Sujecki, S., Abdel-Moneim, N., Tang, Z., Furniss, D., Seddon, A., Bang, O.: Mid-infrared supercontinuum covering the 1.4–13.3 micron molecular fingerprint region using ultra-high NA chalcogenide step-index fiber. Nature Photonics. **8**, 830–834 (2014)
69. Deng, D., Liu, L., Tuan, T.H., Kanou, Y., Matsumoto, M., Tezuka, H., Suzuki, T., Ohishi, Y.,: Advanced Solid State Lasers Conference, paper ATu2A.32, Optical Society of America (2015)
70. Yu, Y., Zhang, B., Gai, X., Zhai, C., Qi, S., Guo, W., Yang, Z., Wang, R., Choi, D.-Y., Madden, S., Luther-Davies, B.: 1.8–10 micron mid-infrared supercontinuum generated in a step-index chalcogenide fiber using low peak pump power. Opt. Lett. **40**(6), 1081–1084 (2015)
71. Li, H., Harris, D.A., Xu, B., Wrzesinski, P.J., Lozovoy, V.V., Dantus, M.: Standoff and arms-length detection of chemicals with single-beam coherent anti-Stokes Raman scattering. Appl. Optics. **48**, B17–B22 (2009)
72. Savitzky, A., Golay, M.: Smoothing and differentiation of data by simplified least squares procedures. Anal. Chem. **44**, 1627–1638 (1964)
73. Ingale, S.V., Sastry, P.U., Patra, A.K., Tewari, R., Wagh, P.B., Gupta, S.C.: Micro structural investigations of TNT and PETN incorporated silica xerogels. J. Sol-Gel Sci. Technol. **54**, 238–242 (2010)
74. Banas, A., Banas, K., Bahou, M., Moser, H.O., Wen, L., Yang, P., Li, Z.J., Cholewa, M., Lim, S.K., Lim, C.H.: Post-blast detection of traces of explosives by means of Fourier transform infrared spectroscopy. Vib. Spectrosc. **51**, 168–176 (2009)
75. Janni, J., Gilbert, B.D., Field, R.W., Steinfeld, J.I.: Infrared absorption of explosive molecule vapors. Spectrochim. Acta A. **53**, 1375–1381 (1997)
76. Harrington, J.A.: Infrared Fibers and Their Applications. SPIE, Bellingham/Washington, DC (2004)
77. Fuller, M.P., Griffiths, P.R.: Diffuse reflectance measurements by infrared Fourier transform spectroscopy. Anal. Chem. **50**, 1906–1910 (1978)
78. Gottfried, J.L., Lucia Jr., F.C.D., Munson, C.A., Miziolek, A.W.: Laser induced breakdown spectroscopy for detection of explosives residues: a review of recent advances, challenges and future prospects. Anal. Bionanal. Chem. **395**, 283–300 (2009)
79. Martinez, R.A., Guo, K., Flanagan, C.L., Chaudhauri, C, Islam, M.N., Hollister, S.J.: 3D Printing of Thermoplastics with Higher Strength Using SWIR-Supercontinuum Laser, Conference on Lasers and Electro–Optics (CLEO), San Jose (2016)
80. Libby, P.: Atherosclerosis: The new view. Sci. Am. **286**, 47–55 (2002)
81. Tillman, K.A., Maier, R.R.J., Reid, D.T., McNaghten, E.D.: Mid-infrared absorption spectroscopy across a 14.4 THz spectral range using a broadband femtosecond optical parametric oscillator. Appl. Phys. Lett. **85**, 3366–3368 (2004)

82. Sorokin, E., Sorokina, I.T., Mandon, J., Guelachvili, G., Picqué, N.: Sensitive multiplex spectroscopy in the molecular fingerprint 2.4 μm region with a Cr2+:ZnSe femtosecond laser. Opt. Express. **15**, 16540–16545 (2007)
83. Holman, H.-Y.N., Bjornstad, K.A., Martin, M.C., McKinney, W.R., Blakely, E.A., Blankenberg, F.G.: Midinfrared reflectivity of experimental atheromas. J. Biomed. Opt. **13**, 030503 (2008)
84. Awazu, K., Nagai, A., Aizawa, K.: Selective removal of cholesterol esters in an arteriosclerotic region of blood vessels with a free-electron laser. Lasers Surg. Med. **23**, 233–237 (1998)
85. Alexander, V.V., Ke, K., Xu, Z., Islam, M.N., Freeman, M.J., Pitt, B., Welsh, M.J., Orringer, J.S.: Photothermolysis of sebaceous glands in human skin ex vivo with a 1,708 nm raman fiber laser and contact cooling. Lasers Surg. Med. **43**, 470–480 (2011)
86. Paluszkiewicz, C., Kwiatek, W.M., Banas, A., Kisiel, A., Marcelli, A., Piccinini, A.: SR-FTIR spectroscopic preliminary findings of non-cancerous, cancerous, and hyperplastic human prostate tissues. Vib. Spectrosc. **43**, 237–242 (2007)
87. Hooper, B.A., Maheshwari, A., Curry, A.C., Alter, T.M.: Catheter for diagnosis and therapy with infrared evanescent waves. Appl. Optics. **42**, 3205–3214 (2003)
88. Gentner, J.M., Wentrup-Byrne, E., Walker, P.J., Walsh, M.D.: Comparison of fresh and post-mortem human arterial tissue: an analysis using FT-IR microspectroscopy and chemometrics. Cell. Mol. Biol. (Noisy-le-Grand). **44**, 251–259 (1998)
89. Alexander, V.V., Deng, H., Islam, M.N., Terry, J.F.L., Pittman, R.B., Valen, T.: Surface roughness measurement of flat and curved machined metal parts using a near infrared super-continuum laser. Opt. Eng. **50**, 113602 (2011)
90. DeGarmo, E.P., Black, J.T., Kohser, R.A.: Materials and Processes in Manufacturing. Prentice-Hall, Englewood Cliffs (1997)
91. Adachi, S., Horio, K., Nakamura, Y., Nakano, K., Tanke, A.: Development of Toyota 1ZZ-FE engine. SAE Technical Paper series, 981087. (1998)
92. Tung, S., McMillan, M.L.: Automotive tribology overview of current advances and challenges for the future. Tribol. Int. **37**, 517–536 (2004)
93. Beckmann, P., Spizzichino, A.: The scattering of electromagnetic waves from rough surfaces. Pergamon, New York (1963)
94. Kumar, M., Islam, M.N., Terry, J.F.L., Aleksoff, C.A., Davidson, D.: High resolution line scan interferometer for solder ball inspection using a visible supercontinuum source. Opt. Express. **18**, 6722–6739 (2010)
95. Ishii, A., Mitsudo, J.: Constant-magnification varifocal mirror and its application to measuring three-dimensional (3-D) shape of solder bump. IEICE Trans. Electron E. **90**, 6–11 (2007)
96. Endo, T., Yasuno, Y., Makita, S., Itoh, M., Yatagai, T.: Profilometry with line-field Fourier-domain interferometry. Opt. Express. **13**, 695–701 (2005)
97. Yasuno, Y., Endo, T., Makita, S., Aoki, G., Itoh, M., Yatagai, T.: Three-dimensional line-field Fourier domain optical coherence tomography for in vivo dermatological investigation. J. Biomed. Opt. **11**, 014015 (2006)
98. Nakamura, Y., Makita, S., Yamanari, M., Itoh, M., Yatagai, T., Yasuno, Y.: High-speed three-dimensional human retinal imaging by line-field spectral domain optical coherence tomography. Opt. Express. **15**, 7103–7116 (2007)
99. Dorrer, C., Belabas, N., Likforman, J.-P., Joffre, M.: Spectral resolution and sampling issues in Fourier-transform spectral interferometry. J. Opt. Soc. Am. B. **17**, 1795–1802 (2000)
100. Choma, M., Sarunic, M., Yang, C., Izatt, J.: Sensitivity advantage of swept source and Fourier domain optical coherence tomography. Opt. Express. **11**, 7748–7755 (2003)
101. Stuart, B.C., Feit, M.D., Herman, S., Rubenchik, A.M., Shore, A.B.W., Perry, M.D.: Optical ablation by high-power short-pulse lasers. J. Opt. Soc. Amer. B. **13**, 459–468 (1996)
102. Stuart, B.C., Feit, M.D., Herman, S., Rubenchik, A.M., Shore, B.W., Perry, M.D.: Nanosecond-to-femtosecond laser-induced breakdown in dielectrics. Phys. Rev. B. **53**, 1749–1761 (1996)
103. Yakovlenko, S.I.: Physical processes upon the optical discharge propagation in optical fiber. Laser Phys. **16**, 1273–1290 (2006)

104. Dianov, E.M., Bufetov, I.A., Frolov, A.A., Mashinsky, V.M., Plotnichenko, V.G., Churbanov, M.F., Snopatin, G.E.: Catastrophic destruction of fluoride and chalcogenide optical fibres. Electron. Lett. **38**, 783–784 (2002)
105. Richardson, D.J., Nilsson, J., Clarkson, W.A.: High power fiber lasers: current status and future perspectives [Invited]. J. Opt. Soc. Am. B. **27**, B63–B92 (2010)
106. Parker, J.M.: Fluoride glasses. Annu. Rev. Mater. Sci. **19**, 21–41 (1989)
107. Harrington, J.A.: Infrared fibers. In: Bass, M., Enoch, J.M., Van Striland, E.W., Wolfe, W.L. (eds.) Handbook of Optics, Vol 3, Classical, pp. 14.1–14.16. Vision & X-ray Optics, Optical Society of America, Washington, DC (2002)
108. Sonntag, E., Borgnakke, C., Van Wylen, G.J.: Fundamentals of Thermodynamics. Wiley, New York (1998)
109. Li, K., Zhang, G., Hu, L.: Watt-level ~2 μm laser output in Tm3þ-doped tungsten tellurite glass double-cladding fiber. Opt. Lett. **35**, 4136–4138 (2010)
110. Lyakh, A., Pflügl, C., Diehl, L., Wang, Q.J., Capasso, F., Wang, X.J., Fan, J.Y., Tanbun-Ek, T., Maulini, R., Tsekoun, A., Go, R., Kumar, C., Patel, N.: 1:6W high wall plug efficiency, continuous-wave room temperature quantum cascade laser emitting at 4.6 μm. Appl. Phys. Lett. **92**, 111110 (2008)
111. Ke, K., Xia, C., Islam, M.N., Welsh, M.J., Freeman, M.J.: Mid-infrared absorption spectroscopy and differential damage in vitro between lipids and proteins by an all-fiber-integrated supercontinuum laser. Opt. Express. **17**, 12627–12640 (2009)
112. Guo, K., Martinez, R.A., Freeman, M., Gurm, H.S., Islam, M.N.: High SNR Glucose Monitoring using a SWIR Super-Continuum Light Source, Conference on Lasers and Electro-Optics (CLEO), San Jose (2016)
113. Smith, J.: The pursuit of non-invasive glucose: hunting the deceitful turkey, 4th edn, (2015) http://www.mendosa.com/articles_testingGlucose.htm

Chapter 5
Specialty Optical Fibers for Raman Lasers

Guanshi Qin

As described in the preceding chapters, the performances of Raman fiber lasers and amplifiers are governed by the parameters (including Raman gain coefficients, Raman shift, transmission window and loss, damage threshold, chemical and thermal stability, etc.) of optical fibers. At present, over 1 kW 1 μm Raman lasers based on silica fibers have been demonstrated by several esteemed groups [1, 2]. Widely tunable Raman lasers from 1 to 2 μm were reported in silica fibers [3]. However, silica fibers are not suitable for constructing mid-infrared (MIR) fiber lasers and amplifiers (>2.5 μm) because of very high material loss in the mid-infrared spectral region [4]. To solve this problem, several types of specialty optical fibers (SOFs) based on fluoride, tellurite, and chalcogenide glasses with low transmission loss in mid-infrared region have been developed. This chapter deals with the optical properties of such fibers, the related Raman fiber lasers and amplifiers. The parameters of those SOFs are described in Sect. 5.1. Section 5.2 focuses on design and construction of Raman laser and amplifiers based on those SOFs. Section 5.3 concentrates on the generation of MIR Raman soliton lasers in those SOFs.

5.1 Parameters of SOFs

The main mechanisms responsible for guiding of light in optical fibers are total internal reflection and photonic bandgap. According to light-guiding mechanism, optical fibers are categorized as follows: step-index fibers, graded-index fibers, photonic bandgap fibers, or microstructured fibers [4]. A step-index fiber, as the

G. Qin (✉)
Jilin University, Changchun, China
e-mail: qings@jlu.edu.cn

© Springer International Publishing AG 2017
Y. Feng (ed.), *Raman Fiber Lasers*, Springer Series in Optical Sciences 207,
DOI 10.1007/978-3-319-65277-1_5

Table 5.1 The compositions of tellurite, fluoride, and chalcogenide glasses (Ref. [4])

Tellurite glasses	Fluoride glasses	Chalcogenide glasses
TeO_2–ZnO	ZrF_4-based (ZBLAN)	As_2S_3
TeO_2–BaO	AlF_3-based	As_2Se_3
TeO_2–WO_3	InF_3-based	Ge–Sb–Se
TeO_2–Bi_2O_3–ZnO–Na_2O	ZnF_2-based	......
(TBZN)	......	
......		

simplest form of optical fibers, consists of a central glass core surrounded by a cladding layer. The refractive index n_c of the cladding layer is slightly lower than that n_1 of the core. There are two parameters that characterize a step-index fiber. One is the relative core-cladding index difference $\Delta = (n_1 - n_c)/n_1$. The other is the so-called V parameter defined as $V = k_0 a(n_1^2 - n_c^2)^{1/2}$, where $k_0 = 2\pi/\lambda$, a is the core radius, and λ is the wavelength of light. The V parameter determines the number of modes supported by the fiber. A step-index fiber is a single-mode fiber If $V < 2.405$.

5.1.1 SOF Materials and Fabrication

This chapter focuses on non-silica SOFs for Raman fiber lasers and amplifiers. The material of choice for this purpose depends on the design of the fiber devices. At present, the commonly used non-silica SOF materials are tellurite, fluoride, and chalcogenide glasses [4]. Table 5.1 gives a summary of the compositions of tellurite, fluoride, and chalcogenide glasses. The detailed description of them will be given below.

Several techniques including rod-in-tube, extrusion, stacking, and drawing have been developed for fabricating non-silica SOFs [4]. For the extrusion technique, the fiber preform is produced by extruding selectively from a glass rod. In this procedure, the molten glass rod is forced through a die containing the designed pattern of holes. In the case of stacking and drawing technique, the fiber preform is made by stacking multiple capillary tubes or glass rods in a designed pattern around a solid glass rod or an air hole. In both cases, the fabricated fiber preform is drawn into a fiber using a standard fiber-drawing tower. The detailed information on them can be found elsewhere [4].

5.1.2 Loss, Damage Threshold, Refractive Index, and Material Dispersion

The operation wavelength of Raman fiber lasers depends on the transmission loss of SOFs. Figure 5.1 shows the transmission spectra of silica, tellurite, fluoride, and

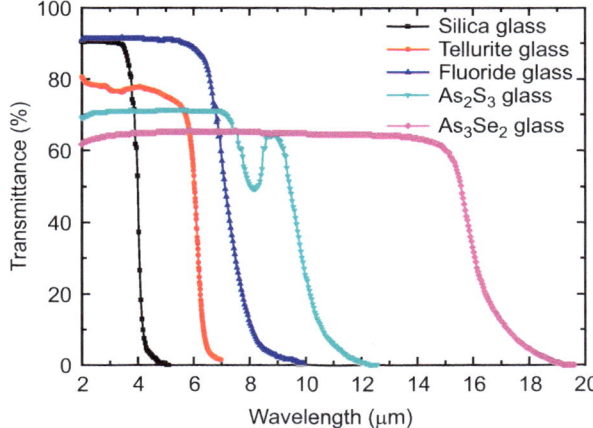

Fig. 5.1 Transmission spectra of silica, tellurite, fluoride, and chalcogenide glasses

Table 5.2 Comparison of the damage threshold, peak Raman shift, peak Raman gain coefficient, and nonlinear refractive index of those fiber materials (Ref. [4–8])

	Damage threshold (GW/cm^2, 1064 nm ps laser)	Peak Raman shift (cm^{-1})	Peak Raman gain coefficient (m W^{-1}, 1 μm)	Nonlinear refractive index (m^2 W^{-1})
Silica	400	440	1×10^{-13}	2.2×10^{-20}
Tellurite	100	750	3.8×10^{-12}	5.9×10^{-19}
Fluoride	200	580	1×10^{-13}	2.1×10^{-20}
As$_2$S$_3$	1	350	7.5×10^{-12}	3×10^{-18}
As$_2$Se$_3$	0.4	250	5.1×10^{-11}	1.1×10^{-17}

chalcogenide glasses [4]. It is seen that the longest possible wavelengths of Raman lasers based on silica, tellurite, fluoride, As$_2$S$_3$, and As$_2$Se$_3$ fibers are about 3, 5, 5.5, 6, and 13 μm, respectively.

In addition to the transmission loss, the maximum available output power of Raman fiber lasers is mainly limited by the damage threshold of SOFs. In general, the damage threshold of the fiber materials is affected by the parameters (including operation wavelength, pulse width, repetition rate, etc.) of the pump light. Here, a comparison of the damage threshold of those fiber materials pumped by a 1064 nm laser with a pulse width of several tens of ps is given in Table 5.2 [5–8]. It is evidently that As$_2$S$_3$ and As$_2$Se$_3$ fibers are not suitable for constructing high-power MIR Raman fiber lasers. Therefore, one needs to choose appropriate SOFs for realizing MIR Raman fiber lasers, as mentioned in the preceding chapters.

Chromatic dispersion, manifesting through the frequency dependence of the refractive index, plays an important role in the propagation of optical pulses inside the fiber [4]. To construct pulsed Raman fiber lasers, chromatic dispersion must be considered for optimizing the laser performances. Figure 5.2a shows the relation of refractive index to wavelength for those fiber materials. The refractive index

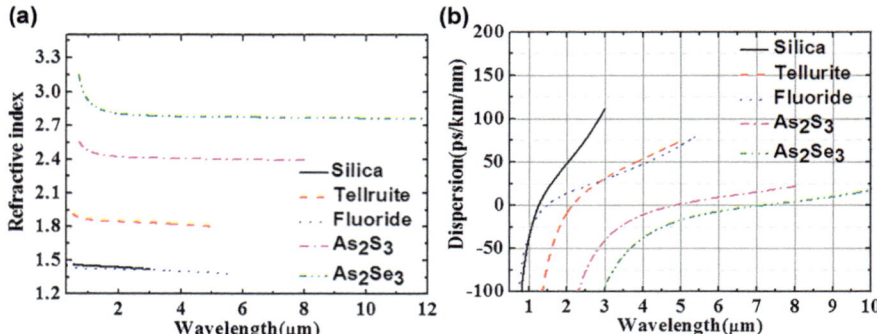

Fig. 5.2 (a) The relation of refractive index to wavelength for those fiber materials. (b) The material dispersion curves of the above fiber materials

decreases with an increase of wavelength. Figure 5.2b shows the material dispersion curves of the above fiber materials. One interesting feature is that the group velocity dispersion value becomes zero at a certain wavelength. Such a wavelength is called by the zero-dispersion wavelength. The zero-dispersion wavelengths for silica, tellurite, fluoride, As_2S_3, and As_2Se_3 glasses are 1.27, 2.15, 1.48, 4.89, and 7.22 μm, respectively. Both material and waveguide dispersion must be considered for investigating optical pulse propagation inside SOFs.

5.1.3 Raman Gain Spectrum

Stimulated Raman scattering (SRS) is one type of nonlinear effects, which results from stimulated inelastic scattering [4]. The Raman gain spectrum is related to the third-order nonlinear susceptibility and the cross section of spontaneous Raman scattering, which is determined by the vibrational modes of the fiber materials. Figure 5.3 shows the normalized Raman gain spectra of several types of fiber materials. Table 5.2 gives the comparison of the Raman shift, Raman gain coefficient, and nonlinear refractive index of those fiber materials. Silica glass has a Raman band of $\sim$440 cm^{-1}, which is ascribed to the Si$-$O vibration. Tellurite glass has several Raman bands. The bands at $\sim$440 and 665 cm^{-1} originate from the stretching vibrations of TeO_4 trigonal bipyramids. The band at $\sim$750 cm^{-1} is due to the stretching vibration of TeO_3 trigonal pyramids [9]. Fluoride glass has one main Raman band of $\sim$580 cm^{-1}, which originates from the symmetric stretching vibration of ZrF_6^{2-} [10]. As_2S_3 glass has a Raman band of $\sim$325 cm^{-1}, which is due to the symmetric stretching vibration of $[AsS_{3/2}]$ regular pyramids [7]. As_2Se_3 glass has a Raman band of $\sim$250 cm^{-1}, which is attributed to the As$-$Se vibration. Among those fiber materials, tellurite fiber has the largest usable Raman shift ($\sim$22.3 THz) and $\sim$16 times higher Raman gain coefficient than that of silica fiber [9]. It means that tellurite fiber can be used as the Raman gain medium for constructing broadband Raman fiber amplifiers and widely tunable Raman fiber lasers.

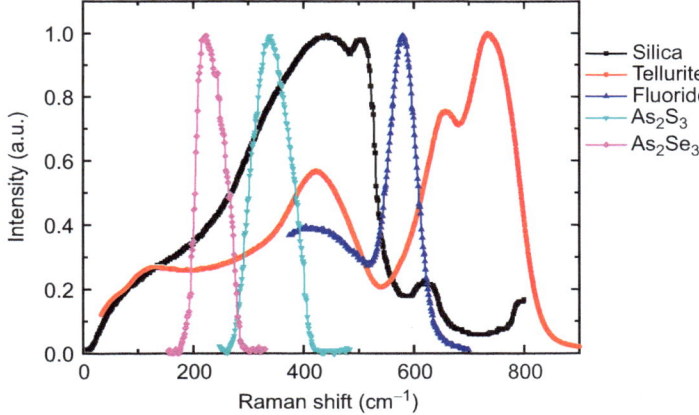

Fig. 5.3 Normalized Raman gain spectra of several types of fiber materials

5.2   Raman Laser and Amplifiers Based on SOFs

5.2.1   Tellurite Fiber Raman Amplifiers and Lasers

5.2.1.1   Ultrabroadband Tellurite Fiber Raman Amplifiers

To meet the growing use of the Internet and reduced data traffic in optical fiber communication networks, wavelength-division multiplexing (WDM) transmission systems with ultralarge transmission capacity have been developed and recognized as the best way to construct ultralarge capacity optical networks. However, the available channels in WDM transmission systems are limited mostly by the gain bandwidth of present optical amplifiers. Therefore, the search for new broadband optical amplifiers has been undertaken worldwide. Among the new broadband optical amplifiers, Raman fiber amplifiers based on stimulated Raman scattering phenomenon are attractive devices in high-capacity and broadband transmission systems/photonic networks because they can provide broad gain bands in any wavelength region required by properly setting the pump wavelengths, which is inaccessible for active ions (rare earth ions or transition metal ions) doped fiber amplifiers.

In 1994, Wang et al. reported a tellurite glass fiber with a background loss of less than 1 dB/m and a comparison of the properties of tellurite, silica, fluoride, and chalcogenide glasses [11]. However, such a background loss is too high for constructing Raman fiber amplifier and lasers. In 1998, Mori et al. used high purity TeO_2 as the raw materials and fabricated a tellurite fiber with a very low loss of $\sim$20 dB/m, which made the realization of fiber devices possible [12]. By using a single-mode tellurite fiber with a core diameter of 2.2 μm, a Δn of 2.2%, and a cutoff wavelength of 1300 nm, its Raman gain spectrum was measured by Mori

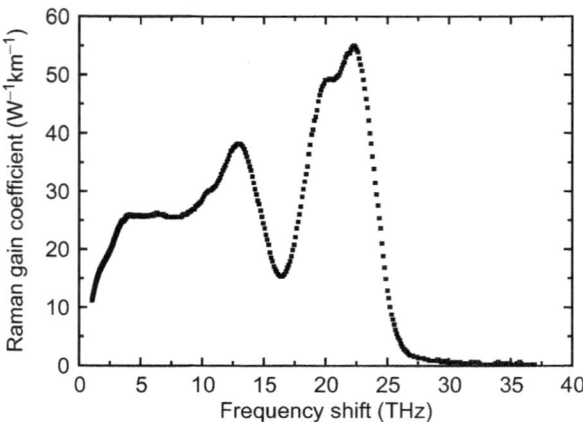

Fig. 5.4 Measured Raman gain coefficients of tellurite fiber pumped at 1460 nm (Ref. [13])

et al. [13]. Figure 5.4 shows the measured Raman gain coefficients of tellurite fiber pumped at 1460 nm. A peak value of Raman gain coefficient of ~55 W⁻¹ km⁻¹ was obtained, which is about 16 times larger than that (~3.5 W⁻¹ km⁻¹) of silica fiber. A usable Raman shift of ~22.3 THz was observed, which is 1.7 times larger than that (~13.1 THz) of silica fiber. Therefore, tellurite fiber Raman amplifiers can provide a gain bandwidth of 170 nm and cover the low-loss region of transmission fiber across the S, C, and L bands. In 2006, Masuda et al. demonstrated a distributed/discrete hybrid tellurite fiber Raman amplifier with an average gain of 22 dB in the wavelength region of 1480 ~ 1607 nm [14]. Figure 5.5 shows the schematic of the amplifier, which is a two-stage amplifier. The first stage is a distributed silica fiber Raman amplifier. The second stage is a discrete tellurite fiber Raman amplifier. Figure 5.6 shows the measured gain spectra with and without gain equalization and the equivalent noise figure. The figure also shows the loss spectrum of 80 km long single-mode silica fiber. The measured gain spectrum after equalization indicates an average gain of ~22 dB in the wavelength region of 1480~1607 nm. Further, experimental results showed that the above amplifier could achieve error-free operation seamlessly across a 124 nm (1485~1609 nm) gain band, which is the widest seamless amplification band yet reported.

Furthermore, ultrabroadband and gain-flattened gain profile can be realized for Raman fiber amplifiers without any gain equalization devices by properly setting the numbers and the power levels of pump lasers. For example, Namiki and Emori reported silica-based FRA with 100 nm of flat gain bandwidth and 0.1-dB flatness over 80 nm [15]. However, by using the silica-based optical fiber, it is impossible to obtain a gain-flattened S+ C+ L band (~20.9 THz) fiber Raman amplifiers due to the limit of inherent Raman gain spectrum (usable Raman shift, ~13.2 THz). In 2007, Qin et al. designed gain-flattened S + C + L ultrabroadband tellurite fiber Raman amplifiers by solving the inverse amplifier design problem

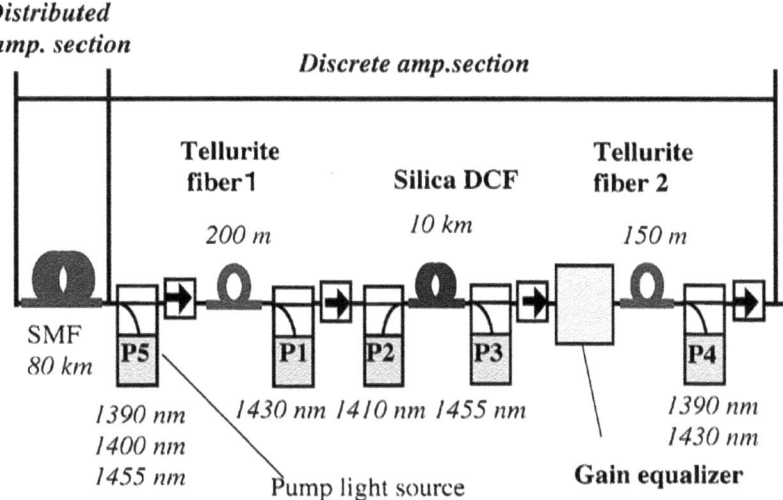

Fig. 5.5 Schematic of a distributed/discrete hybrid tellurite fiber Raman amplifier (Ref. [14])

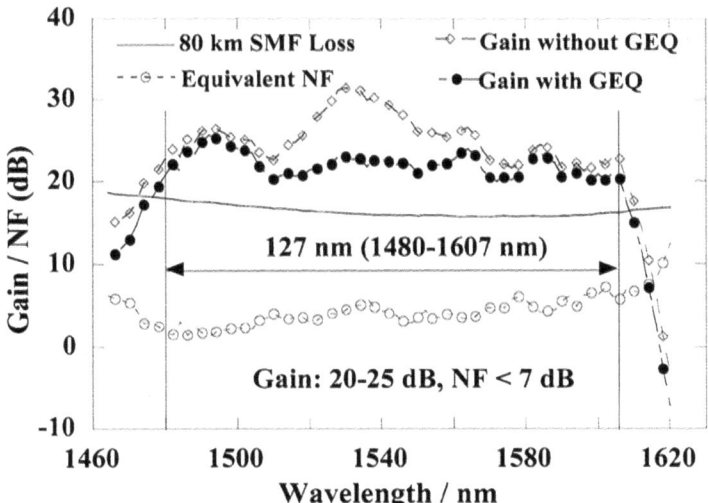

Fig. 5.6 Measured gain spectra with and without gain equalization and the noise figure (Ref. [14])

[16]. Multiwavelength pumped fiber Raman amplifiers can be described by a coupled propagation equation model that includes spontaneous Raman emission and its temperature dependence, amplified spontaneous emission, stimulated Raman amplification (including pump-to-pump, pump-to-signal, and signal-to-signal interactions), and Rayleigh scattering as follows:

$$\pm \frac{dp_k}{dz} = -\alpha_k P_k + \gamma_k P_k + \sum_{j=1}^{k-1} \frac{g_{v_k}(v_j - v_k)}{K_{\text{eff}} A_{\text{eff}}} P_j P_k$$

$$+ 2h v_k \sum_{j=1}^{k-1} \frac{g_{v_k}(v_j - v_k)}{K_{\text{eff}} A_{\text{eff}}} P_k \left[\frac{1}{e^{h(v_j - v_k)/KT} - 1} \right]$$

$$- \sum_{j=k+1}^{m+n} \frac{g_{v_k}(v_k - v_j)}{K_{\text{eff}} A_{\text{eff}}} P_j P_k$$

$$- 2h v_k \sum_{j=k+1}^{m+n} \frac{v_k}{v_j} \frac{g_{v_k}(v_k - v_j)}{K_{\text{eff}} A_{\text{eff}}} P_k P_k \left[\frac{1}{e^{h(v_j - v_k)/KT} - 1} \right]$$

$$\times (k = 1, 2 \ldots, n + m.) \qquad (5.1)$$

Here, P_i, v_i and α_i represent the power, frequency, and attenuation coefficient for the i-th wave, $i = 1, 2 \ldots, n + m$. The frequencies are numerated with decreasing order ($v_i > v_j$ for $i < j$), indexes $k = 1 \ldots$ n correspond to the backward-propagating pump waves (the minus sign on the left-hand side), and indexes $k = n + 1 \ldots n + m$ correspond to the forward-propagating signal waves (plus sign on the left-hand side). The gain coefficients at pump frequency v_i are given by $g_{v_i}(\Delta v) = g_R(\Delta v) \cdot v_i / v_0$, where $g_R(\Delta v)$ is the Raman gain spectrum measured at a reference pump frequency v_0, A_{eff} is the effective area of the fiber, and K_{eff} is the polarization factor. Equation (5.1) is to be solved with the boundary conditions $P_k(L) = P_{k0}(k = 1, 2, \ldots n)$ and $P_k(0) = P_{k0}$ ($k = n + 1, \ldots, n + m$) for pump and signal powers, respectively (L is the fiber length).

The amplification factor for every channel $k = n + 1, n + 2, \ldots n + m$ can be expressed in terms of the power integrals $I_j = \int_0^L P_j(z) dz, j = 1, 2, \ldots, n + m$ as

$$G_k \equiv \frac{P_k(L)}{P_k(0)} = \exp\left(-\alpha_k L + \sum_{j=n+1}^{n+m} g_{jk} I_j\right) \exp\left(\sum_{j=1}^{n} g_{jk} I_j\right) = G_{L,k} G_{G,k} \qquad (5.2)$$

where $G_{L,k}$ represents the effects of fiber attenuation and signal-to-signal Raman scattering (Raman tilt), $G_{G,k}$ the ON–OFF or gross (pump-to-signal) Raman gain experienced by the channel k, $g_{j,k} = g_{v_j}(v_j - v_k)/K_{\text{eff}} A_{\text{eff}}$ for $v_j > v_k$, and $g_{j,k} = -g_{v_j}(v_k - v_j) v_k / v_j K_{\text{eff}} A_{\text{eff}}$ for $v_j < v_k$. Supposing the terms $G_{L,k}$ are known, the complex problem finding the pump combination that yields the flattest net gain profile G_k constant (1 for 0 dB, 100 for 20 dB) breaks into two simpler problems to be solved one after the other. The first problem is to find such a set of frequencies v_j and constants I_j^*, so that the sum $\sum_{j=1}^{n} g_{jk} I_j^*$ is close to $\log G_{L,k}^{-1}$ for all of the frequencies v_k within the specified gain band. The second problem is to find such a set of input pump powers P_{j0} for these frequencies, so that the solutions $P_j(z)$ of Eq. (5.1) with initial conditions P_{j0} have the integrals I_j exactly equal to the optimal values I_j^*.

To solve the first problem, one of all possible sets of n pairs (v_j, I_j^*) is chosen to minimize the relative gain flatness parameters, defined as the difference (in decibels) between the maximal and the minimal gains $\log G_{G,k}$ normalized by their target values $\log G^{-1}{}_{L,k}$ as

$$F_{\text{rel}} \equiv \left[\left(\frac{\log G_{G,k}}{\log G^{-1}{}_{L,k}} \right)_{\text{max}} - \left(\frac{\log G_{G,k}}{\log G^{-1}{}_{L,k}} \right)_{\text{min}} \right] \cdot 100\%. \qquad (5.3)$$

A genetic algorithm in a 2n-dimensional objective space is used to find the optimal values. Using the optimal pair (v_j, I_j^*), we solve Eq. (5.1) to find the pump power for the corresponding wavelength by an iterative and genetic algorithms starting from some reasonable values. In our simulations, the square deviation between the resulting pump power integrals I_j and the target value $I_j^* \sum_{j=1}^{n} (I_j - I_j^*)^2$ for all the cases is less than 0.001. All the simulations are performed by MATLAB software.

Gain-flattened broadband multiwavelength backward pumped tellurite fiber Raman amplifiers operating in S + C + L regions were designed by solving the inverse amplifier design problem. They fixed the fiber length at a practical value of $L = 250$ m due to larger Raman gain coefficient of tellurite-based glass than that of silica and the required net gain at $G = 20$ dB for compensating the span loss of 80 km in a C band optical transmission system. For S + C + L bands, they considered a 20 THz WDM system pumped at 2, 3, 4, and 8 wavelengths. Such a 20 THz WDM system has 200 signal channels with 100 GHz spacing. Figure 5.7a–d shows the net gain-flattened profiles of S + C + L bands tellurite fiber Raman amplifiers pumped at 2, 3, 4, and 8 wavelengths. As one can see from Fig. 5.7d, gain-flattened S + C + L tellurite fiber Raman amplifiers with $F_{\text{rel}} = 11.724\%$ can be achieved using eight-wavelength pumping. With increasing the number of pump wavelengths from 2 to 8, the relative gain flatness is reduced from 39.84% to 11.724% and the corresponding root-mean square deviation (RMSD) for fitting the ideal net gain profile by calculated net gain profile from 0.71865 to 0.0605. Super-overlapped profiles (which corresponds to the gross signal gain index ln $G = [\sum_{j=1}^{n} g_{jk} I_j]$) of S + C + L bands tellurite fiber Raman amplifiers pumped at 2, 3, 4, and 8 wavelengths are presented in Fig. 5.7a–d inset. The results show that the effective bandwidth (defined as full width at half maximum of the superoverlapped profile covering from 1460 nm to longer wavelengths) of S + C + L bands tellurite fiber Raman amplifiers is expanded from 170 to 187 nm with increasing the numbers of pump wavelengths from 2 to 8, which is limited by the internal Raman shift of tellurite glass.

The composite Raman gain coefficient profiles to obtain optimal gain-flattened fiber Raman amplifiers pumped at 2, 3, 4, and 8 wavelengths are shown in Fig. 5.7e–h. Shorter pump wavelength gives larger Raman gain coefficients due to the photon conservation law and the frequency dependence of effective mode area. Figure 5.7i–l represents the evolutions of multiwavelength pump powers inside the tellurite fiber. To obtain a 20 dB 200-channel broadband and gain-flattened S + C + L band tellurite fiber Raman amplifiers pumped at 8 wavelengths,

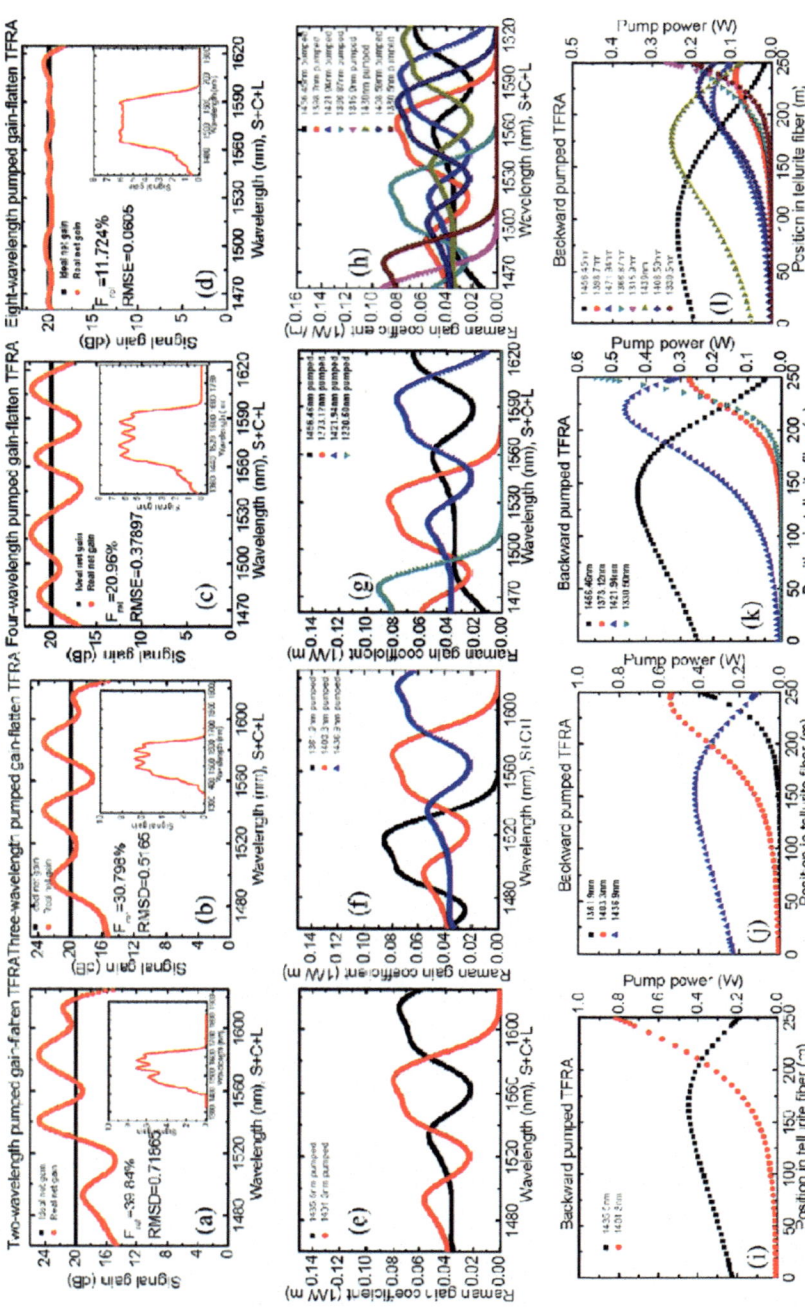

Fig. 5.7 (**a**–**d**) Net gain-flattened profiles and superoverlapped profiles (inset) of S + C + L band TBZN FRA pumped at 2, 3, 4, and 8 wavelengths, respectively. (**e**–**h**) Composite Raman gain coefficient profiles to obtain optimal gain-flattened FRA pumped at $n = 2, 3, 4, 8$ wavelengths, respectively. (**i**–**l**) Evolutions of multiwavelength pump powers inside the TBSNWP fiber (Ref. [16])

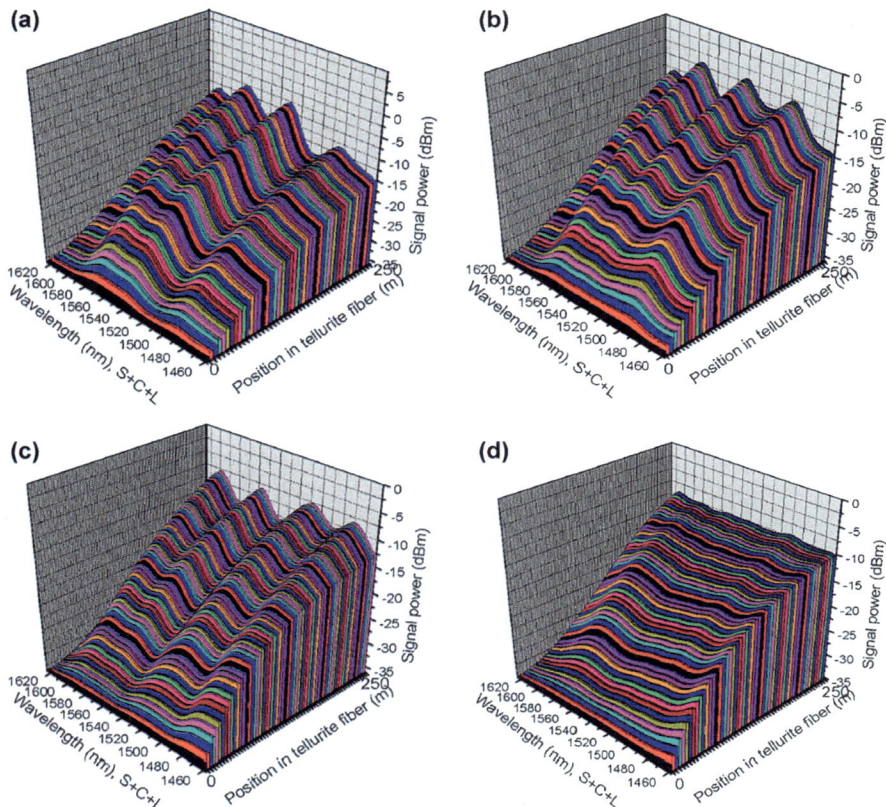

Fig. 5.8 (**a–d**) Evolution of signal powers upon propagation through the TBZN amplifiers for 2, 3, 4, and 8 wavelengths, respectively (Ref. [16])

the required powers for these 8-wavelength pump light (1456.4 nm, 1398.7 nm, 1421.9 nm, 1366.8 nm, 1315.9 nm, 1439 nm, 1408.6 nm, and 1330.5 nm) are 0.0168 W, 0.0799 W, 0.102 W, 0.1882 W, 0.2653 W, 0.0758 W, 0.1421 W, and 0.1963 W, respectively. The total pump power to realize a 20 dB 200-channels ultrabroadband and gain-flattened S + C + L band tellurite fiber Raman amplifiers is 1.0664 W. Figure 5.8a–d shows the evolution of signal powers upon propagation through the tellurite fiber Raman amplifiers for 2, 3, 4, and 8 wavelengths pumping, respectively. Strong Raman tilt effects were also observed in gain-flattened tellurite fiber Raman amplifiers.

5.2.1.2   Tellurite Fiber Raman Lasers

Widely tunable fiber lasers covering the S + C + L + U band have played important roles in optical communications, optical device testing, and optical sensor

Fig. 5.9 Schematic of tunable ring cavity tellurite fiber Raman lasers (Ref. [17])

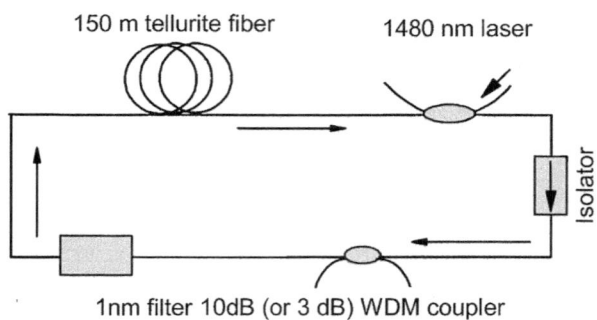

150 m tellurite fiber   1480 nm laser

1nm filter 10dB (or 3 dB) WDM coupler

systems. However, it is difficult to obtain widely tunable fiber lasers covering the S + C + L + U band by using one solely active ion doped fiber laser or silica-based fiber Raman laser. As aforementioned, among those fiber materials, tellurite fiber has the largest usable Raman shift (~22.3 THz) and ~16 times higher Raman gain coefficient than that of silica fiber. Therefore, tellurite fiber could be used as the Raman gain medium for constructing widely tunable fiber lasers. In 2008, Qin et al. reported a widely tunable ring cavity tellurite fiber Raman laser covering the S + C + L + U band and confirmed that the tunable wavelength range of the tellurite fiber Raman laser could cover from 1495 to 1688 nm when pumped at 1480 nm [17]. Figure 5.9 shows the schematic of tunable ring cavity tellurite fiber Raman lasers. A 1480 nm fiber Raman laser with the maximum output power of 5 W was used as the pump source. The pump light was launched into the ring cavity by using a 1480/1550 nm WDM coupler. The pump power was measured by monitoring the output power of one port of the WDM coupler. The Raman gain medium we used was a 150 m long single-mode tellurite fiber, which had a core diameter of 2.2 μm, a Δn of 2.2%, a background loss of ~20 dB/m at 1550 nm, and a cutoff wavelength of 1300 nm. The lasing wavelengths were tuned by using a tunable optical bandpass filter, which has a FWHM of 1 nm and tunable range of 1480–1600 nm. An isolator was inserted into the ring cavity to make sure of unidirectional laser operation. The tunable Raman lasing was output from one port of a 10 dB (or 3 dB) WDM coupler at 1550 nm. Figure 5.10a shows the emission spectra of tunable tellurite fiber Raman laser by tuning the tunable optical bandpass filter and free running 1665 nm Raman laser when the pump power of 1480 nm laser was fixed at 1.4 W. Figure 5.10b shows the power dependence of the tunable tellurite fiber Raman laser. Those results show that they have obtained 1495 nm lasing with increasing the pump power because tellurite fiber has a relatively large Raman gain coefficient, which means that one could obtain lasing even at the valley around 16 THz in the Raman gain spectrum of tellurite fiber. Similarly, one could obtain lasing even at 1688 nm when pumped at 1480 nm if one inserts an optical bandpass filter at 1688 nm into the ring cavity, because the Raman gain coefficient at 1688 nm is 15.5 W^{-1} km^{-1}, which is the same value at 1495 nm when pumped at 1480 nm. Therefore, one solely tunable fiber laser covering the S + C + L + U band could be realized by using tellurite fiber as the Raman gain medium.

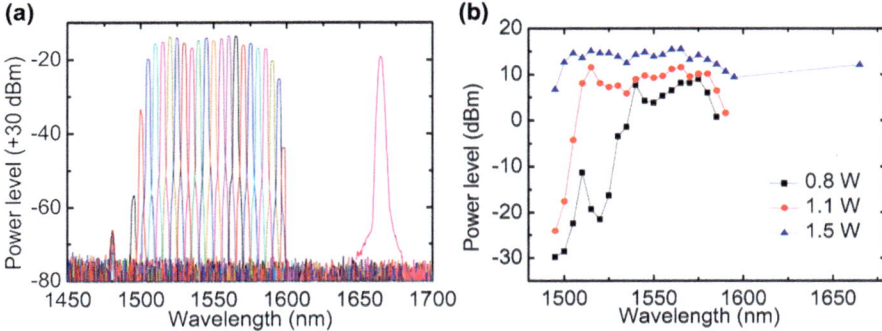

Fig. 5.10 (**a**) Emission spectra of tunable tellurite fiber Raman laser by tuning the tunable optical bandpass filter and free running 1665 nm Raman laser when the pump power of 1480 nm laser was fixed at 1.4 W. (**b**) Power dependence of the tunable tellurite fiber Raman laser (Ref. [17])

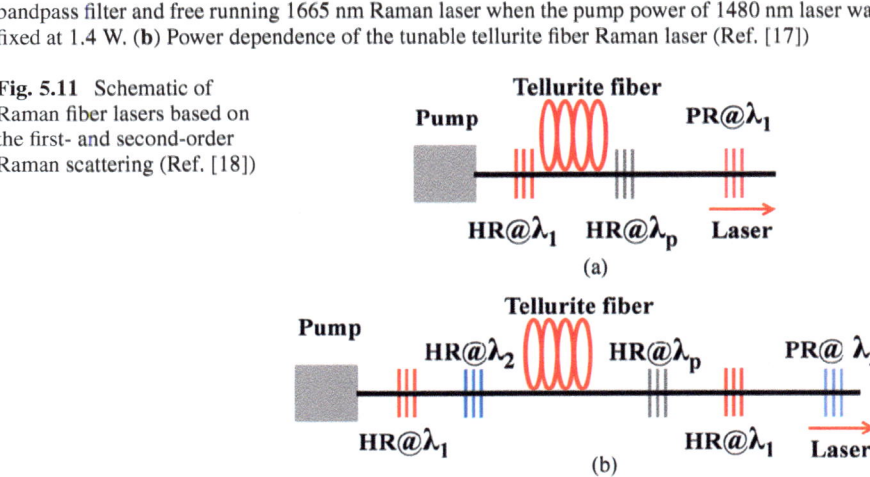

Fig. 5.11 Schematic of Raman fiber lasers based on the first- and second-order Raman scattering (Ref. [18])

Interestingly, since tellurite fiber has a wide MIR transmission window (up to 5 μm) and a large usable Raman shift (~22.3 THz), it could be used as the Raman gain medium for constructing MIR cascaded Raman lasers with long operation wavelengths. In 2015, Zhu et al. showed the possibility of obtaining 10-watt-level 3–5 μm Raman lasers using tellurite fiber through numerical simulations [18]. Figure 5.11 shows the schematics of Raman fiber lasers based on the first- and second-order Raman scattering. The pump laser they used is a 20 W 2.8 μm fiber laser. In all cases, a high-reflectance fiber Bragg grating (FBG) works as the cavity mirror, while a partially reflective acts as the output coupler. In order to increase the efficiency of the Raman fiber laser, a high-reflectance FBG at the pump wavelength is employed to reflect the residual pump back into the gain fiber. In the second-order Raman fiber laser, a pair of a high-reflectance FBGs is utilized to convert the pump light to first-order Raman light most efficiently. The curve 1 in Fig. 5.12 shows output power of the first-order Raman fiber laser as a function of wavelength at a pump power of 20 W when the Raman gain fiber length and the reflectance of the

Fig. 5.12 Output power of
the first- and second-order
Raman fiber laser as a
function of wavelength at a
pump power of 20 W (Ref.
[18])

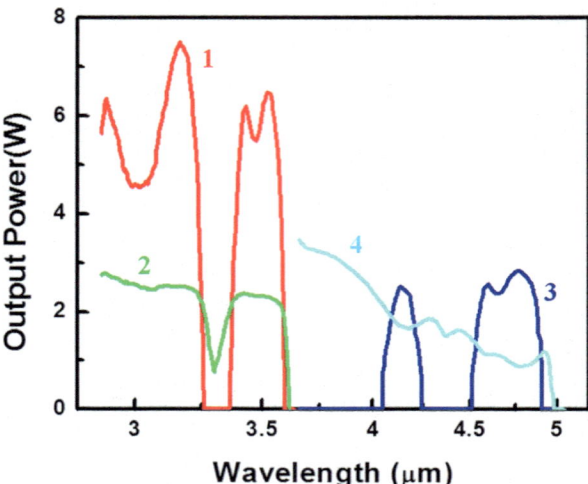

output FBG coupler are fixed at 1 m and 90%, respectively. The curve 2 in Fig. 5.12 shows output power of the first-order Raman fiber laser as a function of wavelength at a pump power of 20 W when the Raman gain fiber length and the reflectance of the output FBG coupler are fixed at 1.25 m and 98%, respectively, and the pump laser is at 2.8 μm. The curve 3 in Fig. 5.12 shows output power of the first-order Raman fiber laser as a function of wavelength at a pump power of 20 W when the Raman gain fiber length and the reflection of the output FBG coupler are fixed at 0.34 m and 40.5%, respectively, and the pump laser is at 2.9 μm. The curve 4 in Fig. 5.12 shows output power of the second-order Raman fiber laser as a function of wavelength at a pump power of 20 W when the Raman gain fiber length and the reflectance of the output FBG coupler are fixed at 0.19 m and 88%, respectively, and the first-order Stokes oscillates at 3.53 μm. The above results show the feasibility of obtaining a fiber laser at any wavelength between 3 and 5 μm by utilizing first-order and second-order Raman scattering in tellurite fiber. Although Zhu et al.'s report is simulated results, it is believed that MIR tellurite fiber Raman lasers will be realized by improving the quality of tellurite fibers in the near future.

5.2.2   Fluoride and Chalcogenide Fiber Raman Lasers

Since fluoride fiber has large damage threshold, wide MIR transmission window, and relatively large usable Raman shift, it can be used to construct high-power MIR Raman lasers. In 2011, Fortin et al. demonstrated the first Raman laser based on a fluoride fiber. The pump laser they used was a 9.6 W Tm^{3+}: silica CW fiber laser operating at a wavelength of 1940 nm [19]. The maximum output power measured was 580 mW at the first Stokes order of 2185 nm. Later, in 2012, they reported a fluoride fiber Raman laser delivering as much as 3.7 W of output power at a wavelength of 2.231 μm [20]. Those results show the potential of fluoride fibers for constructing high-power MIR Raman lasers. Even so, the quality of fluoride fibers needs to be improved for this purpose due to their poor chemical stability and low nonlinearity.

As aforementioned, although tellurite and fluoride fibers have large usable Raman shift and large damage threshold, both of them cannot be used to obtain Raman lasers with an operation wavelength of >5 μm due to very high material loss in the wavelength region of >5 μm. For this purpose, chalcogenide fibers are promising gain media for constructing >5 μm Raman fiber lasers. In 2003, Thielen et al. investigated numerically chalcogenide fiber Raman lasers and proposed the realization of a 8 W MIR fiber Raman laser operating at 6.46 μm pumped by a 10 W 5.59 μm carbon monoxide laser [21]. In 2013, Bernier et al. reported the first demonstration of a chalcogenide fiber Raman laser with an operation wavelength of ~3.34 μm and a maximum average output power of ~47 mW [22]. In 2014, a 3.77 μm chalcogenide fiber Raman laser with a maximum average output power of ~9 mW was reported by the same group [23]. Although the average output power of the reported 3.77 μm laser was low, it indicated that longer wavelength Raman lasers could be obtained by using cascaded Raman gain in chalcogenide fibers. For chalcogenide fiber Raman lasers, the main limitation for power scaling is low damage threshold and poor thermal stability of chalcogenide fibers, which might be relaxed by exploring new fiber materials and reducing the fiber loss.

Note that, most of works on fluoride and chalcogenide fiber Raman lasers are done by Réal Vallée's esteemed group. The detailed description on them can be found in Chap. 3.

5.3   MIR Raman Soliton Lasers Based on SOFs

SRS is an important nonlinear process that can provide optical gain and turn optical fibers into Raman fiber amplifiers and lasers. Interestingly, when optical soliton pulses with a width of ~1 ps or shorter are propagating inside a Raman-active optical fiber with anomalous dispersion at the operation wavelength, the spectral width of optical soliton is large enough that the low-frequency (red) spectral components of the soliton can be amplified via the Raman gain with its high-frequency (blue) spectral components as a pump. The amplification though energy transfer from blue to red spectral components continues along the fiber, causing a continuous redshift in the soliton spectrum. Such a phenomenon is called Raman-induced frequency shift (RIFS) [4]. RIFS can be used to generate tunable Raman soliton lasers. The efficiency of RIFS is related to the parameters of pump light and optical fibers. For the pump light, short pulse width and high peak power are required for RIFS generation, while highly efficient RIFS can be obtained in dispersion-engineered optical fibers.

Such a RIFS has been considered an immutable feature of sub-picosecond soliton propagation in conventional optical fibers for several decades. However, in 2003, Skryabin et al. demonstrated that RIFS could be canceled in a silica microstructured fiber with a negative dispersion slope and tunable redshifted dispersive waves were generated in the silica microstructured fiber [24]. It indicated that tunable redshifted dispersive waves could be generated in dispersion-engineered optical fibers.

Up to know, RIFS based on silica fibers have been widely investigated and tunable Raman soliton lasers from 700 to 2200 nm have been obtained [4]. In this section, we focused on MIR Raman soliton lasers based on SOFs.

5.3.1 MIR Raman Soliton Lasers Based on Tellurite Fibers

To achieve MIR Raman soliton lasers, various types of dispersion-engineered tellurite fibers have been designed and fabricated recently. In 2012, Liu et al. reported a tunable Raman soliton generation from 1.6 to 2.2 μm in a tellurite microstructured fiber with a core diameter of ∼1.1 μm pumped by a 1.56 μm fs fiber laser [25]. In 2015, Koptev et al. demonstrated tunable Raman soliton lasers from 2.1 to 2.65 μm in a tellurite microstructured fiber pumped by a 2 μm fs fiber laser [26]. Figure 5.13 shows the measured spectra at the output of 50 cm piece of tellurite microstructured fiber pumped at 2 μm, the corresponding calculated spectra, and the filtered Raman solitons in the time domain. They obtained Raman solitons up to 2.65 μm with the pump at 2 μm. The spectral widths of the most redshifted solitons correspond to Fourier transform-limited durations of order 100 fs.

Very recently, fluorotellurite microstructured fibers (FTMFs) with low loss in the spectral region of 0.4∼6 μm and good chemical and physical properties compared to fluoride fibers have been developed by us for constructing high-power mid-infrared fiber lasers [27–29]. Furthermore, by introducing the birefringence into FTMFs, two polarization axes including slow and fast axes exist in birefringent FTMFs. Since each polarization axis has its own dispersion profile, two sets of tunable mid-infrared dispersive waves can be generated in birefringent FTMFs. As a result, the tuning range of mid-infrared dispersive waves in birefringent FTMFs is larger than that in non-birefringent FTMFs. In 2016, Yao et al. reported tunable mid-infrared dispersive waves generation in a birefringent FTMF pumped by a 1560 nm femtosecond fiber laser [30]. The FTMFs based on TeO_2-BaF_2-Y_2O_3 glasses were fabricated by using a rod-in-tube method. The FTMF had a birefringence of 3.5×10^{-2} and two zero-dispersion wavelengths (ZDWs) for each polarization axis. As the pump laser was polarized along the fast (or slow) axis of the FTMF, tunable mid-infrared dispersive waves from 2680 to 2725 nm (or from 2260 to 2400 nm) were generated in the FTMF.

In their experiments, they fabricated birefringent FTMFs by using the rod-in-tube method. The glass compositions of the core and cladding of the FTMFs were $70TeO_2$-$20BaF_2$-$10Y_2O_3$ and $65TeO_2$-$25BaF_2$-$10Y_2O_3$, respectively. For the above glass system, the addition of Y_2O_3 was not only to avoid the occurrence of

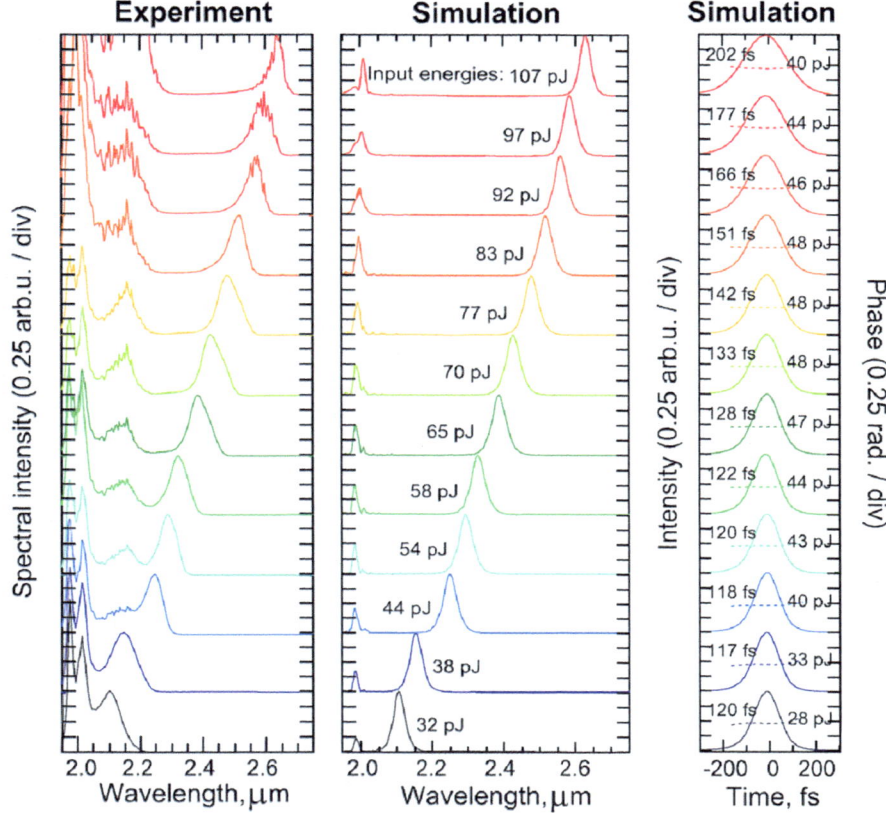

Fig. 5.13 Measured spectra at the output of 50 cm piece of tellurite microstructured fiber pumped at 2 μm, the corresponding calculated spectra, and the filtered Raman solitons in the time domain (Ref. [26])

crystallization during fiber drawing but also to obtain glasses with high transition temperatures (∼424 °C). Such a value was much higher than that of previously reported tellurite or fluorotellurite glasses, which would be preferable for obtaining high-power fiber-based light sources. The absorption coefficient at ∼3.1 μm caused by the residual hydroxyl groups in the glass was ∼0.08 cm^{-1}, which indicated the loss in the mid-infrared range was relatively low for generating mid-infrared light sources. The inset of Fig. 5.14a shows the scanning electron microscope image of the fabricated birefringent FTMF. The fiber had a "wagon wheel" structure, consisting of an unsymmetrical solid core with a size of 1.08 × 0.7 μm surrounded by six air holes. During fiber drawing, dry nitrogen gas was pumped into the holes of the fiber preform. Therefore, they could fabricate FTMFs with varied diameter ratio of the holey region to core by varying the pressure of dry nitrogen gas. The birefringence of the FTMF (shown in the inset of Fig. 5.14a) at 1560 nm

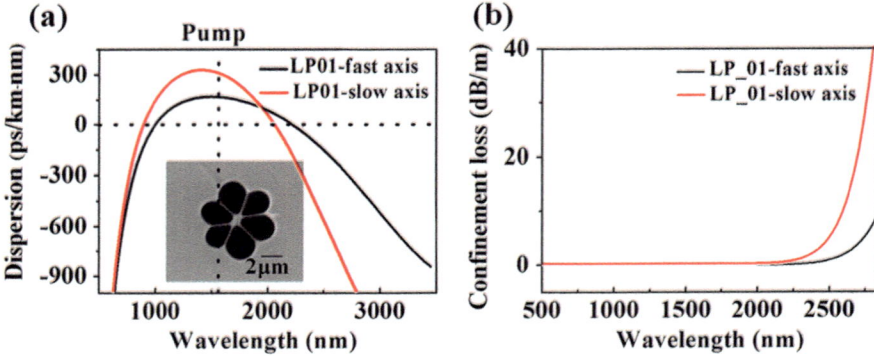

Fig. 5.14 (a) The calculated dispersions of propagating LP01 modes in the fast and slow axes of the FTMF. Inset, scanning electron micrograph of birefringent FTMFs. (b) The calculated confinement losses of the LP01 modes of the fast and slow axes, respectively (Ref. [30])

was calculated to be 3.5×10^{-2} by using the formula $B_m = |n_{x}-n_{y}|$, where the n_x and n_y are the effective indexes of polarization modes for the fast and slow axis, which was large enough for observing the birefringence effect in the above fiber. Figure 5.14a shows the calculated group velocity dispersion (GVD) profiles of the fundamental propagation mode in the FTMF by using the full vectorial finite difference method. The fiber had two zero-dispersion wavelengths (ZDWs) for each polarization axis, which was required for generating tunable redshifted dispersive waves. The ZDWs for the fast axis were 1000 and 2224 nm, respectively. Similarly, the ZDWs for the slow axis were 897 and 2042 nm, respectively. The calculated nonlinear coefficients (γ) at 1560 nm of the fundamental propagation modes were 4451 and 5322 $km^{-1} W^{-1}$ for the fast and slow axes, respectively, by using a nonlinear refractive index of $1.4 \times 10^{-18} m^2 W^{-1}$ for fluorotellurite glasses. Figure 5.14b shows the calculated confinement losses of the fundamental propagation mode for the fast and slow axes by using the full vectorial finite difference method. The confinement losses for both cases increased very much when the operating wavelength was longer than 2.8 μm. The background loss at 1560 nm of the FTMF was measured to be 0.14 dB/cm by using a cutback method.

To clarify the potential of the birefringent FTMF for the generation of mid-infrared dispersive waves, they performed the following experiments, and the experimental setup was shown in Fig. 5.15. A 1560 nm femtosecond fiber laser with a pulse width of ∼150 fs, a repetition rate of ∼50 MHz, and a maximum output power of 500 mW (the corresponding pulse energy: ∼10 nJ) was used as the pump source. The pump laser was launched into the birefringent FTMF through a couple of aspheric lens, and the measured coupling efficiency, defined as the launched power divided by the power incident on the lens, was about 25%. The polarization state of the pump laser was controlled by a polarization controller. The output signals were monitored by using an optical spectrum analyzer (OSA Yokogawa

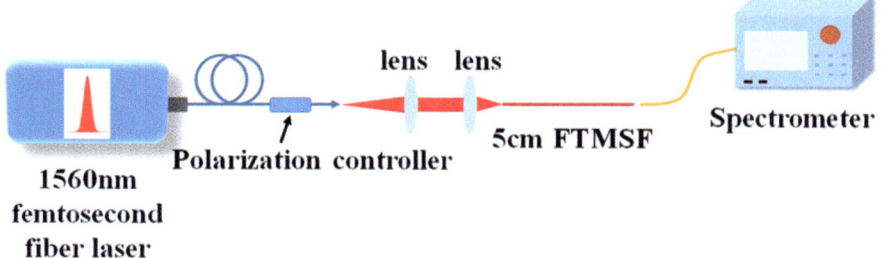

Fig. 5.15 Schematic diagram of the experimental setup for mid-infrared dispersive wave generation (Ref. [30])

6375A, the spectral range: 1200 nm ∼ 2400 nm) and a grating spectrometer with an InSb detector through a large mode field area ZBLAN fiber cable mechanically spliced the output end of the birefringent FTMF.

Figure 5.16a shows the dependence of the measured output spectra from a 5 cm long birefringent FTMF on the launched average power of the 1560 nm femtosecond laser when the pump laser was polarized along the fast axis of the fiber. Since the dispersion for the fast axis was anomalous in the wavelength range of 1000∼2224 nm, the pumping wavelength was located at the anomalous dispersion region of the birefringent FTMF. The group velocity dispersion value (β_2) at 1560 nm for the FTMF was about −218 ps^2/km, and the calculated dispersion length was ∼10 cm. For the 5 cm long birefringent FTMF, the first soliton was generated by the soliton fission as the average pump power reached ∼25 mW. With further increasing the average pump power, the soliton self-frequency shift (SSFS) was clearly observed due to the Raman effect, where the blue spectral part of the soliton pumped the red part, causing a continuous redshift in the soliton spectrum. Interestingly, as the average pump power was increased to 39.9 mW, the first Raman soliton met the second ZDW point with a negative dispersion slope (β_3 <0) for the fast axis, and the SSFS cancelation occurred. Meanwhile, the redshifted dispersive wave with the wavelength at 2725 nm was observed. With further increasing the pump power, the redshifted dispersive wave shifted from 2725 to 2680 nm. Besides, we also found that the second Raman soliton appeared as soon as the average 1560 nm femtosecond pulse power reached 39.7 mW. The above phenomena could be explained as follows. As the Raman soliton shifted into the spectral region in which the dispersion slope of the fiber was negative (β_3 <0), the soliton would emit a radiation band with a wavelength of longer than the second ZDW through the Cherenkov mechanism. Because of the momentum conservation, as the dispersive wave was emitted in the normal GVD regime, the soliton should recoil further into the anomalous GVD regime; the spectral recoil mechanism was responsible for the suppression of SSFS. As far as the spectral recoil was large enough, the SSFS cancelation occurred, and the dispersive wave was amplified with an increase of the pump power. The above experimental results showed that tunable redshifted

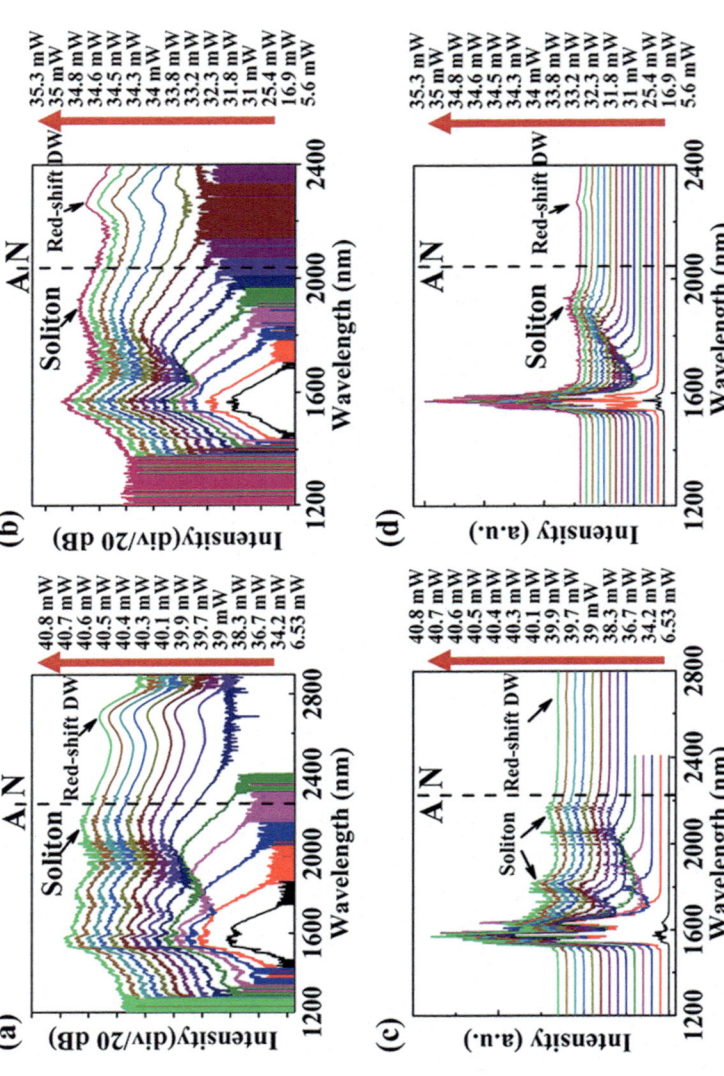

Fig. 5.16 (a) The dependence of the measured output spectra from a 5 cm long birefringent FTMF when the pump laser was polarized along the fast axis of the fiber. (b) The dependence of the output spectra from a 5 cm long birefringent FTMF on the average pump power of the 1560 nm femtosecond laser when the pump laser was polarized along the slow axis of the fiber. (c) The dependence of the measured output spectra plotted in linear scale from a 5 cm long birefringent FTMF on the launched average power of the 1560 nm femtosecond laser when the pump laser was polarized along the fast axis of the fiber. (d) The dependence of the measured output spectra plotted in linear scale from a 5 cm long birefringent FTMF on the launched average power of the 1560 nm femtosecond laser when the pump laser was polarized along the slow axis of the fiber (Ref. [30])

dispersive waves from 2680 to 2725 nm could be obtained in the birefringent FTMF when 1560 nm femtosecond laser was polarized along the fast axis of the fiber.

Figure 5.16b shows the dependence of the output spectra from a 5 cm long birefringent FTMF on the average pump power of the 1560 nm femtosecond laser when the pump laser was polarized along the slow axis of the fiber. Since the dispersion for the fast axis was anomalous in the wavelength range of 897 $\sim$ 2042 nm, the pumping wavelength was also located at the anomalous dispersion region. The β_2 at 1560 nm for the slow axis of FTMF was about -388 ps^2/km, and the calculated dispersion length was $\sim$5.6 cm. As the pump power reached 31 mW, the first Raman soliton was observed, and the redshifted dispersive wave at 2400 nm was observed when the pump power reached 34 mW. With further increasing the pump power to 35.3 mW, the redshifted dispersive wave can be tuned from 2400 to 2260 nm. Figure 5.16c, d shows the dependence of the measured output spectra which were plotted in linear scale on the launched average power of the 1560 nm femtosecond laser when the pump laser was polarized along the fast and slow axes of the fiber, respectively.

The above results showed that two sets of tunable mid-infrared dispersive waves could be generated in birefringent FTMFs by varying the polarization state of the pump laser and the tuning range of mid-infrared dispersive waves in birefringent FTMFs could be larger than that in non-birefringent FTMFs.

In addition, they performed numerical simulations by solving the generalized nonlinear Schrödinger equations. In the simulations, they took the calculated chromatic dispersion data shown in Fig. 5.14a; the aforementioned nonlinear coefficients; the pumping laser with an operating wavelength of $\sim$1560 nm, a pulse width of $\sim$150 fs, a repetition rate of $\sim$50 MHz, and a maximum average pump power of $\sim$40.8 mW (the corresponding pulse energy: $\sim$ 810 pJ); and the Raman response function derived from the Raman gain spectrum of tellurite glass. Figure 5.17a, b showed the dependence of the simulated supercontinuum spectra on the average pump power of the pump laser for the fast and slow axis of the fiber, respectively. The dashed curves of Fig. 5.17c, d show the simulated supercontinuum spectra as the average pump power was 40.8 and 35.3 mW for the fast and slow axes of the fiber, respectively. The solid curves of Fig. 5.17c, d show the corresponding measured output spectra as the average pump power was 40.8 and 35.3 mW for the fast and slow axes of the fiber, respectively. The simulated results agreed with the corresponding measured ones for both the fast and slow axes of the fiber. The discrepancy between the simulated and experimental results might be caused by large loss at long wavelengths (including the confinement loss of the fiber and the materials loss caused by the absorption of residual OH- in the fiber material) of the fiber we used. Figure 5.17e, f showed the simulated spectrograms of output pulse corresponding to the simulation results of Fig. 5.17c, d. The spectrograms display the temporal and spectral characteristics of the soliton and dispersive wave. From the spectrograms, the pulse width of the dispersive waves at 2725 and 2260 nm was estimated to be 1.2 and 1.05 ps, respectively.

Note that, by using the designed birefringent FTMFs with varied ZDWs as the nonlinear media and a 2 μm femtosecond laser as the pump source, the operating

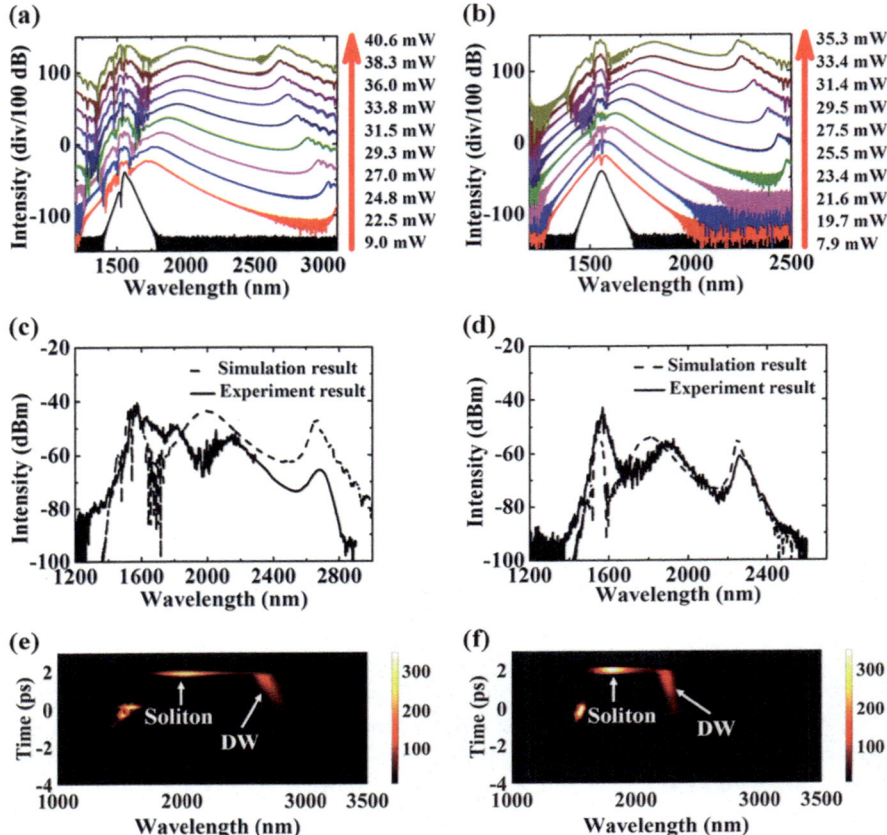

Fig. 5.17 (a) The dependence of the simulated supercontinuum spectra on the average pump power of the pump laser for the fast axis of the fiber. (b) The dependence of the simulated supercontinuum spectra on the average pump power of the pump laser for the slow axis of the fiber. (c) The simulated and measured supercontinuum spectra as the average pump power was 40.8 mW for the fast axes of the fiber. (d) The simulated and measured supercontinuum spectra as the average pump power was 35.3 mW for the slow axes of the fiber. (e, f) The simulated spectrograms of output pulse (Ref. [30])

wavelength of mid-infrared dispersive waves can be extended to longer wavelength (>3 μm) (see Fig. 5.18). To verify the above idea, they designed the FTMFs, which have the similar structure as the current fiber in the inset of Fig. 5.14a and varied core sizes of 1.08×0.7, 1.3×0.77, 1.19×0.84, 1.41×0.91, 1.51×0.98, and 1.62×1.05 μm (named as Fiber 1–6). Figure 5.18a shows the calculated confinement losses of those fibers. It is evidently that the confinement loss can be significantly reduced by increasing the core size of the FTMF. Figure 5.18b shows the calculated GVD profiles of those fibers. The corresponding second ZDWs

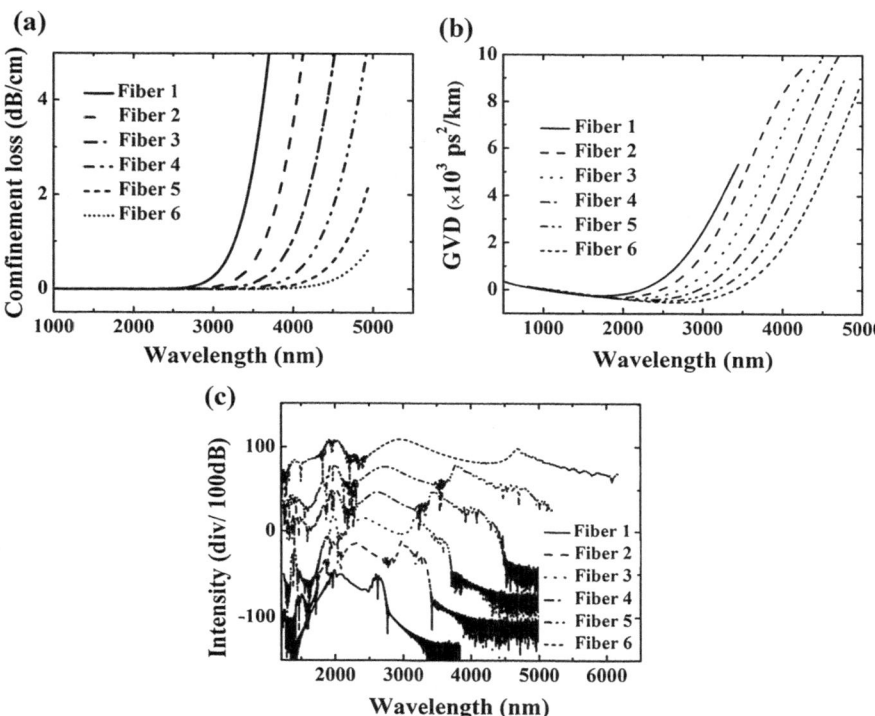

Fig. 5.18 (a) The confinement losses of the designed microstructured fibers with the similar structure and the core diameters of 1.08, 1.3, 1.19, 1.41, 1.51, and 1.62 μm. (b) The calculated GVD profiles of the designed fibers. (c) The simulated supercontinuum spectra when the pump laser is polarized along the fast axis of Fiber 1–6 (Ref. [30])

were 2224, 2468, 2704, 2937, 3171, and 3398 nm for Fiber 1, 2, 3, 4, 5, and 6, respectively. Based on those GVD profiles, we simulated the redshifted dispersive wave generation by solving the generalized nonlinear Schrödinger equation. Here, a 2 μm femtosecond laser with a pulse width of ~120 fs and a peak power of ~1800 W was used as the pump source. Figure 5.18c shows the simulated supercontinuum spectra when the pump laser is polarized along the fast axis of Fiber 1–6. The simulated results showed that, by designing the birefringent FTMFs with varied ZDWs, the operating wavelength of mid-infrared dispersive waves can be extended to longer wavelength (>3 μm).

Another efficient way to extending the operation wavelength of Raman soliton lasers is to use a long wavelength laser as the pump source. In 2016, Kedenburg et al. demonstrated a Raman soliton laser at 4.2 μm by using a W-type index tellurite fiber pumped at 3.2 μm [31]. The output power of the Raman soliton laser was about 130 mW.

5.3.2   MIR Raman Soliton Lasers Based on Fluoride Fibers

The damage threshold reported for fluoride glass pumped by a femtosecond laser was about 5 TW/cm^2, which is nearly half of that of silica glass. Therefore, fluoride fiber is suitable for generating high-power mid-infrared light source (see Chap. 5). In 2014, Liu et al. firstly showed the possibility of obtaining widely tunable mid-infrared Raman soliton source in fluoride fibers and demonstrated a widely tunable Raman soliton source (1.93–3.95 μm) in a fluoride photonic crystal fiber (PCF) pumped by a 1.93 μm fiber laser with a pulse width of 200 fs through numerical simulations [32].

The fluoride microstructured fiber used in their simulation is composed of five rings of air holes with a triangular structure in fiber cladding, which can be fabricated using a stack-and-draw or extrusion method (the inset of Fig. 5.19a). In order to get endlessly single-mode operation (the relative hole size constant is less than 0.45), the relative hole size constant of our fiber is set as 0.4, the corresponding air hole size is 2 μm, and the pitch is 5 μm. Figure 5.19a shows the calculated dispersion data of the fundamental propagation mode in the fluoride fiber. Its zero-dispersion wavelength is about 1.34 μm. The chromatic dispersion value at the pump wavelength ~1.93 μm is about 27.7 ps/km.nm, and the corresponding GVD coefficient is about −54.7 ps^2/km. The calculated nonlinear coefficient at 1.93 μm is about 1.7433 km^{-1}W^{-1}, by using a nonlinear refractive index of 2.1 × 10^{-20} m^2 W^{-1}. Figure 5.19b shows the calculated confinement loss of fluoride PCF. The calculated confinement loss is about 1 dB/m at 4 μm, so it has little effect on Raman soliton generation in the region of 2–4 μm.

Raman soliton generation is induced by Raman gain, so the use of correct Raman gain spectrum is very important for achieving the accuracy of simulations. The Raman gain spectrum is given by

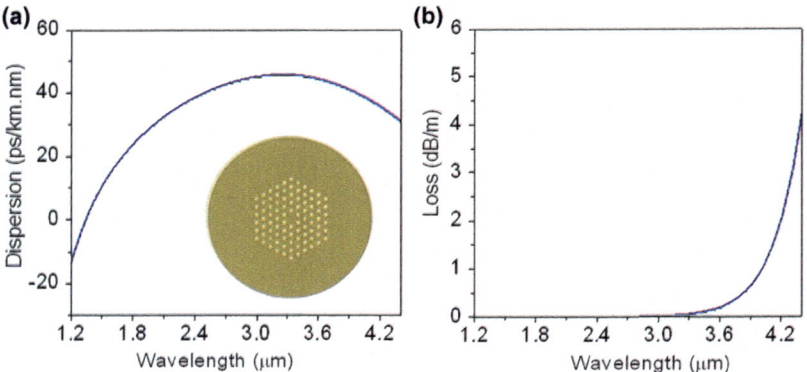

Fig. 5.19 (a) Calculated chromatic dispersion data of the fundamental propagating mode in the fluoride fiber, we used. Inset of Fig. 1(a): the cross section of the fluoride fiber. (b) The calculated confinement loss of the fiber (Ref. [32])

$$g_R (\Delta\omega) = \frac{\omega_0}{cn_0} f_R \chi^{(3)} \, \text{Im} \left[\tilde{h}_R (\Delta\omega) \right] \tag{5.4}$$

Where $\Delta\omega = \omega - \omega_0$ and Im stands for the imaginary part. The real part of $\tilde{h} (\Delta\omega)$ can be obtained from the imaginary part by using the Kramers–Kronig relations. The Fourier transformation of $\tilde{h} (\Delta\omega)$ can provide the Raman response function $h_R(t)$. An approximate analytic form of the Raman response function is given by

$$h_R(t) = \frac{\tau_1^2 + \tau_2^2}{\tau_1 \tau_2^2} \exp(-t/\tau_2) \sin(t/\tau_1) \tag{5.5}$$

For fluoride fibers, the derived values from the Raman gain spectrum of fluoride glasses are $\tau_1 = 9$ fs, $\tau_2 = 134$ fs, and $f_R = 0.1929$.

The initial incident pulse we choose has a hyperbolic-secant field profile, which is common for passively mode-locked fiber lasers.

$$A(0, T) = \sqrt{P_0} \sec h(T/T_0) \tag{5.6}$$

Where P_0 is the peak power and T_0 is the input pulse width.

The split-step Fourier method is used to solve the generalized nonlinear Schrödinger equation (GNLSE). To make sure of the accuracy of numerical simulations, a fourth-order Runge–Kutta algorithm is used in our calculations, and 2^{15} time and frequency discretization points and a longitudinal step size <5 μm were used in our simulations.

In their simulations, the pumping wavelength is set as 1.93 μm, a typical operating wavelength of Tm^{3+}-doped short pulse fiber lasers. The curves in Fig. 5.20a show the simulated normalized output pulse profiles from a 1 m long fluoride microstructured fiber pumped by a 1.93 μm fiber laser with a pulse width of ~200 fs and a peak power of (1) incident pulse, (2) 40 kW, (3) 90 kW, (4) 190 kW, (5) 300 kW, and (6) 800 kW. The corresponding output pulse width are 69 fs, 73 fs, 76 fs, 85 fs, and 99 fs, respectively, when the pump peak power are set as 40 kW, 90 kW, 190 kW, 300 kW, and 800 kW. A temporal window of 190 ps is used in the simulations. As can be seen, such a temporal window is wide enough to preclude the occurrence of spurious interferences between the leading and the trailing edge of the output pulse, even when the pump peak power is increased to 800 kW (This value corresponds to a pump density of ~2 TW/cm^2, which is lower than the damage threshold of fluoride glass).

The curves in Fig. 5.20b show the corresponding simulated RIFS spectra output from a 1 m long fluoride microstructured fiber pumped by a 1.93 μm fiber laser with a pulse width of 200 fs and a peak power of (1) incident pulse, (2) 40 kW, (3) 90 kW, (4) 190 kW, (5) 300 kW, and (6) 800 kW. Widely tunable soliton source (1.93–3.95 μm) was obtained in the fluoride microstructured fiber pumped by a 1.93 μm fiber laser with a pulse width of 200 fs. As we know, the soliton is generated by the balance between group velocity dispersion and self-phase modulation. The calculated dispersion length is about 0.24 m by using a group velocity dispersion

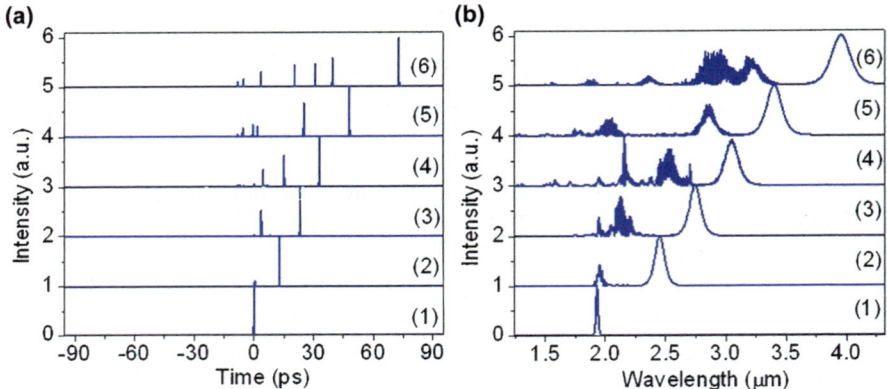

Fig. 5.20 (**a**) Normalized output pulse profiles from a 1 m long fluoride fiber pumped by a 1.93 lm fiber laser with a pulse width of 200 fs and a peak power of (1) incident pulse, (2) 40 kW, (3) 90 kW, (4) 190 kW, (5) 300 kW, and (6) 800 kW. (**b**) Corresponding normalized output spectrum (Ref. [32])

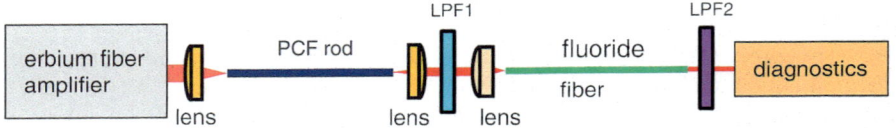

Fig. 5.21 Schematic of the experimental setup for RIFS in the fluoride fiber (Ref. [33])

value of -54.7 ps^2/km (The group velocity dispersion value at the pump wavelength of $\sim$1.93 μm for the fluoride fiber we used is about -54.7 ps^2/km). When the fiber length is 1 m, the soliton can be effectively generated as the pump power increases. For the soliton $A(0,T) = N \operatorname{sech}(T/T_0)$, the soliton order of the input pulse N is determined by both the pulse and fiber parameters through $N^2 = L_D/L_{NL}$. In case of the fundamental soliton ($N = 1$), both the spectral and temporal profiles remain unchanged during propagation, while higher-order solitons ($N \geq 2$) undergo periodic spectral and temporal evolution when the peak power is lower. When the peak power is high enough, the soliton fission occurs, and the input pulses break up into several fundamental solitons, as shown in Fig. 5.20a. Consequently, the generated solitons shift into the red spectral region due to the stimulated Raman scattering effect, as shown in Fig. 5.20b. With further increasing the pump peak power, the soliton collision and interaction processes might occur. The above simulated results showed that a high-power widely tunable mid-infrared Raman soliton source (1.93∼3.95 μm) could be realized in fluoride fibers pumped by short pulse fiber lasers.

As expected, in 2016, Tang et al. reported generation of intense 100 fs Raman solitons tunable from 2 to 4.3 μm in a 2 m long InF$_3$ fiber [33]. Figure 5.21 shows the schematic of the experimental setup for RIFS in the fluoride fiber. The pump

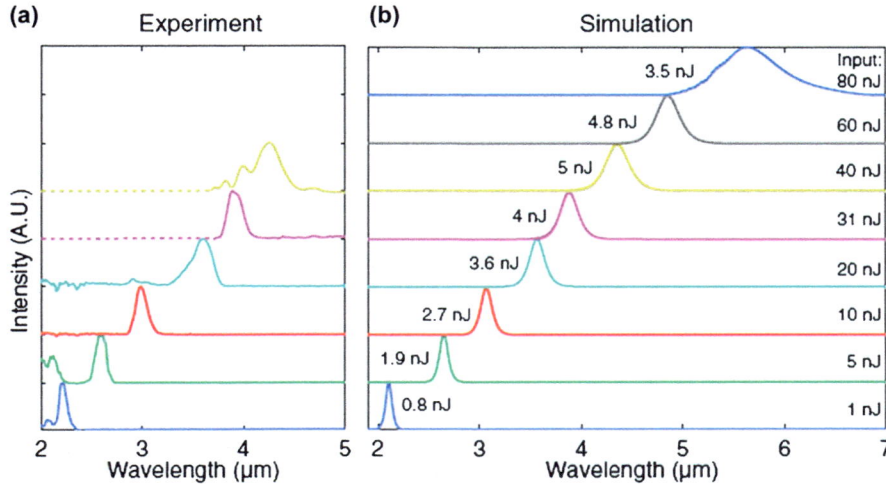

Fig. 5.22 (**a, b**) Measured and simulated spectra of the Raman soliton at different wavelengths (Ref. [33])

laser they used was a 1.9 μm Raman soliton laser with 120 nJ pulse energy and 70 fs duration. The measured and simulated spectra of the Raman soliton at different wavelengths were shown in Fig. 5.22. 100 fs Raman soliton with nanojoule energies, wavelength tunable from 2 to 4.3 μm, were generated by the RIFS in fluoride fibers. With peak powers around 50 kW, these are the most intense MIR pulses generated by a fiber source to date. Subsequently, in 2016, Duval et al. demonstrated watt-level fiber-based Raman soliton laser source tunable from 2.8 to 3.6 μm by using fluoride fibers pumped by a 2.8 μm mode-locked fiber laser [34]. Those fiber-based, broadly tunable source of short pulses should facilitate numerous applications in the important 3–5 μm region.

5.3.3 MIR Raman Soliton Lasers Based on Chalcogenide Fibers

Since the zero-dispersion wavelengths of As_2S_3 and As_2Se_3 glasses are 4.89 and 7.22 μm, respectively, one needs to shift the zero-dispersion wavelength of the chalcogenide fiber to short wavelength region or use the long wavelength fs laser as the pump source for achieving MIR Raman soliton lasers. In 2014, Cheng et al. reported tunable Raman soliton generation from 2.986 to 3.419 μm in a chalcogenide microstructure fiber pumped by an optical parametric oscillator operating at 2.8 μm [35]. In 2016, Cheng et al. demonstrated the generation of MIR dispersive wave at 4.9 μm in an all-solid $AsSe_2$–As_2S_5 microstructured optical fiber with two zero-dispersion wavelengths [36].

Note that, most of works on MIR Raman soliton lasers are related to MIR supercontinuum generation. The detailed description on MIR supercontinuum light sources can be found in Chap. 4.

References

1. Zhang, L., Liu, C., Jiang, H., Qi, Y., He, B., Zhou, J., Xijia, G., Feng, Y.: Kilowatt Ytterbium-Raman fiber laser. Opt. Express. **22**, 18483–18489 (2014)
2. Ma, P., Zhang, H., Huang, L., Wang, X., Zhou, P., Liu, Z.: Kilowatt-level near-diffraction-limited and linear-polarized ytterbium-Raman hybrid nonlinear amplifier based on polarization selection loss mechanism. Opt. Express. **23**, 26499–26508 (2015)
3. Zhang, L., Jiang, H., Yang, X., Pan, W., Feng, Y.: Ultra-wide wavelength tuning of a cascaded Raman random fiber laser. Opt. Lett. **41**, 215–218 (2016)
4. Agrawal, G.P.: Nonlinear Fiber Optics, 5th edn. Academic, New York (2013)
5. Xia, C., Xu, Z., Islam, M.N., Terry Jr., F.L., Freeman, M.J., Zakel, A., Mauricio, J.: 10.5 W time-averaged power mid-IR Supercontinuum generation extending beyond 4 um with direct pulse pattern modulation. IEEE J. Sel. Top. Quantum Electron. **15**, 422–434 (2009)
6. O'Donnell, M.D., Richardson, K., Stolen, R., Seddon, A.B., Furniss, D., Tikhomirov, V.K., Rivero, C., Ramme, M., Stegeman, R., Stegeman, G., Couzi, M., Cardinal, T.: Tellurite and Fluorotellurite glasses for Fiberoptic Raman amplifiers: glass characterization, optical properties, Raman gain, preliminary Fiberization, and fiber characterization. J. Am. Ceram. Soc. **90**, 1448–1457 (2007)
7. Aggarwal, I.D., Sanghera, J.S.: Development and applications of chalcogenide glass optical fibers at NRL. J. Optoelectronics Adv. Mater. **4**, 665–678 (2002)
8. Stuart, B.C., Feit, M.D., Herman, S., Rubenchik, A.M., Shore, B.W., Perry, M.D.: Optical ablation by high-power short-pulse lasers. J. Opt. Soc. Am. B. **13**, 459–468 (1996)
9. Mori, A.: Tellurite-based fibers and their applications to optical communication networks. J. Cerma. Soc. Jpn. **116**, 1040–1051 (2008)
10. Toth, L.M., Quist, A.S., Boyd, G.E.: Raman spectra of zirconium(iV) fluoride complex ions in fluoride melts and polycrystalline solids. J. Phys. Chem. **77**, 1384–1388 (1973)
11. Wang, J.S., Vogel, E.M., Snitzer, E.: Tellurite glass: a new candidate for fiber devices. Opt. Mater. **3**, 187–203 (1994)
12. Mori, A., Kobayashi, K., Yamada, M., Kanamori, T., Oikawa, K., Nishida, Y., Ohishi, Y.: Low noise figure broadband tellurite-based Er^{3+} −doped fiber amplifiers. Electron. Lett. **34**, 887–888 (1998)
13. Mori, A., Masuda, H., Shikano, K., Oikawa, K., Kato, K., Shimizu, M.: Ultra-wideband tellurite-based Raman fiber amplifier. Electron. Lett. **37**, 1442–1443 (2001)
14. Masuda, H., Mori, A., Shikano, K., Shimizu, M.: Design and spectral characteristics of gain-flattened tellurite-based fiber Raman amplifiers. J. Lightw. Technol. **24**, 504–515 (2006)
15. Namiki, S., Emori, Y.: Ultrabroad-band Raman amplifiers pumped and gain-equalized by wavelength-division-multiplexed high power laser diodes. IEEE J. Sel. Topics Quantum Electron. **7**, 3–16 (2001)
16. Qin, G., Jose, R., Ohishi, Y.: Design of ultimate gain-flattened O-, E-, and S+ C+ L ultrabroadband fiber amplifiers using a new fiber Raman gain medium. J. Lightw. Technol. **25**, 2727–2738 (2007)
17. Qin, G., Liao, M., Suzuki, T., Mori, A., Ohishi, Y.: Widely tunable ring-cavity tellurite fiber Raman laser. Opt. Lett. **33**, 2014–2016 (2008)
18. Zhu, G., Geng, L., Zhu, X., Li, L., Chen, Q., Norwood, R.A., Manzur, T., Peyghambarian, N.: Towards ten-watt-level 3-5 μm Raman lasers using tellurite fiber. Opt. Express. **23**, 7559–7573 (2015)

19. Fortin, V., Bernier, M., Carrier, J., Vallée, R.: Fluoride glass Raman fiber laser at 2185 nm. Opt. Lett. **36**, 4152–4154 (2011)
20. Fortin, V., Bernier, M., Faucher, D., Carrier, J., Vallée, R.: 3.7 W fluoride glass Raman fiber laser operating at 2231 nm. Opt. Express. **20**, 19412–19419 (2012)
21. Thielen, P.A., Shaw, L.B., Sanghera, J.S., Aggarwal, I.D.: Modeling of a mid-IR chalcogenide fiber Raman laser. Opt. Express. **11**, 3248–3253 (2003)
22. Bernier, M., Fortin, V., Caron, N., El-Amraoui, M., Messaddeq, Y., Vallée, R.: Mid-infrared chalcogenide glass Raman fiber laser. Opt. Lett. **38**, 127–129 (2013)
23. Bernier, M., Fortin, V., El-Amraoui, M., Messaddeq, Y., Vallée, R.: 3.77 μm fiber laser based on cascaded Raman gain in a chalcogenide glass fiber. Opt. Lett. **39**, 2052–2055 (2014)
24. Skryabin, D.V., Luan, F., Knight, J.C., Russell, P.S.J.: Soliton self-frequency shift cancellation in photonic crystal fibers. Science. **301**, 1705–1708 (2003)
25. Liu, L., Tian, Q., Liao, M., Zhao, D., Qin, G., Ohishi, Y., Qin, W.: All-optical control of group velocity dispersion in tellurite photonic crystal fibers. Opt. Lett. **37**, 5124–5126 (2012)
26. Koptev, M.Y., Anashkina, E.A., Andrianov, A.V., Dorofeev, V.V., Kosolapov, A.F., Muravyev, S.V., Kim, A.V.: Widely tunable mid-infrared fiber laser source based on soliton self-frequency shift in microstructured tellurite fiber. Opt. Lett. **40**, 4094–4097 (2015)
27. Yao, C., He, C., Jia, Z., Wang, S., Qin, G., Ohishi, Y., Qin, W.: Holmium-doped fluorotellurite microstructured fibers for 2.1 μm lasing. Opt. Lett. **40**, 4695–4698 (2015)
28. Yao, C., Jia, Z., He, C., Wang, S., Zhang, L., Feng, Y., Ohishi, Y., Qin, G., Qin, W.: 2.074-μm lasing from Ho^{3+}-doped fluorotellurite microstructured fibers pumped by a 1120-nm laser. IEEE Photon. Technol. Lett. **28**, 1084–1087 (2016)
29. Wang, F., Wang, K., Yao, C., Jia, Z., Wang, S., Wu, C., Qin, G., Ohishi, Y., Qin, W.: Tapered fluorotellurite microstructured fibers for broadband supercontinuum generation. Opt. Lett. **41**, 634–637 (2016)
30. Yao, C., Zhao, Z., Jia, Z., Li, Q., Hu, M., Qin, G., Ohishi, Y., Qin, W.: Mid-infrared dispersive waves generation in a birefringent fluorotellurite microstructured fiber. Appl. Phys. Lett. **109**, 101102–101105 (2016)
31. Kedenburg, S., Steinle, T., Mörz, F., Steinmann, A., Nguyen, D., Rhonehouse, D., Zong, J., Chavez-Pirson, A., Giessen, H.: Solitonic supercontinuum of femtosecond mid-IR pulses in W-type index tellurite fibers with two zero dispersion wavelengths. APL Photonics. **1**, 086101–086110 (2016)
32. Liu, L., Qin, G.-S., Tian, Q.-j., Zhao, D., Qin, W.-P.: Numerical investigation of mid-infrared Raman soliton source generation in endless single mode fluoride fibers. J. Appl. Phys. **115**, 163102–163104 (2014)
33. Tang, Y., Wright, L.G., Charan, K., Wang, T., Xu, C., Wise, F.W.: Generation of intense 100 fs solitons tunable from 2 to 4.3 μm in fluoride fiber. Optica. **3**, 948–951 (2016)
34. Duval, S., Gauthier, J.-C., Robichaud, L.-R., Paradis, P., Olivier, M., Fortin, V., Bernier, M., Piché, M., Vallée, R.: Watt-level fiber-based femtosecond laser source tunable from 2.8 to 3.6 μm. Opt. Lett. **41**, 5294–5297 (2016)
35. Cheng, T., Kanou, Y., Asano, K., Deng, D., Liao, M., Matsumoto, M., Misumi, T., Suzuki, T., Ohishi, Y.: Soliton self-frequency shift and dispersive wave in a hybrid four-hole $AsSe_2$-As_2S_5 microstructured optical fiber. Appl. Phys. Lett. **104**, 121911–121914 (2014)
36. Cheng, T., Tuan, T.H., Liu, L., Xue, X., Matsumoto, M., Tezuka, H., Suzuki, T., Ohishi, Y.: Fabrication of all-solid $AsSe_2$-As_2S_5 microstructured optical fiber with two zero-dispersion wavelengths for generation of mid-infrared dispersive waves. Appl. Phys. Express. **9**, 022502–022504 (2016)

Chapter 6
Distributed Feedback Raman and Brillouin Fiber Lasers

Paul S. Westbrook, Kazi S. Abedin, and Tristan Kremp

6.1 Introduction

Distributed feedback (DFB) lasers are waveguide lasers in which the optical gain region overlaps with a distributed Bragg grating reflector that provides the feedback required to achieve lasing. They were motivated by a desire for compact, integrated, narrow-linewidth lasers for applications in telecom, sensing, and frequency conversion. They were first demonstrated in the early 1970s [1–5] and implemented in planar waveguide semiconductor fabrication platforms. While the first DFB lasers used Bragg gratings with a uniform phase, it was later shown [6, 7] that a single π phase shift in the periodic index modulation of the Bragg grating could define a laser cavity with low threshold and narrow-linewidth performance. In effect, the phase-shifted DFB is the shortest possible distributed Bragg reflector (DBR) waveguide, with a cavity length equal to the effective penetration depth into the Bragg grating. In general, DBR waveguides also use Bragg gratings as the optical feedback mechanism, but these gratings are spatially separated from the gain region, in contrast to DFB lasers. Semiconductor DFBs have been used extensively in telecommunications and other applications that require narrow-linewidth sources such as sensing and frequency conversion.

With the invention of side-written fiber Bragg gratings in the late 1980s [8], it became possible to fabricate DFB structures at any wavelength directly in optical fibers with rare-earth-ion (REI) dopants such as Er and Yb that exhibit optical gain. Fiber Bragg gratings could be fabricated using both phase mask techniques [9, 10] and direct or point-by-point techniques [11, 12]. This allowed for a wide range of possible cavity designs to be fabricated. The first π phase-shifted fiber DFB was reported in 1994 [13]. Since then they have been fabricated in both Er and Yb

P.S. Westbrook (✉) • K.S. Abedin • T. Kremp
OFS Labs, OFS Fitel, LLC, Somerset, NJ, USA
e-mail: westbrook@ofsoptics.com

© Springer International Publishing AG 2017

Y. Feng (ed.), *Raman Fiber Lasers*, Springer Series in Optical Sciences 207,
DOI 10.1007/978-3-319-65277-1_6

fibers for a range of applications. Their primary advantages are lower noise and narrower linewidth in comparison to semiconductor DFBs, as well as the usual fiber attributes of compactness, and an all glass structure free of electrical contacts and thus insensitive to any electrical interference. Also, their in-fiber design allows for efficient coupling to fiber amplifiers and other fiber components. These lasers have been commercialized for high-performance applications in sensing and frequency conversion.

Due to the very high Q-factor of a fiber DFB cavity and its low intrinsic loss, only very small gain is required to reach the lasing threshold. A typical fiber DFB laser using Er or Yb requires only 5–10 cm of fiber for the cavity, and pump thresholds are in the mW range. Moreover, the flexibility of fiber grating fabrication schemes allows for optimization of cavity performance through tailoring of the refractive index modulation [14]. One limitation of rare-earth-doped fibers, though, is the limited range of wavelengths for which optical gain is possible. A more flexible gain mechanism would rely on intrinsic mechanisms such as stimulated Raman and Brillouin scattering. As has been discussed in other chapters of this book, stimulated Raman scattering is a flexible gain mechanism allowing for frequency conversion of a pump source over a very large range of wavelengths. As with fiber DFBs, fiber Bragg gratings played a significant role in harnessing optical fiber Raman gain for generation of high-power laser output at a range of wavelengths. A laser cavity defined by a pair of fiber Bragg gratings whose resonant wavelength falls within the Raman gain spectrum results in significant conversion of the pump light to the Raman-shifted signal resonating in the cavity. The flexibility of fiber Bragg grating fabrication allows for several such cavities to share the same length of Raman gain fiber, thereby allowing for a large number of Raman conversions, resulting in a cascade of laser lines with the lowest order carrying significant output power. Such Raman fiber lasers were first demonstrated as cascading wavelength conversion devices in the early 1990s [15]. Since the Raman gain is much smaller than REI gain, high pump powers and long lengths (typically >100 m) are required to generate sufficient amplification for efficient lasing. As a result, a conventional Raman laser operates on a large number of longitudinal modes and has a linewidth greater than 10 GHz.

The first proposal for a Raman fiber DFB was discussed by Perlin and Winful in 2001 [16, 17]. These authors considered a uniform fiber Bragg grating of 1 m length. Using a linear analysis, they showed that the threshold could be below 1 W. To compute the laser output, they reported a full-time domain solution of the nonlinear coupled mode equations (NLCME) including the grating, Kerr nonlinearity, and the Raman effect. A significantly improved design was proposed by Hu and Broderick [18] who showed with numerical simulations that a 20 cm long DFB cavity in practical germanosilicate fibers could give a threshold close to 1 W if the Bragg grating (of uniform amplitude) has a π phase shift close to the longitudinal center of the grating. The fact that the phase shift is located slightly offset from the center increases the output power on the side that is closer to the phase shift. Moreover, further offsetting the phase shift could provide unidirectional lasing as had been demonstrated for DFBs using REI doping [19]. A subsequent study by Shi and Ibsen

[20] considered the effect of grating imperfections on the lasing characteristics. Further improvements in the simulation schemes were reported in a study that included the effect of two-photon absorption (TPA) on Raman DFB lasers [21].

The first experimental demonstration of a Raman fiber DFB used an offset phase shift design in 124 mm of standard Raman gain fiber [22]. An improved Raman fiber DFB of similar length in a highly Ge-doped nonlinear fiber was also shown [23]. These demonstrations used a cascaded Raman resonator pump with six Raman orders extending from Yb gain wavelengths near 1 μm to the Er band near 1.5 μm. Therefore, these results demonstrated that a standard chain of Raman conversions could be terminated with a Raman DFB cavity and converted to MHz linewidth light, resulting in a reduction of linewidth by more than three orders of magnitude. Subsequently, 30 cm center phase-shifted Raman DFBs with only a single Raman conversion were also demonstrated [24, 25]. Pumped directly with both unpolarized and linearly polarized Yb lasers, significant improvements in slope efficiency, threshold power, and linewidth were demonstrated. These results showed that Raman DFBs could have performance similar to that of highly efficient REI fiber DFBs used in commercial applications. In addition to these Raman fiber DFB results, it has been shown that the output linewidth of Raman lasers may be greatly narrowed by using a short DBR cavity [26]. This laser used a linearly polarized pump and polarization-maintaining (PM) cavity and operated with a single Raman conversion from the Yb emission band. These authors showed an output linewidth of 60 pm for an effective cavity length of 17 cm. They subsequently used the unabsorbed pump power in a Raman amplifier to increase the total output power [27]. Such a Raman fiber master oscillator/power amplifier (MOPA) configuration is also applicable to Raman fiber DFBs, due to the relatively high fraction of unabsorbed pump power that passes through these very short cavities.

An important extension of the Raman fiber DFB is the Brillouin fiber DFB [28]. In fact, the typical design for a Raman fiber DFB makes such a cavity suitable for use with Brillouin gain as well. The primary requirement is a very long and thus spectrally very narrow cavity. The reflection spectrum of a Raman DFB grating can be so narrow that it may be pumped by a signal which is only one Brillouin Stokes shift away from the DFB lasing wavelength, where the linear transmission of the grating is already very close to unity, such that there are no significant cavity reflections of the pump wave. Because Brillouin gain, much like Raman gain, is always present in optical fibers, it is thus possible to use either or both gain mechanisms with a single cavity. As we discuss in this chapter, Brillouin DFB lasers have shown the potential for line narrowing of the incoming pump beam, as well as forming the basis for novel pulsed light sources.

In this chapter, we begin with a theoretical description of Raman DFB lasers. We show how they can be modeled using a set of nonlinear coupled mode equations. In agreement with a closed-form approximation to the threshold gain, time domain simulations reveal the dependence of threshold and slope efficiency on cavity parameters such as gain, loss, the specifics of the grating profile, and nonlinear effects such as two-photon absorption. We then review the realizations of narrow-linewidth Raman fiber lasers. We show how different pump schemes and cavities affect the performance and discuss possibilities for improvements. Finally, we describe the Brillouin DFB laser and compare its performance with that of a Raman DFB laser made with the same cavity.

6.2 Modeling and Simulation

In this section, we describe a theoretical model of the Raman DFB laser. It describes the temporal dynamics of the pump wave as well as the backward and forward signal waves. A numerical implementation of this model can be used to predict important quantities such as the output power and the slope efficiency of the laser, assuming that all parameters of the DFB cavity are known. A fit of measured data to the corresponding prediction from the model can also be used to determine cavity parameters. Examples of cavity parameters that are often not exactly known due to fabrication tolerances are the grating strength and the grating phase shift. To handle the prediction or data fitting tasks with sufficient accuracy even at high input powers, the model also takes into account the most relevant nonlinear effects. At the other extreme of the possible range of input powers, i.e., at low powers, analytic approximations are given to determine the threshold power and detuning, as a fully dynamic model is not well suited to this regime.

The first mathematical model of a Raman DFB laser was published in 2001 by Perlin and Winful [16]. The same equations were used in 2009 by Hu and Broderick [18] for the first simulations of a Raman DFB laser with a π phase shift, which lowers the threshold power and increases the efficiency of the laser. While only the nonlinear Kerr effect was included in these publications, we here describe a more general model [21] that also takes into account nonlinear absorption effects such as two-photon absorption. Temperature effects can be incorporated as well.

6.2.1 Nonlinear Coupled Mode Equations (NLCME)

In this subsection, we present the nonlinear coupled mode equations (NLCME) that describe the temporal evolution of the pump and signal fields. To use an intuitive notation, the indices p and s refer to the pump wavelength λ_p and the signal wavelength λ_s. To simplify the presentation, we assume that the waveguide is single-moded in this wavelength range. We denote the transverse mode profile of this mode at wavelength λ by $w(x, y, \lambda) = w_x(x, y, \lambda)e_x + w_y(x, y, \lambda)e_y$, with the complex-valued mode profile functions w_x and w_y. The transverse direction vectors e_x and e_y are normalized to $1 = \|e_x\| = \|e_y\|$ (in the standard Euclidean norm) and are perpendicular to each other and to the waveguide axis, which is defined to be identical with the z-direction. In the case of a multimode waveguide, the fields at both wavelengths would need to be written as a sum over all modes that are relevant at these wavelengths.

In a Bragg grating, e.g., a DFB grating, the refractive index $n_0(x, y)$ is longitudinally perturbed with a periodic modulation of period Λ, such that the total refractive index is $n_0(x, y) + g(x, y)\left[\Delta n_{\mathrm{dc}}(z) + \Delta n_{\mathrm{ac}}(z)\cos\left(\frac{2\pi z}{\Lambda} + \phi(z)\right)\right]$, with the period-averaged ("dc") index change $\Delta n_{\mathrm{dc}}(z)$ and the "ac" modulation amplitude $\Delta n_{\mathrm{ac}}(z)$ and the index modulation phase $\phi(z)$. The dimensionless transverse profile $g(x, y)$ of the perturbation is typically zero outside the usually doped, photosensitive core region of the waveguide. The effective index $n_{\mathrm{eff},s}$ at the signal wavelength λ_s is then also modulated with the same grating period Λ. The total effective index can thus be written as $n_{\mathrm{eff},s} + \eta\left[\Delta n_{\mathrm{dc}}(z) + \Delta n_{\mathrm{ac}}(z)\cos\left(\frac{2\pi z}{\Lambda} + \phi(z)\right)\right]$, with the dimensionless number η describing the relative overlap of the transverse profiles of the waveguide mode and the grating. If the maximum index perturbation $|g(x, y)|(|\Delta n_{\mathrm{dc}}| + |\Delta n_{\mathrm{ac}}|)$ is sufficiently small such that its impact on the transverse mode profile w is negligible, the overlap coefficient v can be approximately calculated [29, Eq. (18-5)] by a simple integral ratio according to $\eta \approx \iint_{-\infty}^{\infty}|w_s|^2 g \, dx dy / \iint_{-\infty}^{\infty}|w_s|^2 dx dy$, with the notation $|w_s|^2 := \|w(x, y, \lambda_s)\|^2$ and omitting the arguments (x, y) for simplicity. Denoting the speed of light in vacuum as c_0 and the time coordinate as t, we write the forward and backward propagating electrical fields at the signal wavelength λ_s as $E_f(t, x, y, z) = u(t, z)\,e^{\mathrm{i}2\pi(c_0 t - n_{\mathrm{eff},s}z)/\lambda_s}\cdot w(x, y, \lambda_s)$ and $E_b(t, x, y, z) = v(t, z)\,e^{\mathrm{i}2\pi(c_0 t + n_{\mathrm{eff},s}z)/\lambda_s}\cdot w(x, y, \lambda_s)$, respectively. Their envelopes u and v are coupled by such a grating if λ_s lies sufficiently close to the Bragg wavelength $\lambda_B = 2n_{\mathrm{eff},s}\Lambda$. In contrast, the pump wavelength λ_p is assumed to be sufficiently far away from the Bragg wavelength λ_B that it does not interact with the grating, i.e., there is only a forward-propagating pump field $E_p(t, z) = p(t, z)\,e^{\mathrm{i}2\pi(c_0 t - n_{\mathrm{eff},p}z)/\lambda_p}\cdot w(x, y, \lambda_p)$, with the effective index $n_{\mathrm{eff},p}$ at the pump wavelength λ_p.

The group indices at λ_p and λ_s are $n_p = \frac{\mathrm{d}(n_{\mathrm{eff},p}/\lambda_p)}{\mathrm{d}(1/\lambda_p)} = n_{\mathrm{eff},p} - \lambda_p\frac{\mathrm{d}n_{\mathrm{eff},p}}{\mathrm{d}\lambda_p}$ and $n_s = n_{\mathrm{eff},s} - \lambda_s\frac{\mathrm{d}n_{\mathrm{eff},s}}{\mathrm{d}\lambda_s}$, and the corresponding group velocities are $\frac{c_0}{n_p}$ and $\frac{c_0}{n_s}$, respectively. For convenience, we normalize the envelopes of the pump, forward and backward signal fields such that their powers (in unit watts) are $|p|^2$, $|u|^2$, and $|v|^2$. Thus, the NLCME [21, Eqs. (6.1)–(6.3)] read

$$\left(\frac{n_p}{c_0}\frac{\partial}{\partial t} + \frac{\partial}{\partial z}\right)p = \left[-\frac{\alpha_p}{2} - \frac{g_p}{2}\left(|u|^2 + |v|^2\right)\right.$$

$$\left. +\mathrm{i}\left(\gamma_p + \mathrm{i}\frac{b_p}{2}\right)\left(|p|^2 + 2|u|^2 + 2|v|^2\right)\right]p(t, z), \qquad (6.1)$$

$$\left(\frac{n_s}{c_0}\frac{\partial}{\partial t} + \frac{\partial}{\partial z}\right)u = \left[-\frac{\alpha_s}{2} + \frac{g_s}{2}|p|^2\right.$$

$$\left. +\mathrm{i}\left(\gamma_s + \mathrm{i}\frac{b_s}{2}\right)\left(2|p|^2 + |u|^2 + 2|v|^2\right)\right]u(t, z) + \kappa(t, z)\,v(t, z),$$

$$\qquad (6.2)$$

$$\left(\frac{n_s}{c_0}\frac{\partial}{\partial t} - \frac{\partial}{\partial z}\right) v = \left[-\frac{\alpha_s}{2} + \frac{g_s}{2}|p|^2\right.$$

$$\left. + i\left(\gamma_s + i\frac{b_s}{2}\right)\left(2|p|^2 + 2|u|^2 + |v|^2\right)\right] v\,(t,z) - \kappa^*\,(t,z)\,u\,(t,z).$$

$$(6.3)$$

The left-hand sides of Eqs. (6.1)–(6.3) describe the longitudinal derivatives of the envelopes p, u, and v in a reference frame that is moving with group velocity at their individual wavelengths. In other words, these are the envelope changes experienced by a virtual observer riding along the waveguide axis at the same speed as the envelope of the wave. The first term on the right-hand sides of Eqs. (6.1)–(6.3) describes linear attenuation, with the linear loss coefficients α_p and α_s, respectively, both having the unit of inverse length. The following two product terms describe Raman gain (for the signal envelopes u and v) or Raman loss (for the pump envelope p), Kerr nonlinearity, and nonlinear losses such as two-photon absorption (TPA). The Raman gain efficiencies g_p and g_s, the Kerr efficiencies γ_p and γ_s, as well as the TPA efficiencies b_p and b_s all have the same unit $W^{-1}\,m^{-1}$. Finally, the last term in Eqs. (6.2) and (6.3) describes grating-induced linear counterpropagating mode coupling. This effect is driven by the complex-valued mode coupling coefficient (here using the sign convention from [30]):

$$\kappa(z) := \frac{\pi\eta}{i\lambda_B}\Delta n_{ac}(z)\exp\left(-i\left[\phi(z) - \frac{4\pi\eta}{\lambda_s}\int_0^z \Delta n_{dc}\,(\zeta)\,d\zeta\right]\right), \quad (6.4)$$

which describes the coupling of the two counterpropagating waves. We note that other sign conventions (e.g., $i\kappa$ or $-i\kappa^*$ instead of κ) are also common in the literature on fiber Bragg gratings [31–33].

6.2.2 Threshold: Linear Steady State with Zero Incident Fields

The threshold power of the DFB cavity is the minimum incident pump power at which the gain of the cavity at the signal wavelength λ_s equals the loss. At such low power levels, the nonlinearities in Eqs. (6.1)–(6.3) can be neglected. Thus, Eq. (6.1) becomes completely decoupled from Eqs. (6.2) and (6.3) and has the solution $p\,(t,z) = p\left(t - \frac{n_p}{c_0}z, 0\right)e^{-z\alpha_p/2}$. For the following linear analysis, we assume the time harmonic case $u\,(t,z) = \tilde{u}(z)e^{i\Delta\beta c_0 t/n_{eff,s}}$ and $v\,(t,z) = \tilde{v}(z)e^{i\Delta\beta c_0 t/n_{eff,s}}$ with the complex-valued wavenumber offset $\Delta\beta$ (with respect to the reference wavenumber $2\pi n_{eff,s}/\lambda_B$). The real part of $\Delta\beta$ describes the detuning, and its imaginary part describes the gain. Thus, we obtain from the remaining terms of Eqs. (6.2) and (6.3) the linear coupled mode equations in frequency domain:

$$\frac{\partial}{\partial z}\begin{pmatrix}\tilde{u}\\\tilde{v}\end{pmatrix} = \begin{pmatrix}-\frac{\alpha_s}{2}-\mathrm{i}\Delta\beta & \kappa(z)\\ \kappa^*(z) & \frac{\alpha_s}{2}+\mathrm{i}\Delta\beta\end{pmatrix}\begin{pmatrix}\tilde{u}(z)\\\tilde{v}(z)\end{pmatrix}. \tag{6.5}$$

Equation (6.5) describes the linear propagation of the harmonic field envelopes $\tilde{u}$ and $\tilde{v}$ along the grating. To model a laser cavity (e.g., of length L), Eq. (6.5) must be subject to the boundary condition of vanishing incident fields:

$$\tilde{u}(0) = \tilde{v}(L) = 0. \tag{6.6}$$

Thus, we obtain the complex-valued threshold wavenumber offset [30]

$$\Delta\beta_{\text{thresh}} = \delta_{\text{thresh}} + \mathrm{i}\frac{g_{\text{thresh}}}{2}, \tag{6.7}$$

with the real-valued threshold gain g_{thresh} and threshold detuning $\delta_{\text{thresh}} = 2\pi n_{\text{eff},s}$ $\left(\frac{1}{\lambda_s} - \frac{1}{\lambda_B}\right)$. The incident pump power $|p(t,0)|^2$ that is required to reach the threshold gain g_{thresh} is (see also Eqs. (6.2) and (6.3))

$$P_{\text{p,thresh}} = \frac{g_{\text{thresh}}}{g_s}. \tag{6.8}$$

6.2.2.1   Uniform Grating

In the case of a strong uniform Bragg grating of length L, i.e., a grating having a constant coupling coefficient κ with $|\kappa|L \gg 1$, Eqs. (6.5/6.6) yield the threshold gain and threshold detuning for the first longitudinal mode (cf. [16] [3, Eqs. (28)/(29)]),

$$g_{\text{thresh}} \approx \alpha_s + \frac{2\pi^2}{|\kappa|^2 L^3}, \tag{6.9}$$

$$\delta_{\text{thresh}} \approx |\kappa| + \pi^2 \frac{1 - \frac{1}{|\kappa|^2 L^2}}{2|\kappa|L^2}. \tag{6.10}$$

The threshold gain from Eq. (6.9) obviously decreases polynomially with increasing grating strength and length.

6.2.2.2   Grating with Approximate π Phase Shift

We may also consider a grating with a phase shift. A threshold gain that decreases exponentially with increasing grating strength and length can be achieved by a grating with an approximate π phase shift located close to the center of the grating, as illustrated in Fig. 6.1.

Fig. 6.1 DFB grating of
length L with phase shift
$\varphi \approx \pi$ at position $(L + \Delta L)/2$

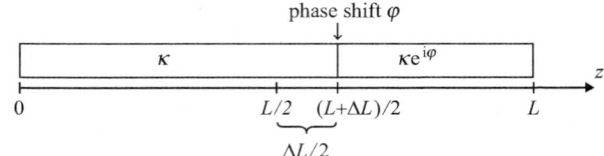

If such a grating has a uniform amplitude $|\kappa(z)| = |\kappa|$ with $|\kappa|L \gg 1$ and a single discrete phase shift φ with $|\cos(\varphi) + 1| \ll 1$ (preferably $\varphi = \pi$) at position $z = (L + \Delta L)/2$, Eqs. (6.5)/6.6) yield [30, Eqs. (24/25)]

$$g_{\text{thresh}} \approx \alpha_s + 2\,|\,\kappa\,|\,r\left[1 + r - r^2\frac{\widetilde{L}\,|\kappa| - 3}{2}\right.$$

$$\left. + \sin^2(\varphi)\left(\frac{\widetilde{L}\,|\kappa| - 3}{8} + r\frac{6\widetilde{L}\,|\kappa| - 9}{8} - r^2\frac{9\widetilde{L}^2\,|\kappa|^2 - 62\widetilde{L}\,|\kappa| + 51}{16}\right)\right],$$

$$(6.11)$$

$$\delta_{\text{thresh}} \approx \frac{|\kappa|}{2}\sin(\varphi)\left[1 + r - r^2\frac{2\widetilde{L}\,|\kappa| - 5}{2} + \sin^2(\varphi)\frac{1 + r\left(\widetilde{L}\,|\kappa| - 1\right)}{8}\right],$$

$$(6.12)$$

with the effective length $\widetilde{L} := L - \Delta L\tanh(\Delta L|\kappa|)$ and the spatial offset factor $r := \frac{\cosh(\Delta L|\kappa|)}{\cosh(L|\kappa|)}$. Hence, the threshold gain from Eq. (6.12) decreases exponentially with increasing grating strength and length.

We note that Eqs. (6.11/6.12) are a generalization of [34, Eq. (17a)] to nonzero offsets $\Delta L \neq 0$ and including higher-order terms of φ. In the case $\varphi = \Delta L = 0$, Eq. (6.11) yields the approximation $g_{\text{thresh}} \approx \alpha_s + 2\,|\,\kappa\,|\,\frac{\cosh(L|\kappa|)+1}{\cosh^2(L|\kappa|)} \approx \alpha_s + 4\,|\,\kappa\,|$ $e^{-L|\kappa|}$. The latter one is equivalent to [35, Eq. (1)] and [36, Eq. (18)], and the prior one is more accurate than [37, Eq. (6)] where only first-order terms were considered in the derivation.

In the case of negligible loss $\alpha_s \approx 0$ and a phase shift $\varphi = \pi$ at the center of the grating, the ratio of the threshold gains from Eqs. (6.9) and (6.11) is approximately $|\kappa|^3 L^3/(\pi^2\cosh(|\kappa|L))$. Already for a relatively short and/or weak grating with $|\kappa|L = 4$, this is a ratio of 0.237, which means that the π phase-shifted grating requires only 23.7% of the threshold power of a similar uniform grating [see also 6, Fig. 6.3] or [13, Fig. 6.5]. For longer or stronger gratings, this ratio becomes even smaller, e.g., 0.92% for $|\kappa|L = 10$. The reason for the reduced threshold power is the exponential decay of the signal intensity toward the ends of the grating (see Fig. 6.2), corresponding to a higher Q-factor [30, Eq. (28)] of the cavity.

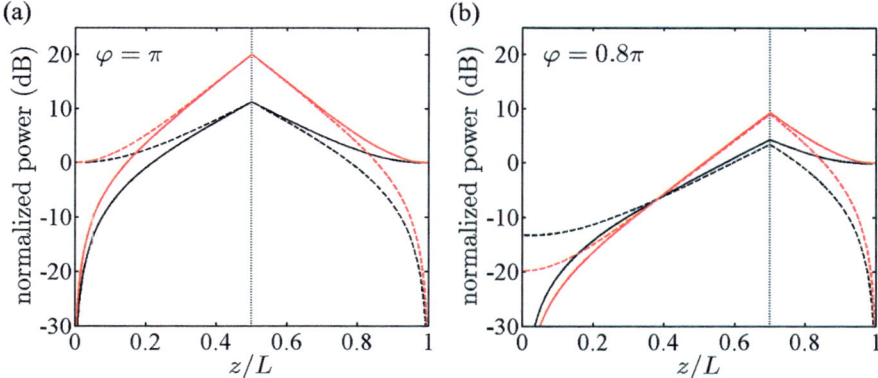

Fig. 6.2 Normalized forward and backward powers $|u(z)|^2/|u(L)|^2$ (——) and $|v(z)|^2/|u(L)|^2$ (——) at threshold, with $|\kappa|L = 4$ (*black*) and $|\kappa|L = 6$ (*red*). (**a**) $\varphi = \pi$, $\Delta L = 0$. (**b**) $\varphi = 0.8\pi$, $\Delta L = 0.4L$

6.2.3   Dynamical Behavior: Numerical Solution

We next consider simulations of the dynamic behavior of the DFB. In the following, we assume equal group indices $n := n_p = n_s$ in Eqs. (6.1)–(6.3) and introduce the following characteristic variables [39, Eq. (4)] and their partial derivatives:

$$\xi := \frac{\frac{c_0 t}{n} + z}{2}, \qquad \frac{\partial}{\partial \xi} = \frac{n}{c_0}\frac{\partial}{\partial t} + \frac{\partial}{\partial z}, \tag{6.13}$$

$$\eta := \frac{\frac{c_0 t}{n} - z}{2}, \qquad \frac{\partial}{\partial \eta} = \frac{n}{c_0}\frac{\partial}{\partial t} - \frac{\partial}{\partial z}. \tag{6.14}$$

Inserting Eqs. (6.13)/(6.14) in the quasilinear hyperbolic system Eqs. (6.1)–(6.3), we obtain a system of ordinary differential equations [21, Eq. (5)]:

$$\begin{pmatrix} \frac{\partial}{\partial \xi} & 0 & 0 \\ 0 & \frac{\partial}{\partial \xi} & 0 \\ 0 & 0 & \frac{\partial}{\partial \eta} \end{pmatrix} \begin{pmatrix} p \\ u \\ v \end{pmatrix} = (\mathbf{L} + \mathbf{N}) \begin{pmatrix} p \\ u \\ v \end{pmatrix}, \tag{6.15}$$

with the linear and nonlinear operators:

$$\mathbf{L} := \begin{pmatrix} -\frac{\alpha_p}{2} & 0 & 0 \\ 0 & -\frac{\alpha_s}{2} & \kappa \\ 0 & -\kappa^* & -\frac{\alpha_s}{2} \end{pmatrix}, \tag{6.16}$$

$$\mathbf{N} := \begin{pmatrix} N_{11} & 0 & 0 \\ 0 & N_{22} & 0 \\ 0 & 0 & N_{33} \end{pmatrix}, \tag{6.17}$$

$$N_{11} := -\frac{g_p}{2}\left(|u|^2 + |v|^2\right) + i\left(\gamma_p + i\frac{b_p}{2}\right)\left(|p|^2 + 2|u|^2 + 2|v|^2\right), \tag{6.18}$$

$$N_{22} := \frac{g_s}{2}|p|^2 + i\left(\gamma_s + i\frac{b_s}{2}\right)\left(2|p|^2 + |u|^2 + 2|v|^2\right), \tag{6.19}$$

$$N_{33} := \frac{g_s}{2}|p|^2 + i\left(\gamma_s + i\frac{b_s}{2}\right)\left(2|p|^2 + 2|u|^2 + |v|^2\right). \tag{6.20}$$

The most common method to solve Eq. (6.15) is a fourth-order Runge-Kutta method [39, Eq. (7)]. However, since this is an implicit algorithm, it requires time-consuming iterations. In contrast, an explicit method that does not require iterations is the exponential split-step method (see, e.g., [40, p. 212]). With the step size h, we use a symmetric implementation [21, Eq. (9)], which is more stable and accurate than explicit nonsymmetric implementations and faster than implicit methods:

$$\begin{pmatrix} p\left(t + \frac{hn}{c_0}, z\right) \\ u\left(t + \frac{hn}{c_0}, z\right) \\ v\left(t + \frac{hn}{c_0}, z\right) \end{pmatrix} := e^{\frac{h}{2}\mathbf{L}} \begin{pmatrix} \widetilde{p}(z) \\ \widetilde{u}(z) \\ \widetilde{v}(z + h) \end{pmatrix}, \tag{6.21}$$

$$\begin{pmatrix} \widetilde{p}(z) \\ \widetilde{u}(z) \\ \widetilde{v}(z) \end{pmatrix} := e^{h\mathbf{N}(A)}\mathbf{A}, \qquad \mathbf{A} := e^{\frac{h}{2}\mathbf{L}}\begin{pmatrix} p\left(t, z - h\right) \\ u\left(t, z - h\right) \\ v\left(t, z\right) \end{pmatrix}. \tag{6.22}$$

At each point in time, Eq. (6.21) is performed for all $N + 1$ discrete positions $z_m := mh$, $m = 0, \ldots, M = L/h$ along the grating length L, with the boundary conditions Eq. (6.6) and a known incident pump envelope $P_p(t) = p(t, 0)$.

We note that heating-induced nonlinearities can be included in the simulation by solving the heat equation in the cavity with the pump and signal power-dependent heat load at the current time t and then using the material and grating parameters for the found temperature distribution in the next iteration of Eq. (6.21).

6.2.4   Example: 12.4 cm Grating in OFS Highly Nonlinear Fiber

As an example, we consider the Raman DFB laser from [21, 23] with a pump wavelength $\lambda_p = 1480$ nm and a signal wavelength $\lambda_s = 1583.5$ nm using OFS highly nonlinear fiber (HNLF; see DFB2 in Table 6.1) with linear loss coefficients $\alpha_p = \alpha_s = 0.0064/m$ (0.028 dB/m), effective area $A_{eff} = 11.4\ \mu m^2$, Raman gain coefficient $g_R = 5.7 \cdot 10^{-14}$ m/W, Raman gain efficiency $g_s = \frac{g_R}{A_{eff}} = \frac{5}{W\ km}$, Kerr efficiency $\gamma_s = \frac{2\pi n_2}{\lambda_s A_{eff}}$, nonlinear index $n_2 = 3.29 \cdot 10^{-20}\ m^2/W$, two-photon absorption efficiencies $b_p = b_s = 0$, and the relation $\frac{\lambda_s}{\lambda_p} = \frac{g_p}{g_s} = \frac{\gamma_p}{\gamma_s}$. The DFB grating of length $L = 12.4$ cm has a strength $|\kappa| = 90/m$ and a perfect $\varphi = \pi$ phase shift at the position $z = \frac{L+\Delta L}{2} = 0.58L = 7.192$ cm.

Inserting these values in Eq. (6.11), we expect a threshold gain $g_{thresh} \approx 0.022/m$, corresponding to a threshold power $P_{p,thresh} = \frac{g_{thresh}}{g_s} = 4.31$W. This value is in perfect agreement with the observed pump threshold of approximately 4.3 W recorded in Table 6.1.

In the example of an input pump power $P_p = |p(t,0)|^2 = 20$W that enters the DFB cavity from the left, we simulate the turn-on behavior of this Raman DFB laser, using the symmetric exponential split-step method from Eqs. (6.21) and (6.22). As shown in Fig. 6.3a, the laser reaches a steady state within about 2 μs, independently of the exact type and amount of spontaneous emission noise that is used at the beginning of the simulation. A very important effect is the pump depletion, i.e., the difference between the input and output pump powers (solid and dashed black lines), which starts to become noticeable at $t \approx 1.8\mu$s. The resulting steady-state power distribution inside the cavity is shown in Fig. 6.3b. The signal intensity increases exponentially in the neighborhood of the π phase shift at $z = \frac{L+\Delta L}{2} = 7.192$cm, and the pump is being depleted as it passes this very high signal intensity.

The higher the input pump power $P_p = |p(t,0)|^2$, the steeper is the slope in Fig. 6.3a, and the faster the convergence of the numerical solution from Eq. (6.21) toward a steady state as shown in Fig. 6.3b. On the other hand, the smaller the difference $P_p - P_{p,thresh}$, the slower the convergence of any numerical scheme that solves the coupled mode Eqs. (6.1)–(6.3) in the time domain. For $P_p \leq P_{p,thresh}$, a time domain simulation does not converge at all, i.e., the exact threshold quantities such as the threshold power need to be determined from the approximations in Eqs. (6.11) and (6.12).

Performing such simulations for a range of input pump powers $P_p = |p(0)|^2$, the output characteristics of the Raman DFB laser can be determined; see Fig. 6.4a. These simulation results are in perfect agreement with the analytical threshold power from Eq. (6.11), which is also included. The spectrum of the output signal $u(L)$ as well as the transmission spectrum of the DFB grating is shown in Fig. 6.4b. In the absence of a pump wave p, i.e., at pump power $P_p = 0$, the linear transmission spectrum of the grating is perfectly symmetric and has the typical very narrow transmission spike in the center. For pump powers $P_p > P_{p,thresh}$, the Kerr effect induces a nonlinear shift of the effective index in the neighborhood of the position of the phase shift. Thus, the higher the pump power, the more asymmetric the transmission spectrum of the grating, see Fig. 6.4b. This detuning of the grating is shown in Fig. 6.4c as a function of the input pump power P_p. The intracavity pump power $|p(z)|^2$ is plotted in linear scale in Fig. 6.4d. Since the Raman-induced loss of the pump is proportional to the square of the signal power in Eq. (6.1), the depletion of the pump is more pronounced at higher power levels.

Table 6.1 Properties of narrow-linewidth Raman fiber lasers discussed in the text

Reference	Westbrook 2011, [23] DFB1	Westbrook 2011, [23] DFB2	Shi 2012a, [24] R-DFB1	Shi 2012a, [24] R-DFB2	Shi 2012b, [25]	Siekra 2012, [26]	Units
Cavity type	DFB 8% offset	DFB 8% offset	DFB centered	DFB centered	DFB centered	PM DBR	
Cavity length	124	124	300	300	300	170	mm
Coupling constant	70	90	37	30	30		1/m
Loading	Deuterium	Deuterium	None	None	None		
Fiber type	OFS Raman	OFS HNLF	PS980	UHNA4	UHNA4	OFS PM Raman	
Core dopants	Ge	Ge	B/Ge	Ge	Ge	Ge	
Raman gain (signal)	0.47	0.57	0.86	1.5	1.5	1.24	10^{-13} m/W
Effective area (signal)	19	12	30	5.4	5.4	7.3	μm^2
NA	0.23	0.29	0.13	0.35	0.35		
Signal wavelength	1584	1583.4	1118	1110	1110	1151	nm
Pump wavelength	1480	1480	1068	1068	1064	1100	nm
Pump polarization	Unpolarized	Unpolarized	Unpolarized	Unpolarized	PM polarized	PM polarized	
Raman order	6	6	1	1	1	1	
Threshold power	39	4.3	2	1	0.44	4.1	W
Slope efficiency (incident pump)	0.2	1.2	14	14	13.4	9	%
Slope efficiency (absorbed pump)	3	6	74	93	40		%
Maximum power	150	350	1500	1600	120	700	mW
Linewidth	7.5	4			<0.0025	14000	MHz

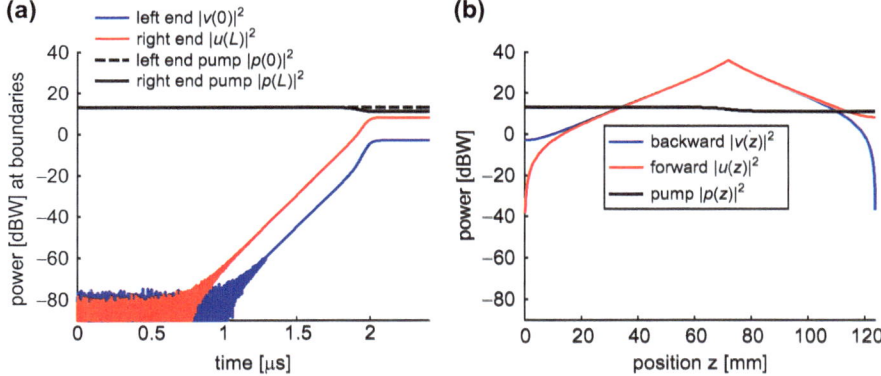

Fig. 6.3 Raman DFB laser simulation with HNLF fiber (DFB2 in Table 6.1) at input pump power $P_p = |p(t, 0)|^2 = 20$ W. (**a**) Temporal dynamics of the transmitted (unabsorbed) pump power $|p(L)|^2$, output backward signal power $|v(0)|^2$, and output forward signal power $|u(0)|^2$. (**b**) Steady-state power distribution. Note that the input pump power enters from the left, i.e., the pump envelope $p(z)$ and the forward signal envelope $u(z)$ travel from left to right, while the backward signal envelope $v(z)$ travels from right to left

6.2.5   Two-Photon Absorption

Finally, we consider the effect of two-photon absorption on Raman DFB operation. While nonlinear absorption processes are negligible in silica, they can become significant in highly nonlinear materials that are used for optical fibers, such as tellurite (TeO_2, [41]) or chalcogenide ($As_2 Se_3$, [38]). In the case of two-photon absorption (TPA), which is the dominant and lowest-order nonlinear absorption process, two photons are simultaneously absorbed to create an excited state of an atom or molecule in the material. Hence, the TPA-induced absorption of a propagating wave is proportional to the power (square of the modulus of the envelope). With the TPA absorption coefficient α_{TPA}, the TPA efficiencies that are being multiplied with the powers in Eqs. (6.1)–(6.3) are $b_s = \frac{\alpha_{TPA}}{A_{eff,s}}$ and $b_p = \frac{\alpha_{TPA}}{A_{eff,p}}$.

Since both the TPA absorption and the nonlinear index shift are proportional to the optical power, we define the dimensionless nonlinear *figure of merit* [21]:

$$\text{FOM} := \frac{n_2}{\alpha_{TPA}\lambda_s} = \frac{\gamma_s}{2\pi b_s} = \frac{\gamma_p}{2\pi b_p}, \tag{6.23}$$

again assuming the relation $\frac{\lambda_s}{\lambda_p} = \frac{g_p}{g_s} = \frac{\gamma_p}{\gamma_s} = \frac{b_p}{b_s}$. A favorable material for a Raman DFB laser should have an FOM $\gg 1$, i.e., a large ratio of the nonlinear index shift (which is typically large in materials with a high Raman gain coefficient) in comparison to the TPA absorption coefficient.

For a first set of TPA simulations, we use the chalcogenide fiber parameters from [38]: pump wavelength $\lambda_p = 1560$ nm, signal wavelength $\lambda_s = 1619$ nm, linear

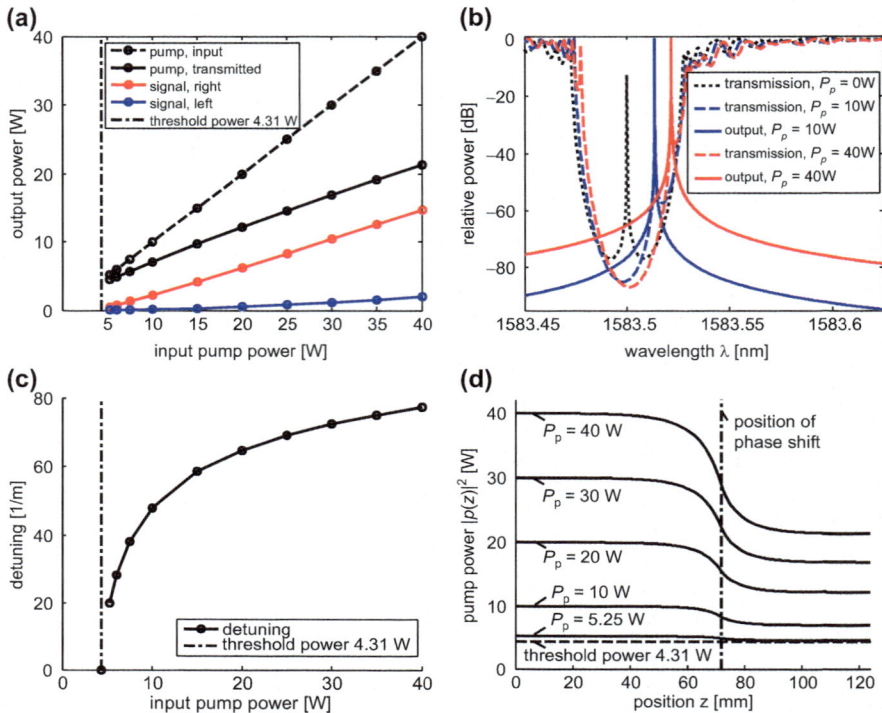

Fig. 6.4 Steady-state quantities of the Raman DFB laser with HNLF fiber for various input pump powers P_p. (**a**) Output signal power. (**b**) Normalized spectrum of output signal $P_s(z = L)$ (*solid*) and linear transmission spectrum of the cavity (*dashed*). (**c**) Detuning $\Delta\beta$. (**d**) Pump power distribution $P_p(z)$

loss coefficients $\alpha_s = \alpha_p = 0.23/m$ (1 dB/m), effective area $A_{eff} = 21\ \mu m^2$, Raman gain coefficient $g_R = 2 \cdot 10^{-11}$ m/W, and nonlinear index $n_2 = 7.5 \cdot 10^{-18}$ m^2/W. The DFB grating is assumed to have a length $L = 10$ cm with a perfect π phase shift at position $\frac{L+\Delta L}{2} = 5.8$ cm (i.e., $\Delta L/L = 0.16$ in Fig. 6.1).

The intracavity power distribution is shown in Fig. 6.5a for the case of an input power $P_p = |p(t,0)|^2 = 500$ mW and a grating strength $|\kappa| = 10/m$. Already for FOM $= 20$, the nonlinear losses keep the intensity much lower than in the case FOM $= \infty$ without TPA. The corresponding effect on the output power $|u(L)|^2$ can be observed in Fig. 6.5b, where a shorter and stronger grating ($L = 5$ cm, $|\kappa| = 200/m$, dashed) is more severely affected by high TPA than a longer and weaker grating ($L = 20$ cm, $|\kappa| = 50/m$, dotted). In the absence of TPA, the output power depends (approximately) only on the product $L|\kappa|$. As shown in Fig. 6.5c, an efficient laser operation requires FOM $\gtrsim 100$. Hence, the low FOM $= 1.16$ (due to $\alpha_{TPA} = 4 \cdot 10^{-12}$ m/W) of this As$_2$Se$_3$ fiber is not suitable for a phase-shifted Raman DFB laser. We note that this is also the case if the phase shift is spatially distributed as in [18].

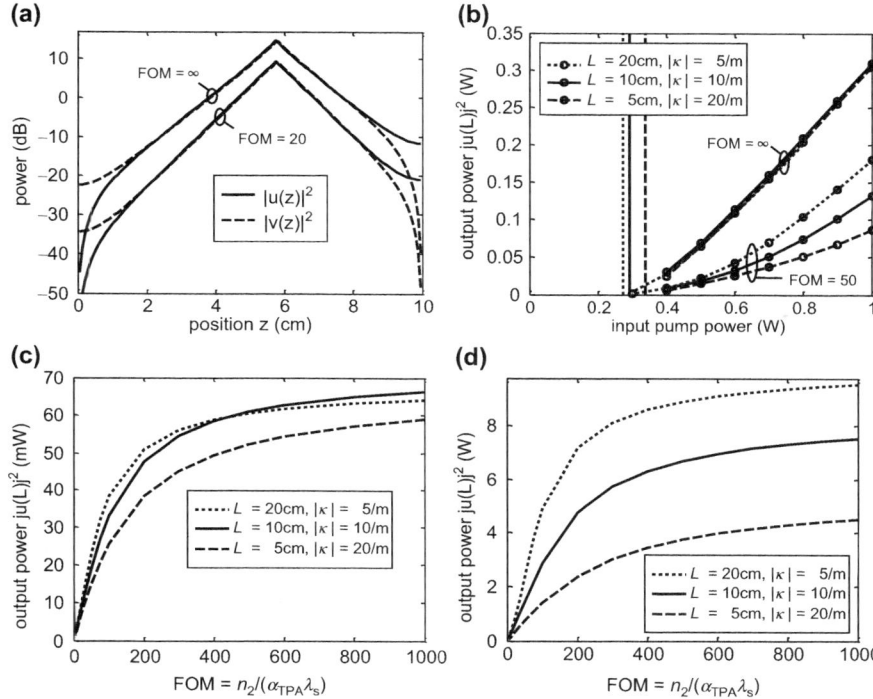

Fig. 6.5 As$_2$Se$_3$ fiber [38]: (**a**) Intracavity power. (**b**) Output power vs. input power (*vertical lines* denote threshold powers). (**c**) Output power vs. FOM for input power $|p(0)|^2 = 500$mW. (**d**) HNLF silica fiber [23] with input power $|p(0)|^2 = 30$W

In a second set of simulations, we again use the HNLF silica fiber from Figs. 6.3 and 6.4 and [23]: $\lambda_p = 1480$ nm, $\lambda_s = 1583.5$ nm, $\alpha_s = \alpha_p = 0.0064$/m (0.028 dB/m), $A_{eff} = 11.4$ μm^2, and $n_2 = 3.29 \cdot 10^{-20}$ m^2/W. Even though the Raman gain coefficient $g_R = 5.7 \cdot 10^{-14}$m/W is by a factor of 350 lower than for As$_2$Se$_3$, the resulting dependence on the FOM in Fig. 6.5d is very similar to Fig. 6.5c: Again, FOM $\gtrsim 100$ or higher is required for efficient laser operation.

6.3   Experimental Results

In this section, we describe three different realizations of narrow-linewidth Raman fiber lasers. Figure 6.6 compares schematics for each laser. Figure 6.6a illustrates a Raman DFB laser using a cascaded Raman laser pump operating near 1480 nm [Westbrook 2011, 23]. Figure 6.6b shows a longer DFB cavity pumped through only a single Raman order. This configuration was used with both unpolarized [Shi 2012a, 24] and polarized [Shi 2012b, 25] pumps. The polarized pump configuration

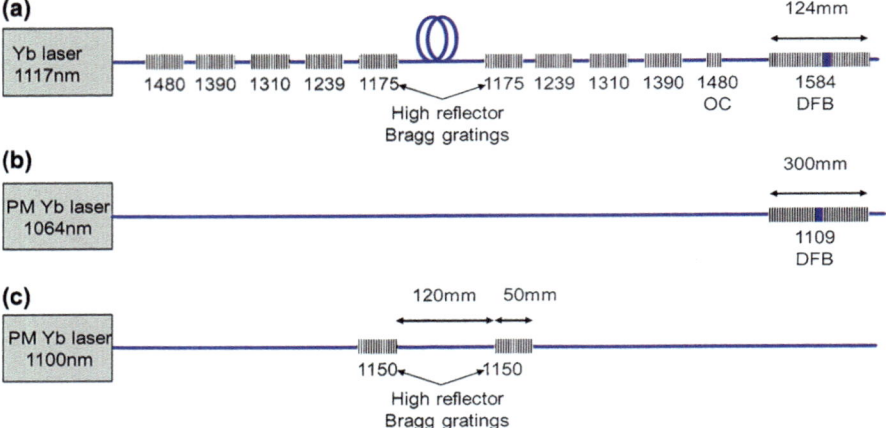

Fig. 6.6 Schematics of narrow-linewidth Raman fiber lasers discussed in Table 6.1 and in the text. (**a**) Cascaded Raman resonator laser terminated at the sixth order with 124 mm DFB cavity [Westbrook 2011]. (**b**) High-performance single-order Raman DFB with very high efficiency and kHz linewidth [Shi 2012b]. Location of DFB π phase shifts indicated with blue mark. (**c**) Short cavity Raman DBR laser [Siekiera 2012]. Components such as isolators and WDMs to access the backward signal are left out for clarity

is shown. Finally, Fig. 6.6c presents a very short Raman DBR laser, again with a single Raman shift and a polarized pump [Siekiera 2012, 26]. Table 6.1 compares various design and performance parameters for each Raman laser. Most of the experimental details are similar for these lasers, and so we present detailed figures for Fig. 6.6a while indicating the differences and improvements observed in the other configurations.

The first demonstration of a Raman fiber DFB laser used the configuration (Fig. 6.1a) [Westbrook 2011, 23]. The aim of this work was to prove that the well-known Raman laser frequency conversion schemes using Raman gain and fiber Bragg gratings could be extended from the very long Raman gain cavities of conventional Raman lasers to very short Raman DFB cavities. By presenting a Raman DFB laser pumped through many Raman orders, this result therefore shows that the typically very large bandwidth of a Raman laser could be converted to a narrow linewidth for a large range of wavelengths above the fundamental pump wavelength.

Two different fibers were considered in this work. Firstly a commercial Raman gain fiber, identical to that used in the Raman fiber laser, was used. This DFB laser was designated DFB1. Clearly such a fiber would be of interest since it could allow a narrowed Raman laser with the DFB portion included in the set of output gratings defining the cascaded Raman resonator cavity. Secondly, a highly nonlinear fiber with smaller effective area and higher Raman gain efficiency was used in order to improve the performance. This device was designated DFB2.

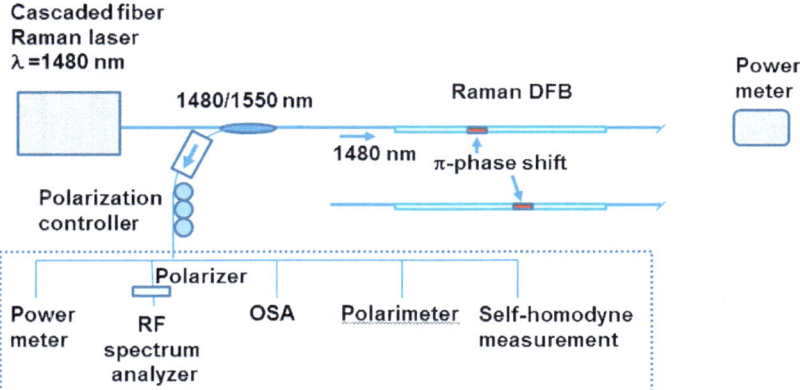

Fig. 6.7 Experimental apparatus used to characterize the DFBs of Westbrook 2011. The self-homodyne linewidth measurement is shown in Fig. 6.12

The experimental setup used to pump and characterize these DFB lasers is shown in Fig. 6.7. The pump laser was a high-power cascaded Raman fiber laser that produced a maximum of 81 W at 1480 nm [42]. The pump output was spliced to a 1480/1550 nm WDM, which was then spliced to the DFB fiber, to observe the backward lasing power. An isolator in the 1550 nm path ensured low feedback to the DFB cavity from the backward direction, while the DFB fiber in the forward direction was angle cleaved to decrease back reflection. As indicated in Table 6.1, DFB1 used OFS Raman fiber with a numerical aperture (NA) of 0.23, effective area of 19 μm^2, and Raman gain of 0.47×10^{-13} m/W (Raman gain efficiency for unpolarized light of 2.5/W/km). DFB2 used a variant of OFS highly nonlinear fiber (HNLF) with NA = 0.29, effective area 12 μm^2, and Raman gain of 0.57×10^{-13} m/W (Raman gain efficiency for unpolarized light of 5/W/km). The fibers were loaded with deuterium to increase photosensitivity. Gratings of length 124 mm and pitch of 547.30 nm (DFB1) and 544.22 nm (DFB2) were inscribed using a CW direct write system with a 244 nm inscription wavelength. The gratings were annealed after inscription to stabilize the refractive index modulation defining the grating cavity. The resulting Bragg wavelengths were near 1584 nm (see Fig. 6.8). During the measurements, the DFB cavities had less than 0.5 m of gain fiber pigtail on either side, and the bare end of the fiber in the forward direction was angle cleaved to minimize Fresnel reflection. The grating profile was uniform, and a π phase shift was placed 8% from the center of the grating. Such an offset results in a preferential lasing in one direction. The grating transmission spectra measured with a commercial high-resolution spectrometer (LUNA OVA) are shown in Fig. 6.8. A visual fit of the grating transmission width to a uniform 124 mm phase-shifted grating gave estimated index modulations of $\approx 3.5 \times 10^{-5}$ (DFB1) and $\sim 4.5 \times 10^{-5}$ (DFB2). The corresponding grating coupling constants were $\kappa = 70(1/m)$ and $90(1/m)$, respectively. While DFB1 shows the characteristic cavity notch, DFB2 is too strong to show the notch in this figure. A second

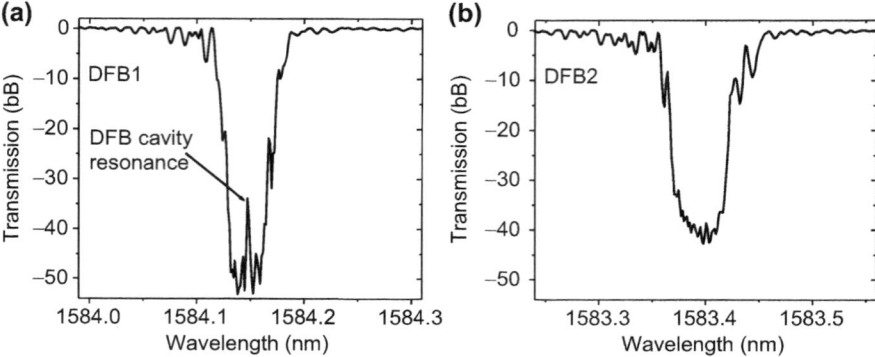

Fig. 6.8 Transmission spectra of 124 mm DFB cavities used in Westbrook 2011. (**a**) DFB1, (**b**) DFB2. Figure 6.16 shows a higher-resolution spectrum of DFB2 indicating the cavity resonance

higher-resolution measurement of the transmission spectrum of DFB2 performed using an external cavity tunable laser is shown in Fig. 6.16, indicating the narrow notches of the cavity resonance. Each DFB grating was placed on a metal bar and surrounded by a thermally conductive paste to reduce thermal variations. Like REI fiber DFBs, Raman DFBs are very sensitive to external conditions. Proper thermal and mechanical packaging can therefore improve the optical performance. The backward output of the laser was coupled to one of several measurements as indicated in Fig. 6.7. These measurements were power, optical spectrum, linewidth (delayed self-homodyne interferometer + RF spectrum analyzer; see Fig. 6.12b), a polarizer and RF spectrum analyzer, and a polarimeter. During the measurements, very little pump was absorbed, making the pump power at the output of the lasers very large compared to the forward signal. Additional filtering was not sufficient to measure the forward power of the laser accurately. Thus the laser was oriented with its phase shift to lase in the backward direction (toward the WDM).

The backward signal power vs. pump power is shown in Fig. 6.9 for DFB1 and DFB2. Pump power was calibrated to the 1480 nm laser output pigtail fiber. The signal power was corrected for the losses of the WDM and isolator (~3.6 dB). The pump power used for DFB2 was kept below 35 W as a precaution because of heating in a splice connecting the pump to the DFB. Figure 6.9 also shows the transmitted pump power. At maximum pump power, both gratings had more than 75% of the pump transmitted. Because of the large transmission of pump power, two slope efficiencies are estimated for each laser: a slope efficiency vs. incident pump power as well as a slope efficiency vs. absorbed pump (see Table 6.1). DFB1 exhibited a nonlinear relationship in the output power vs. pump. However, one can estimate that these two slope efficiencies were 0.2% and 3%, respectively. DFB2 showed a more typical linear relationship and the efficiencies were estimated to be 1.2% and 6%, respectively.

The very small converted pump power is common to all of the lasers presented here and is seen in the simulations as well (see Fig. 6.4d). The large transmitted

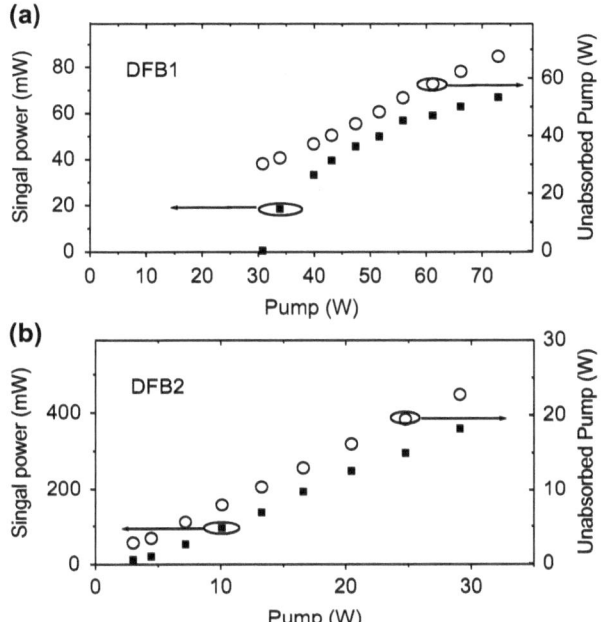

Fig. 6.9 Output power and unabsorbed pump power for the Raman DFB lasers of [Westbrook 2011, 21]. (**a**) DFB1, (**b**) DFB2

pump power presents the possibility of a master oscillator/power amplifier (MOPA) configuration. Both signal and transmitted pump could be coupled into additional Raman gain fiber. We review one MOPA result below applied to a narrow-linewidth DBR laser.

Figure 6.10 shows a wideband spectrum of the pump and forward Raman DFB spectrum for DFB2. The Yb pump and all of the orders of the full Raman pump laser are visible in the spectral range shown. The DFB order at 1583.4 nm shows significant narrowing with respect to the various Raman orders in the pump, particularly the high-power pump order at 1480 nm. Figure 6.11 shows the optical spectrum (resolution 0.01 nm) below and above threshold for DFB1. Below threshold, the backward power spectrum matches the grating spectrum of Fig. 6.8a, since the DFB grating is reflecting Raman ASE generated in the fiber connecting the pump and DFB. Above threshold, the lasing peak is resolution limited, indicating a linewidth less than 0.01 nm. The dependence of the lasing signal on the wavelength is similar to the simulated curves in Fig. 6.4a. To obtain a more accurate linewidth, the DFB outputs were sent through a delayed self-homodyne interferometer with a 20 km fiber delay in one arm. Figure 6.12a shows the RF spectrum measured through the delayed self-homodyne interferometer for the two lasers. Measurements were recorded at both full power and near threshold. From the 3 dB widths of these spectra, the optical linewidths at full power were

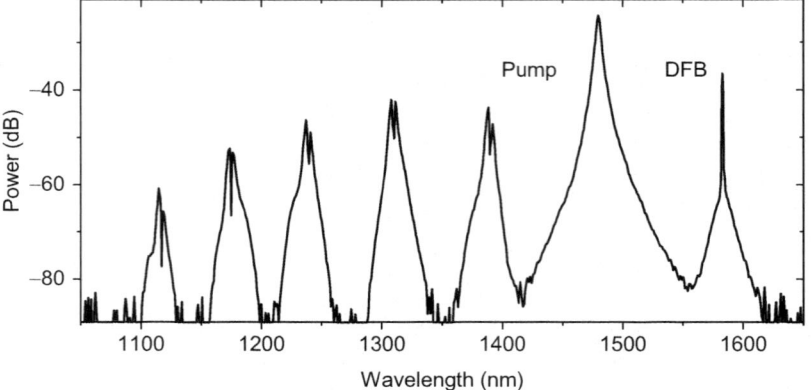

Fig. 6.10 Wideband spectrum of DFB2 showing all orders of the cascaded Raman pump laser, including the strong pump at 1480 nm. The substantially narrowed DFB line is observed at the sixth order at 1583 nm

Fig. 6.11 Backward spectrum of Raman DFB1 below and above threshold

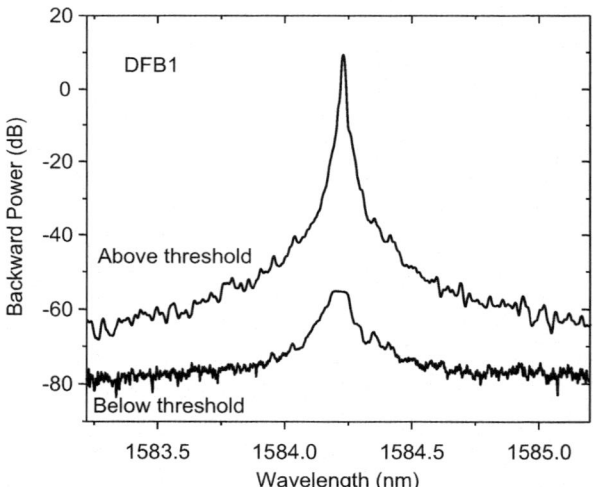

estimated to be 7.5 MHz (DFB1) and 4 MHz (DFB2). Near threshold, these linewidths were somewhat reduced. The RF spectrum was also measured on a larger frequency scale. This measurement showed a beat note at 100 MHz (DFB1) and 200 MHz (DFB2), indicating the presence of a second lasing mode. The most likely explanation for this peak is dual-polarization lasing. Fiber DFBs have been shown to lase on two polarizations [13]. To verify that the beat note was due to an additional polarization state, polarization controllers followed by a polarizer were placed in front of the RF spectrum analyzer. This measurement showed that the beat note could be nulled by adjusting the polarization with the paddles. Figure 6.13 shows the RF beat note from DFB1 at 100 MHz. Both the maximum strength

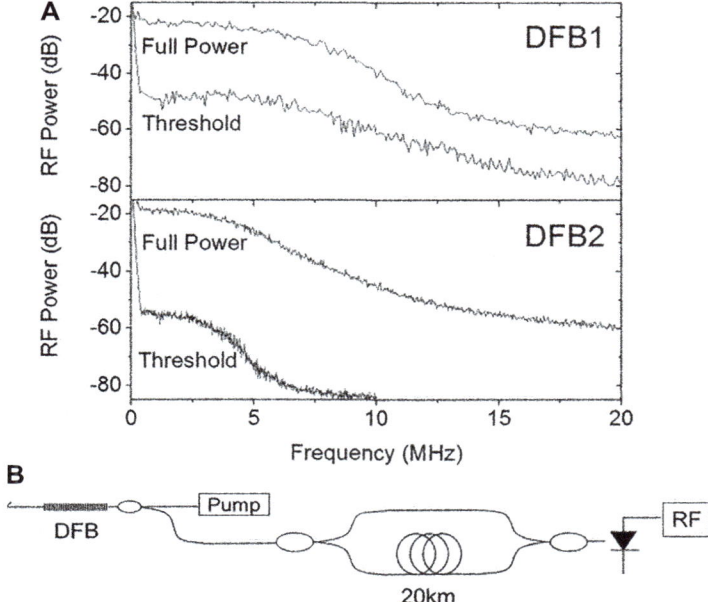

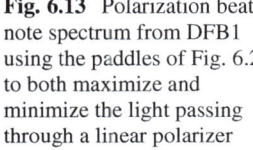

Fig. 6.12 Linewidth measurement for DFB1 and DFB2. (**a**) RF spectrum recorded with the delayed self-homodyne interferometer shown in (**b**). Measurements of both near threshold and at full power (as indicated in Fig. 6.9) are shown, indicating some increase with pump power

Fig. 6.13 Polarization beat note spectrum from DFB1 using the paddles of Fig. 6.2 to both maximize and minimize the light passing through a linear polarizer

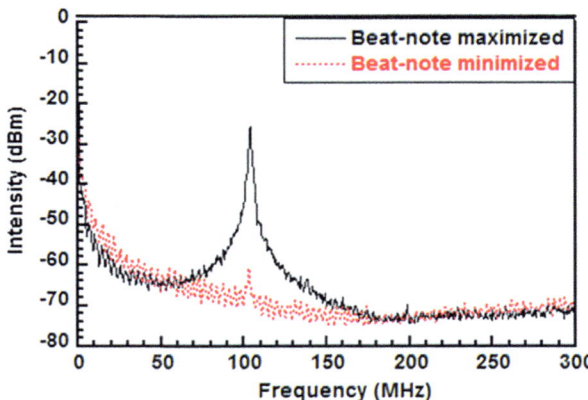

and minimum strength and paddle adjustment are shown. These measurements are a strong indication that these lasers were operating on the two polarizations of a single longitudinal mode. A measurement of the degree of polarization (DOP) of the signal showed it to be near 100%. This was consistent with the very low value of the minimized polarization beat note power observed in Fig. 6.13. Finally, as shown in the high-resolution/high-dynamic range spectrum of DFB2 (Fig. 6.16), two resonances separated by 220 MHz were visible in the spectrum, consistent with the appearance of the beat note near 200 MHz observed for this laser.

These early demonstrations of multi-Raman order DFB lasers showed far lower pump absorption and output power than observed in simulations of Sect. 2 and subsequent demonstrations [Shi 2012a, Shi 2012b]. While the simulated threshold power is consistent with experimental observations, the measured output power and slope efficiency were much lower than predicted by simulation. Moreover, the MHz linewidths were much larger than expected and also much larger than subsequent observations in [Shi 2012b]. One explanation for the discrepancy is thermal variations induced as the pump increased in power. The UV-exposed and deuterium-loaded fiber used in DFB1 and DFB2 is known to have increased absorption in the IR. An estimate of the thermal sensitivity gives the order of magnitude for temperature changes that would affect the DFB lasing. The DFB grating bandwidths were roughly 50 pm. Since the local Bragg condition can change by $\sim$11 pm/C, temperature fluctuations of a few degrees C could distort the grating profile enough to affect the optical spectrum and hence the lasing output. Dynamic temperature fluctuations arising from pump fluctuations may also have affected the laser performance. We note that the 4 MHz linewidth of DFB2 corresponds to a temperature change of only 3mK. Moreover, the simulations of DFB2 presented in Sect. 2 indicate a laser turn-on time on a μsec timescale. If such a temperature fluctuation were to occur on a timescale similar to that of the laser power buildup, the effect would be an increased linewidth and a low slope efficiency, both of which were observed for DFB1 and DFB2. Furthermore, the linewidth broadening would increase with pump power as was observed in Fig. 6.12. As we discuss below, DFB2 had a significantly higher slope efficiency when being tested as a Brillouin DFB laser. Since the typical pump powers for Brillouin lasing are 100× lower than for Raman operation, these results indicate that the performance of DFB2 is improved with lower pump power. A precise explanation for the performance of these cascaded Raman resonator terminated Raman DFBs requires additional measurements. However, these considerations indicated that significant improvements should be possible in cavities with lower absorption and lower threshold power. Moreover, the use of a lower-noise, single-polarization pump should also improve performance.

Such improvements were observed in a subsequent set of Raman fiber DFB demonstrations. These results were reported in [Shi 2012a, 24] and [Shi 2012b, 25] and are summarized next to those of Westbrook 2011 in Table 6.1. A schematic of the Raman DFB laser of [Shi 2012b] and its pump is shown in Fig. 6.6b. In the study of [Shi 2012a], the pump source was an unpolarized Yb laser operating at 1068 nm. Only a single order of Raman gain was considered, thus eliminating the intervening Raman gain conversions of the previous scheme [Westbrook 2011, 23]. Two fibers were chosen as summarized in Table 6.1. The first fiber had a relatively large mode field diameter and a Boron-Germanium codoped core, while the second fiber had a very small mode field diameter and much larger Raman gain. These cavities had larger Raman gain than [Westbrook 2011, 23] due to both their core dopants and the operation wavelength near 1.1 μm. The cavity for these demonstrations was 30 cm long, and the phase shift was placed in the center of the cavity to minimize the threshold pump power. The DFB cavity grating was inscribed using a CW direct writing system operating at 244 nm similar to that of Westbrook 2011. However, no deuterium loading was applied to the fiber for photosensitization. Moreover, the index modulation amplitude of the gratings was less than half as strong as for those of [Westbrook 2011, 23]. While the 1.1 μm operating wavelength would have more UV-induced loss in the grating, the cavity design and fabrication parameters would be expected to minimize the loss within the grating.

As is evident in Table 6.1, these authors reported substantial performance improvements over the results of [Westbrook 2011, 23]. Threshold powers were observed to be 2 W and 1 W, respectively. Slope efficiencies vs. incident pump power were above 13% for both lasers, almost an order of magnitude improvement over [Westbrook 2011, 23]. Much as in [Westbrook 2011, 23], only a limited amount of pump power was converted into narrow-linewidth Raman signal. Considering only the absorbed pump power, these lasers showed slope efficiencies of 74% and 93%. The latter result is close to the theoretical maximum. Simulations by these authors using methods similar to those presented in this work predicted similar performance.

These authors subsequently published another set of still more significant improvements [Shi 2012b]. The laser cavity for this work was the same as the second laser from [Shi 2012a]. In this work, the pump was a linearly polarized Yb laser operating at 1064 nm. With the polarized pump, the laser threshold was less than half of that with an unpolarized pump, a result that is predicted from theory. This demonstrated for the first time that a Raman DFB could have sub-watt threshold power, making the technology much more attractive. Slope efficiency vs. input pump was 13%, similar to that of [Shi 2012a].

Significantly, in [Shi 2012b], the linewidth was also measured using the delayed self-heterodyne technique. A linewidth of less than 2.4 kHz was observed. Such a narrow linewidth is typical of fiber-optic DFB lasers in REI-doped fibers. Because such narrow linewidths are a key advantage of fiber-optic DFB lasers over semiconductor lasers, this observation is a significant advance for Raman fiber DFBs showing that it is possible to obtain a performance similar to that observed in conventional REI DFBs. We note that this linewidth is three orders of magnitude smaller than that observed in [Westbrook 2011, 23]. The slope efficiency is also much higher than in the case of DFB1 and DFB2. It is likely that both of these effects are related to a difference in the laser cavity loss as well as different noise and polarization characteristics of the pump lasers. We also note that the difference in wavelength is expected to result in significant differences in effective area, gain, and cavity performance. The results of [Shi 2012a and 2012b] represent a significant advance for the Raman fiber DFB approach. Future studies will aim to achieve these impressive lasing characteristics with the cascading Raman wavelength conversion effects exploited in the demonstration of [Westbrook 2011, 23].

Lastly, we discuss another approach to a narrow-linewidth Raman laser, namely, the use of a short DBR laser [Siekiera 2012, 26]. We include these more conventional Raman laser results as a baseline for comparison with Raman DFBs. Moreover, these authors also tested their narrow-linewidth source in a MOPA configuration as we describe below.

The DBR laser of [Siekiera 2012, 26] is shown in Fig. 6.6c. As in [Shi 2012a, b] a single Raman order was used in this work. The cavity was formed by inscribing two narrow-linewidth FBGs in a PM Raman gain fiber. The FBGs were inscribed over 5 cm using a phase mask and a 248 nm excimer laser. Their bandwidths were 108 pm for the high reflector and 94 pm for the weaker output coupler. The reflectivities were near 100% and 96% for the HR and OC, respectively. The gratings were separated by 120 mm. The pump source was a Yb laser operating at 1100 nm. As in [Shi 2012b], the pump source was polarized.

This laser showed a threshold power of 4.1 W and a maximum power of 0.1 W at 8 W of pump for a conversion efficiency vs. incident pump of 9%, similar to that observed by [Shi 2012a, b]. The measured bandwidth of the output was 60 pm, indicating no more than 24 longitudinal cavity modes and corresponding to a linewidth of roughly 14GHz. Analysis of the RF spectrum of the laser output indicated the presence of at least six longitudinal modes. The cavity mode spacing was found to be 607 MHz corresponding to an effective cavity length of 170 mm. Since the spacing between the gratings was only 120 mm, this indicated that light propagation into the gratings contributed significantly to the effective cavity length. Thermal fluctuations were measured over 1 h and observed to be as high as 40%, indicating the possible influence of thermal effects. As in the case of Raman DFB cavities, thermal management has a significant effect on the performance of this laser.

As discussed previously, the low conversion efficiency of these various Raman lasers may be improved by using the excess pump to power a subsequent Raman amplifier. Multi-watt Raman amplification of narrow-linewidth sources, such as semiconductor DFBs, has been reported [43, 44]. The same approach was examined by Siekiera et al. using their short Raman DBR lasers [26, 27]. Both pump and signal from the Raman DBR in Fig. 6.6c were directed into 55 m of Raman gain fiber identical to that used to fabricate the Raman DBR. A conversion efficiency as high as 56% was achieved, with an output power of more than 5 W with an 8 W pump. However, the bandwidth of the signal increased from less than 100 pm to more than 340 pm at the highest power. The initial narrow-linewidth spectrum of the master oscillator (MO) was broadened with a substantial continuous bandwidth outside of the MO's 60 pm output. This broadening sets in at a pump power of 6 W and was attributed to four-wave mixing among the modes of the fiber.

6.4 Brillouin Fiber DFB Lasers

In this section, we consider the performance of a fiber DFB laser using Brillouin gain. As stated previously, Raman DFB cavities can be configured to operate with Brillouin gain because of their very narrow reflection spectra. Therefore, we also review Brillouin laser operation and some applications. We then describe the first observation of a fiber DFB laser using Brillouin gain. The particular cavity we discuss is the same as DFB2 from Westbrook 2011 [23], whose Raman operation was discussed in detail in Sect. 3. Since the Brillouin gain efficiency $g_s = \frac{g_B}{A_{eff}} = \frac{4.2}{W\,m}$ (Brillouin gain coefficient $g_B = 5 \cdot 10^{-11}$ m/W, effective area $A_{eff} = 12\ \mu m^2$ in HNLF) is about three orders of magnitude larger than the Raman gain efficiency (see Sect. 2.3), the same DFB cavity is expected to lase at a much lower pump power level in the mW range, making it possible to study the lasing operations with minimal thermal and nonlinear distortions.

Brillouin lasing is based on stimulated Brillouin scattering (SBS), which occurs when narrowband laser radiation propagates through an optical fiber. It is useful for various applications such as narrowband amplification, lasing, distributed sensing, phase conjugation, and slow light generation. Light that is scattered in the backward direction (known as the Stokes wave) experiences a frequency downshift by an amount equal to $\nu_B = 2n\nu_A/\lambda_p$, where ν_A is the acoustic velocity, n is the refractive index, and λ_p is the wavelength of the incident radiation (pump). SBS can be used to amplify a counterpropagating signal with wavelength near the Stokes wave. Taking advantage of Brillouin gain, fiber lasers, in both ring and Fabry-Pérot configurations, have been realized to generate single- as well as multiple-order Stokes waves [45–53]. SBS fiber lasers operating in a single longitudinal mode are particularly attractive for optical microwave signal processing [48, 49] and also for suppression of amplitude/phase noise of optical radiation [50–52].

Despite the narrow SBS gain bandwidth (a few tens of MHz) in silica fibers, it is difficult to realize single-frequency lasing. As with Raman fiber lasers, Brillouin lasers require fiber lengths of hundreds of meters to allow for sufficient gain to reach threshold, causing the cavity to operate with multiple longitudinal modes. Short cavity SBS lasers with active feedback stabilization were used to make the cavity doubly resonant to both pump and Stokes, allowing low threshold [46, 51, 52]. Also, single-frequency SBS lasers have been demonstrated using only a few meters of chalcogenide glass fiber, which has a Brillouin gain coefficient about two orders of magnitude higher than that of silica fibers [53]. However this chalcogenide glass fiber has a number of drawbacks, which include low damage threshold; difficulty in handling, in particular splicing to silica fiber; and high transmission loss. SBS was also reported in a chalcogenide photonic chip (length of $\approx$4cm) resonator where the Fresnel reflection, $\approx$17%, at the end facets gives rise to a Fabry-Pérot cavity [54]. Multiple Stokes waves were generated using a pulsed pump source with peak power of 1.5 W.

Large Brillouin amplification has also been demonstrated in silicon by engineering waveguide structures. Although silicon waveguides exhibit large Kerr and Raman nonlinearities, coupling between photons and acoustic phonons in typical silicon waveguide is very weak. However, using a membrane-suspended waveguide, Kittlaus et al. demonstrated Brillouin amplification greater than 5 dB for modest pump powers and a low threshold (5 mW) for net amplification, which promises realization of efficient Brillouin amplifiers and lasers for potential applications [55].

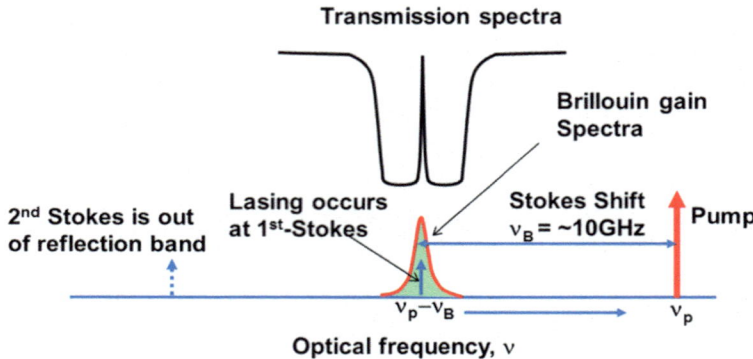

Fig. 6.14 Schematic spectrum of the optical signal (pump) and the DFB fiber Bragg grating

A new approach to overcome these problems is the Brillouin fiber DFB laser [28]. The Brillouin DFB cavity requires a phase-shifted Bragg grating whose reflection bandwidth is less than the Brillouin shift. Narrow linewidth laser radiation with appropriate wavelength can then be launched into the DFB as a pump wave without reflecting off of the cavity grating. The pump is tuned so that its optical frequency is higher than the resonant transmission frequency of the grating by the Stokes frequency shift ν_B in the fiber. In such an arrangement, as shown in Fig. 6.14, the pump can transmit through the grating to the phase shift region and provide Brillouin gain that results in lasing at the Stokes wavelength within the grating. The DFBs have a high Q-factor enabling low lasing threshold and narrow bandwidth that ensures operation in a single longitudinal mode.

The offset phase-shifted Raman DFB2 from Sect. 3 was used in these experiments. The DFB had a threshold as low as 30 mW and slope efficiency vs. incident pump of more than 20%, while higher-order Stokes waves were suppressed by more than 20 dB. In contrast to the "optical" phonons (out-of-phase movements of neighboring atoms) involved in Raman scattering, the "acoustic" phonons (in-phase movements of neighboring atoms) involved in Brillouin scattering carry a significant crystal momentum. Hence, conservation of momentum implies that Brillouin gain is only experienced by a backward propagating signal. However, the DFB laser could lase in either the forward or backward direction, simply by changing the orientation of the offset phase shift with respect to the pump propagation direction. As we discuss below, the laser could operate over a very broad range of pump frequencies >1.2 GHz, which is 50 ~ 60 times larger than the SBS gain bandwidth.

A typical Brillouin DFB laser setup is shown in Fig. 6.15. The DFB fiber Bragg grating was the same as DFB2 from Table 6.1. Light from a CW external cavity semiconductor laser operating near 1583 nm, with a typical linewidth of approximately 150 kHz, was amplified by an L-band amplifier to a maximum power of 143 mW and launched into the DFB grating through a circulator. The wavelength of the pump in relation to the grating transmission spectrum is shown in Fig. 6.16, which shows a high-resolution spectrum of DFB2. The reflection spectrum of the

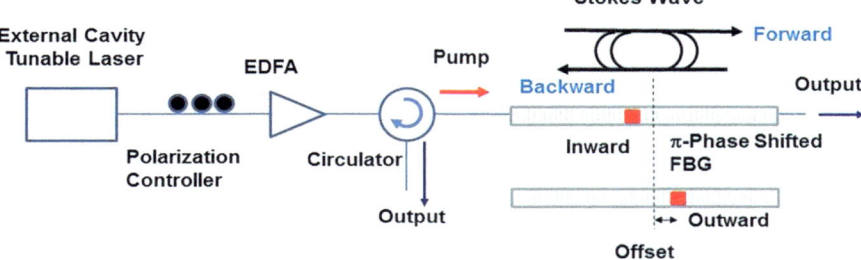

Fig. 6.15 Schematic of a DFB Brillouin fiber laser

Fig. 6.16 Transmission
spectrum of DFB2
measured with a
high-resolution/high-dynamic
range laser scan. The
Brillouin pump and lasing
frequencies are indicated by
vertical lines

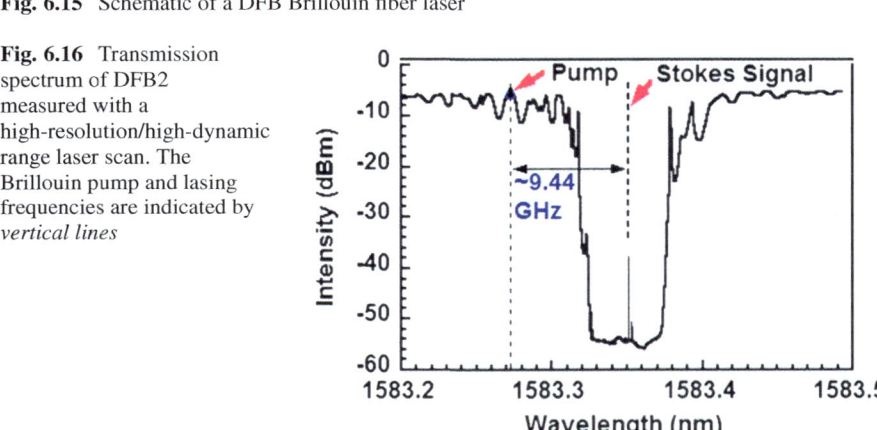

grating has a half-width of about 3.7 GHz (30 pm). The pump was tuned so that
its optical frequency was higher than the resonant transmission frequency of the
grating by about 9.4 GHz, which was equivalent to the Stokes shift in the nonlinear
fiber used in the experiment. Therefore, the pump is located outside of the grating's
reflection band and can thus transmit through the grating and provide Brillouin gain.
Note that this π phase-shifted grating has a pair of resonant transmission peaks near
1583.35 nm, separated by 220 MHz, which is due to birefringence of the fiber.
The ripples shown in the edges of the transmission spectra are due to the fact
that the grating was not fully apodized. Apodization of the grating can make the
transmission edge much smoother and flatter, which would be useful for efficient
pumping of the laser. The spectrum of spontaneous Brillouin scattering measured in
highly nonlinear fiber (OFS) is shown in Fig. 6.17. The Stokes shift was 9.415GHz,
while the bandwidth was 20 MHz.

Figure 6.18 shows SBS DFB laser output characteristics. Stable lasing at the
Stokes wavelength was achieved by tuning the optical frequency of the pump to
overlap the SBS gain spectrum with the narrowband resonance of the grating. Figure
6.18a shows the spectra of the output in the backward and forward directions,
when the phase shift offset was inward. In the backward direction, a strong Stokes

Fig. 6.17 Spontaneous Brillouin emission spectrum in OFS HNLF

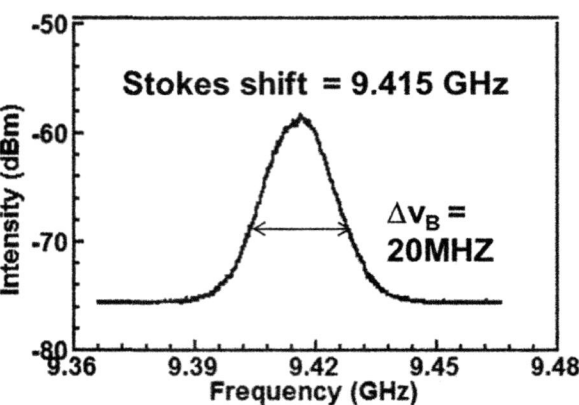

component (red-shifted by ≈80 pm) along with a fraction of pump backreflected by the uniform grating was observed. Ideally, the DFB grating should be transparent at the pump wavelength; however, due to ripples in the transmission spectrum, some part of the pump was reflected by the grating. The backward Stokes output was about 20 dB higher than that in the forward Stokes output. Moreover, since the second-order Stokes wavelength was out of resonance, subsequent conversion to higher-order Stokes waves was greatly suppressed. Both forward and backward second-order Stokes waves were found to be more than 20 dB below the (backward) first-order Stokes wave. Figure 6.18b shows the power of the backward Stokes wave and residual pump wave in the forward (transmitted through the DFB) and backward (reflected from the grating) direction plotted against the incident pump. The slope efficiency of conversion from incident pump to backward Stokes wave was about 23%, which is found to be an order of magnitude higher than the Raman DFB slope efficiency (1.2%) of [Westbrook 2011, 23] demonstrated using the same cavity.

Outputs from the DFB both in the forward and backward directions were recorded using an optical spectrum analyzer and a power meter. The beating between the Stokes signal and a small part of the backreflected pump was measured using a high-speed photodiode and an electrical spectrum analyzer. The electrical spectrum of the beat frequency of the pump and Stokes waves (Fig. 6.19) clearly showed narrow, single-frequency Stokes generation. The peak observed at ≈9.44 GHz was in agreement with the Stokes shift measured from spontaneous Brillouin scattering in this fiber for a pump at 1583 nm. The DFB Brillouin fiber laser oscillated in only one polarization state, because the birefringence in the DFB grating caused polarization splitting that is about 200 MHz, which is much larger than the Brillouin gain linewidth.

When the DFB grating was operated with the phase shift facing outward, efficient lasing of the Stokes wave in direction of the pump was also achieved. Although the Stokes wave experiences gain in the DFB resonator only while propagating in the direction opposite to the pump, the laser output could be obtained in forward direction due to the resonance provided by the DFB grating, with the output coupling determined only by the position of the phase shift.

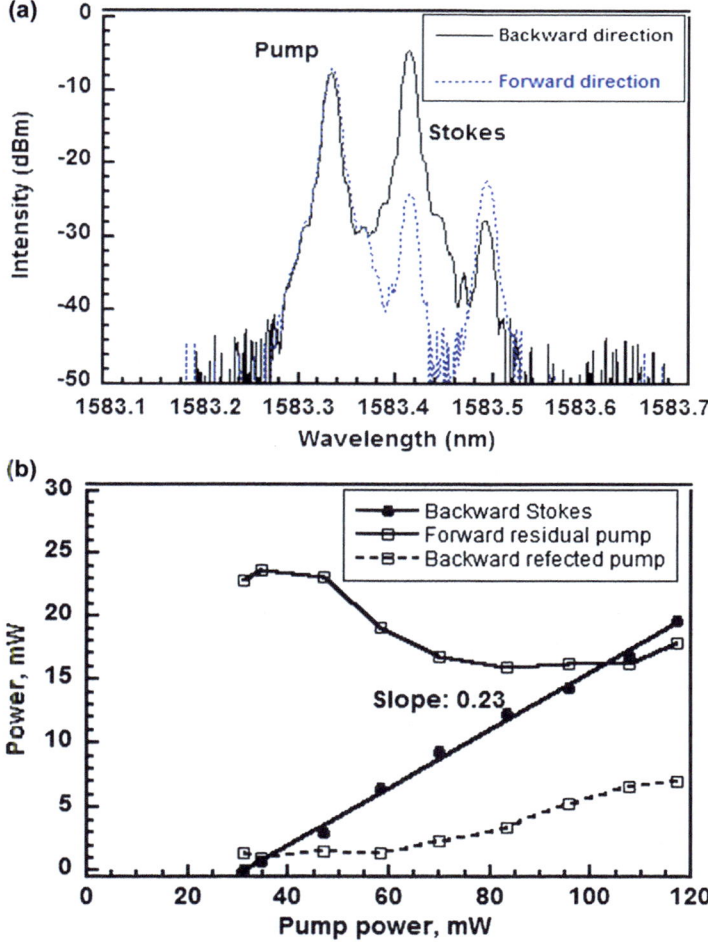

Fig. 6.18 Optical spectra of the output in the forward and backward direction (**a**). Powers of Stokes wave, residual pump and reflected pump plotted as a function of incident pump power (**b**). The discrete phase shift is facing inward

The optical spectrum of the output in the forward direction is shown in Fig. 6.20. The Stokes wave with intensity stronger than the pump was generated in the forward direction when the pump wavelength was optimally tuned. Stokes output was 3 dB stronger than the residual pump, indicating an internal conversion efficiency (ignoring the backreflected pump) greater than 67%. The slope efficiency of conversion from incident pump to forward Stokes was 27%, which could be improved by optimizing the grating design to suppress the backreflection of pump radiation.

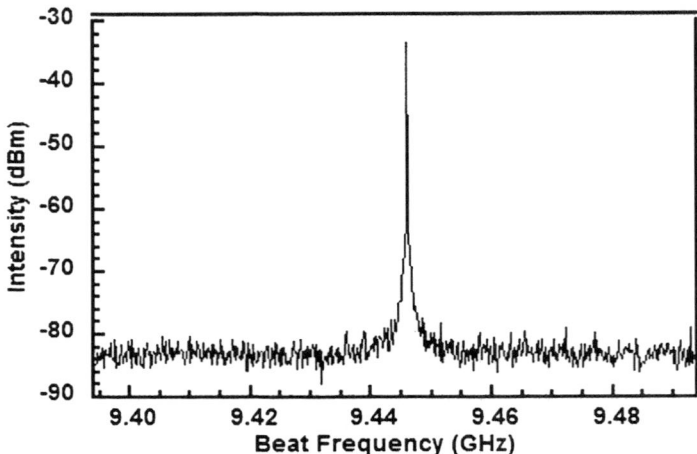

Fig. 6.19 The electrical spectrum of the pump-signal beat note

An unexpected property of the Brillouin DFB was that it lased for a range of pump frequencies much larger than the Brillouin gain bandwidth. This effect is related to hysteresis in the dependence of signal power vs. pump frequency detuning. To show the hysteresis, the pump power was held constant at 120 mW, and starting from no lasing, the Stokes output power and the RF beat frequency between the pump and Stokes wave were measured as the optical frequency of the pump was gradually reduced, bringing it closer to the DFB resonant frequency. As shown in Fig. 6.21a, the intensity of the Stokes wave continued to grow as the pump frequency was reduced until it suddenly ceased lasing. While the beat frequency remained relatively constant at about 9.44GHz, the range of pump tuning over which lasing occurred was 1.2 GHz, about 60 times larger than the Brillouin gain bandwidth. In contrast, lasing was not observed to initiate when the pump frequency was tuned from low to high frequency.

The reason for this behavior is explained using Fig. 6.21b. As the overlap of the Brillouin gain with the narrow resonance of DFB cavity increases, the Stokes signal strengthens. The refractive index inside the DFB, predominantly near the discrete phase shift, can be increased by the intracavity Stokes field due to both the Kerr effect [19] and nonuniform heating resulting from absorption [56]. This causes the DFB cavity resonance to shift to a lower frequency. The lasing frequency thus continues to decrease as the pump frequency is further decreased. This continues until the peaks of both SBS gain and the lasing wavelength coincide. At this pump frequency detuning, the output power is maximized. Beyond this point, a further decrease in pump frequency results in smaller gain, and thus reduced Kerr effect and heating, shifting the cavity resonance to higher frequency. Since the direction of the shift of the lasing frequency is now opposite to that of the pump detuning, the Stokes field no longer tracks the pump (in region III), which results in abrupt termination of

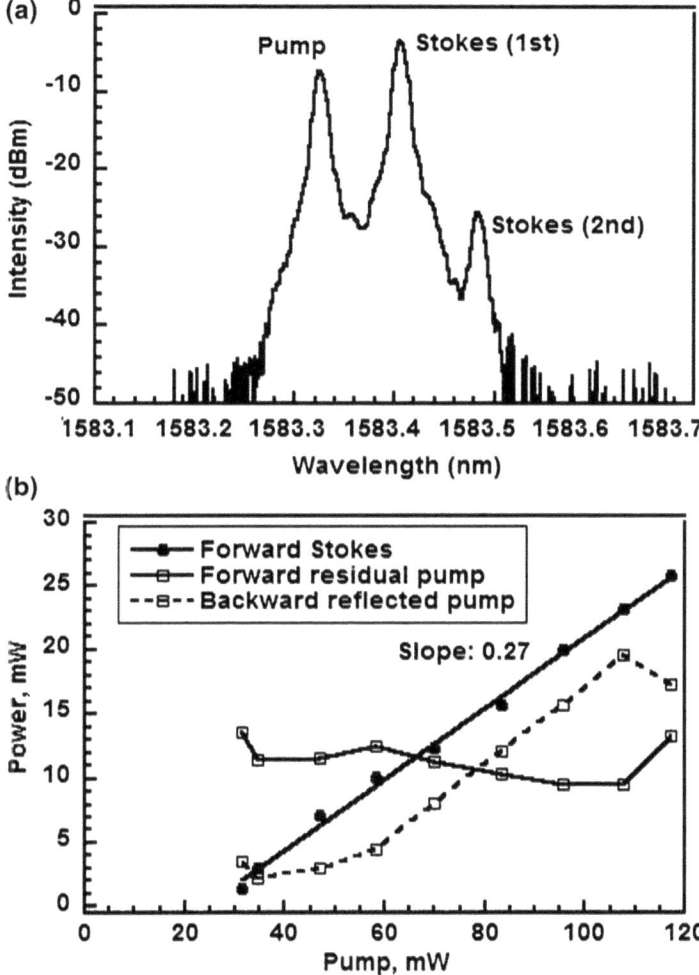

Fig. 6.20 Optical spectrum of the output in the forward direction, with the discrete phase shift facing outward (**a**). Powers of Stokes wave and residual pump plotted as a function of incident pump power (**b**)

lasing. It was observed that the Brillouin DFB laser can be turned on by decreasing the pump frequency. In contrast, the laser did not turn on when the pump frequency was increased. As the pump was detuned by as much as 1.2 GHz, the beat frequency between pump and Stokes wave changed by no more than 30 MHz (similar to the Brillouin gain bandwidth), indicating that the Stokes frequency was pulled along with the pump frequency. This is in stark contrast with Brillouin fiber ring lasers such as [51] that relied on active stabilization because of small tolerance (few tens of MHz only) against pump frequency fluctuations.

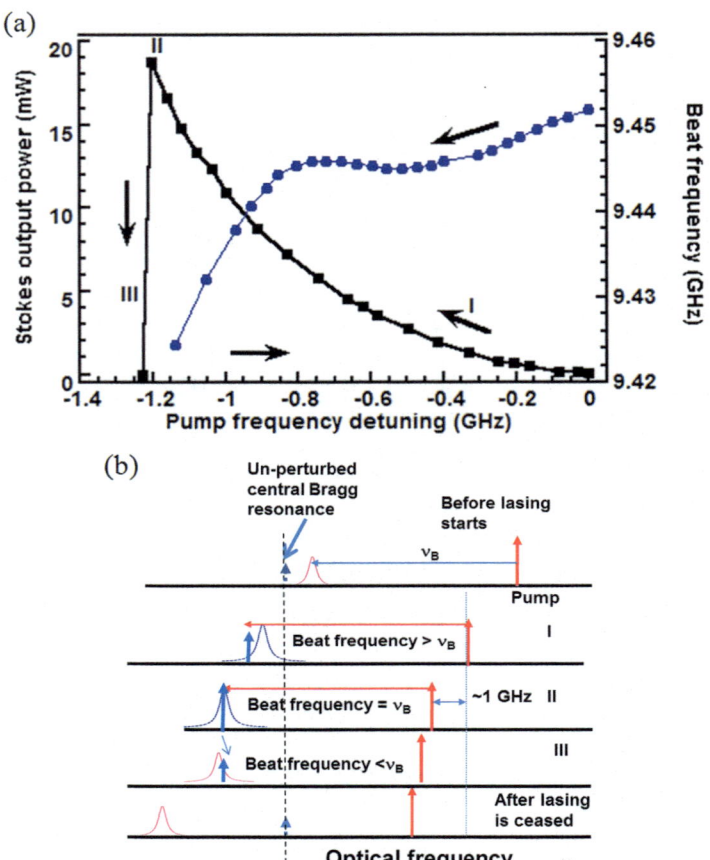

Fig. 6.21 (**a**) Output power dependence on relative pump frequency. Zero detuning corresponds to the pump frequency when the laser begins to oscillate. (**b**) Heuristic explanation of the Brillouin fiber DFB hysteresis effect

   Although the Brillouin pump was a narrow-linewidth source, it is also possible to use wider bandwidth sources while still obtaining a narrow-linewidth DFB output. In another set of experiments, such linewidth narrowing was shown using the same DFB cavity [57]. The DFB was pumped by a semiconductor DFB laser with a 30 MHz linewidth. The temperature of the grating was controlled using a Peltier cooler that allowed fine tuning of the spectrum of the fiber DFB grating. Optical heterodyne measurements of the DFB Brillouin laser were conducted by mixing with the output of an external cavity semiconductor (reference) laser with a linewidth of 150 kHz. The RF spectrum (Fig. 6.22a) shows three distinct frequency components, resulting from beating between the residual pump (semiconductor DFB), the Stokes wave, and the reference laser. Relative values of the optical frequencies shown in Fig. 6.22b were $\nu_{Pump} > \nu_{Ref} > \nu_{Stokes}$. The beat frequency

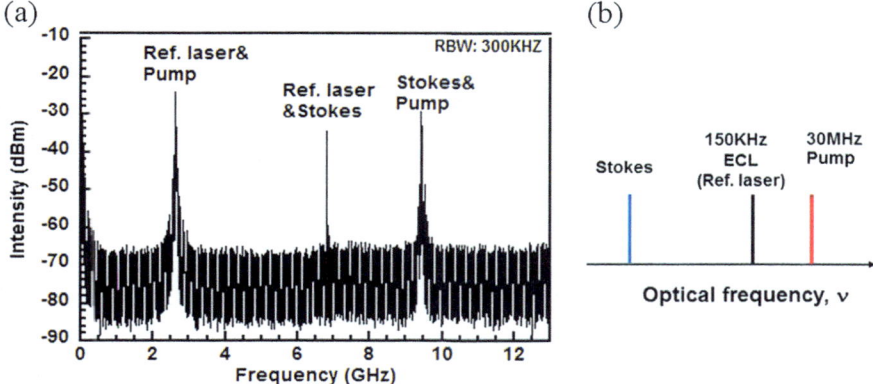

Fig. 6.22 (a) RF heterodyne spectrum of the laser output mixed with a narrow-linewidth semiconductor tunable reference laser (linewidth, ~150 kHz). (b) Relative optical frequencies

component between the reference laser and Stokes output was found to be narrower than that which resulted from beating between the reference laser and the pump. This suggested that optical spectra narrowed as the pump was converted to a Stokes wave, in a manner similar to that reported previously on single-frequency ring-cavity Brillouin laser [52].

Recently, Brillouin lasing has also been demonstrated in a π phase-shifted DFB gratings, 10–20 cm in length, inscribed in standard silica fiber. The laser exhibited similar dependence on pump frequency detuning, but to a lesser extent, due to reduced nonlinear resonance shifts [58].

While rare-earth DFB fiber lasers have the tendency to oscillate with two polarization modes, the DFB Brillouin fiber laser was observed to oscillate in only one polarization state. The birefringence in the DFB grating caused polarization splitting that is about 200 MHz, much larger than the Brillouin gain linewidth, thus making it possible to restrict lasing to only one polarization state. Moreover, since Brillouin gain becomes largest for a signal with the same polarization as the pump, the forward Stokes wave exhibits the same polarization as the pump wave.

Winful et al. have recently developed a model for numerical simulation of distributed feedback Brillouin fiber lasers by taking into account the pump wave, the Stokes wave, and the acoustic wave, and the results of simulation are found to be in good agreement with the results of our experiments [59].

6.5   Conclusion

In this chapter, we have described advances in narrow-linewidth fiber Raman lasers. We showed how they can be modeled using a set of nonlinear coupled mode equations and presented closed-form approximations to the threshold gain and detuning. An efficient numerical method was developed that can be used to reveal the dependence of the slope efficiency on cavity parameters such as gain, loss, the specifics of the grating profile, and nonlinear effects such as two-photon absorption.

Recent results have shown that the well-known cascaded resonator Raman fiber laser may be used to pump a fiber DFB laser using only Raman gain. Output linewidths are reduced by more than three orders of magnitude from that of traditional Raman lasers. Thus the flexible wavelength generation properties of the Raman laser approach are extended to applications requiring narrow linewidth. It has also been shown that such Raman DFBs can be exploited in applications requiring kHz linewidths. While these advances were obtained with only a single Raman conversion, they nonetheless point the way to future Raman DFB lasers with the full wavelength flexibility of Raman lasers and the superlative linewidth performance of fiber DFB lasers.

We have also demonstrated operation of a single-frequency Brillouin DFB fiber laser using the same π phase-shifted DFB gratings. The Brillouin gain coefficient being three orders of magnitude higher than the Raman gain coefficient in silica glass lasing could be initiated with only a few tens of milliwatts of pump power. Lasing with an order of magnitude higher slope efficiency along with a narrowing of the Stokes wave compared with the pump has been observed. Such lasers are expected to have potential application in communication, signal processing, as well as sensing.

References

1. Kogelnik, H., Shank, C.V.: Stimulated emission in a periodic structure. Appl. Phys. Lett. **18**, 152–154 (1971)
2. Shank, C.V., Bjorkholm, J.E., Kogelnik, H.: Tunable distributed-feedback dye laser. Appl. Phys. Lett. **18**, 395–396 (1971)
3. Kogelnik, H., Shank, C.V.: Coupled-wave theory of distributed feedback lasers. J. Appl. Phys. **43**, 2327–2335 (1972)
4. Nakamura, M., Yariv, A., Yen, H.W., Somekh, S., Garvin, H.L.: Optically pumped GaAs surface laser with corrugation feedback. Appl. Phys. Lett. **22**, 515–516 (1973)
5. Scifres, P.R., Burnham, R.D., Streifer, W.: Distributed-feedback single heterojunction GaAs diode laser. Appl. Phys. Lett. **25**, 203–206 (1974)
6. Haus, H.A., Shank, C.V.: Antisymmetric taper of distributed feedback lasers. IEEE J. Quantum Electron. **12**, 532–539 (1976)
7. Utaka, K., Akiba, S., Sakai, K., Matsushima, Y.: λ/4-shifted InGaAsP/InP DFB lasers by simultaneous holographic exposure of positive and negative photoresists. Electron. Lett. **20**, 1008–1010 (1984)
8. Meltz, G., Morey, W.W., Glenn, W.H.: Formation of Bragg gratings in optical fibers by a transverse holographic method. Opt. Lett. **14**(15), 823–825 (1989)
9. Hill, K.O., Malo, B., Bilodeau, F., Johnson, D.C., Albert, J.: Bragg gratings fabricated in monomode photosensitive optical fiber by UV exposure through a phase mask. Appl. Phys. Lett. **62**, 1035–1037 (1993)
10. Anderson, D.Z., Mizrahi, V., Erdogan, T., White, A.E.: Production of in-fibre gratings using a diffractive optical element. Electron. Lett. **29**, 566–568 (1993)

11. Asseh, A., Storøy, H., Sahlgren, B.E., Sandgren, S., Stubbe, R.A.H.: A writing technique for long fiber Bragg gratings with complex reflectivity profiles. J. Lightwave Technol. **15**, 1419–1423 (1997)
12. Loh, W.H., Cole, M.J., Zervas, M.N., Barcelos, S., Laming, R.I.: Complex grating structures with uniform phase masks based on the moving fiber-scanning technique. Opt. Lett. **20**, 2051–2053 (1995)
13. Kringlebotn, J.T., Archambault, J.-L., Reekie, L., Payne, D.N.: Er3+:Yb3+—codoped fiber distributed-feedback laser. Opt. Lett. **19**, 2101–2103 (1994)
14. Yelen, K., Zervas, M.N., Hickey, L.M.B.: Fiber DFB lasers with ultimate efficiency. J. Lightwave Technol. **23**, 32–43 (2005)
15. Grubb, S.G., Erdogan, T., Mizrahi, V., Strasser, T., Cheung, V.Y., Reed, W.A., Lemaire, P.J., Miller, A.E., Kosinski, S.G., Nykolak, G., Becker, P.C., Peckham, D.W.: 1.3 µm cascaded raman amplifier in germanosilicate fibers. Paper presented at optical amplifiers and their applications conference, PD3 1994
16. Perlin, V.E., Winful, H.G.: Distributed feedback fiber Raman laser. IEEE J. Quantum Electron. **37**, 38 (2001)
17. Perlin, V.E., Winful, H.G.: Stimulated Raman scattering in nonlinear periodic structures. Phys. Rev. A. **64**, 043804 (2001)
18. Hu, Y., Broderick, N.G.R.: Improved design of a DFB Raman fibre laser. Opt. Commun. **282**, 3356–3359 (2009)
19. Lauridsen, V.C., Povlsen, J.H., Varming, P.: Design of DFB fibre lasers. Electron. Lett. **34**, 2028–2030 (1998)
20. Shi, J., Ibsen, M.: Effects of phase and amplitude noise on π phase-shifted DFB Raman fibre lasers. Paper presented at bragg gratings, photosensitivity and poling in glass waveguides, JThA30, 2010
21. Kremp, T., Abedin, K.S., Westbrook, P.S.: Simulation of two-photon absorption in Raman DFB lasers. Paper presented at advanced photonics congress, OSA technical digest (Optical Society of America), paper BW3E.5, 2012
22. Westbrook, P.S., Abedin, K.S., Nicholson, J.W., Kremp, T., Porque, J.: Demonstration of a Raman fiber distributed feedback laser. Paper presented at CLEO, PDPA11, 2011
23. Westbrook, P.S., Abedin, K.S., Nicholson, J.W., Kremp, T., Porque, J.: Raman fiber distributed feedback lasers. Opt. Lett. **36**, 2895–2897 (2011)
24. Shi, J., Alam, S., Ibsen, M.: Highly efficient Raman distributed feedback fibre lasers. Opt. Express. **20**, 5082–5091 (2012)
25. Shi, J., Alam, S., Ibsen, M.: Sub-watt threshold, kilohertz-linewidth Raman distributed-feedback fiber laser. Opt. Lett. **37**, 1544–1546 (2012)
26. Siekiera, A., Engelbrecht, R., Nothofer, A., Schmauss, B.: Short 17-cm DBR Raman fiber laser with a narrow Spectrum. IEEE Photon. Technol. Lett. **24**, 107–109 (2012)
27. Siekiera, A., Engelbrecht, R., Nothofer, A., Schmauss, B.: Characterization of a narrowband Raman MOPA with short master oscillator. Paper presented at Fiber Lasers IX: Technology, systems, and applications, edited by Eric C. Honea, Sami T. Hendrow Proc. of SPIE **8237**, 82371I (2012)
28. Abedin, K.S., Westbrook, P.S., Nicholson, J.W., Porque, J., Kremp, T., Liu, X.: Single-frequency Brillouin distributed feedback fiber laser. Opt. Lett. **37**, 605–607 (2012)
29. Snyder, A.S., Love, J.D.: Optical Waveguide Theory. Kluwer Academic Publishers Group, Boston (1983)
30. Kremp, T., Abedin, K.S., Westbrook, P.S.: Closed-form approximations to the threshold quantities of distributed-feedback lasers with varying phase shifts and positions. IEEE J. Quantum Electron. **49**, 281–292 (2013)
31. Kashyap, R.: Fiber Bragg Gratings. Academic Press, San Diego (1999)
32. Othonos, A., Kalli, K.: Fiber Bragg Gratings: Fundamentals and Applications in Telecommunications and Sensing. Artech House, Boston/London (1999)
33. Erdogan, T.: Fiber grating spectra. J. Lightwave Technol. **15**, 1277–1294 (1997)

34. Foster, S.: Spatial mode structure of the distributed feedback fiber laser. IEEE J. Quantum Electron. **40**, 884–892 (2004)
35. Løvseth, S.G., Rønnekleiv, E.: Fundamental and higher order mode thresholds of DFB fiber lasers. J. Lightwave Technol. **20**, 494–501 (2002)
36. McCall, S.L., Platzman, P.M.: An optimized $\pi/2$ distributed feedback laser. IEEE J. Quantum Electron. **21**, 1899–1904 (1985)
37. Barmenkow, Y.O., Kir'yanov, A.V., Perez-Millan, P., Cruz, J.L., Andres, M.V.: Threshold of a symmetrically pumped distributed feedback fiber laser with a variable phase shift. IEEE J. Quantum Electron. **44**, 718–723 (2008)
38. Tuniz, A., Brawley, G., Moss, D.J., Eggleton, B.J.: Two-photon absorption effects on Raman gain in single mode As_2Se_3 chalcogenide glass fiber. Opt. Express. **16**, 18524–18534 (2008)
39. de Sterke, C.M., Jackson, K.R., Robert, B.D.: Nonlinear coupled mode equations on a finite interval: a numerical procedure. J. Opt. Soc. Am. B. **8**, 403–412 (1991)
40. Taha, T.R., Ablowitz, M.J.: Analytical and numerical spects of certain nonlinear evolution equations II. Numerical nonlinear Schrödinger equation. J. Comput. Phys. **55**, 203–230 (1984)
41. Qin, G., Liao, M., Suzuki, T., Mori, A., Ohishi, Y.: Widely tunable ring-cavity tellurite fiber Raman laser. Opt. Lett. **33**, 2014–2016 (2008)
42. Nicholson, J.W., Yan, M.F., Wisk, P., Fleming, J., DiMarcello, F., Monberg, E., Taunay, T., Headley, C., DiGiovanni, D.J.: Raman fiber laser with 81 W output power at 1480 nm. Opt. Lett. **35**, 3069–3071 (2010)
43. Feng, Y., Taylor, L.R., Calia, D.B.: Multiwatts narrow linewidth fiber Raman amplifiers. Opt. Express. **16**, 10927–10932 (2008)
44. Feng, Y., Taylor, L.R., Calia, D.B.: 25 W Raman-fiber-amplifier-based 589 nm laser for laser guide star. Opt. Express. **17**, 19021–19026 (2009)
45. Stokes, L.F., Chodorow, M., Shaw, H.J.: All-fiber stimulated Brillouin ring laser with submilliwatt pump threshold. Opt. Lett. **7**, 509–511 (1982)
46. Smith, S.P., Zarinetchi, F., Ezekiel, S.: Narrow-linewidth stimulated Brillouin fiber laser and applications. Opt. Lett. **16**, 393–395 (1991)
47. Al-Mansoori, M.H., Kamil Abd-Rahman, M., Mahamd Adikan, F.R., Mahdi, M.A.: Widely tunable linear cavity multiwavelength Brillouin-erbium fiber lasers. Opt. Express. **13**, 3471–3476 (2005)
48. Loayssa, A., Benito, D., Garde, M.J.: Optical carrier-suppression technique with a Brillouin-erbium fiber laser. Opt. Lett. **25**, 197–199 (2000)
49. Norcia, S., Tonda-Goldstein, S., Dolfi, D., Huignard, J.-P.: Efficient single-mode Brillouin fiber laser for low-noise optical carrier reduction of microwave signals. Opt. Lett. **28**, 1888–1890 (2003)
50. Geng, J., Staines, S., Jiang, S.: Dual-frequency Brillouin fiber laser for optical generation of tunable low-noise radio frequency/microwave frequency. Opt. Lett. **33**, 16–18 (2008)
51. Geng, J., Jiang, S.: Pump to-stokes transfer of relative intensity noise in Brillouin fiber ring lasers. Opt. Lett. **32**, 11–13 (2007)
52. Geng, J, Staines, S, Wang, Z., Zong, J., Blake, M., Jiang, S.: Actively stabilized Brillouin fiber laser with high output power and low noise. Paper presented at optical fiber communication conference, OThC4, 2006
53. Abedin, K.S.: Single-frequency Brillouin lasing using single-mode As2Se3 chalcogenide fiber. Opt. Express. **14**, 4037–4042 (2006)
54. Pant, R., Li, E., Choi, D.Y., Poulton, C.G., Madden, S.J., Luther-Davies, B., Eggleton, B.J.: Cavity enhanced stimulated Brillouin scattering in an optical chip for multiorder stokes generation. Opt. Lett. **36**, 3687–3689 (2011)
55. Kittlaus, E.A., Shin, H., Rakich, P.T.: Large Brillouin amplification in silicon. Nat. Photonics. **10**, 463–468 (2016)
56. Rokhsari, H., Vahala, K.J.: Observation of Kerr nonlinearity in microcavities at room temperature. Opt. Lett. **30**, 427–429 (2005)

57. Abedin, K.S., Westbrook, P.S., Kremp, T., Zhu, B., Nicholson, J.W., Porque, J., Liu, X.: Highly efficient distributed feedback Brillouin fiber laser. Paper presented at advanced photonics congress OSA Technical Digest paper BW3E.3, 2012
58. Loranger, S., Lambin-Iezzi, V., Wahbeh, M., Kashyap, R.: Stimulated Brillouin scattering in ultra-long distributed feedback Bragg gratings in standard optical fiber. Opt. Lett. **41**, 1797–1800 (2016)
59. Winful, H.G., Kabakova, I.V., Eggleton, B.J.: Model for distributed feedback Brillouin lasers. Opt. Express. **21**, 16191–16199 (2013)

Chapter 7
Random Distributed Feedback Raman Fiber Lasers

Sergey A. Babin, Sergey I. Kablukov, Ekaterina A. Zlobina, Evgeniy V. Podivilov, Sofia R. Abdullina, Ivan A. Lobach, Alexey G. Kuznetsov, Ilya D. Vatnik, Dmitry V. Churkin, and Sergei K. Turitsyn

7.1 Introduction

Random lasers represent now a rapidly growing class of light sources, in which a conventional optical cavity is substituted by a multiple-scattering feedback in a disordered gain medium, such as laser crystal or semiconductor powders (see [1–3] for a review). Recent developments in this field include improvements of the random laser performances as well as demonstrations of lasing in disordered media of new types. So, low-threshold surface-plasmon-enhanced lasing is demonstrated in a matrix of randomly distributed gold nano-islands coated by a wave-guiding layer of a dye-doped polymer [4] or in a semiconductor active medium (ZnO nanorods) with graphene oxide nanoflakes [5]. Fluidic paper-based random laser devices are fabricated by conventional soft lithography techniques on a usual paper

S.A. Babin (✉) • E.V. Podivilov • I.D. Vatnik
Institute of Automation and Electrometry SB RAS, 630090, Novosibirsk, Russia

Novosibirsk State University, 630090, Novosibirsk, Russia
e-mail: babin@iae.nsk.su; podivilov@iae.nsk.su; ilya.vatnik@gmail.com

S.I. Kablukov • E.A. Zlobina • S.R. Abdullina • I.A. Lobach • A.G. Kuznetsov
Institute of Automation and Electrometry SB RAS, 630090, Novosibirsk, Russia
e-mail: kab@iae.nsk.su; ZlobinaKaterina@gmail.com; sonka@ngs.ru; ivan.lobach@gmail.com; leks.kuznecov@gmail.com

D.V. Churkin
Novosibirsk State University, 630090, Novosibirsk, Russia
e-mail: churkin@nsu.ru

S.K. Turitsyn
Aston Institute of Photonic Technologies, Aston University, Birmingham, UK

Novosibirsk State University, 630090, Novosibirsk, Russia
e-mail: s.k.turitsyn@aston.ac.uk

© Springer International Publishing AG 2017
Y. Feng (ed.), *Raman Fiber Lasers*, Springer Series in Optical Sciences 207,
DOI 10.1007/978-3-319-65277-1_7

[6]. Random lasing may be obtained in such exotic media as cold-vapor atoms [7] or biological tissues including active dye-infiltrated bones [8], butterfly wing with semiconductor ZnO nanoparticles [9], and even a single cell [10]. These results initiate the development of advanced technologies toward the realization of biocompatible and implantable active photonic components [9, 10], bio-imaging of a new type including mapping of malignant tumors [11], diagnostics/dynamics of granular [12] or turbid [13] media with a great potential in pharmacology, as well as the development of low-coherence sources suitable for speckle-free full-field microscopy or digital light projector systems [14].

For the development of new light sources, a competitive device performance becomes quite a challenge. In this sense, fiber-based random lasers [15] are recognized as light sources superior to random lasers of other types and in some cases to conventional lasers. The fiber waveguide structure is nearly one-dimensional forming an output beam of high quality (single transverse mode with a Gaussian beam profile) in a desired direction by using fiber flexibility. For random lasing, even conventional telecom fibers are suitable. As the fiber material (silica glass) is highly transparent for radiation, especially in the telecom spectral window around 1.5 μm, the gain and feedback mechanisms here are quite different from those in bulk random lasers. The fiber gain is induced by inelastic stimulated Raman scattering (SRS) of the pump light by vibrating SiO_2 molecules in a glass lattice, whereas the feedback is provided by elastic Rayleigh scattering of the SRS-induced Stokes wave on sub-micron irregularities of the glass structure, with a small part ($\sim 10^{-3}$) of scattered light coming back into the fiber. Though the feedback is very weak, it is sufficient for lasing in a kilometer-long passive fiber given that the integral Raman gain is proportional to the fiber length and pump power.

As shown recently [13], a high-power non-resonant pumping enables Raman lasing in non-active bulk random materials (e.g., $BaSO_4$ [13]) as well, which makes random lasing possible in almost any white powder, thus offering a new direction in the development of devices and diagnostic techniques. Nevertheless, fiber-based random Raman lasers demonstrate at the moment the highest efficiency of pump-to-Stokes wave conversion exceeding 70% both for the first [16–18] and for the second Stokes order [19], with the output beam power up to 200 W [20]. Such random Raman fiber lasers (RRFLs) generate a quasi-continuous mode-free spectrum with the resulting shape defined by the Schawlow-Townes narrowing of amplified SRS emission near the generation threshold and nonlinear broadening at high generated powers [15, 21]. Fiber-based spectral filters can be relatively simply imbedded into the low-power part of RRFLs providing flat tuning within the entire Raman gain spectral range of >35 nm [22], as well as power-equalized multiwavelength generation [23], and order-of-magnitude spectral width reduction defined by the filter characteristics [24]. RRFLs also grant configurations/regimes similar to those in conventional fiber lasers, such as direct pumping by inexpensive and powerful multimode diodes [25], internal intensity modulation [26], pulsed operation via active [27] or passive [28] Q-switching, etc.

In this chapter we briefly review the basic principles of Raman fiber lasers with random distributed feedback, including details of feedback mechanism, various cavity designs and corresponding power and spectral characteristics, as well as their statistical properties both in experiment and theory. Then we compare performances of the random Raman fiber lasers with that for conventional Raman fiber lasers (RFLs) with linear cavity based on two reflectors, whose role may play fiber loop mirrors, fiber Bragg gratings, or Fresnel reflection from cleaved fiber end. The performance analysis includes polarization management, optimization of conversion efficiency, cascaded generation of higher Stokes orders, opportunities for short-wavelength generation via direct pumping by high-power laser diodes, or frequency doubling of random Raman fiber laser radiation. Pulsed operation of random Raman fiber lasers via active or passive Q-switching will be also analyzed.

The analysis shows that the output characteristics of Raman fiber lasers with random distributed feedback reached to the moment already outperform in many aspects those for conventional Raman fiber lasers, in particular, higher efficiency of the first-order and especially cascaded generation in cavity-free RRFL, much broader tuning range and higher flatness of multiwavelength generation, narrower bandwidth and higher conversion efficiency of frequency doubling in nonlinear crystal, efficient active Q-switching, etc. Other types of random fiber lasers especially those based on active fibers and random feedback based on random FBGs are also present. The unique performance of random fiber lasers opens the door to their application in advanced technologies, such as long-distance amplifier-free transmission and remote sensing, low-coherence IR and visible sources for bio-imaging and others, which are reviewed at the end of the chapter.

7.2 Basic Principles of Random RFLs

7.2.1 Gain and Feedback Mechanisms in Random Fiber Lasers

The basic scheme of the random distributed feedback (DFB) fiber laser is very simple comprising only a pump laser and a piece of a passive fiber playing the role of both gain medium and a cavity. Here and further in this section, we focus on the properties of the random DFB fiber laser operating via Raman gain (other gain mechanisms are discussed in Sect. 7.3.7). The random DFB fiber laser converts the pump radiation into new spectral range defined by the pump wavelength and the Stokes shift of a particular Raman gain media. The distributed random feedback in optical fibers is originated from the Rayleigh scattering on a natural disorder of the fiber core's refractive index. Although a randomness in other random systems usually means a stochastic behavior of output beam in temporal, spatial, and spectral domains, the radiation of the random DFB fiber laser is well confined in spectrum and stable over time. Amazingly, random DFB fiber lasers can even outperform conventional fiber lasers of a similar kind in several aspects, as it will be demonstrated below.

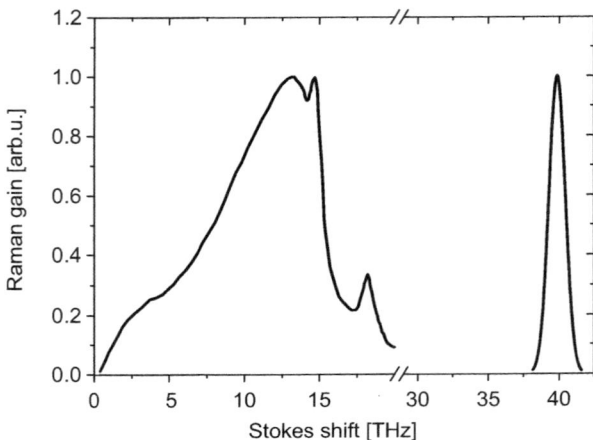

Fig. 7.1 A Raman gain spectrum of a germanosilica-based optical fiber (left peak) and phosphorus-doped silica fibers (both peaks)

The gain mechanism which is widely used to obtain random lasing in fibers is based on the stimulated Raman scattering. A pump light of frequency ν_p incident on a medium excites a quantum of molecular vibrations of the silica glass during an inelastic scattering process and loses some small part of its energy. Residual energy is carried out by a so-called Stokes photon possessing lower frequency ν_s than the pump photon. The frequency difference between pump and Stokes photons is called a Stokes shift. The Stokes shift value is determined by a structure of vibration levels of the host media. In the amorphous medium of silica glass with a variety of different collective vibrations, Stokes phonons of the wide energy range following the Raman scattering spectral profile could be spontaneously emitted. If there are some spontaneously scattered photons in the media, the stimulated Raman scattering can take place being a coherent process. The stimulated Raman scattering rate depends on the pump wave, P_p, and the Stokes wave, P_s, powers and can be described as $dP_s/dz = g_R P_p P_s$, where g_R is the frequency-dependent Raman gain coefficient. Concrete Raman gain spectral profile depends on a structure of a fiber used. While all silica fibers possess broad peak around 13 THz corresponding to amorphous silica vibration bands, phosphosilicate fibers having P_2O_5 molecules in the core as a dopant demonstrate additional peak arising around 40 THz (Fig. 7.1). The total amplification acquired by the Stokes wave in the fiber span of length L can be found as $\exp(g_R P_p L)$, meaning that the total gain exponentially increases over the fiber length. However, in practice, the gain is limited by the pump wave attenuation, $P_p(z) \sim \exp(-\alpha_p z)$. As a result, the total gain is limited by $\exp(g_R P_p L_{\text{eff}})$, where $L_{\text{eff}} = 1/\alpha_p$ is the approximate expression for the effective length in the very long fibers; so $L_{\text{eff}} \sim 20$km in SMF-28 fiber with $\alpha_p \simeq 0.05\,\text{km}^{-1}$ at $\sim 1.5\,\mu\text{m}$. At pump power level of 1 W, the total gain could be as high as 10^5–10^6.

The Rayleigh scattering (RS) used in the random DFB laser as a feedback source is an elastic scattering process. While propagating in the fiber, the light scatters on the random (in strength and position over the fiber but unchanging in time) fluctuations of the refractive index in the fiber core obeying the Rayleigh law. Such fluctuations are "frozen" during the fiber's manufacturing process. The Rayleigh scattering contributes sufficiently to the total linear losses α of optical fiber in the wavelength range below $1.55\,\mu m$. Small part of the light scattered at angles close to π is recaptured by the fiber waveguide and propagates in direction opposite to the direction of the incident light. The backscattered part of the radiation is equal to $\varepsilon = \alpha_s \cdot Q \sim 5 \times 10^{-5}\ km^{-1}$ being extremely small. Here the geometrical factor $Q \sim 0.001$ is defined by a numerical aperture and geometrical dimensions of the fiber [29].

Despite the Rayleigh backscattering is extremely small, the backscattered signal can be detected that is widely used in optical time domain reflectometry since the 1970s [30]. In the 1990s it was shown that the presence of the RS feedback in a Brillouin fiber lasers improves laser's performances, namely, its output line width is narrowed by three orders of magnitudes [31]. In the 2000s the effect of double Rayleigh scattering was observed in long-haul fiber-optic transmission lines [32]. The double Rayleigh scattering manifested itself as irregular spikes of lasing that occurred at high values of the distributed Raman gain in the system. In 2009 it was pointed out for the first time that the Rayleigh scattering based random feedback may be sufficient for lasing in ultra-long fiber lasers ($\simeq270\,km$), [33]. In cavities of such length, the Rayleigh scattering based random feedback has a lower threshold compared to the lasing threshold of the corresponding FBG-based linear cavity. After that, intensive endeavors directed at the development of self-consistent Rayleigh scattering based lasing in passive fibers have resulted in the invention of a new class of lasers—random DFB Raman fiber lasers or random RFLs.

7.2.2   Cavity Designs of Random RFLs

There are two different basic configurations of the random DFB Raman fiber laser which are dramatically different in power properties. The first one is the so-called forward-pumped configuration, as the generation wave out-coupled from the laser is co-propagating with the pump wave (Fig. 7.2a). This configuration could be realized in a fiber span of length L and one pump laser used with a mirror (e.g., a fiber loop mirror or fiber Bragg grating) placed at $z = 0$, thus providing half-open cavity.

The pump wave could be coupled from fiber ends, i.e., at $z = L$ (Fig. 7.2b). This scheme is called the backward-pumped configuration of the random DFB (RDFB) fiber laser, as the output generation wave is counter-propagating with the pump wave. Note that both schemes have symmetrical modifications with doubling all elements and symmetrically reflecting the scheme against point $z = 0$ with similar characteristics [34]. Symmetrical modifications were used in first demonstration of RDFB laser [15] and could be considered for telecommunication applications

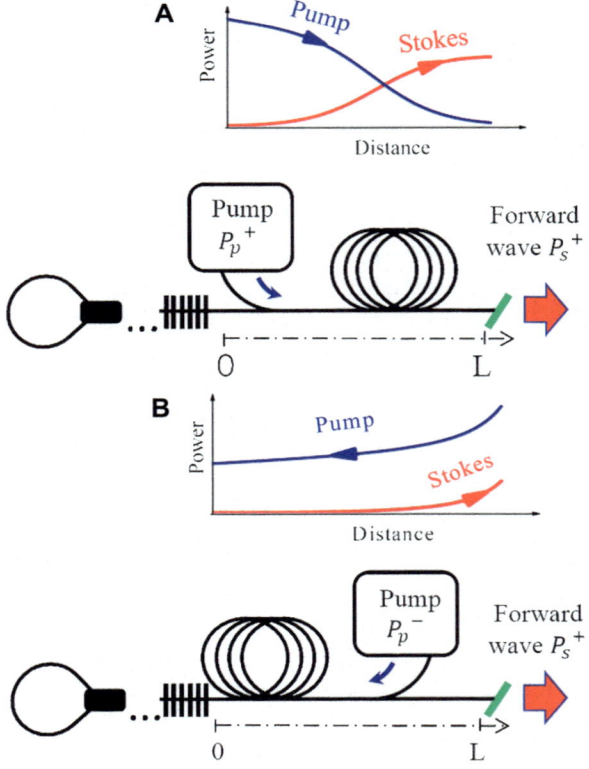

Fig. 7.2 Random distributed feedback fiber laser configurations: (**a**) forward-pumped laser, (**b**) backward-pumped laser

[35]. Note that there is a third scheme comprising a fiber span of length L with one pump laser coupled at point $z = 0$ with no any point-based reflector, so the laser is operated via the random distributed feedback only. This so-called single-arm open-cavity configuration demonstrates similar power performance with backward scheme [34], but has higher threshold level, and thus is usually out of scope.

7.2.3 Power Characteristics

The random DFB Raman fiber laser has a clear generation threshold in all configurations. As an example, the power performance in the forward-pumped configuration of Fig. 7.2a is shown in Fig. 7.3a. The key laser features—a generation threshold and the rapid growth of the output power above the threshold—are clearly seen. The experimentally measured laser threshold in the forward-pumped laser based on 850 m of phosphosilicate fiber [16] is close to a value of 5.5 W for one pump. Near the generation threshold, more than 2 W of output power is generated from only 0.5 W of pump power excess over the generation threshold. At high pump

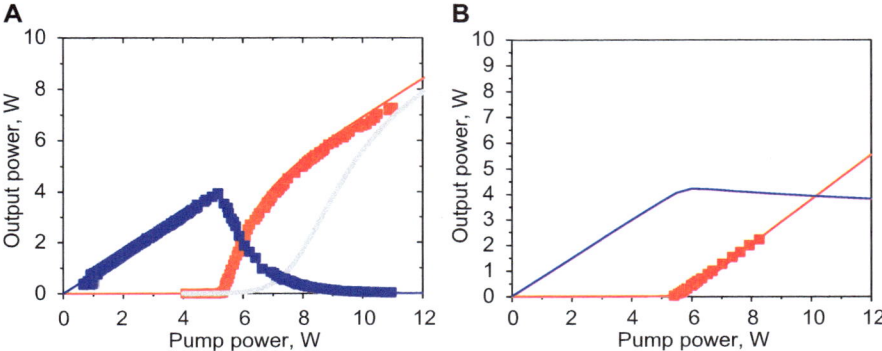

Fig. 7.3 Typical output power performances of the laser in (**a**) forward-pumped scheme, (**b**) backward-pumped scheme. *Squares* are experimental data, and *lines* are numerical results. *Blue* is for pump and *red* for generation. *Gray line* is numerical results for generation in case of absence of Rayleigh feedback

power, the quantum conversion efficiency defined as a ratio of generated and pump photons (in absense of lasing) at the laser output exceeds 100%. This is due to the specific longitudinal power distribution along the cavity (see Fig. 7.2a) and lower loss for the generated wave compared to the pump wave. The scheme can be used in high-power generation for moderate levels of excess over threshold [16, 18].

Backward-pumped scheme has a very different power properties with linear output dependence of generation power, with slope efficiency achieving quantum limit and with much more weaker dependence on fiber length and other parameters [36].

In general, power characteristics of the random DFB lasers are the consequence of the presence of Rayleigh feedback, and in an imaginary case of generation without influence of feedback predicted power properties dramatically deviate from the RDFB case (see Fig. 7.3a).

7.2.4   Spectral Characteristics

A random DFB Raman fiber laser generates a well-confined optical spectrum with a typical width of 1 nm. When the pump power is below the generation threshold, the spectrum is broad and corresponds to the amplified spontaneous emission spectrum. However, above the threshold, the generation spectrum becomes much narrower (1 nm) than the spectral profile of the Raman gain ($\sim$10 nm) (Fig. 7.4a). The abrupt narrowing of the spectrum above the threshold is the important criteria for that the real lasing is achieved in contrast to the amplified spontaneous emission. This narrowing is similar to the classical Shawlow-Townes line narrowing, but there is a difference defined by different mechanisms of narrowing. In the conventional

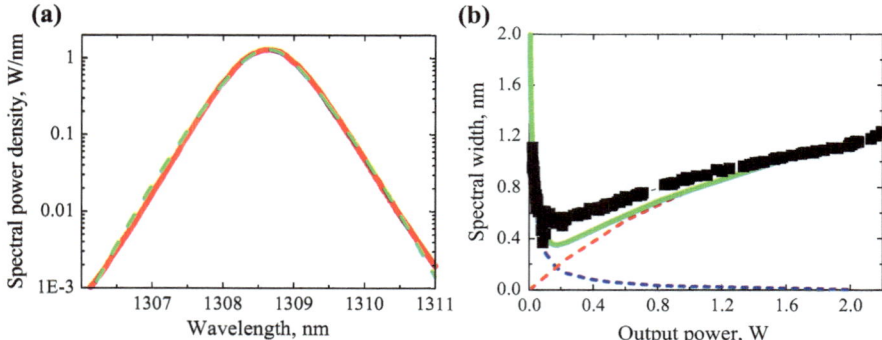

Fig. 7.4 Typical spectra of RDFB generation [21] obtained in phosposilicate fiber (**a**) experimentally measured (*red*) and theoretically predicted by nonlinear kinetic theory (*green*) optical spectrum in backward-pumped scheme of RDFB laser. (**b**) Spectrum width as a function of the laser's output power in theory and experiment. Experimental data are shown by *black squares*. The prediction for spectrum broadening from nonlinear kinetic theory is shown by *green dashed line* and is a sum of nonlinear (*red*) and linear (*blue*) contributions (After Ref. [21])

laser having cavity modes, the spectral profile of an individual mode providing its reflection at each round trip leads to narrowing after many round trips, whereas in the random DFB fiber laser with broadband RS reflection, only the gain spectral profile defines narrowing at each round trip. So one can effectively treat this laser as a single mode one with a continuous spectrum of the width and central frequency defined by the gain profile [37].

Well above the threshold, the generated spectrum exhibits power broadening: the higher is the pump power, the broader is the generation spectrum (Fig. 7.4b). The observed spectral broadening is somehow similar to the spectral broadening in conventional Raman fiber lasers [38]. Kinetic theory describing the spectral behavior including linear narrowing at the threshold and spectral broadening above the threshold was introduced in [21] and proved to be in good agreement with experimental data (see Fig. 7.4b).

To experimentally check the predictions of the nonlinear kinetic theory of spectral broadening and linear kinetic theory of spectral narrowing, a specific experimental setup was designed in [21].

The random DFB fiber laser is designed to have a short length of only 850 m. Phosphosilicate fiber was used because it has a single Raman gain peak with a spectral shape close to parabolic one (of full width at half maximum of about 8 nm).

High above the generation threshold, the measured and calculated within the nonlinear kinetic theory spectral shapes agree well (Fig. 7.4a). The description of the spectral broadening law is good as well at high power (Fig. 7.4b, red line). However, one should take into account both nonlinear spectral broadening and linear spectral narrowing. The sum of linear and nonlinear terms describes well the experimentally measured laser spectrum width in all power range (Fig. 7.4b, green line).

Thus, the spectrum of the random fiber laser could be quantitatively described within the nonlinear kinetic theory which takes into account the nonlinear Kerr effect leading to self-phase modulation of light generated in random fiber laser.

Since both the Raman gain and the RS-based random feedback are broadband, it is challenging to implement multiwavelength or tunable operation of the random DFB Raman fiber laser.

The first realization of a multiwavelength RDFB laser was based on the backward-pumped configuration of the RDFB Raman fiber laser (Fig. 7.2b) where array of 22 FBG reflectors operating at different wavelengths is implemented instead of the single FBG [23]. This configuration has been directly compared with the conventional Raman laser realized in the same fiber using broadband 4% Fresnel reflection from the normally cleaved fiber end. In contrast to the normal linear cavity characterized by competition and nonuniform power distribution between laser lines, the RDFB laser exhibits higher efficiency (up to 40%) and highly stable competition-free multiwavelength generation with a very flat power-equalized distribution of the generated 22 wavelengths in the range of 1552–1570 nm with spacing of ∼0.8 nm. Another approach to obtain a multiwavelength generation is to use an all-fiber Lyot filter [39] based on two 45° tilted fiber Bragg gratings inscribed into a polarization-maintaining (PM) fiber along its principle axis and a polarization-maintaining fiber as cavity. The filter has both filtering and polarizing functions, and its bandwidth and free spectral range (FSR) could be adjusted by changing the length of the PM cavity. A Lyot filter with a 0.2 nm bandwidth and 0.4 nm FSR was incorporated in the symmetric forward-pumped random DFB fiber laser [15] thus providing multiple wavelengths simultaneously in the whole range of the Raman gain spectral profile. Each line has a typical spectrum width of about 0.07 nm, which is 3 times less than a spectral width of the Lyot filter transmission peak and much less than a spectral width in the case of FBG-based multiwavelength random DFB fiber laser. In papers [40, 41], multiwavelength generation was obtained in the forward-pumped configuration (Fig. 7.2a), using a Sagnac loop mirror with an insertion of high-birefringence (Hi-Bi) photonic crystal fiber in the loop, which works as a multiwavelength filter and results in a modulation of the mirror reflection spectrum. This element allows to obtain multiwavelength generation with mode spacing varying with photonic crystal fiber length. Up to 16 nm span around 1550 nm was filled by multiple lines with a spacing of 0.09 nm only.

An idea of the tunable generation in random DFB Raman fiber lasers was first implemented in [22]. A tunable acousto-optic filter of 1.5 nm spectral width was placed at the central point of the symmetrical forward-pumped scheme [15]. Since the generation power at this point is very low even at the multi-watts output power level, low-power wavelength selection/tuning elements could be used to control the wavelength of high-power random fiber lasers. As a result, a radiation tunable within the range of 1530–1575 nm has been achieved with high pump-to-Stokes conversion efficiency. Another impressive result here is that the power variations within the tuning range 1535–1570 nm do not exceed 3%. The obtained flatness of the tuning curve is much better than that in the conventional linear cavity that is caused by the fact that the total generation efficiency of random DFB fiber lasers barely depends on the wavelength and the value of the Raman gain.

A similar approach was implemented in [42], where the forward-pumped configuration with a fiber loop mirror was used (Fig. 7.2a). For wavelength selection a FBG-based tunable filter of 0.2 nm spectral width was used at some point within the fiber loop mirror. A small part of the generated radiation propagating in a backward direction was out-coupled by using a 1% coupler. Tuning over 21 nm within a range of 1550–1571 nm was achieved with power variations of less than 3 dB. However, because most of the radiation in this scheme is generated in a forward direction and due to the small coupling ratio of output coupler, the power efficiency was relatively low in this scheme.

The generation linewidth can be also reduced using wavelength selection/tuning elements. In paper [24], a narrow linewidth as low as 0.05 nm is achieved by employing narrowband filters in the central point of the symmetrical forward-pumped scheme [15]. Two types of filters have been used: FBG (spectral width is 0.05 nm) or fiber-coupled Fabry-Perot filter (FPF) having a finesse of 486 and a free spectral range of 623.60 GHz that corresponds to a spectral width of 10 pm for every FFP transmission peak. With a FBG as a spectral filter, spectral width of the generated wave is almost constant at a level of around 0.05 nm and follows the spectral width of used FBG up to the pump power of 1.2 W. With FPF as a spectral filter, narrowband generation of spectral width below 0.02 nm is generated.

In general, the random DFB fiber laser can be operated in any mix of tunable, multiwavelength, and narrowband regimes. As an example, tunable multiwavelength generation of small spectral width of each individual line was demonstrated in [43]. The authors used a Mach-Zehnder interferometer based on two long-period FBGs as a tunable broadband filter (tuning is provided by bending the interferometer). At the same time, a Fabry-Perot filter was used for a fine selection of 12 different wavelengths within the Mach-Zehnder interferometer bandwidth. As a result, the multiwavelength generation with tuning range 1553.9–1565.4 nm and individual line spectral width of only 0.034 nm was achieved at mW level of power.

7.2.5 Statistical Properties

Another important feature of laser radiation to be considered is its statistical properties. In terms of generated intensity dynamics and therefore of statistical properties, random fiber lasers belong to the class of quasi-continuous wave lasers like ytterbium-doped fiber lases [44] and Raman fiber lasers [45, 46].

In simplest consideration, the generated radiation of quasi-CW laser could be described within the assumption of the Gaussian statistics for generated complex amplitudes. Within the approach of the wave kinetics, spectral properties of the quasi-CW turbulent generation of both conventional Raman fiber laser [45] and random fiber laser could be described [21]. The statistical properties of the radiation cannot be revealed in the kinetic description as it is based on the assumption of Gaussian statistics. However, it is known from experiments that the complete Gaussian statistics is not always the case [47]. In particular, the generation of different types of rogue waves is observed in conventional Raman fiber lasers [48, 49].

Statistical properties could be calculated in direct numerical simulations. For quasi-CW fiber lasers with conventional mirror-based resonator, this could be done within the full equation set based on coupled nonlinear Schrodinger equations [50–54].

Similar model could be adapted for random fiber laser by including additional terms describing random Rayleigh scattering in a way that the feedback is provided only by energy income [55]. It has been shown numerically that the intensity probability density function (pdf) deviates from exponential law, meaning that the statistics of the radiation is not pure Gaussian. It has been also found that statistical properties of the random fiber laser do depend on generation power and fiber coefficients (nonlinearity and dispersion). This could indicate the existence of correlations between the generated spectral components in spite of the seemingly random and turbulent nature of the radiation.

In experiment, studying statistical properties of the quasi-CW lasers including random ones is a challenging task. The reason of that is optical bandwidth of the generated radiation which is usually much larger than the electrical bandwidth of measurement setups. Indeed, the typical spectral width is around 1 nm, i.e., 100–300 GHz at 1–$1.5\,\mu$m, while the bandwidth of state-of-the-art real-time oscilloscope does not usually exceed 60 GHz. Different experimental approaches are proposed to study temporal and spectral properties of radiation of such a type. A time microscopy based on mixing of the source under test with a chirped fs pulse in a BBO crystal proved to be powerful tool to study statistical properties of quasi-CW radiation [56–58].

This technique requires a complicated optical setup which limits its applicability. Simpler way which could qualitatively reveal if the statistics is Gaussian or not is to perform a set of bandwidth-limited measurements with variable electrical bandwidth [59]. This technique has been applied for random fiber lasers [47]. It has been found that for a random fiber laser, the statistics deviates from Gaussian. Despite intensity pdfs which could be approximated by exponential function at high intensities, a decay factor is not equal to those characteristics for the completely Gaussian statistics. It was also found that the lower the generation power, the more probable is a generation of events of extreme intensities (up to 20 times higher than average intensity). However, the technique does not allow the quantitative description of the statistical properties.

The more detailed study of statistical properties of random fiber laser has been performed recently on an example of multiwavelength random fiber [60]. In this paper, a single-armed, forward-pumped configuration is realized with an all-fiber Lyot filter placed at the one end of the cavity to provide multiwavelength generation. The generation spectrum was a set of narrowband lines covering the total span up to 13 nm (Fig. 7.5). By filtering out different individual lines from the total multiwavelength spectrum, the time dynamics and then statistical properties

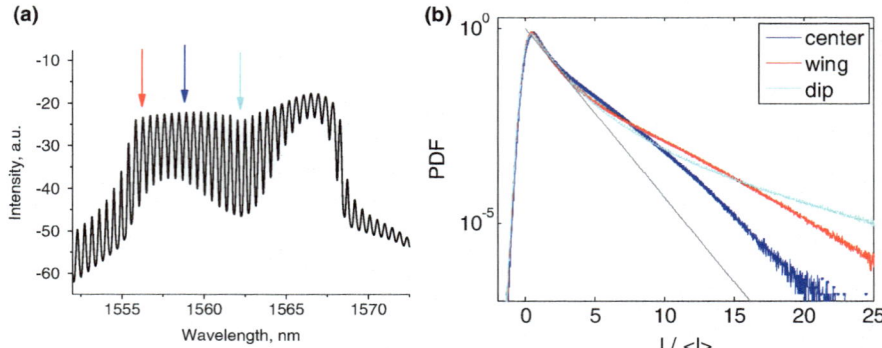

Fig. 7.5 (**a**) Generation spectrum and (**b**) intensity probability density function of the multiwavelength random fiber laser. Each line has a bandwidth less than 0.25 nm. Lines located near local Raman gain maxima exhibit Gaussian statistics, while lines generated on the edge of the gain profile exhibit non-Gaussian statistics with increased probability of intense events

have been measured. Note that to filter individual lines, a rectangular spectral filter with width 0.25 nm (optical frequency bandwidth 31 GHz at wavelength 1560 nm) was used which corresponds to the electrical bandwidth of used real-time oscilloscope (33 GHz). In this way, the direct study of statistical properties has been performed, and measurements were not affected by limited electrical bandwidth of measurement setup. It has been found that lines near the local maxima of the Raman gain (blue line in Fig. 7.5) exhibit almost exponential intensity pdf. This means that such a line could be thought as a sum of statistically independent (i.e., random in phases) frequency components with Gaussian statistics for amplitudes of each component. Lines located at spectrum edge (red and cyan lines in Fig. 7.5) exhibit increased probability of high-intensity events and reveal non-Gaussian statistics.

It is worth to note that the envelope of the generated multiwavelength spectrum does not repeat the Raman gain spectral profile. This could be an evidence of cross-lines spectral correlations. To check this, the total multiwavelength spectrum has been simulated within a model describing a set of individual spectral components with Gaussian statistics and pre-given phase relations between them. It was found that introduced correlations between lines affect substantially the total spectral shape, and the observed spectrum could be explained within the phenomenological model.

The existence of cross-line correlations leads to practically important fact of possibility to achieve a flat power distribution between multiple components in multi-wavelegnth random fiber laser despite the Raman gain is not flat and has a double-peak structure. Further, the cross-correlations between lines could be enhanced via implementation of different pumping schemes and cavity configurations. For instance, the use of erbium-doped fiber allows to achieve a multiwavelength generation with increased operating bandwidth and flatness of power distribution between different lines [61].

7.3   Performance of Random RFLs

7.3.1   Polarization Management

RRFLs employing conventional single-mode fibers under depolarized pumping generate depolarized (or randomly polarized) light [15–28] that limits range of RRFL applications. The first attempts to govern the polarization state of the RRFL output radiation faced some problems. Implementation of linearly polarized pumping [62] in the RRFL scheme results in generation of partially polarized output radiation. The laser characteristics (threshold, output power, efficiency, and degree of polarization) appear to be significantly influenced by the polarization state of pump radiation. Furthermore, the results indicate that the efficiency of lasing is reduced considerably compared to that for depolarized pumping. In the opposite case, depolarized forward pumping of a 500 m polarization-maintaining (PM) fiber in the half-open cavity configuration with a highly reflecting FBG connected to one fiber end [63] leads to generation of linearly polarized Stokes radiation with a 14 dB polarization extinction ratio (PER) at low output powers (<3 W) if special measures (strongly coiled piece of fiber inserted) are taken for the polarization selection. Herewith, the polarization state almost fully degrades (PER < 3 dB) while the power approaches 9.5 W, at that the maximum conversion efficiency is also rather low (~40%).

In [64], new scheme of a random fiber laser based on an all-PM configuration with linearly polarized pumping has been proposed. In this scheme it appears possible to generate stable linearly polarized Stokes radiation with high PER in the full range of the obtained powers. Moreover, it exhibits superior stability and efficiency of the generated Stokes radiation with relatively narrow spectrum.

The laser scheme is shown in Fig. 7.6. The pumping was provided by a CW linearly polarized single-mode all-fiber master-oscillator power-amplifier (MOPA)

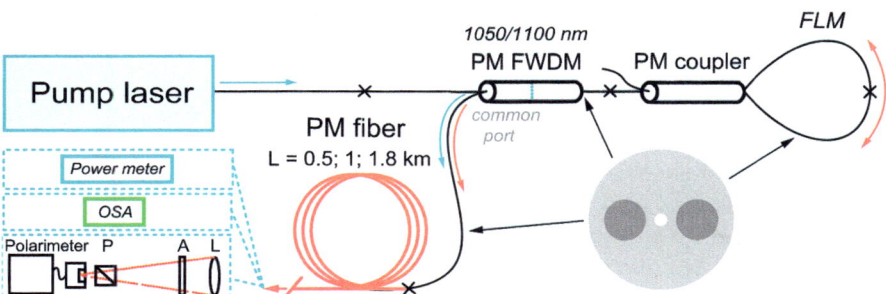

Fig. 7.6 Experimental setup: PM coupler, polarization-maintaining fused fiber coupler with splitting ratio 50/50 at 1050 nm; PM FWDM, polarization-maintaining filtered wavelength-division multiplexer; OSA, optical spectrum analyzer; L, lens; A, broadband attenuator; P, Glan-Thompson polarizer

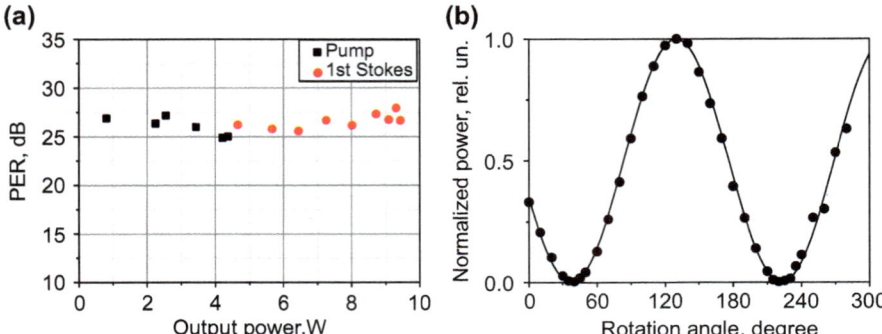

Fig. 7.7 (**a**) Polarization extinction ratio vs. output power for the transmitted pump (*squares*) and generated first Stokes wave (*circles*). (**b**) Variation of the normalized power with increasing polarizer rotation angle at $P_s = 9.4$ W

source. It consisted of a multimode laser-diode pumped ring-cavity Yb-doped fiber laser (YDFL) with depolarized output radiation and a two-stage polarization-maintaining Yb-doped fiber amplifier (YDFA). A fiber polarization beam splitter being set at the YDFL output extracted the linearly polarized component of radiation, which was then launched into the YDFA. As a result, the pump laser generated linearly polarized radiation at 1054.6 nm with a power of up to 14 W and linewidth of <0.2 nm in the entire power range.

The pump radiation was launched into a 500-m-long PM passive fiber (Fujikura SM98-PS-U25D) via a 1050 nm port of a high-power 1050/1100 nm PM filtered WDM (FWDM). The PM fiber loop mirror (FLM) formed by a fused PM fiber coupler with a splitting ratio of 50/50 at 1050 nm was spliced to the other port of the FWDM. The output fiber end was cleaved with an angle of >10° to eliminate Fresnel reflection. The lasing in the scheme occurred via pump-induced Raman gain at the Stokes wavelength and broadband feedback provided by the reflection from the loop mirror and Rayleigh backscattering distributed along the fiber span. The output laser power and spectra were measured by a power meter and optical spectrum analyzer Yokogawa AQ6370, respectively.

Polarization measurements. The polarization properties of the pump and RRFL output radiation were measured by means of a special scheme consisting of a lens (L), a broadband attenuator (A), a Glan-Thompson polarizer (P), and a free-space polarimeter with a PAN5710IR2 external sensor head (Thorlabs). The lens and attenuator were used to slightly focus and attenuate the output radiation in front of the polarizer. The polarization extinction ratio was defined as $10 \log(P_{max}/P_{min})$, where P_{min} and P_{max} correspond to the minimum and maximum power transmitted by the polarizer during the rotation of its optical axis. The transmitted power was measured by a polarimeter which provides a large dynamic range (from -60 to 10 dBm) at 1000–1350 nm spectral regions.

Figure 7.7a shows the measured polarization extinction ratio (PER) for the transmitted pump (squares) and generated first Stokes (circles) waves. It appeared that the PER values for the pump and Stokes waves were nearly the same slightly varying between 25 and 28 dB. Figure 7.7b presents the variation of normalized power as a function of the polarizer rotation angle at the maximum output power $P_s = 9.4$ W. The experimental data (points) were accurately fitted by the sine function (solid curve) indicating that the PER measurements were correct. Presented PER values were much higher than in [63], because discriminating pump-induced gain and the FWDM used for polarization selection in [64] did not insert sufficient losses and provided the PER values close to that for the pump (>25 dB) at all power levels.

7.3.2 High-Efficiency/High-Power Generation

As balance model [21] shows, the reduction of fiber length results in the generation threshold increase, but the RRFL efficiency is also increased. High-power pumping of RRFLs based on a relatively short (≤ 1 km) conventional passive fiber enables high-power (tens of watts) generation with high optical efficiency of pump-to-Stokes wave conversion [16–18, 20, 64–66]. The best performance characteristics (low threshold, high stability, and efficiency of single-side output beam) are observed for half-open forward-pumped cavity configuration with a fiber loop mirror at one fiber end and an angle cleaved facet at the other end (Fig. 7.2a [16, 18, 64–66]).

The highest conversion efficiency of the pump power into the Stokes power at the lowest generation threshold was demonstrated in all-PM RRFL configuration [64]. Figures 7.8 and 7.9 show obtained in [64] experimental output power and corresponding output spectra as a function of the incident pump power. The output power behavior was typical for single Stokes and cascaded lasing in short random fiber lasers [16–19, 65]. Below the lasing threshold, only the transmitted pump wave was present at the laser output. It was slightly attenuated in the 0.5-km-long passive fiber. The transmitted power grew linearly with increasing input pump power up to the first Stokes generation threshold (5.2 W pumping). Above the threshold, the Stokes wave power started to grow very rapidly approaching linear extrapolation for the transmitted pump power. The Stokes power increased up to the second Stokes generation threshold (11 W). At the same time, the pump power became depleted since it was almost fully converted into the Stokes generation. The absolute optical efficiency of pump-to-Stokes wave conversion calculated as a ratio of the first Stokes wave power to the input pump power reached 87%. This value is close to the quantum limit estimated as the ratio of generated Stokes photon and input pump photon numbers ($\approx 95\%$) and exceeded the record efficiency of 85% obtained for conventional unpolarized Raman fiber lasers (RFL) with a 30 m linear cavity [67]. Besides, demonstrated efficiency was two times higher than that for conventional RFLs with linearly polarized output [68, 69]. Comparable ultra-high

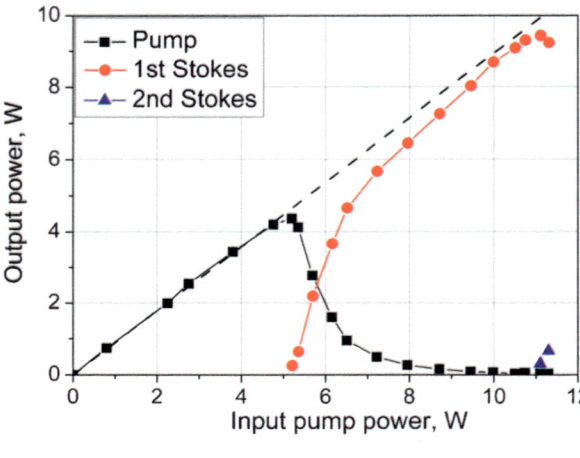

Fig. 7.8 Measured output power for the transmitted pump (*squares*) and the generated first Stokes (*circles*) and second Stokes (*triangles*) components versus input pump power. The *dashed straight line* is a linear extrapolation for the undepleted pump

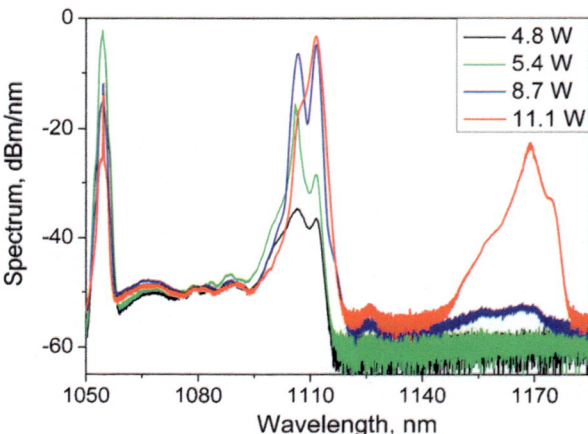

Fig. 7.9 Output spectra of the linearly polarized random fiber laser at different pump powers

optical efficiency of 89% was later demonstrated also for unpolarized random RFL at much higher pumping of shorter fiber [20].

The output spectra measured at different pump powers in [64] demonstrate appearance of the first ($1.11\,\mu m$) and the second ($1.17\,\mu m$) Stokes components above the corresponding thresholds in addition to the transmitted pump radiation at $1.055\,\mu m$ (see Fig. 7.9). The spectra were stable and smooth, in contrast to those in the random fiber laser based on a short non-PM fiber at linearly polarized pumping, where random sharp peaks in the output spectra of the first Stokes component were observed [62]. Such smooth spectra are typical for RRFLs based on short non-PM fibers with depolarized pumping [16–18, 65].

As the Raman gain profile in germanosilicate fibers has two nearly equal peaks shifted by ~440 and $\sim490\,cm^{-1}$ relative to the pump wave [70], the first Stokes output spectrum consisted of two generation lines at 1106 and 1111 nm, correspondingly. The first line was mainly generated at low pump powers, whereas the second one became dominating at high powers. This behavior was observed also in the RRFLs with randomly polarized output [15, 18, 62, 71]. Although the width of the output spectrum near the generation threshold was much smaller than the gain-peak bandwidth exhibiting Schawlow-Townes narrowing, the laser line broadens from 1.3 nm up to 1.8 nm with increasing output power due to nonlinear processes [21]. The obtained linewidth values were similar to that observed in a polarized RRFL with the PM fiber of the same length and FBG-based half-open cavity [63] and was slightly larger than the linewidth in a linearly polarized RFL with an FBG-based linear cavity of shorter length [69]. Besides, the generated line is narrower at the same power level as compared to the depolarized radiation of random lasers in a similar half-open configuration [16, 18], because equal polarization states of the pump and Stokes waves eliminate the effects of cross-phase modulation (XPM) between the orthogonal polarization components, thus resulting in considerable reduction of spectral broadening. Another interesting spectral effect was the reduction of the linewidth for transmitted pump radiation in the regime of its strong depletion; it is discussed in details in the next section.

Power scaling of Raman fiber lasers is restricted by high-order Stokes waves whose generation threshold is directly proportional to the fiber length. High output power of the first-order Stokes wave can be achieved by shortening fiber length, but the threshold of a RRFL will increase dramatically. First RRFL with output power of several tens of watts was reported by H. Zhang et al. [17]. They used 300-m-long Corning SMF-28 fiber directly core pumped by an 1120 nm YDFL from one side with output power of up to 100 W. The whole system was assembled in an all-fiber configuration. The pump source and Raman Stokes waves were separated by a wavelength-division multiplexing (WDM) in the backward direction. All the fiber free ends of the system were angle cleaved to avoid Fresnel reflection. Due to the short fiber length, the generation threshold of such a mirrorless laser was as high as 50 W. The maximum generated power at 1184 nm at the pump power of $\approx$100 W reached 58 W and 15 W in forward and backward direction, respectively. A 3 dB bandwidth at the maximum power was 4 nm due to nonlinear broadening. Further power increase was limited by the second-order Stokes wave. At the maximum power, the total optical efficiency was as high as 75%. The output was in single mode operation although the cutoff wavelength of the passive fiber was 1260 nm.

The same group tried also 1090 nm pumping from YDFL with output power of up to 150 W. They used 320-m-long passive fiber with 10 μm core and open or half-open cavity configuration of RRFL [72]. Open cavity contained WDM to detect the generated backward Stokes wave (relates to the pump direction). The laser generation threshold here was as high as 92 W, and total power of CW RRFL at 1146 nm reached 124 W with 75 and 49 W in forward and backward direction, respectively. The optical efficiency and 3-dB-generated linewidth were 79% and 3.5 nm, respectively, which were close to those in [17]. Output radio frequency spectrum was free of resonant signal corresponding to the mode spacing; therefore, the lasing was random. In half-open forward-pumped cavity configuration, the WDM was replaced by FBG centered at 1150 nm, whose reflectivity was 99% and 3 dB bandwidth was about 1.5 nm. So the feedback was provided both by the point reflection and distributed Rayleigh scattering, which reduced the pump power threshold down to 32 W. Output power of 112 W at 1150 nm was achieved when the pump power was 132 W. Therefore, the optical efficiency reached the value of 85%. Taking into account 9% system loss, which can be estimated from the transmitted pump power below the threshold, such efficiency was close to the quantum limit (94.8%). Second-order Stokes wave around 1214 nm limited the further power increase. The 3 dB bandwidths at the maximum power were 2 nm, which is larger than that for applied FBG.

The developed RRFL was applied as a pump source for 23-W-level Ho-doped fiber laser (HDFL) at 2050 nm [72]. Obtained slope efficiency 33% was comparable with one demonstrated in high-power HDFL pumped by conventional Raman fiber laser (37%) [73].

Further increase of the output power was achieved by the shortening of passive fiber. Thus, X. Du et al. obtained 194 W power at 1173 nm in half-open cavity RRFL based on 120-m-long G.625 passive fiber forward-pumped by YDFL at 1120 nm [20]. This value corresponds to 89% of optical efficiency. High reflectivity FBG was used as a point reflector, and the passive fiber was evenly winded on the outside surface of an air-cooled metal cylinder (diameter of 300 mm) for appropriate heat management. Due to the short cavity design with 120 m passive fiber, the thresholds for the first-and second-order Raman Stokes waves were at high levels of about 80 and 170 W, respectively. The FWHM of the peak centered at 1173 nm wavelength was broadened to 6 nm at maximum output power due to nonlinear Kerr effect. It was also demonstrated that RRFL maintains satisfying temporal stability at high output powers making this laser promising as a new seed source for power amplifiers.

7.3.3 High-Order Cascaded Generation at Long Wavelengths

A next important step concerns high-order random Raman lasing in a polarization-maintaining fiber. It appears that the cascaded generation of the RRFL in the all-fiber PM configuration exhibits no degradation of output characteristics with increasing Stokes order [66].

7.3.3.1 Experiment

The experimental setup in [66] was the same as presented in Fig. 7.6 with the fiber length increased to 1 km. The FLM reflection coefficient R amounted to 91% at 1.11 μm (first Stokes), 66% at 1.17 μm (second Stokes), 36% at 1.23 μm (third Stokes), and 12% at 1.3 μm (fourth Stokes) and was as low as 6% at 1.4 μm in correspondence with the coupling ratio at these wavelength. As the linearly polarized pumping was used whose axis coincides with a chosen (slow) axis of the PM fiber, the Raman gain for another polarization component was strongly discriminated, and generation of one (slow) linearly polarized component was demonstrated, similar to conventional Raman fiber lasers with polarized pumping [68, 69].

In the experiment, cascaded random lasing was observed starting from the first (1.11 μm) to higher Stokes orders, which appeared consecutively in the output spectrum with increasing input pump power measured ahead of the 1-km-long PM fiber [66] (see Fig. 7.10). The first Stokes output spectrum consisted of two generation lines at 1106 and 1111 nm according to the Raman gain profile; at low powers the first line was mainly generated, whereas the second one becomes dominating at high powers (see Fig. 7.10). For the higher Stokes orders, the first peak was always dominating. A probable reason is that the Stokes wave playing a role of the pump for the next order has sufficiently larger linewidth as compared with the YDFL pump, which leads to the smoothing of the relatively narrow second peak in the Raman gain spectrum (see the lowest spectrum in Fig. 7.10c characterizing the amplified spontaneous emission). The detailed structure of the RRFL generation spectrum arising from the two-peak shape of the Raman gain spectral profile in germanosilicate fibers has been treated in [74].

Figure 7.11a shows individual power data for the residual pump (squares), first Stokes (triangles), second Stokes (circles), and third Stokes (stars) orders at the output as functions of the input pump power. Evolution of the output power was typical for RRFLs. Below the lasing threshold, only the residual pump wave slightly (by $\sim$15%) attenuated in the 1 km passive fiber was present at the fiber output. Its power grew proportionally to the input pump power up to the first Stokes threshold (2.6 W). Then the first Stokes power started to grow, and the output pump power was depleted as long as almost all the pump power was converted into the Raman lasing. The first Stokes power grew up to the second Stokes threshold (5.9 W) and then started to deplete and so on for the higher Stokes orders. The absolute optical efficiency of pump-to-Stokes conversion calculated as a ratio of the corresponding j-order Stokes power P_{Sj} ($j = 1, 2, 3$) to the input pump power P_{in} exceeded 75% for the first and second Stokes waves and approached 70% for the third Stokes line. Those values were close to the corresponding quantum limits (95, 90, and 86%, respectively) and set record values for the second and the third Raman Stokes waves in random fiber lasers. The generated power was also high amounting to 4.4, 7.4, and 9.1 W for $j = 1, 2, 3$, respectively. The high efficiency was demonstrated for all components in spite of reduced reflectivity of the end mirror for higher orders because the output power in RRFLs with a forward-pumped half-open cavity is only weakly dependent on the FLM reflectivity [71], as the Stokes power at the terminating mirror (and its loss accordingly) is lower by several orders of magnitude than the power at the output end [16, 19].

The measured polarization extinction ratio of the transmitted pump power and all generated Stokes lines was more than 22 dB and did not degrade with the generated power and Stokes order. Moreover, the intensity of the output Stokes order was quite stable in the time scale longer than 0.1 ms (see the averaged intensity dynamics in Fig. 7.11b) being fully stochastic in the scale shorter than 1 ns in accordance with the

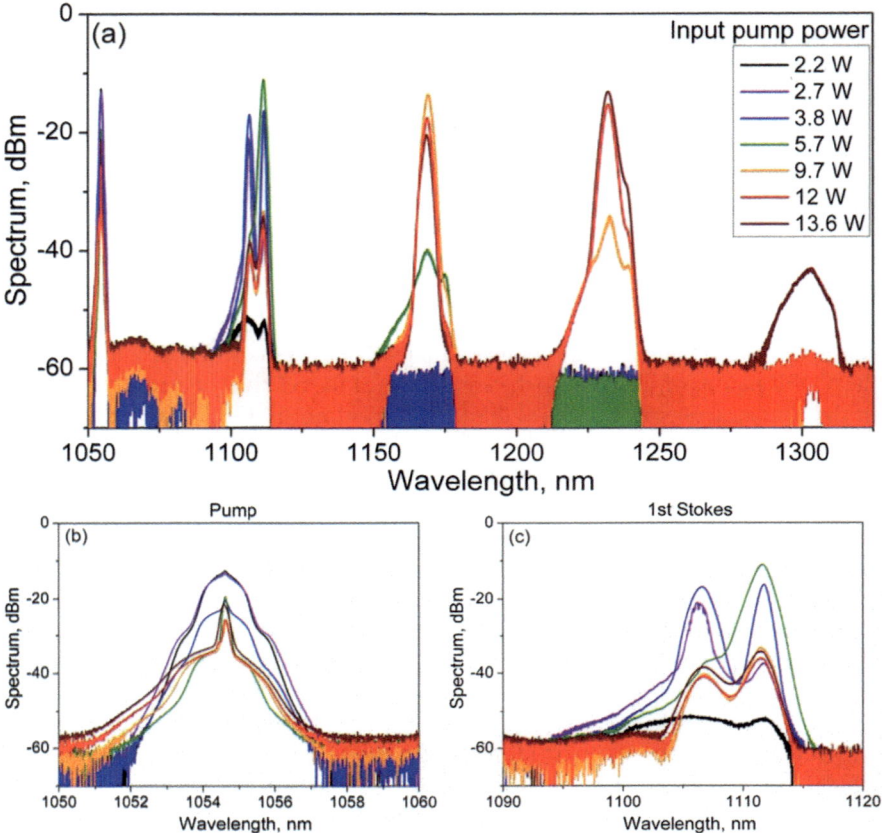

Fig. 7.10 Measured output spectra of the cascaded PM RRFL at different input pump powers at 1.05 μm: (**a**) wide spectral range; (**b**) transmitted pump; (**c**) first Stokes wave

generated RRFL spectrum consisting of random frequencies with random phases of the Gaussian statistics [21, 34].

Figure 7.12 demonstrates the linewidth of the first peak of the all Stokes orders and the second peak for the first Stokes wave as a function of the power in the corresponding line. All spectral lines behaved similarly exhibiting the Schawlow-Townes narrowing near the generation threshold and slight broadening with increasing generation power. The generation linewidth for the $440\,\mathrm{cm}^{-1}$ Raman peak varied in the range of 1.1–1.5 nm, 1.4–2.5 nm, and 2.3–3.4 nm for the first, second, and third Stokes components, correspondingly, whereas the $490\,\mathrm{cm}^{-1}$ peak for the first Stokes wave was considerably narrower (0.5–1.2 nm). The evolution of the spectral linewidth comprising rather large constant value at the threshold and small power-variable part was principally different from theory [21] especially

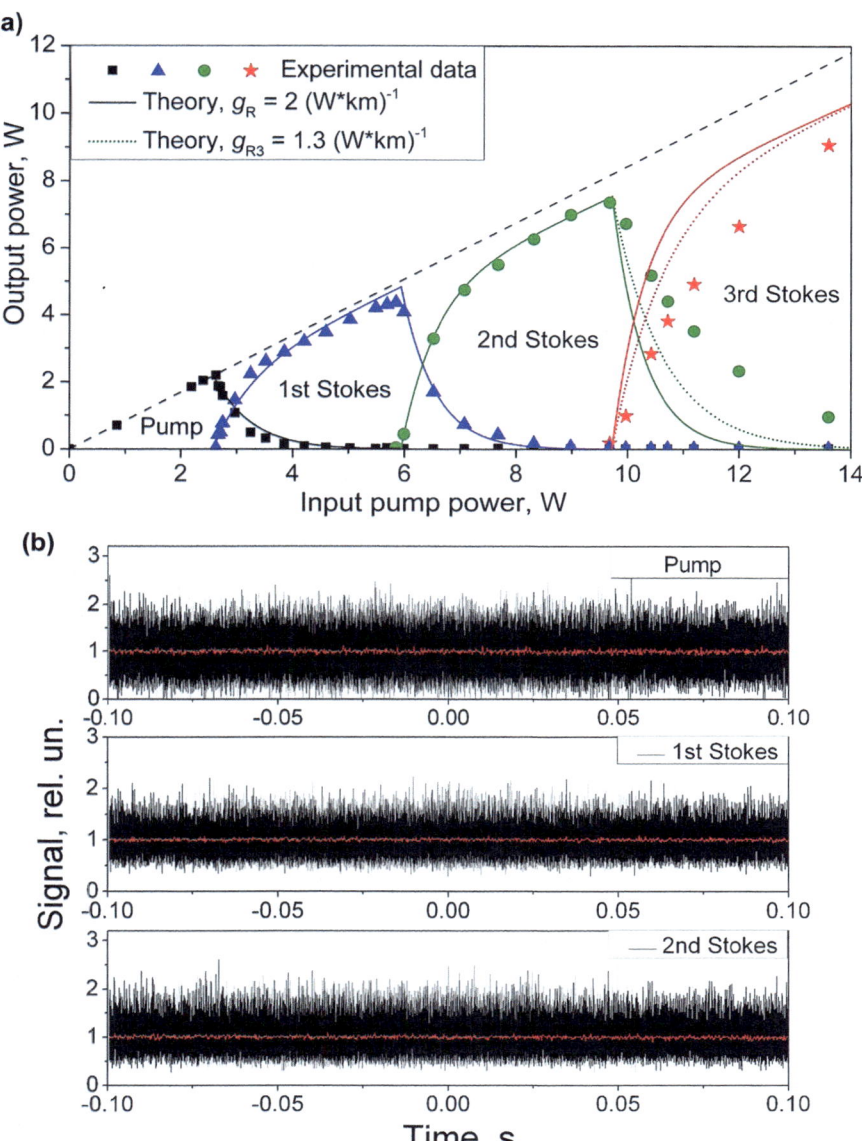

Fig. 7.11 (a) Output power of the cascaded PM RRFL as a function of input pump power. Points correspond to experimental data for transmitted pump (*squares*) and generated first Stokes (*triangles*), second Stokes (*circles*), and third Stokes (*stars*) orders. *Solid* and *dotted lines* show the analytical model of power distribution calculated at $g_R = 2$ and $g_R = g_{R3} = 1.3\,\mathrm{W}^{-1}\,\mathrm{km}^{-1}$, respectively. (b) Intensity dynamics of the pump and the first and the second Stokes waves measured with 400 ps resolution. *Red line* shows the intensity dynamics averaged over 0.2 ms. It is normalized to unity

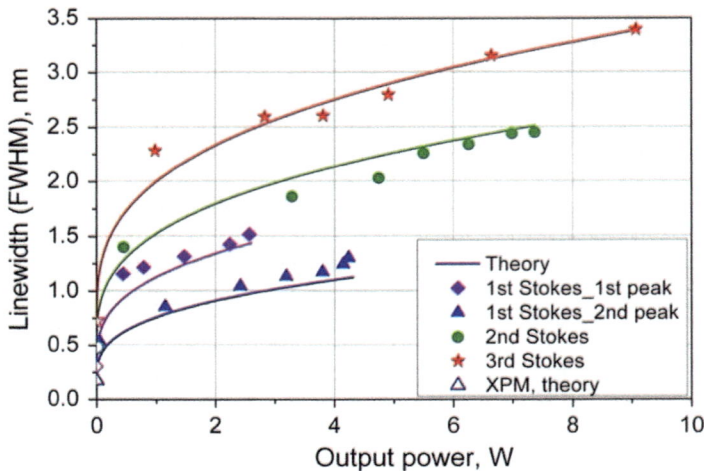

Fig. 7.12 Spectral widths (FWHM) of the individual Stokes line as a function of its power. Points correspond to experimental data for the first Stokes wave generated at $440 \, \text{cm}^{-1}$ (*diamonds*) and $490 \, \text{cm}^{-1}$ (*triangles*) Raman shift, second Stokes (*circles*) and third Stokes (*stars*) components. *Solid lines* show the analytical model of spectral bandwidth

for the higher Stokes orders. The rather unique features of the cascaded linearly polarized lasing in random RFL are discussed in the next section.

Another interesting spectral effect observed in the cascaded RRFL [66] was the reduction of the linewidth for transmitted pump radiation in the regime of its strong depletion by the cascaded Raman generation as shown in Fig. 7.10b. Taking that the Raman conversion predominantly occurs for high-intensity peaks existing in the stochastic time-domain trace (see Fig. 7.11b), which correspond to the broad spectral wings of the integral spectrum, the two-scale pump wave spectrum is differently affected by the Raman process. So, the broadband spectral tails acquired by the pump wave during its propagation (via SPM) are efficiently converted into the Stokes waves, whereas the rest narrowband input part with the minimum intensity fluctuations survives better and becomes dominating at the output fiber end at strong pump depletion. This is only a qualitative explanation, which requires a more detailed study.

When the fiber length is increased up to 1.8 km in the configuration shown in Fig. 7.6, cascaded generation of the Stokes components from the first up to the fifth orders is observed at the output, when the input pump varies in the range from 1 to 14 W. Figure 7.13a–c shows the generated spectra at the maximum input pump power of 14.1 W (a–b) or 15.2 W (c) for different pump wavelengths, 1040.6, 1064.4, and 1080.1 nm, respectively. Pump wavelength was changed from 1040 to 1080 nm by setting appropriate high-reflectivity FBG inside the YDFL cavity in Fig. 7.6. Generated Stokes lines have narrow bandwidth. But when the next Stokes order approaches the vicinity of zero dispersion wavelength (ZDW) of $\sim 1.4 \, \mu\text{m}$ of the long passive PM fiber, a broadband spectrum generation is observed similar to the one obtained in the depolarized cascaded RRFL [75]. Figure 7.3d–f shows the

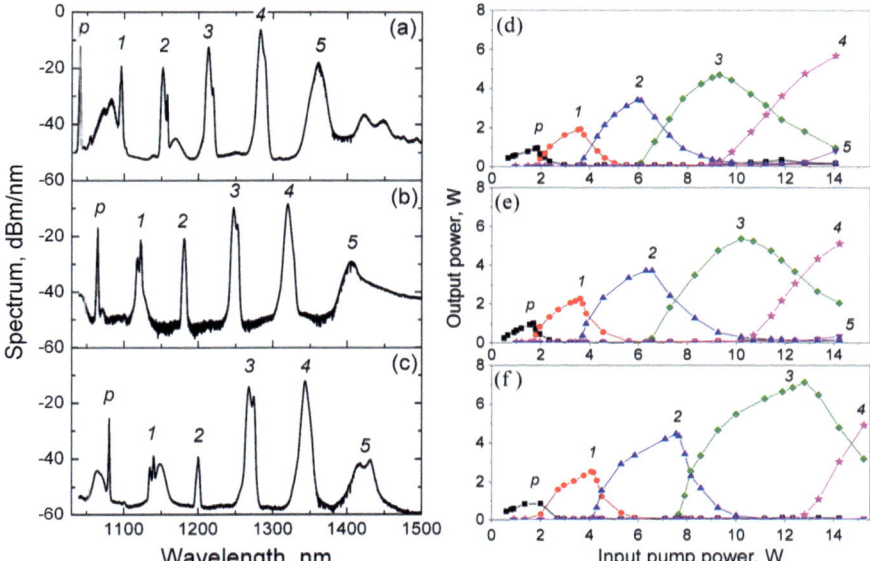

Fig. 7.13 (**a**)–(**c**) Output spectra of the all-PM 1.8-km-long random Raman fiber laser at different pump wavelengths—1040.6, 1064.4, and 1080.1 nm, respectively; input pump power is 14.1 W in (**a**)–(**b**) and 15.2 W in (**c**); (**d**)–(**f**) the output power for different Stokes components together with residual pump power as a function of the input pump power for pump wavelengths 1040.6, 1064.4, and 1080.1 nm, respectively

power of residual pump (squares), first Stokes (circles), second Stokes (triangles), third Stokes (diamonds), fourth Stokes (stars), and fifth Stokes (inverted triangle) as a function of the input pump power at the entrance of the PM fiber. The behavior of generated powers up to the 4th order is similar to that one observed for cascaded lasing in the short random RFLs. However, in the vicinity of the ZDW ($\sim$1.4 μm), the fifth-order Stokes wave starts to generate despite the previous Stokes order has not achieved its maximum value. Low threshold generation for the highest Stokes order can be related to the influence of the parametric four-wave mixing (FWM) process, which becomes much more efficient near the ZDW. Competition between the nonlinear effects results in significant reduction of the conversion efficiency to the higher Stokes orders and gives rise to the spectral continuum in the anomalous dispersion region. Figure 7.13d–f shows that changing the pump wavelength from 1040 to 1080 nm results in an increase of the threshold power for almost all Stokes components, especially at the higher orders. It can be explained by combination of several factors: decrease in the Rayleigh backscattering and Raman gain coefficients, and reflectivity of the fiber loop mirror. As a result, the output power of the appropriate Stokes orders is much higher for the 1080 nm pump wavelength.

The absolute optical efficiency of pump-to-Stokes waves conversion calculated as a ratio of the corresponding Stokes order power to the input pump power was

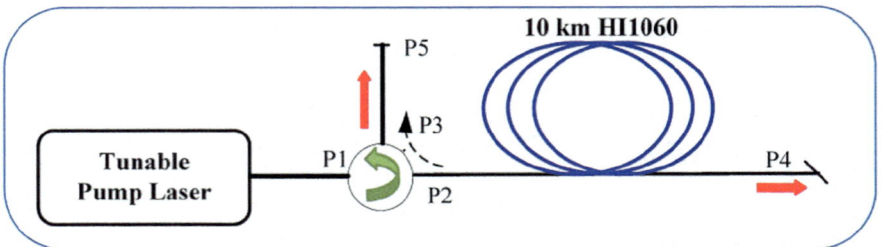

Fig. 7.14 Schematic diagram of the tunable cascaded Raman random fiber laser

$\sim$62%, $\sim$59%, and $\sim$54% for the first, the second, and the third orders, respectively, for the cases (e)–(f) in Fig. 7.13 and slightly lower in the case (d). Generated power varies from 2 to 7 W for different Stokes orders and reaches 5 W at 1.34 μm for the fourth order (Fig. 7.13c, f) and 0.8 W at 1.36 μm for the fifth order—the longest generated wavelength (Fig. 7.13a, d). In addition, all-PM configuration allows to generate stable and linearly polarized radiation as was demonstrated in [64, 66] previously.

The high-order Raman generation was also demonstrated in non-PM configuration with long passive fiber [75]. Figure 7.14 illustrates the schematic diagram of tunable cascaded Raman random fiber laser investigated in [75]. Tunable Yb-doped fiber laser was used as a pump source. It had MOPA configuration based on Yb-doped fibers with a core diameter of 10 μm. The seed laser had ring cavity geometry with tunable bandpass filter based on the thin film which was used to select the operating laser wavelength. The filter was tunable from 1000 to 1099 nm with a tuning resolution of 0.02 nm and a bandwidth of 1 nm at 3 dB. The amplifier was optimized for the 1020–1080 nm wavelength range. The whole pump laser system was all fibered and polarization maintained. The output power of the YDFL was over 45 W from 1020 to 1080 nm (the maximum output power reached 58 W at 1030 nm) and the 3 dB linewidth of all the wavelengths kept less than 0.5 nm.

The tunable pump laser was injected into the Raman random laser through a broadband (1020–1100 nm) circulator. A piece of 10-km-long HI1060 fiber (from YOFC) was used as the gain fiber, which supplies the distributed Rayleigh scattering and the Raman gain simultaneously. The output end of this fiber (P4) was cleaved at an angle of >8° to suppress the backward reflection. P3 of the circulator was 0° cleaved to increase the feedback. The isolation of the circulator from P3 to P2 was 20 dB, so the effective reflection for the Raman fiber laser from the circulator side was about 4×10^{-4} at the specified circulator wavelength range.

Figure 7.15 shows the typical output spectra at different pump powers with the pump wavelength fixed at 1045 nm. With the increase of pump laser power, the first-to fifth-order Raman emissions are observed successively. At long wavelength near the zero dispersion wavelength (1400 nm) of the Raman gain fiber, FWM process broadens the laser linewidth significantly prohibiting cascaded generation at the longer wavelengths.

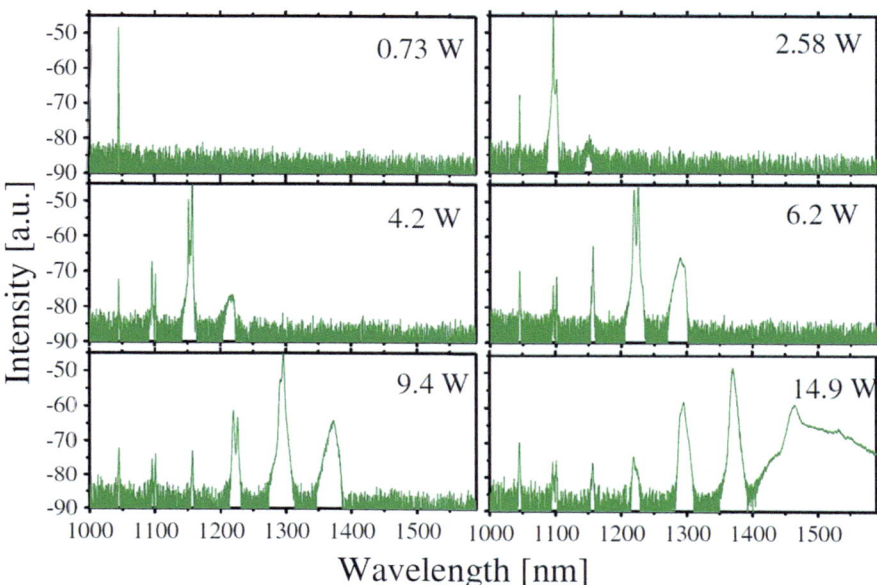

Fig. 7.15 The output spectra of the non-PM Raman random fiber laser at different pump powers

By adjusting the pump laser wavelength and power together, wavelength tuning over 300 nm (from 1070 to 1370 nm) was achieved with the RRFL. The highest output power at 1360 nm reached 1.8 W, corresponding to an optical efficiency of 15% from 1080 nm. These values are less than the all-PM RRFL discussed above because of long length (10 km) of the fiber used in non-PM RRFL.

Two experiments demonstrate that efficient higher-order Stokes generation at telecom wavelength region of 1.3–1.4 μm is possible in the normal dispersion region of the long passive fiber. Besides, narrowband Stokes spectrum is obtained by pumping in the anomalous dispersion region ∼1.45 μm [15]. But close to the ZDW (1.4 μm), competition between the nonlinear effects results in a growth of spectral continuum and decrease of the conversion efficiency. One should solve this problem in order to expand the generation range further to the longer wavelengths, for example, to 1.45 μm.

One solution is to add narrowband FBG with reflectivity at the wavelengths of the fifth- or sixth-order Stokes components in the RRFL scheme. Then the narrowband spectrum of the FBG should assist in the generation of the high-order Stokes wave. Such modification was performed in the all-PM RRFL configuration (see Fig. 7.6), but the output spectrum in the vicinity of the ZDW was not changed. It is explained by the fact that the continuum is generated at a single pass in the fiber. The other possible techniques for increasing the number of the Stokes orders are shortening the fiber length together with the significant increase of the input pump power and using a passive fiber with the longer zero dispersion wavelength The last approach allows generating tunable wavelength from 1 μm up to 1.9 μm [76].

7.3.3.2 Analytical Model

The unique features of the linearly polarized cascaded random Raman fiber laser can be described by an analytical model first introduced in [66].

Balance model for the power. In the SRS process, every pump quantum with a frequency ν_0 is absorbed by the medium giving birth to the Stokes wave quantum with a frequency ν_1 and the vibration quantum of the medium with a frequency $\nu_v = \nu_0 - \nu_1$, regardless of the initial frequency ν_0. The same process occurs during the Raman scattering of the Stokes wave into the next-order Stokes wave with a frequency $\nu_2 = \nu_1 - \nu_v$. The balance equations for the cascaded SRS process converting the power of the pump wave (P_0) into the j-th order Stokes waves ($P_j^{\pm}$), which co(+)- and counter(−)-propagate with the pump wave along the fiber axis x, are written as follows [34]:

$$\frac{dP_0}{dx} + \alpha_0 P_0 = -\left(\frac{\nu_0}{\nu_1}\right) g_{R1} P_0 \left(P_1^+ + P_1^-\right), \tag{7.1}$$

$$\pm \frac{dP_j^{\pm}}{dx} + \alpha_j P_j^{\pm} = g_{R(j)} P_j^{\pm} \left(P_{j-1}^+ + P_{j-1}^-\right) \tag{7.2}$$

$$- g_{R(j+1)} \left(\frac{\nu_j}{\nu_{j+1}}\right) P_j^{\pm} \left(P_{j+1}^+ + P_{j+1}^-\right) + \varepsilon P_j^{\mp}.$$

Here ν_j, α_j, and $g_{R(j)}$ are the frequency, attenuation, and Raman gain coefficient for j-th order Stokes wave, respectively, and α_0 is pump attenuation coefficient. A rather simple analytical solution for the system of differential equations (7.1)–(7.2) is possible under the following assumption. The pump wave is launched in the fiber at point $x = 0$, thus $P_0(0) = P_{\text{in}}$. In addition, the reflecting mirror is also placed at $x = 0$, thus $P_j^-(0) = P_j^+(0)$. The distributed feedback provided by the Rayleigh backscattering (with coefficient ε) of the co-propagating wave (P_j^+) into the counter-propagating wave ($P_j^-(0)$) can be substituted by a local reflector with a reflection coefficient $R_{\text{eff}(j)} \ll 10^{-4}$ placed at the point $x = L_{\text{RS}j}$ where $P_j^+(x)$ reaches its maximum value:

$$P_j^-(L_{\text{RS}j}) = R_{\text{eff}(j)} P_j^+(L_{\text{RS}j}). \tag{7.3}$$

The Rayleigh scattering of the P_j^- wave into the P_j^+ wave can be neglected in comparison with the mirror reflection. In the approximation of point reflectors, P_j^- can be expressed via P_j^+ as $P_j^-(x)/P_j^-(0) = P_j^+(0)/P_j^+(x)$. Therefore, the counter-propagating wave is much weaker than the co-propagating wave, and the values of P_j^- can be neglected in the right side of Eqs. (7.1)–(7.2). In addition, the boundary condition in Eq. (7.3) is rewritten as $P_j^+(0) = \sqrt{R_{\text{eff}(j)}} P_j^+ (L_{\text{RS}j})$. The simplified equations can be integrated under the assumption that the absorption coefficient is independent of the wavelength $\alpha_0 = \alpha_j = \alpha$, and Raman gain coefficient has the model frequency dependence:

$$g_{R(j)} = g_{R1} \frac{\nu_0}{\nu_{j-1}} \tag{7.4}$$

Using this relation, the power threshold for the all Stokes orders obtained from the gain and loss balance can be expressed in the iteration form with single gain coefficient $g_{R1} = g_R$:

$$P_j^{\text{th}} = P_{j-1}^{\text{th}} - \ln(r_j) / (g_R \tilde{L}) = -\ln \left(\prod_{i=1}^{j} r_i \right) / (g_R \tilde{L}), \tag{7.5}$$

where $r_j = \sqrt{R_{\text{eff}(j)}} \exp(-\alpha L_{\text{RS}j})$, L and $\tilde{L} = [1 - \exp(-\alpha L)]/\alpha$ are the total and effective fiber lengths, respectively, and g_R is the Raman gain coefficient of the first Stokes wave which is set for higher orders too.

If the input pump power P_{in} exceeds the generation threshold for the j-th Stokes wave, the Stokes power distribution along the fiber becomes inhomogeneous:

$$P_0(x) = (P_{\text{in}}/2) \exp(-\alpha x) (1 - \tanh[g_R P_{\text{in}} (\tilde{x}(x) - x_1)/2]), \tag{7.6}$$

$$P_j(x) = (P_{\text{in}}/2) (\lambda_0/\lambda_j) \exp(-\alpha x) \tag{7.7}$$
$$\times (\tanh[g_R P_{\text{in}} (\tilde{x}(x) - x_j)/2] - \tanh[g_R P_{\text{in}} (\tilde{x}(x) - x_{j+1})/2]),$$

$$P_k(x) = (P_{\text{in}}/2) (\lambda_0/\lambda_k) \exp(-\alpha x) (1 + \tanh[g_R P_{\text{in}} (\tilde{x}(x) - x_k)/2]). \tag{7.8}$$

Here $P_0(x)$ and $P_j(x)$ are the longitudinal distributions involving a specific hyperbolic tangent function for the pump (0) and j-th Stokes (j) waves at the k-th cascade ($j = 1 \ldots k$); λ_0 and P_{in} are the wavelength and input power of the pump wave, respectively, α is the average attenuation coefficient, and the coordinate

$$x_j = \frac{P_j^{\text{th}}}{P_{\text{in}}} \tilde{L} - \frac{\ln \left(1 - \exp\left[-g_R \tilde{L} \left(P_{\text{in}} - P_j^{\text{th}} \right) \right] \right)}{g_R P_{\text{in}}} \tag{7.9}$$

is the point, where $P_{j-1}(x_j) = P_j(x_j) (\nu_{j-1}/\nu_j)$. The validity of approximate solutions (7.6), (7.7), and 7.8) was checked up to the third stage of the cascaded RRFL, and their deviation from exact solutions is negligibly small. Simplified solutions for output power P_j^{out} of the generated j-th order Stokes wave can be derived from the exact solutions in Eqs. (7.6), (7.7), and (7.8) at $x = L$ under conditions $x_j \approx \tilde{L} P_j^{\text{th}} / P_{\text{in}}$ and $\exp\left[-g_R \tilde{L}\left(P_{\text{in}} - P_j^{\text{th}}\right)\right] < 1$:

$$P_j^{\text{out}} = P_{\text{in}}(\lambda_0/\lambda_j) \exp(-\alpha L) \left(1 - \exp\left[-g_R \tilde{L}\left(P_{\text{in}} - P_j^{\text{th}}\right)\right]\right) \qquad (7.10)$$

Here P_j^{th} is the power threshold for the j-th generation Stokes wave ($P_0^{\text{th}} = 0$ as there is no threshold for the pump wave). The output power of the generated Stokes wave is exponentially approaching the maximum value of the input pump power (if $j = 1$) or of the previous Stokes component playing the role of a pump at $j > 1$. Herewith, the power of the pump wave (or previous Stokes component) starts to decrease exponentially with increasing input pump power above the threshold:

$$P_{j-1}^{\text{out}} = P_{\text{in}}(\lambda_0/\lambda_{j-1}) \exp\left[-g_R \tilde{L}\left(P_{\text{in}} - P_j^{\text{th}}\right) - \alpha L\right] \qquad (7.11)$$

The power curves (solid lines) calculated from Eqs. (7.10) and (7.11) are compared in Fig. 7.11 with the experimental data for the output pump and Stokes waves of the first, second, and third orders. The experimental values for the threshold powers and parameters $L = 1\,\text{km}$, $\alpha = 0.15\,\text{km}^{-1}$, and $g_R = 2\,\text{W}^{-1}\,\text{km}^{-1}$ are used in the calculations.

A comparison shows that the derived formulae agree quite well with the experimental data for the first and second orders of the cascaded Raman generation. The difference becomes noticeable for the third-order generation whose wavelength is longer by 17% than that for the input pump. The model parameters start deviating noticeably from the experimental ones. If the model gain coefficient g_R in the formula for the third-order output power is replaced by the real value ($g_{R3} = 1.3\,\text{W}^{-1}\,\text{km}^{-1}$), the agreement between the theory and experiment becomes appreciably better. Moreover, this parameter affects only the intermediate power domain, whereas the maximum generated power remains unchanged. It is clear that the power growth above the threshold and the corresponding depletion of previous component are adequately described by the exponential functions predicted by the theory for all orders.

The power/efficiency behavior of the cascaded random RFL is quite different from that for conventional RFLs, either linearly polarized [68, 69] or non-polarized [45, 77], exhibiting a nearly linear power growth with limited efficiency because of high losses for intermediate Stokes components in the cavity. As was mentioned, in RRFL the generated power above the threshold exponentially approaches the maximum value being close to the quantum limit, independently of the Stokes order, i.e., almost all input pump photons are converted into the highest-order Stokes photons. This feature of the RRFL is defined by the specific power distributions along the fiber. The longitudinal power distributions calculated from Eqs. (7.6), (7.7), and (7.8) are shown in Fig. 7.16 for two values of the input pump power $P_{\text{in}} = 6$ and $9.7\,\text{W}$, which nearly correspond to the maximum output of the first and second Stokes waves. It is seen that the transition regions in the power distribution

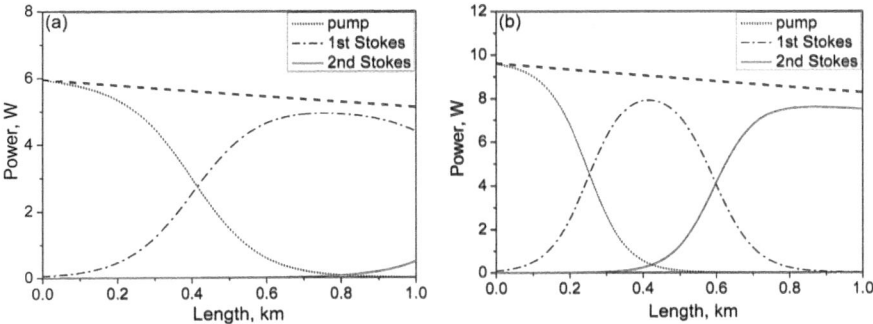

Fig. 7.16 Longitudinal power distribution for different components calculated at the input pump power of 6 W (**a**) and 9.7 W (**b**). *Dashed line* shows attenuation of the pump power along the fiber

are described by the steep hyperbolic tangent function predicted by the theory and different waves are almost fully separated in space. The pump distribution is concentrated near the left end ($x = 0$), whereas the output Stokes order is at the right end ($x = L$), and intermediate ones are in between. Thus, the intermediate components have zero intensity at the fiber ends, while the highest Stokes order has the maximum output power. This is quite different from RFLs with conventional cavity, where the intermediate components are reflected from the cavity mirrors at the fiber ends thus experiencing losses [77]. The Raman conversion in the RRFL occurs inside the passive fiber, so only the Rayleigh scattering losses are present here for all waves. They are proportional to the fiber length and nearly equal for the pump and low-order Stokes components. So, the maximum quantum efficiency is limited by the fiber transmission amounting to about 0.85 in the scheme with a 1 km PM fiber. As a result, almost all transmitted pump photons are converted into the output Stokes order, and the quantum efficiency of converting the pump radiation (1.05 μm) to the output Stokes wave is around 80%, independent of the Stokes order, which is close to the transmission coefficient of the 1 km fiber.

Kinetic model for spectra. It is assumed that the effective dispersion length in the studied RRFL is much larger than the integral nonlinearity and Raman gain, which allows to neglect the dispersion effect in the kinetic equations [21, 78] describing the effect of self-phase modulation (SPM) for the generated wave. Taking the spectral intensity of the Stokes components in the form

$$I_j(\omega, x) = F_j(\omega)P_j(x), \quad \frac{1}{2\pi}\int F_j(\omega)\,d\omega = 1, \tag{7.12}$$

where ω is detuning from the line center, and solving equations (S5.2-S5.4) from supplementary information of [66], one can write the kinetic equation as

$$\int_0^L dx\, g_R P_{j-1}(x) \frac{\Delta_{\text{RMS}(j)}^2 + \frac{d^2}{dt^2}}{\Delta_{g(j)}^2}\, F_j(t) = \tag{7.13}$$

$$\frac{\left(\gamma_{\text{SPM}} P_j(L)\right)^2}{g_R^2 P_{\text{in}}^2 + \beta_{(j)}^2 \Delta_{\text{RMS}(j)}^4}\, \left(F_j(t) - F_j(t)^3\right),$$

where γ_{SPM} is the Kerr nonlinear coefficient for the SPM processes, β is the second-order dispersion coefficient, and Δ_{RMS} is the spectral half width. The left side describes spectral filtration over the round trip (by means of the gain spectral function) leading to the Schawlow-Townes spectral narrowing, whereas the right side describes SPM-induced broadening. The solution of this equation is

$$F_j(t) = \frac{1}{\cosh\left(\Delta_{\text{RMS}(j)} t\right)} \tag{7.14}$$

and the corresponding spectral density of the j-th Stokes component in the case of $g_R P_{\text{in}} \ll \beta \Delta_{\text{RMS}}^2$ is similar to that for the first component under the same condition [78]:

$$I_j(\omega, L) = \frac{\pi P_j(L)}{\Delta_{\text{RMS}(j)} \cosh\left(\omega \pi / 2\Delta_{\text{RMS}(j)}\right)} \tag{7.15}$$

with

$$\Delta_{\text{RMS}(j)} = \left[\frac{\gamma_{\text{SPM}} P_j^{\text{out}} \Delta_{g(j)}}{\beta_{(j)} \sqrt{2 g_R \tilde{L} \left(P_j^{\text{th}} - P_{j-1}^{\text{th}}\right) (\lambda_0 / \lambda_{j-1})}}\right]^{1/3}.$$

Thus, the resulting FWHM linewidth at the $-3\,\text{dB}$ level is expressed as a cubic root function of the output power of the j-th Stokes wave P_j^{out}:

$$\Delta_{\text{SPM}(j)} = \frac{4}{\pi} \ln\left(2 + \sqrt{3}\right) \Delta_{\text{RMS}(j)}, \tag{7.16}$$

which is supplemented by the pump-induced cross-phase modulation effect. The XPM contribution can be estimated at the corresponding threshold, similar to the case of the first Stokes wave generation in RFLs [21, 45]:

$$\Delta_{\text{XPM}(j)} = 2 \sqrt{\frac{2 \ln 2 \gamma_{\text{XPM}} P_{\max(j-1)} \Delta_{g(j)}}{\Delta \beta_j \sqrt{g_R P_{\max(j-1)} \tilde{L}}}}. \tag{7.17}$$

Table 7.1 Parameters of the Stokes waves used in the calculations

j	λ, nm	P^{th}, W	P_{max}, W	β, ps^2 km^{-1}	$\Delta\beta$, ps m^{-1}	Δ_g, $\times 10^{12}$ rad s^{-1}
0	1055	0	2.6			
1 (first peak)	1106	2.6	2.5	21	1.74	4.8
1 (second peak)	1111	2.6	4.4	20.8	1.9	1.7
2	1169	5.9	7.4	17.3	1.43	6.5
3	1232	9.7	9.1	13.5	1.12	7

Here γ_{SPM} and γ_{XPM} are the Kerr nonlinearity coefficients for the SPM and XPM processes, $\beta_{(j)}$ and $\Delta\beta_j$ are the second-order dispersion and dispersion walk-off coefficients, respectively, $P_{max(j-1)}$ is the maximum power of the $(j-1)$-th order Stokes wave (or pump wave if $j=1$) corresponding to the j-th Stokes generation threshold, and $\Delta_{g(j)}$ is the HWHM width of the Raman gain spectral peak. The solid curves in Fig. 7.12 show the total linewidth $\Delta_{FWHM} = \Delta_{SPM} + \Delta_{XPM}$ calculated with the nonlinear coefficients $\gamma_{SPM} = \gamma_{XPM}/2 = 6$ W^{-1} km^{-1} and experimental values for the power and Raman widths $\Delta_{g(j)}$ estimated from the ASE spectrum at low powers (see Fig. 7.10a), which are collected in Table 7.1 together with the dispersion coefficients for all orders. As a result, good agreement of these formulae with the experiment is demonstrated.

The experiment and calculations show that the XPM effect arising from the pump (or previous Stokes order) defines the minimum bandwidth of the generated spectrum amounting to 0.17–0.3, 0.48, 0.72, and 1.03 for the first, second, third, and fourth Stokes waves, respectively, growing nearly proportionally to the pump power at corresponding threshold. At the same time, the SPM linewidth is a slowly growing cubic root function of the generated power for all orders. This is quite different from the results obtained for the first Stokes generation under conditions of weak dispersion and nonlinearity as compared to the Raman gain, when the linewidth grows almost linearly with generated power [21]. In this case, the linewidth is nearly constant in the broad range of generated powers but is increasing proportionally to the Stokes order. This behavior is different from the case of conventional Raman fiber lasers, where the minimum spectral width is defined by the FBG (or other filtering element) bandwidth [45, 68, 69], whereas its power broadening behaves as a linear or square root function of the generated power depending on the ratio between nonlinearity and dispersion [45].

A complete description of the cascaded Raman lasing in fibers enables development on this base of high-performance devices, which offer wide opportunities for advanced applications, especially in telecommunications and sensing based on fiber-optics links, in which all-fiber PM RRFLs can be easily integrated. Linear polarization and a high power at a relatively narrow bandwidth (which may be further reduced and made tunable by insertion of a spectral filter similar to [22–24]) also offer an efficient second harmonic generation that will transfer the generated spectrum in visible range (0.5–0.8 μm is feasible), thus enabling implementation of this source in bio-imaging and display technologies. In the next sections, two options for RRFL short-wavelength generation (direct LD pumping and second harmonic generation) are treated.

7.3.4 Short-Wavelength Generation

Raman lasing is usually obtained in conventional single-mode telecom fibers (e.g., SMF-28) as well as other passive fibers pumped by, e.g., rare-earth (RE)-doped fiber lasers in robust all-fiber configurations (see e.g., [79] for a review). Such single-mode pumping induces in a passive fiber the distributed Raman gain that may be higher than linear attenuation in a fiber together with losses in splice/connector points already at watts level of pumping. Using highly reflective fiber Bragg gratings as cavity mirrors, one can easily achieve positive feedback enough for lasing at the first Stokes wavelength, as well as at higher-order Stokes wavelengths usually achieved in cascaded FBG cavities [80], which makes possible generation in all-fiber configuration at various wavelengths in $1–2\,\mu m$ spectral range [81]. Since the Raman laser is based on the pump-induced Raman gain in passive fibers, it has fundamental difference in lasing properties compared with RE-doped fiber lasers, namely, small quantum defect, fast response of the gain saturated due to pump depletion only, low background spontaneous emission, as well as absence of photo-darkening effect that is a problem in RE fiber lasers especially at short wavelengths [82]. At the same time, a relatively low Raman gain as compared with the gain value in active fibers requires much larger length of passive fiber exceeding 100 m in conventional RFLs and kilometers in random DFB schemes of RFLs. That is why RFLs are usually core-pumped by high-power Yb- or Er-doped fiber lasers with single-mode output. Operation of RFLs in $\sim1\,\mu m$ spectral range is possible if one uses for pumping Yb-doped fiber lasers (YDFL) with high power output around $1.06\,\mu m$. As an example, CW output power as high as 150 W at 1120 nm has been obtained for the YDFL-pumped Raman fiber laser with 85% conversion efficiency [67].

Short wavelength ($\leq1\,\mu m$) operation of RFLs seemed to be difficult because of a lack of high-power single-mode laser sources operating well below $1\,\mu m$. The lower limit for YDFLs is ~980 nm, while the photodarkening effect increases with the wavelength shortening [83]. In this direction, there exist two perspective approaches that offer opportunity to explore multimode laser diodes for RFL pumping. First one consists of using multimode graded-index (GRIN) fibers with coupling of multimode pump beam directly into its relatively large core [84]. As a result, the Raman fiber laser based on a multimode GRIN fiber pumped by the multimode Nd-YAG laser generating nearly single-mode output at 1116 nm has been demonstrated. Much better quality of the RFL output beam in comparison with that for multimode pump LDs provided due to the effect of beam cleanup typical for stimulated Raman scattering (SRS) in fibers with graded index core [85]. This effect was also explored in similar RFL scheme for combination of multimode pump beams and their conversion into three times brighter beam with conversion efficiency of $\sim56\%$ [85]. Then T. Yao et al. demonstrated a possibility to increase the RFL output power up to 20 W and slope efficiency up to 80% using shorter (1.5-km-long) GRIN fiber and higher-power multimode pumping reached by means of two 975 nm LD beams combination [86]. Y. Glick et al. obtained 80 W of output power at 1020 nm in the similar experimental scheme by using only 0.5-km-long GRIN fiber [87].

Second approach concerns implementation in RFLs of a cladding pumping that has been proposed and realized in [88] demonstrating the RFL based on the double-clad single-mode passive fiber being end-pumped by a multimode Er:Yb-doped fiber laser in MOPA configuration with up to 10 W output power at 1660 nm from the single-mode fiber end. Later on, up to 100 W at 1120 nm with 71% slope efficiency [89] from the RFL based on the double-clad passive fiber forward-pumped by a multimode YDFL at 1064 nm has been obtained by the same group. At that, $M^2 = 1.35$ at low output power was demonstrated. It slightly degraded to $M^2 = 1.6$ at 80 W of generated power. This research group also investigated concept of multimode LD pumping in double-clad passive fiber obtaining 6 W of output power at 1019 nm and nearly five times the brightness enhancement [86].

The described approaches offer an opportunity to use multimode laser diodes for RFL pumping, just like in cladding-pumped high-power Yb-doped fiber lasers [82]. At that both the conventional linear cavity, consisting of two mirrors, and half-open cavity with random distributed feedback may be explored.

7.3.4.1 LD-Pumped Raman Fiber Laser Operating at 980 nm

Availability of high-power multimode laser diodes pigtailed by a 105 μm core fiber which generates more than 100 W at 915 and 980 nm makes it possible to obtain efficient Raman lasing at the short wavelengths (< 1 μm). Thus, RFL generation at the wavelength as short as 835 nm at 806 nm LD pumping was demonstrated in pulsed regime [90]. At the same time, CW RFL generation at the wavelength of 980 nm with direct 938 nm LD pumping into GRIN fiber was obtained in [91].

The experimental setup of the RFL used in [91] is illustrated in Fig. 7.17a. Commercially available multimode LD with power of up to 75 W at 935–940 nm delivered into 105 μm core fiber pigtail was chosen as a pump source. The LD beam collimated by lens L_1 at fiber output was then focused by lens L_2 into the 62.5 μm core of the multimode Corning 62.5/125 CPC7 GRIN fiber with coupling coefficient of $\sim 60\%$. The normally cleaved end of the fiber from one side and highly reflective fiber Bragg grating inscribed in the multimode GRIN fiber core by high-power UV laser [92] formed linear cavity of RFL. Output beam collimated with lens L_3 was then separated by the dichroic mirror (DM) onto pump and generated Stokes radiation. Three power meters (PM_1, PM_2, PM_3) were used to control power in back and forward directions. The optical spectra were measured by Yokogawa AQ6370 optical spectrum analyzer (OSA).

The fiber length was chosen on the base of the laser threshold calculations using a balance model [93]. The threshold power can be calculated using the following formula:

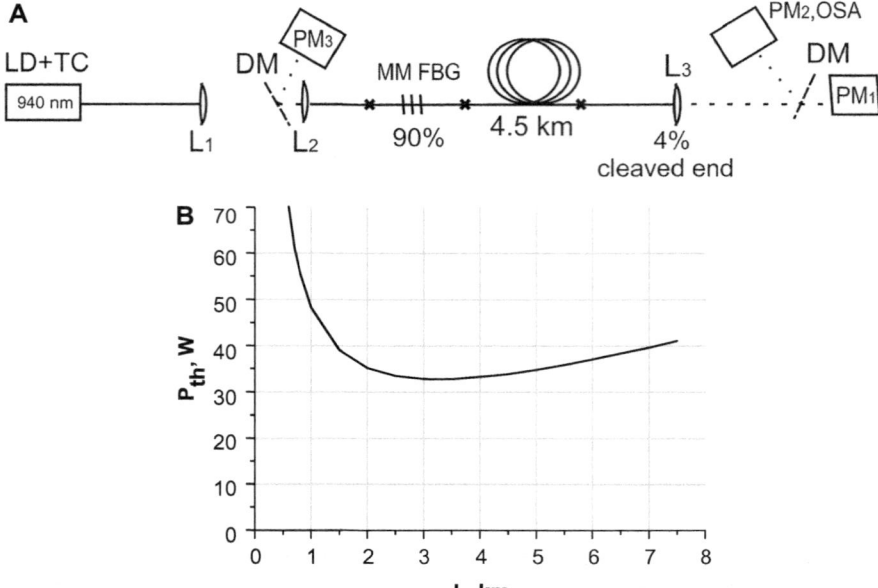

Fig. 7.17 (a) Scheme of the experimental setup. (b) Calculation of the threshold power versus fiber length in the cavity formed by mirrors with $R_1 = 90\%$ and $R_2 = 4\%$

$$P_{\text{th}} = \frac{2\alpha_s L - \ln(R_1 R_2)}{2g_R L_{\text{eff}}} \tag{7.18}$$

where L_{eff} is effective absorption length:

$$L_{\text{eff}} = \frac{1 - \exp(-L\alpha_p)}{\alpha_p} \tag{7.19}$$

Here $\alpha_s = 1.7\,\text{dB km}^{-1}$ and $\alpha_p = 2\,\text{dB km}^{-1}$ are the signal and the pump loss coefficients and $g_R = 0.23\,\text{dB km}^{-1}\,\text{W}^{-1}$ is the Raman gain coefficient, which has been measured by standard technique of small-signal attenuation and special technique based on analysis of spectral dependence for amplified Raman scattering at different pump powers [94], correspondingly. The effective reflection coefficients were taken, $R_1 \approx 0.9$ and $R_2 \approx 0.04$. Value of the effective length was estimated as $L_{\text{eff}} \sim 1.9\,\text{km}$. The calculated threshold power variation with increasing fiber length is presented in Fig. 7.17b. Following the results of calculations, the threshold power P_{th} took the minimum value in the range of cavity lengths between 2.5 and 4.5 km. Since the dependence of threshold was steepen to shorter lengths, 4.5-km-long spool was chosen in the first experiment. The calculated value of threshold power at that length amounted to $P_{\text{th}} \sim 33\,\text{W}$. The experimental threshold power was 35 W.

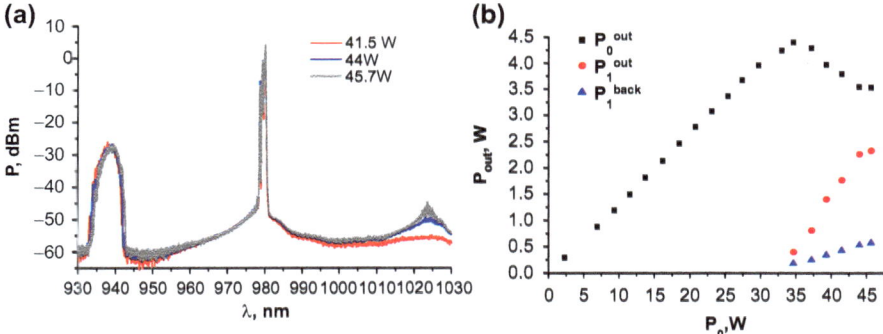

Fig. 7.18 (**a**) RFL output spectrum at different pump powers. (**b**) Generated Stokes wave power (at 980 nm) in forward P_1^{out} (*circles*) and backward P_1^{back} (*triangles*) directions and residual pump power P_0^{cut} (*squares*) versus coupled pump power P_0 (at $\sim$ 938 nm)

Output spectrum above the threshold contained two lines: pump spectrum centered at $\lambda_0 \approx 938$ nm and generated first Stokes wave spectrum at $\lambda_1 \sim 980$ nm (see Fig. 7.18a). The Stokes power rapidly grew with increasing pump power above the threshold value and amounted to $P_1^{out} = 2.3$ W (measured by PM$_2$) at coupled pump power of $P_0 = 46$ W, see Fig. 7.18b. The residual pump power started decreasing above the threshold from 4.4 to 3.3 W due to the pump-to-Stokes wave conversion. The sufficient Stokes power ($P_1^{back} \sim 0.6$ W at maximum pumping) was observed in opposite direction, so the total output power reached $\sim$2.9 W. Moreover, a clear saturation of the first Stokes wave was observed which was induced by the second Stokes wave generation at 1025 nm like in conventional cascaded RFLs [77, 95] (see Fig. 7.18a).

Low value of the threshold for the second Stokes wave was explained by that the only low-index transverse modes of the first Stokes wave were generated due to cleanup effect in graded-index fibers [84, 85]. A relatively good quality of the generated beam and detailed structure of the generated spectrum are seen in (Fig. 7.19a). Output spectrum at 980 nm consisted of tree spectral components which were mainly defined by multimode FBG used in the experiments. The width (FWHM) of the individual peak was within 0.5–1 nm range at different output powers, and the main part of the power was concentrated in the peak centered at 980 nm during generated power increasing. Wavelength difference $\Delta\lambda$ between two neighboring groups of the transverse modes in the GRIN fiber can be estimated by simple formula on $\Delta\lambda$ [91]

$$\Delta\lambda = \lambda^2 NA/ \left(\pi \, d \, n_1^2 \right) \tag{7.20}$$

where λ is wavelength of the fundamental mode, NA is numerical aperture, d is core diameter, and n_1 is cladding refractive index. Using GRIN fiber characteristics NA $= 0.275$, $d = 62.5\,\mu$m, $n_1 = 1.45$, and wavelength $\lambda = 980$ nm, the value

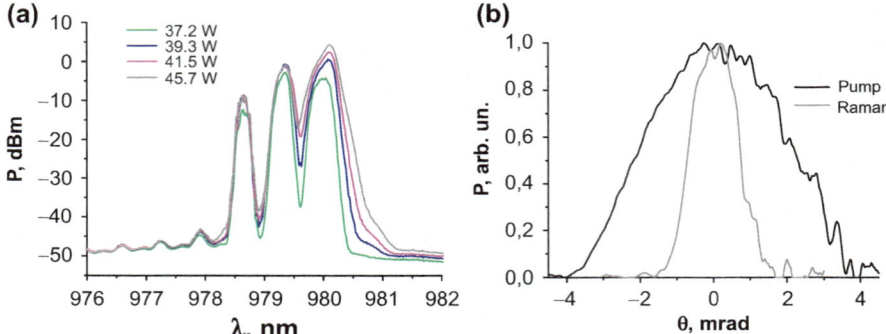

Fig. 7.19 (a) Detailed output spectrum of the generated Stokes wave at different pump powers. (b) Output beam profile for the pump (*black*) and Raman (*gray*) lasers

of wavelength difference 0.6 nm was obtained which was in close agreement with the distance between two lines observed in the RFL generation spectrum (see Fig. 7.19a). Therefore, those lines corresponded to neighboring groups of low-index transverse modes which were reflected by FBG. Since intermediate peaks were not observed in the generated spectrum, the intermodal coupling within the multimode grating did not take place. The far-field profile of the collimated output beam measured by Coherent BeamMaster BM-7 (Fig. 7.19b) demonstrated reduction of beam divergence by three times compared to the pump beam.

The RFL slope efficiency (calculated for forward output radiation only) was not as high ($\eta \sim 25\%$) as that in conventional single-mode RFLs pumped by fiber lasers [67, 79, 80] because of the large GRIN fiber length. Moreover, effective reflectivity of the initially highly reflective FBG became relatively low at high-power lasing of the RFL, $R \to 40\%$, as was estimated by formula [67]

$$P_1/P_2 = (R_2/R_1)^{1/2}(1 - R_1)/(1 - R_2) \tag{7.21}$$

where $R_2 = 0.04$ is the Fresnel reflection coefficient and P_1 and P_2 are backward and forward Stokes powers. The method of efficiency enhancement was applied in [91] by adding a bulk mirror placed behind the dichroic mirror DM (see Fig. 7.17a) in order to send the residual pump power back to the cavity. As a result, 1.4 times increase in output power (3.3 W) and efficiency ($\eta \sim 35\%$) was achieved.

7.3.4.2 Random Fiber Laser Directly Pumped by a High-Power Laser Diode

Random fiber laser directly pumped by a high-power laser diode was also demonstrated without conventional resonator [25]. In this case the RFL used positive feedback for the Stokes wave provided by a randomly distributed Rayleigh backscat-

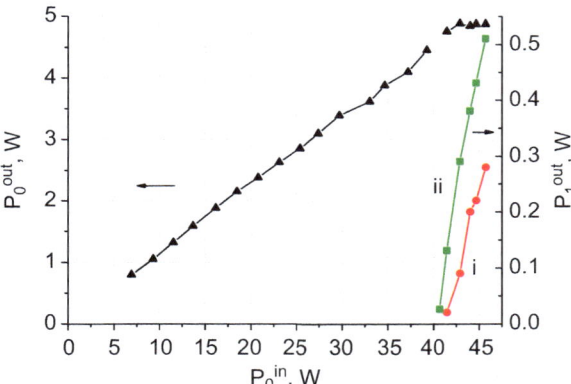

Fig. 7.20 Residual pump at 938 nm (P_0^{out}) and generated output power at 980 nm (P_1^{out}) versus input pump power (P_0^{in})

tering which is also stronger than that in conventional step-index single-mode fibers. Experimental study of the random lasing was performed on the base of RFL scheme applied in [91] (see Fig. 7.17a), where the narrowband highly reflective input FBG was replaced by a broadband bulk mirror, placed instead of power meter head PM3, and output GRIN fiber end was cleaved with angle $>15°$ to avoid Fresnel reflection. The pump power of 45 W launched into 62.5/125 GRIN fiber was enough for the Stokes wave generation in the half-open cavity configuration involving highly reflective bulk mirror ($R = 99\%$ at 980 nm) and RS-based distributed feedback along the GRIN fiber. The measured output powers for the pump (P_0^{out}) and the Stokes (P_1^{out}) beams are shown in Fig. 7.20. After the RDFB laser threshold ($P_0^{in} \sim 42$ W), the generated 980 nm Stokes radiation starts growing rapidly reaching $P_1^{out} \sim 0.3$ W (curve i) at maximum available pump power ($P_0^{in} = 45.7$ W), whereas the residual pump power P_0^{out} becomes fixed at level of 5 W. If an additional mirror reflecting back the residual pump power was inserted, the generated power increases to $P_1^{out} > 0.5$ W (curve ii).

Corresponding spectra of the output radiation measured near and above the threshold are shown in Fig. 7.21a. In addition to the pump peak at 938 nm, the amplified SRS peak at ∼980 nm was seen that was switched above the threshold to the generation of the Stokes wave. At the maximum Stokes wave power, the second-order Stokes peak appeared in the spectrum at ∼1020 nm. Though the Raman gain profile was broad, the generation peak was narrow, but its position fluctuated in time during the generation looking after the averaging by OSA as a comb near the gain profile maximum at ∼980 nm (upper trace in Fig. 7.21a). Corresponding time trace for the Stokes wave consisted of pulses which were fully stochastic in nanosecond to microsecond scales but exhibited some regularity in ms scale (see Fig. 7.21b). With increasing power, a steady component tended to increase, while characteristic modulation period tended to decrease. Thus, the LD-pumped multimode random

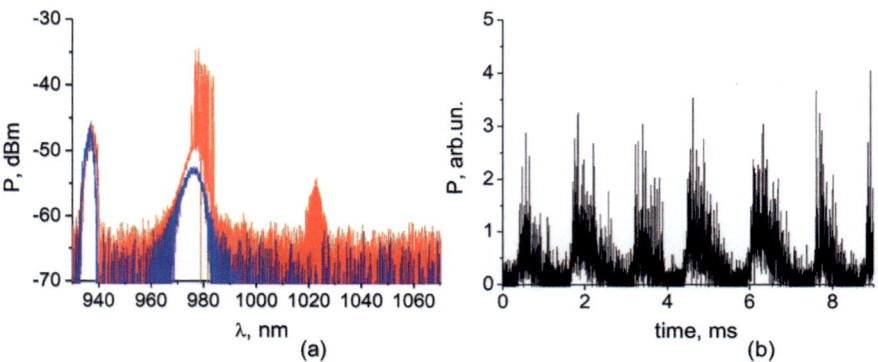

Fig. 7.21 Experimental RDFB spectra at (**a**) $P_0^{in} = 41$ and 45 W and (**b**) corresponding time trace at $P_0^{in} = 45$ W

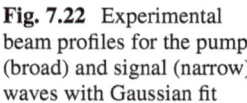

Fig. 7.22 Experimental beam profiles for the pump (broad) and signal (narrow) waves with Gaussian fit

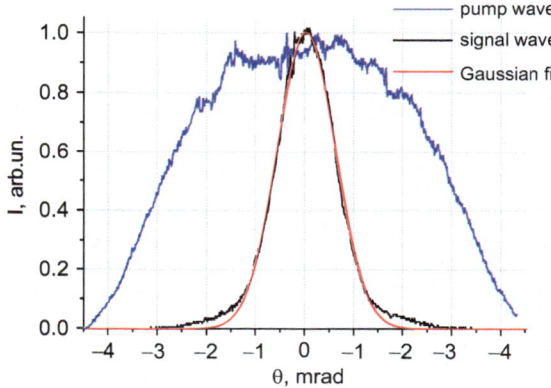

RFL had some similarity with single-mode ones [15, 22, 23] exhibiting near-threshold narrow instantaneous spectra and self-Q-switched pulsations due to the combined effect of Rayleigh scattering feedback and stimulated Brillouin scattering (SBS) [96], which transform well above the threshold into the quasi-continuous regime with broad stationary spectrum. However, the sufficient excess over the threshold was not reached because of a very high threshold value (see Fig. 7.20). The 42 W threshold is defined by the Rayleigh backscattering coefficient estimated as 1.5×10^{-3} km^{-1} (or ~0.7% total reflection) for the used GRIN fiber.

One of the most interesting features of the presented laser concerned of the generated beam profile. Though the pump beam from the multimode LD had a very low beam quality and a high divergence as a consequence, the quality of the generated beam became much better. The far-field profiles measured by CCD camera (see Fig. 7.22) demonstrate a relative reduction of the generated beam divergence by 4.5 times as compared with the pump beam. Taking that the pump beam quality factor was $M^2 \sim 20$, such a reduction indicated a tendency of the laser to generate nearly single transverse mode. Moreover, profile of the Stokes signal was perfectly fitted by the Gaussian function. Besides, an additional influence of the SBS [97] and the RS distributed feedback made the cleanup effect much stronger than that in the LD-pumped RFL scheme with point-action reflectors [91]. Using passive fibers without reflectors may eliminate some principal drawbacks of conventional high-power fiber lasers based on active fibers with resonators, although relatively low efficiency could pose for the application of the proposed solution.

7.3.4.3   954 nm Raman Fiber Laser with Multimode Laser Diode Pumping

Optimization of the GRIN fiber length at the 915 nm pumping provided generation of the CW LD-pumped RFL at the shorter wavelength (954 nm) [98] with higher output power and slope efficiency as compared with characteristics of 980 nm CW RFL [91].

The experimental scheme of such LD-pumped RFL is shown in Fig. 7.23. It was similar to the one in Fig. 7.17a. Commercially available multimode LD with output power of up to 65 W at 912–918 nm was used as a pump source. The LD radiation which emerged from 105 μm core fiber pigtail was launched by collimating lenses *L1* and *L2* into the 62.5 μm core of multimode Corning 62.5/125 GRIN fiber with coupling efficiency of ∼60%. The linear cavity of the laser was formed by high-reflection fiber Bragg grating *FBG1*, and normally cleaved fiber end provided weak (∼4%) Fresnel feedback, or low-reflection output fiber Bragg grating *FBG2*. The multimode FBGs were written by a high-power UV argon laser in the same multimode GRIN fiber and had Bragg wavelengths of ∼954 nm corresponding to the Raman gain spectral maximum. The output laser beam collimated with lens *L3* was splitted by the dichroic mirror *M2* onto the residual pump beam and the Stokes radiation generated at ∼954 nm. Dichroic mirror *M1* was used to protect the *LD* from the backward Raman radiation. Mirrors *M3*-*M5* splitted the output RFL beam between the optical spectrum analyzer *OSA* (Yokogawa AQ6370), Beam Master *BM*

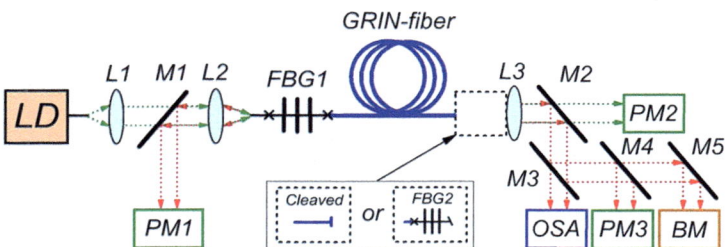

Fig. 7.23 Scheme of the experimental setup: *LD* high-power multimode laser diode, *L1*, *L2*, *L3* collimating lenses, *M1*, *M2*, *M3*, *M4* dichroic dielectric mirrors, *M5* aluminum mirror, *FBG1* and *FBG2* are high- and low-reflection multimode fiber Bragg grating, respectively, *GRIN* fiber, multimode graded-index fiber, *PM1*, *PM2*, *PM3* power meters, *OSA* optical spectrum analyzer, *BM* beam master

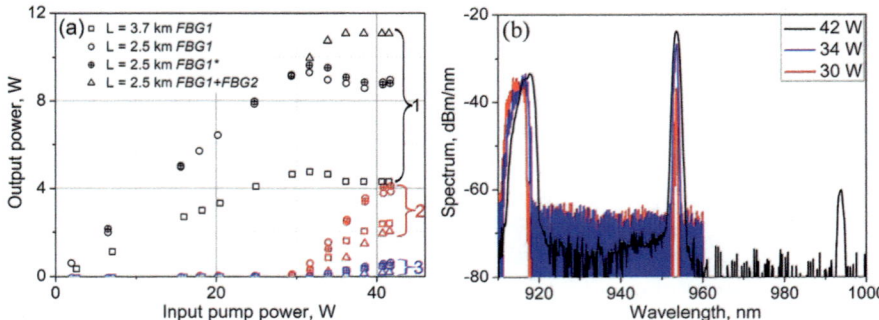

Fig. 7.24 (a) Residual pump power (*1*) and generated Stokes wave power at 954 nm in the forward (*2*) and backward (*3*) directions versus input pump power at $L = 3.7$ km (*squares*) and $L = 2.5$ km in the cavity formed by *FBG1* and cleaved fiber end (*circles*) or output *FBG2* (*triangles*). Data for another highly reflective *FBG1** are indicated with *cross* inside *circle*. (**b**) Output spectrum of RFL with 2.5-km-long *GRIN* fiber at different input pump powers

(Coherent BM-7), and power meter *PM3* (Coherent PM10). Power meters *PM1* and *PM2* were used to control power of backward Stokes and residual pump radiation, respectively.

In order to increase the generated power and conversion efficiency in comparison with [91], the authors shortened the GRIN fiber length from 4.5 km down to $L = 3.7$ and 2.5 km. Figure 7.24a shows residual pump (1) and Stokes generated power (2, 3) at central wavelength of 954 nm versus input (coupled into GRIN fiber) pump power for two fiber lengths in the scheme with normally cleaved output fiber end. When the coupled pump power exceeded the generation threshold, the Stokes wave power started to grow and the residual pump power started decreasing due to the pump-to-Stokes wave conversion. The experimental power threshold of ∼30 W was nearly the same for two fiber lengths, in correspondence with the calculated threshold power curve in Fig. 7.17b. At the same time, the differential efficiency was higher for the shorter length which amounted to 37% and 26% at $L = 2.5$ and 3.7 km, respectively. The maximal Stokes power was provided at shorter cavity length and reached 3.8 W in forward direction. Taking into account the backward Stokes power of 0.4 W, the total generated power reached 4.2 W. Optimization of highly reflective grating did not lead to considerable power growth. With the optimal grating *FBG1** (crosses inside circles in Fig. 7.24a), the output power (differential efficiency) increased up to 4.1 W (41%) in forward direction.

The effective reflectivity R_1 of the *FBG1* estimated by the Eq. (7.21) was $R_1 = 65\%$ and 45% for 2.5- and 3.7-km-long GRIN fibers at $P_{in} = 38$ W. Since the value of R_1 was about 40% at $L = 4.5$ km [91], one can conclude that the effective reflection coefficient of *FBG1* is reduced with the fiber length. The reason may be the following. The transverse modes of multimode GRIN fiber can be classified into groups with nearly the same propagating constant [99]. Fiber imperfections, however, allow power to couple from one mode to another during propagation. At

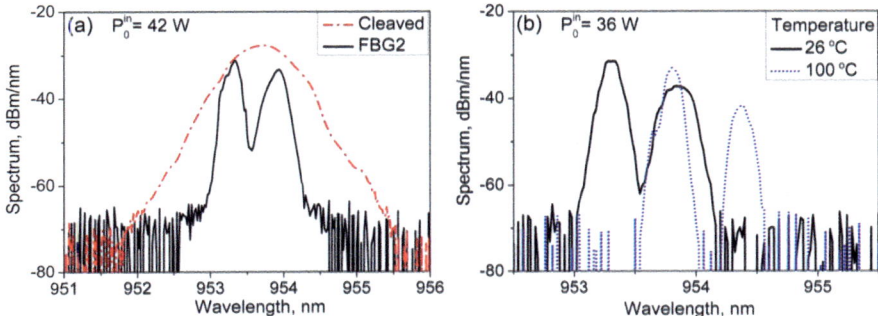

Fig. 7.25 Output spectra of the Stokes wave for $L = 2.5$ km fiber (**a**) with normally cleaved fiber end (*dash-dot line*) or with the *FBG2* (*solid line*) (**b**) at different temperatures of the *FBG2*

that, the coupling between different mode groups is by an order of magnitude lower than the coupling within one group, and in both cases it increases with the increase of fiber length [100]. From the other hand, reflection coefficient and reflection wavelength of multimode FBG depend on transverse mode numbers, and reflection wavelengths do not coincide for different mode groups [101]. Thus, the increasing mode coupling in the longer fiber cavity results in the reduction of the effective reflectivity of multimode *FBG1*.

In addition, a clear saturation of the first Stokes wave because of second Stokes wave generation at 993 nm was observed in 2.5-km-long cavity at high-input pump powers (see Fig. 7.24a). The rather low value of the second Stokes threshold was explained by a high quality of the first Stokes beam, because only low-order transverse modes of the first Stokes wave were generated due to the beam cleanup effect in the GRIN fiber with relatively narrowband ($\sim$0.7 nm) *FBG1*. Good quality of the Stokes beam was confirmed also by the generation spectrum in Fig. 7.24b which is much narrower than the pump one.

The cascaded generation of higher Stokes orders limits potential power for the first Stokes wave. Thus, in order to increase the second Stokes threshold, the cleaved fiber end was substituted by an output *FBG2* with low reflection at 954 nm. The second Stokes line was not observed in the spectrum for that scheme; however, the generation power was also reduced almost twice in comparison with the normally cleaved output end (see triangles in Fig. 7.24a). Effective reflectivity R_2 of the *FBG2* estimated from Eq. (7.21) with $R_1 = 65\%$ turned to be 3%, which was comparable with the Fresnel reflection of the normally cleaved end.

Figure 7.25a shows the output Stokes spectrum measured for the scheme with (solid line) and without (dash-dot line) output *FBG2*. Without *FBG2*, the spectrum of the first Stokes wave was smooth with 3 dB linewidth of 0.7 nm defined by the *FBG1* bandwidth. With the *FBG2*, it consisted of two components spaced by $\sim$0.62 nm with individual linewidths of 0.15–0.2 nm. Distance between the peaks was equal to wavelength difference $\Delta\lambda$ between two neighboring groups of the transverse modes in the 62.5/125 GRIN fiber, so those lines corresponded to two

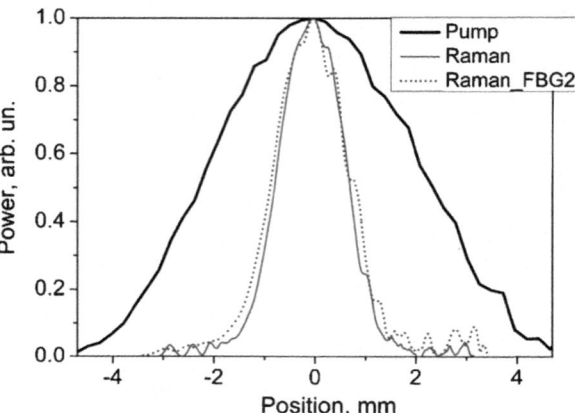

Fig. 7.26 Output beam profile for the pump LD (*black*) and the Raman laser for $L = 2.5$ km fiber with cleaved fiber end (*gray solid line*) or the FBG2 (*gray dotted line*)

neighboring groups of low-index transverse modes reflected by *FBG2*. If *FBG2* was heated by 74°, the two-peak structure synchronously shifted by 0.6 nm (see Fig. 7.25b). This behavior is similar to that observed in single-mode lasers with broadband highly reflective FBG and narrowband output FBG. When temperature is changed, narrowband reflective spectrum of the *FBG2* is tuned over relatively broadband reflective spectrum of *FBG1*. And the generated power is reduced due to the FBG reflective spectra mismatch. When *FBG1* was substituted with a bulk mirror (like in [25]), similar two-peak spectrum or broad (∼2 nm) single-peak spectrum was generated depending on output *FBG2* or reflection from normally cleaved fiber end, correspondingly.

The far-field profile of the collimated output beam in the scheme with and without *FBG2* presented in Fig. 7.26 indicates that generated beam spot size was reduced by three times as compared with the pump beam, similar to [91].

Note that much better results (higher power and efficiency) as well as better mode selection (close to single mode with $M^2 \sim 1.2$) have been recently obtained for output FBG inscribed by fs pulses in central part of graded-index fiber core [102].

Thus, a new class of Raman fiber lasers, namely, RFL pumped by multimode diodes generating high-quality short-wavelength radiation which may compete in <1 μm spectral range with traditional 980 nm LD-pumped Yb-doped fiber lasers [82], is described. They can be also implemented in all-fiber configuration by replacing the bulk coupling scheme by fiber pump combiner which can combine multiple high-power pump beams [103]. Accompanied by the fiber length reduction and two-pass pumping scheme similar to [86], the RFL efficiency of about 80% seems feasible in the all-fiber configuration. Moreover, RFL may operate at any shorter wavelength by using shorter wavelength pump LDs, e.g., 805 nm. High output beam quality of the GRIN-fiber RFL that appears one order brighter than that for multimode LDs opens a way for efficient second harmonic generation thus converting LD-pumped RFL radiation to wavelength range of 400–500 nm applied in biomedical diagnostics such as confocal microscopy and flow cytometry.

7.3.5 Frequency Doubling

An expansion of the spectral region of RFLs operation toward shorter wavelengths can be made by nonlinear frequency conversion such as second harmonic generation (SHG). In particular, one can reduce generation wavelength of near-IR radiation (1.1–1.7 μm) to a visible range involving green, yellow, and red [104], which is attractive for laser guide stars, laser displays, or biomedical applications. About 3 W of yellow radiation at 589 nm, for example, was generated in a periodically poled lithium niobate (PPLN) crystal pumped by a CW narrowband linearly polarized RFL [105]. Recently the power level is increased up to 14 W level in a 20 mm long MgO-doped stoichiometric periodically polled lithium tantalate (MgO:sPPLT) crystal at higher pumping and narrower pump RFL linewidth [106].

The conversion efficiency was found to be limited by the RFL spectral broadening [104]. Another way to increase the conversion efficiency is to implement a narrowband (< 3 MHz) oscillator with high-power Raman fiber amplifier and SHG in an external cavity [107]. Otherwise, the spectral linewidth of the RFL can grow with increasing power up to several nanometers because of the nonlinear effects inside the laser. Such a broad spectrum considerably exceeds the phase-matching bandwidth of SHG crystals. On the other hand, the SHG efficiency for multifrequency RFL radiation increases when compared to that of single-frequency one, owing to the contribution of sum-frequency mixing between different longitudinal modes of the RFL [108].

The spectral width of a random RFL may be easily controlled using intracavity spectral selectors such as FBGs (see Sect. 7.2.4). Furthermore, a wide wavelength tuning range without generation efficiency degradation is easily achieved in RRFLs, which exceeds the performance of conventional RFLs [22, 34]. Frequency doubling of a modeless Raman fiber laser operating via random distributed feedback is of great interest in terms of both investigation of sum-frequency processes in the case of continuous spectrum and extension of the range of RRFL applications.

Paper [109] reports on the experimental demonstration of SHG for a random fiber laser in two configurations (without and with spectral filter) and compares characteristics of the generated visible radiation with SHG for a conventional Raman fiber laser with a linear cavity formed by two mirrors in the same fiber.

Figure 7.27 shows a schematic of the setup for frequency doubling of the RRFL radiation in a nonlinear MgO:PPLN crystal. The laser is based on a piece of 0.85-km-long phosphosilicate fiber (PF) that is terminated by a highly reflective Sagnac mirror providing feedback at one side and a fiber pigtail with angled physical contact connector (FC/APC) preventing Fresnel back reflection at the other side (half-open RRFL cavity configuration). Distributed feedback from the open cavity side is provided by Rayleigh backscattering along the PF. The laser is forward pumped by an ytterbium-doped fiber laser (YDFL) with a maximum output power of 12 W at a wavelength of 1115 nm through a wavelength-division multiplexer (WDM) inserted near the Sagnac mirror. This random fiber laser configuration is similar to one explored in [16] for high-efficiency generation of Stokes radiation at 1308 nm.

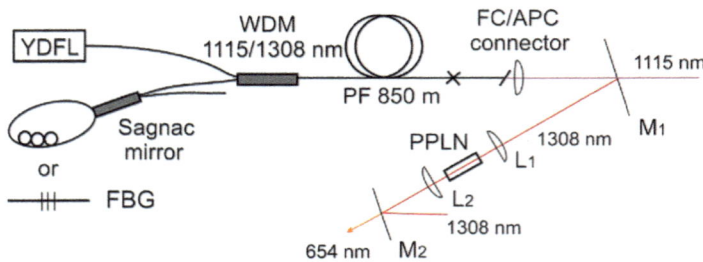

Fig. 7.27 Experimental setup for frequency doubling of RDFB laser

Such a long generation wavelength is defined by the large Raman shift in PF (about $1330\,cm^{-1}$) [81].

Instead of a Sagnac mirror having a rather broad ($\sim 100\,nm$) reflection spectrum, an FBG with bandwidth of 0.15 nm and a reflection coefficient of 80% could also be used. Note that the RRFL with a half-open cavity is only weakly sensitive to the mirror's reflectivity [71], since the integral reflectivity of the output fiber end is very low.

The half-open cavity configuration with the fiber loop mirror also could be easily transformed into a traditional linear cavity configuration. In this case, a narrowband FBG with reflection coefficient of $\sim 14\%$ and spectral linewidth of ~ 0.2 nm is added at the splicing point between the PF and output FC/APC pigtail shown by the cross in Fig. 7.27. To compare SHG efficiency for the laser configurations under study, the laser output radiation coming from the FC/APC pigtail was collimated and then focused into the periodically poled MgO-doped LiNbO$_3$ nonlinear crystal (PPLN) under the same conditions (beam waist radius inside the crystal is $\sim 18.5\,\mu m$). The length of periodically poled structure in the crystal was 8 mm, and the domain period was $\Lambda \sim 12.7\,\mu m$. The radiation coming out of the crystal was collimated and divided by another selective mirror M_2 into the fundamental and SHG components. The optical surfaces of the lenses $L_{1,2}$ and PPLN crystal have antireflection coating for the fundamental and second-harmonic radiation.

Figure 7.28 demonstrates the power and spectral characteristics of the three laser configurations. Conventional-cavity RFL has a low generation threshold ~ 1.5 W, and its output power grows nearly linearly (Fig. 7.28a), whereas both RDFB configurations have a high generation threshold of ~ 5.5 W and exhibit rapid growth of output power above the threshold. The RDFB laser power reaches nearly the same values as that in a conventional RFL when the pump power exceeds 8 W. Note that the output power of the conventional RFL saturates at the ~ 7 W level. The saturation is associated with the second Stokes wave generation that becomes apparent in the output spectrum. These results show that one of the advantages of RDFB-based RFL configurations is a higher threshold of the second Stokes wave generation, which leads to the higher output power for the first Stokes wave.

Another advantage of the RDFB Raman lasers is the relatively narrow spectral width. Figure 7.28b compares the generation spectra for three laser configurations

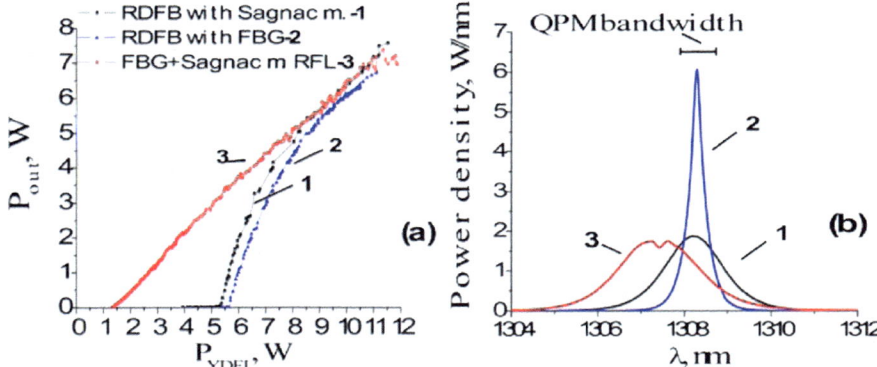

Fig. 7.28 (a) Output Raman laser power at 1308 nm versus input YDFL pump power at 1115 nm for different cavity configurations: (1) RDFB with Sagnac mirror, (2) RDFB with FBG, and (3) conventional RFL with output FBG and Sagnac mirror; (b) their spectra at 7 W pump power

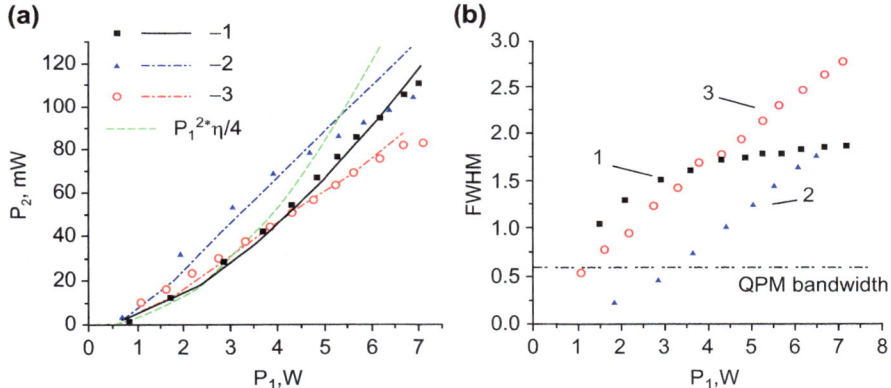

Fig. 7.29 (a) SHG power P_2 versus fundamental power P_1 for three types of lasers: RDFB with Sagnac mirror (*1*) or FBG (*2*) respectivly, conventional RFL (*3*). (b) Corresponding 1308 nm spectral linewidths

at 7 W pump power. It is clear that an RDFB laser with FBG can have a sufficiently narrower spectrum and higher spectral power density compared to a conventional RFL. On the other hand, an RDFB with a Sagnac mirror gives a similar spectrum but without a central dip typical for conventional linear-cavity RFLs with a narrowband output FBG [108, 110]. Figure 7.29b shows the detailed dependence of the laser linewidth on the generated power, demonstrating that the studied laser configurations have different spectral broadening behaviors that in turn influence the SHG efficiency. The conventional-cavity RFL has a relatively wide spectrum with nearly linear low power broadening (Fig. 7.29b). The RDFB laser with a Sagnac mirror also has a wide generation spectrum at low power; at higher powers, its spectral broadening tends to saturate, which results in a narrower spectrum than that for a conventional RFL. The RDFB laser with an FBG instead of a Sagnac mirror attains a narrower linewidth in accordance with the narrower FBG reflection spectrum but only in a limited power range. At high generation powers, both RDFB lasers have close values.

Random fiber laser linewidth was well described by the kinetic theory [21] in the case of a half-open cavity with a broadband reflector, when the nonuniform spectral shape of the Raman gain defined the linear (Shawlow-Townes) line narrowing over the round trips. On the other hand, nonlinear broadening due to self-phase modulation (SPM) is dominant at high powers. However, the kinetic theory is not directly applicable to the case of a narrowband reflector such as FBG providing additional spectral filtering at each random round trip. Since this configuration offers an opportunity to narrow the spectrum and to increase the spectral density at relatively high powers (up to 5 W), it is potentially interesting to further develop the kinetic theory of random fiber lasers thus making possible its spectral optimization by means of intracavity filtering.

Now let's compare the laser linewidths with the quasi-phase matching (QPM) bandwidth amounting to 0.6 nm for the PPLN crystal (shown by the horizontal bar in Fig. 7.28b and by the dash-dot-dot line in Fig. 7.29b). The comparison shows that the linewidths for all configurations exceed the QPM bandwidth in a broad power range. As a result, the measured SHG spectral linewidth is fixed at level of ~0.3 nm and is almost independent of the pump power.

Figure 7.29a shows the measured SHG power as a function of input fundamental power for each laser configuration. Comparing Fig. 7.29a, b, one can find that the narrower the spectral linewidth of the fundamental radiation, the larger the SHG power. The RDFB laser with FBG demonstrates maximum SHG efficiency in a broad power range except at the highest available fundamental power. Then, the SHG power of two RDFB laser configurations approaches each other in accordance with their spectral linewidth equalization. At that point, the power of the red radiation at 654 nm reaches 110 mW, whereas the SHG power for conventional-cavity RFL saturates at a level of ~80 mW because of the broader spectrum and the onset of second Stokes wave generation.

The lines in Fig. 7.29a show the results of the calculations, which were performed for quantitative analysis. The dashed line corresponds to the SHG power for single-frequency radiation calculated according to the well-known formula

$$P_2 = \eta P_1^2 / 4,$$

where $P_{1,2}$ are the fundamental and SHG powers, coefficient 1/4 takes into account the fact that fundamental radiation is nonpolarized and only one linear polarization component is involved in the SHG process, and η is the SHG coefficient for single-frequency linearly polarized radiation [111].

The dash-dot and solid lines in Fig. 7.29a correspond to calculation results for broadband radiation, taking into consideration a finite QPM bandwidth and concrete spectral shapes for all of the lasers. The calculations were performed in

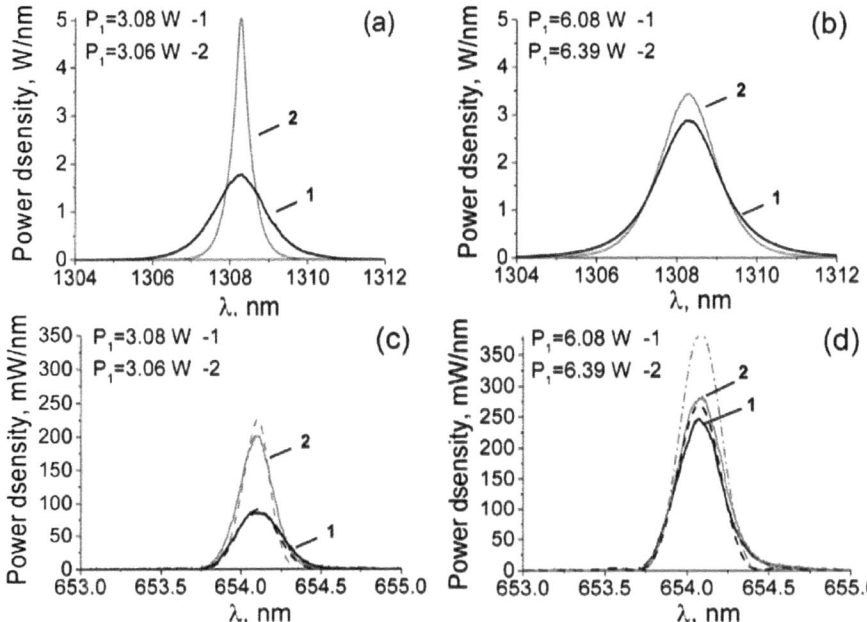

Fig. 7.30 Spectra of RDFB laser with (*1*) Sagnac or (2) FBG mirror for (**a**) ∼3 W, (**b**) ∼ 6 W power at 1308 nm, and (**c**), (**d**) corresponding SHG spectra at 654 nm in experiments (*solid lines*) and calculations (*dashed lines*)

approximation of Gaussian statistics for modes with random phases. The difference between the quasi-continuous spectrum of the RDFB laser and the discrete modes in a conventional RFL was neglected because the intermode spacing is very narrow (∼120 kHz) in the case of an 850 m cavity. For further simplification, the fundamental spectrum was divided into equal intervals of 10^{-3} nm (∼0.18 GHz). The processes of sum-frequency generation between the intervals and direct frequency doubling were considered for all frequency combinations, taking into account QPM conditions. It was confirmed that the calculated integral value of SHG power changes by less than 0.4% with further interval reduction. The detailed calculation method can be found in [108]. Figure 7.30 shows some examples of fundamental and second-harmonic spectra for two RDFB laser configurations with corresponding solid and dashed lines. Left and right columns correspond to the fundamental powers of ∼3 W and ∼6 W, respectively.

Figures 7.29a and 7.30 demonstrate a relatively good agreement between the measured and calculated results. It is well known that the SHG power has quadratic dependence on the fundamental power for single-frequency radiation. Linear SHG power growth for a conventional-cavity RFL is associated with the linear spectral linewidth broadening for the laser when the QPM bandwidth is exceeded [108]. For the SHG, the power increases because of the increase in the number of sum-frequency mixing processes within the RFL generation spectrum, in spite of the absence of spectral power density growth for fundamental radiation.

The RDFB laser with a Sagnac mirror behaves similarly, but at high powers the SHG efficiency becomes higher than that for the conventional-cavity RFL because of the narrower linewidth. In the case of the random laser with a narrowband FBG instead of a broadband Sagnac mirror, the spectral power density is much higher in almost all of the power range due to the FBG filtering effect. At the highest available power, the linewidth in the two RDFB laser configurations is nearly the same, which equalizes the SHG power.

It is well known that the SHG conversion efficiency for a narrow (within the phase-matching bandwidth), non-synchronized multimode spectrum is two times higher than that for a single longitudinal mode [112]. When the multifrequency spectrum becomes wider than the phase-matching bandwidth of the crystal, conversion efficiency starts to decrease, which was confirmed experimentally in the case of multimode RFL [108]. The same is true for a RDFB fiber laser with a quasi-continuous modeless spectrum. As Fig. 7.29a shows, the SHG efficiency is higher for random lasers with a narrower spectrum. In the case of RDFB laser with FBG, it exceeds the calculated efficiency for single-frequency radiation by ~ 2 times for powers <4 W when the fundamental spectrum is narrower than the QPM bandwidth. It can even exceed the theoretical limit. One can conclude that not only the spectral shapes of the lasers but also various radiation statistics influence the SHG power behavior observed in the experiment, whereas the theoretical curves are calculated under the assumption of Gaussian statistics.

Direct comparison of the SHG efficiency in the whole power range for the Raman laser configurations with and without an output cavity reflector demonstrates a higher efficiency for the random fiber lasers without the reflector. The type of highly reflective mirror also influences the conversion efficiency, resulting in ~ 2 times higher SHG power for the RDFB laser with FBG at laser powers < 4 W. Note that the SHG efficiency in the case of a conventional RFL may be improved by the combined optimization of the PF length and the output FBG reflectivity, taking into account the available pump power to obtain a narrow linewidth at a high output power.

Finally, one should note that generated quasi-continuous radiation with nm-wide spectrum filled with random frequencies has great potential for imaging and other applications that require low-coherence visible light [14].

7.3.6 Actively Q-Switched and Other Pulsed Regimes

Conventional RFLs operate in continuous-wave (CW) mode, but pulsed operation is also possible that is usually achieved by pulsed pumping (see, e.g., [90, 113]). All-fiber pulsed Raman fiber source emitting at 1254 nm was realized applying pulsed Yb:Bi codoped fiber laser as a pump and P-doped fiber as an active medium of the Raman converter [114]. The slope conversion efficiency of 70% and the peak power

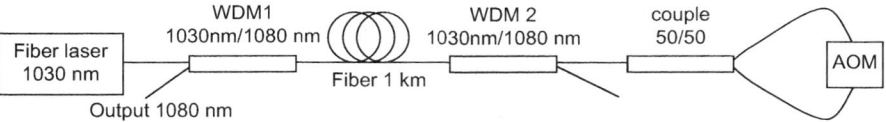

Fig. 7.31 Pulsed RRFL scheme

of 12 W were observed. More complicated intensity modulation mechanisms were exploited in YDFL-pumped RFLs with a common cavity [115, 116] that involve intracavity nonlinear interaction of pump and Stokes waves (via SRS-induced pump depletion), which may be treated as passive self- Q-switching for YDFL pump and pulsed pumping of RFL, correspondingly. Combined effect of Rayleigh scattering (RS) and stimulated Brillouin scattering (SBS) processes has been also explored as a nonlinear passive self-Q-switching mechanism in fiber lasers [96, 117], but pulsed operation in this case is characterized by unstable repetition rates, amplitudes, and pulse widths. Later on, stable continuous action exploiting RS-based feedback has been achieved in a long cavity-free RFL [15]. Various configurations and regimes of such Raman fiber laser with random distributed feedback are implemented (see [34] for a review). Absolute optical efficiencies of such lasers reach ~70% at relative quantum efficiencies of ~100% and output power level of several tens of watts [16, 17]. It has been also shown that they can be internally modulated providing modulated intensity patterns. Unlike conventional lasers where the length of the laser resonator is an essential parameter, in such cavity-free Raman fiber laser, the modulation frequency can be customized without distortions or self-mode-locking effects that has been tested for frequencies up to 12 GHz [26].

It is interesting to explore an opportunity to obtain actively Q-switched operation of the random Raman fiber laser without regular cavity with a goal of generating high-energy pulses in this scheme. A proof-of-principle experiments confirmed by theory of such a Q-switched Raman fiber laser have been performed in [27], which are described below.

The experiment was performed with Yb-doped fiber laser generating CW radiation at wavelength of 1030 nm and maximum optical power of 6 W used for pumping of a passive fiber. The pump radiation is coupled through WDM1 coupler (Fig. 7.31) to OFS TrueWave single-mode fiber having a normal dispersion and length of about 1 km. On the other end of the fiber line, there is another WDM2 coupler that separates pump (1030 nm) and Raman Stokes (1080 nm) waves into different ports. The Raman Stokes component with wavelength of 1080 nm is guided to a highly reflecting fiber loop mirror based on 50:50 coupler with acousto-optical modulator (AOM) inserted into it. All fiber ends were terminated by APC (angled cleaved) connectors to eliminate Fresnel reflection. Initially AOM is switched off, and there is no back reflection from the fiber loop mirror, preventing generation of the Raman Stokes wave in fiber line. After sending an electrical signal to AOM, it closes the fiber loop for a required time interval, and optical square pulse (pedestal) is formed from spontaneous emission at 1080 nm and starts propagating toward the pump laser. Amplified pulsed signal goes out through WDM1 and is registered with a high-speed photodetector. The time duration of electric pulse that drives AOM is varied in the range of 5–20 μs with repetition range of about 30 kHz. At that, the pump power at 1030 nm is always on, and maximum power is $P = 6$ W. The generated average power approaches half-Watt level that corresponds near ten percent efficiency.

Let us first explain basic principles of Q-switched operation in the proposed scheme. The laser has no regular cavity with well-defined length. The feedback is provided by the highly reflective loop mirror and Rayleigh backscattering in the passive fiber. The fiber loop mirror FLM is switched on/off by means of acousto-optic modulator (AOM) thus changing Q-factor of the distributed cavity. The desired generation of pulses at 1080 nm in the Q-switched regime occurs, when the laser becomes to be above the threshold by means of switching on the FLM, whereas it is below the threshold when the FLM is switched off. At the moment of FLM switching, the amplified spontaneous emission of the Stokes wave (at 1080 nm) is reflected by the FLM and starts propagating in backward direction as related to the pump wave. Due to high integral Raman gain, $G = 10 \log [\exp(gPL)] \sim 30$ dB, the Stokes pulse is amplified at propagation reaching high peak powers so that the pump power becomes depleted behind the pulse. At that, the front of pulse experiences much higher amplification (defined by undepleted pump) than its tail. As a result, the pulse is shortened at propagation. During one round trip time ($\tau = 2nL/c \sim 10$ μs), the Stokes pulse goes out of the fiber, while the pump power inside the fiber line is recovered tending laser to CW operation. So, after that, AOM should switch off the FLM thus suppressing lasing, and the pulse generation process can be repeated with the period longer than the round trip time. In this regime, maximum Stokes pulse energy is defined by the total pump energy stored in the cavity that may be converted in SRS process, i.e., $E_p = PLn/c \sim 30$ μJ in our case. Proof-of-principle experiments confirm this qualitative picture.

Figure 7.32a shows typical shape of output pulses in conditions of different pump levels. It is seen that relatively short pulses (1–2 μs) are formed with delay of about 5 μs against the electric pulse in correspondence with pulse propagation time along the fiber. It is noticeable that with the increasing pump, the pulse becomes shorter. Herewith the level of after-pulse pedestal (corresponding to CW lasing) is gradually depleting, and in time interval <17 μs there is no pedestal in case of maximum (6 W) pump power. After 17 μs, the pump wave restores, and the second round trip of CW Raman signal with some relaxation oscillations is registered. To prevent the second round trip of laser wave, the switching time of AOM has been changed from 20 to 5 μs (dashed line on Fig. 7.32a), in this case there is no CW lasing at all. Inset shows the output spectrum of the Q-switched laser at pump power of 6 W. It is relatively narrow (1–2 nm) at wavelength of ~1080 nm, and its shape is typical for Raman fiber lasers [67, 91].

The Q-switched operation can be described by the balance model. The powers of interacting pump wave $P(z, t)$ and signal (Stokes) wave $I(z, t)$ are described by the set of quasi-linear equations, where we neglect linear losses:

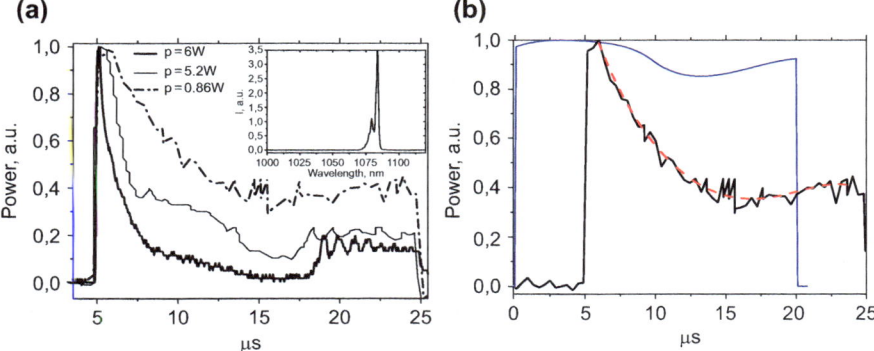

Fig. 7.32 (a) Pulse formation at different pump levels and output spectrum at 6 W pump power (inset); (b) experimental output pulse (*bold line*), approximation by polynomial at pump power 0.86 W (*dashed line*), calculated input signal (*thin line*)

$$(\partial_z + \beta_2\partial_t)\,I\,(z,t) = gP\,(z,t)\,I\,(z,t)\,, \tag{7.22}$$

$$(\partial_z + \beta_1\partial_t)P\,(z,t) = -\widetilde{g}P(z,t)I(z,t), \tag{7.23}$$

where $\beta_{1,2}$ is the inverse group velocities of these waves, $\widetilde{g} = g\lambda_I/\lambda_P$ is interaction constants (Raman gain coefficients), and $\lambda_{I,P}$ are their wavelengths.

For fibers with normal dispersion, $\beta_2 < \beta_1$ and general solution of these equations can be found by means of substitution

$$I = (\partial_z + \beta_1\partial_t)\frac{\Phi(z,t)}{\widetilde{g}}, \tag{7.24}$$

$$P = (\partial_z + \beta_2\partial_t)\frac{\Phi(z,t)}{g}. \tag{7.25}$$

As a result, wave equation with function $\exp(\Phi)$ is obtained, solution of which is function

$$\Phi = \ln(G(t - \beta_1 z) + F(t - \beta_2 z)), \tag{7.26}$$

where $F(t)$ and $G(t)$ are random positive monotonically increasing functions and dash means derivative. Solution of Cauchy problem with border conditions

$$P(0,t) = P_0(t), \quad I(0,t) = I_0(t) \tag{7.27}$$

can be found in general form:

$$F(t) = A\int_{-\infty}^{t} d\hbar \frac{\widetilde{g}I_0(\tau)}{\beta_1 - \beta_2}\exp\int_0^{\tau} dt'\left(\frac{\widetilde{g}I_0(t') + gP_0(t')}{\beta_1 - \beta_2}\right), \tag{7.28}$$

$$G(t) = A \int_{-\infty}^{t} dfi \frac{gP_0(\tau)}{\beta_1 - \beta_2} \exp \int_{0}^{\tau} dt' \left(\frac{\widetilde{g}I_0(t') + gP_0(t')}{\beta_1 - \beta_2} \right). \tag{7.29}$$

When pump and Stokes waves propagate in opposite directions, the solution has the following form:

$$I(z,t) = \frac{I_0(t + \beta_2 z) \exp \left(\int_{t-\beta_1 z}^{t+\beta_2 z} dt' \frac{\widetilde{g}I_0(t') - gP_0(t')}{\beta_1 + \beta_2} \right)}{1 + \frac{\widetilde{g}}{\beta_1 + \beta_2} \int_{t-\beta_1 z}^{t+\beta_2 z} dfi I_0(\tau) \exp \left(\int_{t-\beta_1 z}^{\tau} dt' \frac{\widetilde{g}I_0(t') - gP_0(t')}{\beta_1 + \beta_2} \right)}, \tag{7.30}$$

$$P(z,t) = \frac{P_0(t - \beta_1 z)}{1 + \frac{\widetilde{g}}{\beta_1 + \beta_2} \int_{t-\beta_1 z}^{t+\beta_2 z} dfi I_0(\tau) \exp \left(\int_{t-\beta_1 z}^{\tau} dt' \frac{\widetilde{g}I_0(t') - gP_0(t')}{\beta_1 + \beta_2} \right)}. \tag{7.31}$$

In this case direct solving of the equations is not possible, but inverse problem still can be solved: if one knows output pulse, input signal defined by switching of the loop mirror (stepwise in our case) could be recovered.

Numerical simulations were performed to calculate input signal shape using experimental data for output pulses. The following parameters in (7.30) were chosen to make simulations: $\beta_1 = 1.463/c$, $\beta_2 = 1.464/c$, $g = 1\,W^{-1}km^{-1}$, and c is the light velocity. Figure 7.32b shows the output pulse at pump power $P = 0.86\,W$ approximated by a polynomial. It is seen that after pulse formation, a slight decrease of the input signal is observed in calculations. Obviously the reason of the mismatch between calculated input signal and rectangular pedestal in experiment is pump depletion. In spite of that, the theoretical model describes quite good the time interval of $5\,\mu s$ corresponding to half round trip time.

Thus, it has been proposed and experimentally demonstrated an actively Q-switched random Raman fiber laser based on modulated feedback from fiber loop mirror at one end and distributed feedback via Rayleigh backscattering at the other end. Since Q-switched operation in this scheme is based on different laser states when the mirror is switched off (below lasing threshold) and switched on (above lasing threshold), the feedback from the output end should be as low as possible. From this point of view, the Rayleigh backscattering in fiber itself provides the lowest possible value. Nevertheless, the laser will also operate with point-action output reflector such as Fresnel reflection from normally cleaved end, but lower CW lasing threshold in this case will limit pump power for the Q-switch regime.

The demonstrated scheme is principally different from conventional Q-switched lasers, where pump energy is stored in the active media with rather long relaxation time of population on upper laser level (millisecond for rare earth-doped fiber). In the case of Raman laser with amplification based on the inelastic scattering of light on vibrations of atoms in silica structure, the energy cannot be stored in the media since the relaxation of excited vibrations is very fast (of picosecond scale). So, we can convert the pump power into the Stokes wave only locally. In case of backward propagation of the Stokes pulse, we utilize all the pump power distributed along the cavity, so the maximum generated pulse energy is defined by the product of pump power and fiber length. In our case, available pump energy is 30 μJ that already exceeds pulse energy level for conventional Q-switch master oscillator in MOPA fiber laser systems (see, e.g., [118]). It can be further increased by means of fiber lengthening, but not too much, as CW threshold defined by Rayleigh scattering feedback decreases with lengthening. Nevertheless, at least one order increase seems feasible. Herewith, the pulse duration is expected to be at least one order shorter than now. The pulse energy may be also increased by using large-mode area passive fibers for Raman amplification and single-mode or multimode ones with direct LD pumping like in [91]. One of the main advantages of Raman lasers is their broad range of operating wavelength which cannot be covered by conventional fiber lasers that is important for applications.

A transformation of linear scheme of actively Q-switched RFL (Fig. 7.31) into a ring one and switching of the acousto-optic modulator (AOM) synchronously with the pulse round trip provides mode-locked operation of it [119] with pulse duration down to 2 ns. Adding to active mode-locking, a saturable absorber based on nonlinear polarization evolution (NPE) brings to a pulse shortening of the output signal down to 50 ps [120]. At the same time, the output pulses of such laser could not be further compressed by the grating compressor since they have a noisy component.

In [121], authors demonstrated polarization modulation (PM) scheme to generate pulsed output of a random RFL (Fig. 7.33). It was shown that the output power of the PM RFL is sensitive to the state of polarization (SOP) for the Stokes light in a fiber loop that is used as a mirror, especially at the vicinity of the lasing threshold. Using polarized pump and by switching the SOP of the Stokes light in a fiber loop, the Raman gain is modulated, resulting in the modulation of the output power. The gain and the Q-value of the cavity could be also driven up and down by the SBS process. In [28], a feedback from Rayleigh scattering is used to define the random cavity modes and stimulated Brillouin scattering (SBS) to improve the Q-value of the cavity quickly and significantly, thus efficiently combining random lasing and Q-switching in a simple fiber configuration (Fig. 7.31). As a result, generation of giant pulses with peak power about 2.5 kW and pulse duration down to 42 ns is successfully demonstrated. A multiwavelength Q-switched random fiber laser with erbium-doped fiber as the gain medium was demonstrated in [122]. Q-switched pulses are generated with threshold as low as 27 mW by combining random cavity resonances and the Q-value modulation effect induced by stimulated Brillouin scattering.

In [123] a graphene-based pulse generation with cavity-free random fiber lasers is reported. In this scheme a CW random fiber laser (Fig. 7.34a) is polarized and polarization-rotated by a polarization rotator (PR), followed by intensity modulation with the graphene-coated D-shaped fiber (GDF). A rotation speed of PR is tunable from 1 kHz to 3 MHz. As soon as graphene is featured by a polarization dependent transmission, it serves as a natural polarizer, and a sinusoidal-like single polarized temporal wave after passing the GDF is observed. With the launched power increasing from 13.8 mW to 2.88 W, the sinusoidal-like waveform with full width at half maximum (FWHM) more than 180 ns is gradually compressed to be narrow

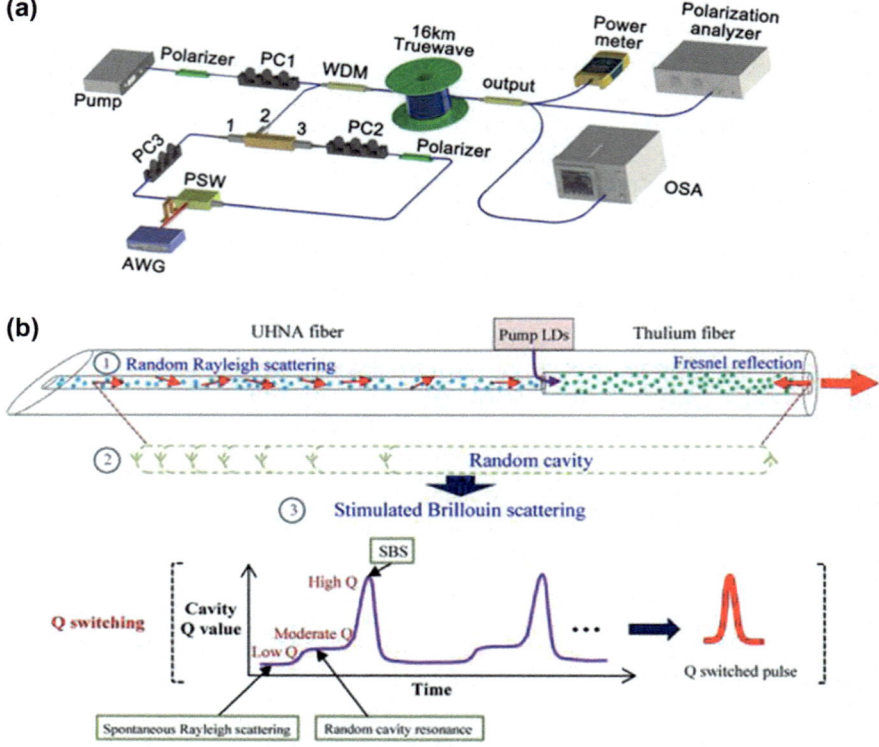

Fig. 7.33 (**a**) Experimental setup of the PMRFL (*PC* polarization controller, *PSW* polarization switch, *AWG* arbitrary waveform generator, *OSA* optical spectrum analyzer); (**b**) apparatus and operating principle of a random Q-switched fiber laser

pulses with duration ∼900 ps (Fig. 7.34b). With increasing repetition rate, the pulses narrow, and the sinusoidal-like pedestal is being suppressed dramatically. So the pulse width of the laser could be tuned with increasing the repetition rate of PR rotation as well as with varying the launched power.

7.3.7 Other Types of Random Fiber Lasers

7.3.7.1 DFB Fiber Lasers Based on Random FBGs

Recently a great deal of attention is attracted to fiber lasers with random distributed feedback (RDFB) based on artificial non-regular refractive index structures. In the literature, various types of RDFB fiber lasers are reported.

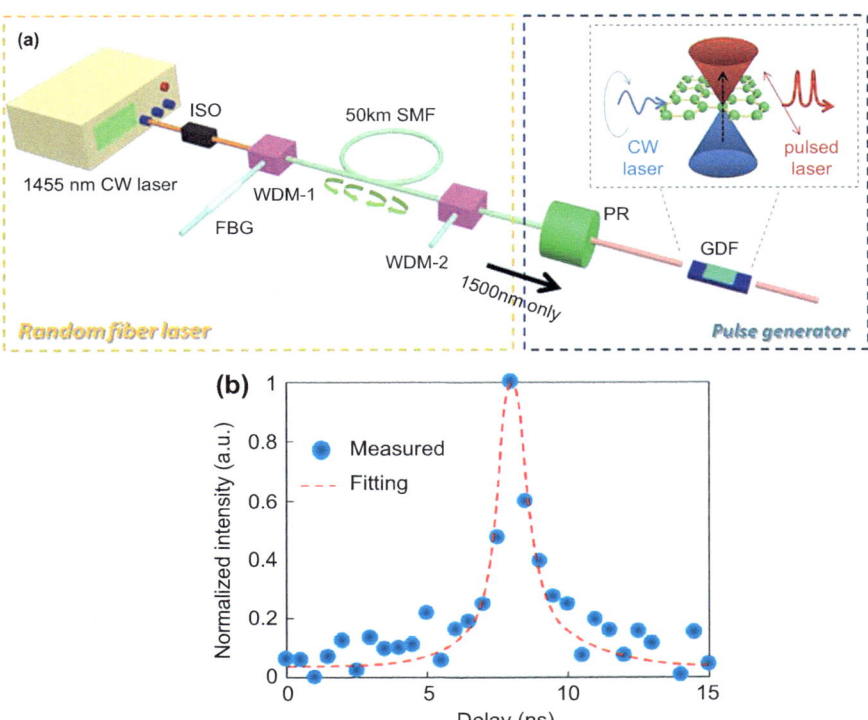

Fig. 7.34 (a) Schematic of the short pulse generation; (b) output pulse at 2.88 W

RDFB lasers are based on the array of FBGs, inscribed in an active or passive fiber with random intervals which form random phase shifts between the FBGs (subgratings) or on the continuous FBG with arbitrary phase shifts between subgratings. A number of subgratings counts up to several tens. A number of generated longitudinal modes depends on the length of random grating and the pump power. To the utmost, it is principally possible to obtain single-frequency regime.

So, in paper [124], a random Er-doped fiber laser generating at the wavelength of ~1.535 μm was proposed and demonstrated. In 150- cm-long fiber, an array of 7–31 gratings was inscribed, each of which has the length of 5 mm and the reflection coefficient of 7–8%. The distance between gratings was randomly chosen from 4.2 to 5.8 mm interval. The generation spectrum consisted of several competing longitudinal modes whose number varies from 2 to 7 depending on the gratings number (see Fig. 7.35).

The positions of the generated lines remain fixed, while power distribution between them changes at the pump power variation. The spectral width of the lines appears to be fairly independent on the number of gratings.

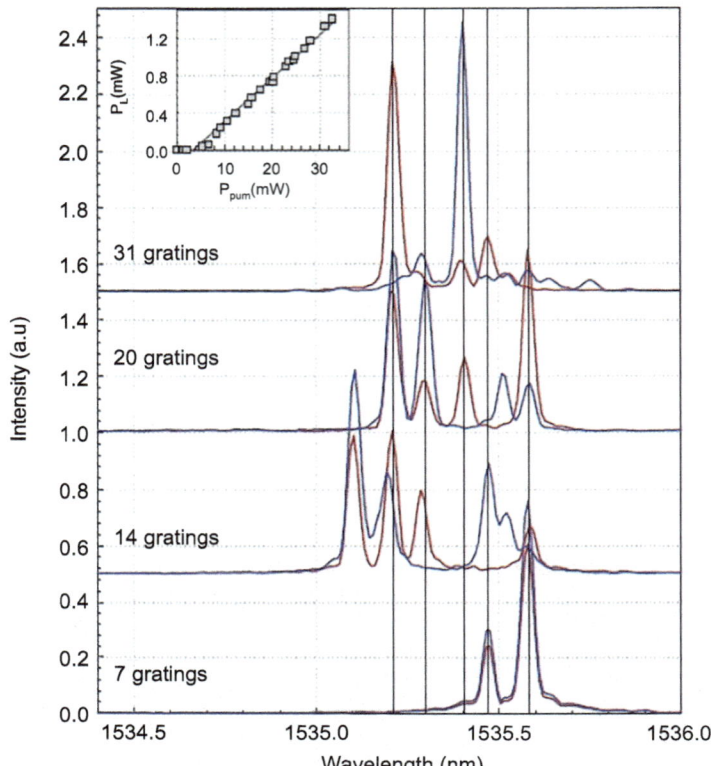

Fig. 7.35 Output spectra of the random lasers with 7, 14, 20, and 31 gratings for a pump of 20 mW (*red curve*) and 40 mW (*blue curve*). The inset: total emitted power as a function of the pump power for the laser with 31 gratings

Figure 7.36 shows the calculated spatial distributions of the intensities of the light for the four wavelengths. The field intensity that corresponds to the wavelength λ_4 decays exponentially along the RFBG, and the transmittance of the system at this wavelength is exponentially small. The distributions that correspond to the wavelengths λ_1, λ_2, and λ_3 have maxima. These distributions are associated with resonance modes. There are two curves that correspond to the wavelength λ_2. The lower curve is for an array of $N = 27$ gratings, and the upper one is for an array of $N = 31$ gratings. One can see that the intensity inside the cavity grows with the increase of the length of the array, but its shape changes only slightly.

In paper [125], another random Er-doped fiber laser generating at the wavelength of $\sim$1.534 μm is described. In contrast to [124], instead of the array with rather large intervals between gratings, a continuous grating with random phase shifts between subgratings of variable length is explored. Polarization-maintaining (PM) active fibers of 30 and 20 cm length were tested. The high degree of randomness

Fig. 7.36 The spatial distribution of the intensity of the light inside the grating array for four different wavelengths. The two plots for the wavelength λ_2 correspond to the arrays with 31 (upper curve) and 27 gratings (lower curve)

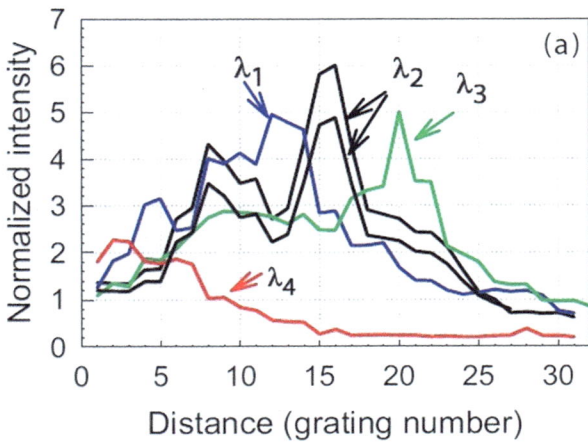

Fig. 7.37 Laser emission spectrum of the 30 cm RFL for a 120 mW pump power at 976 nm

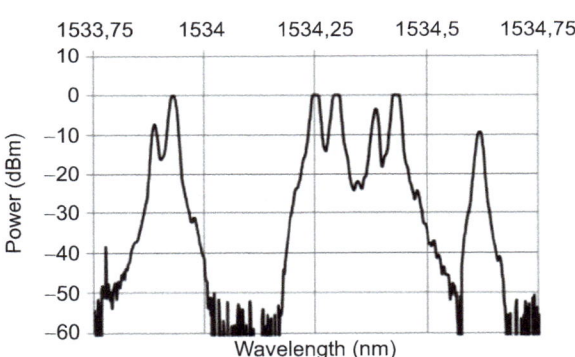

obtained in this scheme allows the RFLs to exhibit the lowest threshold (~3 mW) and linewidth (~0.5 pm) compared to previously reported schemes, e.g., [124]. It appeared that the lasing regime depends on the fiber length and pump power varying from multi- to single longitudinal mode. The 30 cm RFL is single mode up to a pump power of 10 mW, while the 20 cm RFL shows such features up to a pump power of 40 mW. The number of modes can reach 7 for the 30 cm RFL at higher pump power (see Fig. 7.37).

While the relative power distribution and the overall emitted power are both fluctuating in time, the wavelengths of the different lasing modes remain constant.

In paper [126], a tunable Er-doped fiber laser generating near ~1.55 μm is demonstrated. Random DFB is provided by a 10 cm piece of passive SMF-28 fiber, in which a random grating consisting of 100 CO_2 laser-irradiated points with random intervals between two adjacent points (0.5–1 mm) was inscribed. In this configuration the linewidth as low as 2.4 kHz was obtained with the central wavelength tuning within 0.5 nm.

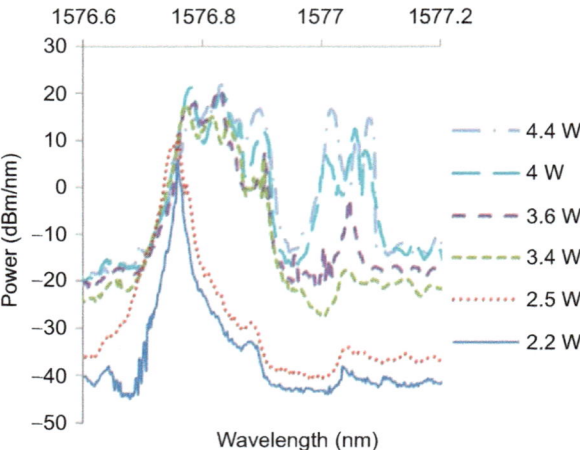

Fig. 7.38 Forward emission spectra of the random FBG-RRFL at different pump powers

In paper [127], a Raman fiber laser is demonstrated which generates at 1.576 μm due to the random DFB provided by a non-regular grating with 1500 random phase shifts inscribed in a 1-m-long passive fiber (random FBG).

Figure 7.38 shows the emitted forward spectra for various pump powers. Single line is observed until a pump power of 2.5 W, and then a low-intensity side mode appears ∼ 0.33 nm away from the peak wavelength, corresponding to another transmission dip. Up to seven dominant modes were observed at fixed frequencies, always on the long wavelength side of the originally emitted wavelength. The single-frequency linewidth amounts to 430 kHz. Contrary to the previous Er-doped random laser [125], the emitted spectrum is stable (being limited by pump power variations) and CW.

As compared to Raman fiber lasers operating via RDFB arising from Rayleigh backscattering on natural refractive-index fluctuations [15, 34], the implementation of random FBG allows to greatly shorten the length of passive fiber and to obtain the single-frequency regime of such random DFB Raman fiber laser.

Thus, the most promising type of RDFB laser is based on implementation of multiple random phase shifts into long homogeneous gratings. In [128] introduction of additional random variations of the grating amplitude is proposed to have an entirely random FBG. Designed RDFB laser is based on FBG with randomly changing phase and amplitude of the subgratings formed in a short (4 cm) Yb-doped fiber. The obtained output characteristics appear to be even better as compared to the regular DFB Yb-doped fiber laser of the same length [129].

Testing of the random FBG with laser diode pumping has demonstrated a single-frequency operation of such RDFB fiber laser. Figure 7.39a shows output spectra of the random laser measured by optical spectral analyzer (OSA) at different pump powers. Two separated peaks correspond to two orthogonal polarization modes. At low pump powers, the spectrum contained an addition of another longitudinal mode which disappeared with the pump power increase. The laser spectrum shifts

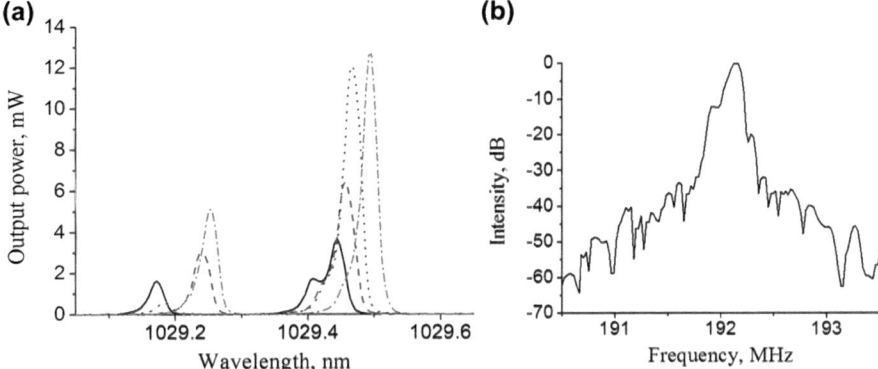

Fig. 7.39 (**a**) Random DFB laser spectra at the pump power of 145 mW (*black solid line*), 255 mW (*red dash line*), 310 mW (*blue dot line*), and 470 mW (*green dash dot line*). (**b**) RDFB and DFB lasers beat spectra at power of hundreds of microwatts

to longer wavelengths with increasing pump power, because of fiber heating. The measured linewidth corresponds to the OSA instrument width (0.02 nm).

For a precise laser linewidth measurement, a frequency beating method was employed. Figure 7.39b shows the radio-frequency spectra of beating of the RDFB laser with the single-frequency DFB laser whose frequency difference is about 190 MHz. The measured linewidth of beat spectra at −3 dB level was about 120 kHz that is comparable to the linewidth of beat spectra of two identical DFB lasers (about 100 kHz). So, the realized RDBF laser is single frequency with linewidth close to that for the regular DFB laser inscribed in the PANDA fiber.

Figure 7.40c shows the RDFB and DFB laser [129] output power as a function of the pump power. In the treated range, the dependences are close to linear, and at that the slope value is 3.5 times higher for random DFB laser (and maximum power, accordingly). The random DFB laser power reaches ∼25 mW at 250 mW pump power, i.e., absolute efficiency is about 10%, while the slope efficiency exceeds this value.

Figure 7.40a, b shows the measured sideward emission in RDFB laser and DFB laser [129] at the pump wavelength (black upper line) and lasing wavelength (red line for RDFB, blue line for DFB).

For the both lasers, the longitudinal intensity distribution at the lasing wavelength measured in the same setup has its maximum in the central region of the grating, because of light localization effect that is known to occur in regular grating with one π-shift, as well as in random gratings [124, 125, 130]. The lasing intensity is reduced at the points ± 10 mm apart from the central maximum by 15–20 dB in the DFB laser (for different sides) and by 12–15 dB in the RDFB laser.

The increase in the generation efficiency could be connected with wider localization area of the generated radiation for random grating in comparison to the grating with one π-shift, herewith the dip in saturated pump radiation distribution

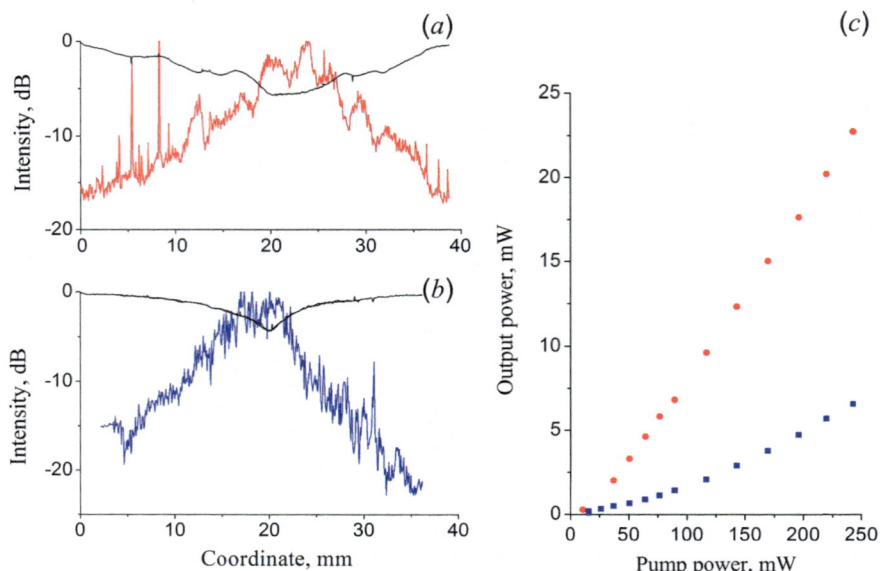

Fig. 7.40 Scattered light intensity distribution along the RDFB (**a**) and DFB (**b**) lasers at the pump (*black upper line*) and lasing wavelength (*red line* for RDFB, *blue line* for DFB). (**c**) The DFB (*blue squares*) and RDFB (*red circles*) laser output power as a function of pump power

is also wider and deeper that indicates a more effective conversion of pump to laser radiation in the case of random DFB laser.

In order to compare the experimental results with theory, a numerical simulation of random FBG spectra and longitudinal distributions of the light intensity inside the FBG at wavelengths corresponding to its transmission peaks was performed using software worked out in paper [131]. The simulation results qualitatively agree with the experimental data, although absolute value of intensity decrease in the experiment is lower than the calculated value that could be explained by imperfections of real grating.

The comparison of the calculated intensity distributions for different random gratings shows that for the FBG in which both the phases and amplitudes of the segments are random, the most pronounced localization with maximum intensity in the center is obtained resulting in single-frequency generation of the random DFB laser.

Thus, FBGs with randomly changing phase and amplitude are first inscribed in a PM Yb-doped fiber. Spectral characteristics of such FBGs are measured using single-frequency DFB Yb-doped fiber lasers with close wavelength. With an appropriate pumping, the random FBG provides lasing at ~1.03 μm due to the random DFB.

Recently Raman DFB fiber laser based on FBGs with randomly changing phase and amplitude was obtained. Random FBG consists of ~50 4-cm-long subgratings with 18 cm pieces of non-stripped fiber between them to decrease fragility. FBGs were inscribed using a reliable phase-mask writing scheme [128]. Between the inscription of adjacent subgratings, the fiber was moved by pulling it manually. The laser was pumped by a MOPA source at 1045.5 nm. Similar MOPA has been described in [66]. Single-frequency generation of around 1092.5 nm is observed at pump power of 2 W with much higher efficiency than that in [102]. The details will be published in prospective paper.

7.3.7.2 Random Fiber Lasers Based on Active Fibers with Dopants

Traditionally the Raman gain is used for development of random fiber lasers. Since the gain factor is proportional to pump power for the Raman process and the total backward Rayleigh scattering due to inhomogeneities within the glass structure of the fiber is extremely weak, the threshold of laser operation is rather high (usually several watts). The laser signal can be amplified in active fiber with different dopants (such as Er, Yb, and so on). In this case, in addition to the decrease of lasing threshold, the expansion of spectral range of laser generation can be expected because of wider gain profile for dopants gain than for Raman one. Such lasers utilizing random FBG for feedback have been described in previous paragraph.

The first demonstration of stable CW random lasing based on Rayleigh scattering feedback and erbium-doped fiber is presented in [132], wherein the spool of 50 km single-mode fiber is used as distributed reflector in half-opened cavity design and 980 nm pumped Er-doped fiber is used as gain. The similar half-open scheme with 2-m-long Er-doped fiber but pump wavelength of 1480 nm is investigated in more detail in [133]. The threshold power of the laser is as low as 10 mW, which is two orders lower in magnitude than that of the reported conventional random fiber laser amplified by Raman effect. The maximum pump efficiency is up to 14%, which is comparable to that of normal erbium-doped fiber lasers. Using spectral selection of two different FBGs, the dual-wavelength random lasing operation is demonstrated. Later on, the same authors used a tunable fiber Fabry-Perot interferometer filter combined with optical circulator-based loop mirror for spectral selection [134] (Fig. 7.41a). A broad wavelength tuning range up to 40 nm (from 1525 to 1565 nm) with maximum power fluctuation of only 1.5 dB is demonstrated (Fig. 7.41b). The tunable operation of random fiber laser based on 5-m-long ytterbium-doped fiber in similar configuration with tunable fiber Bragg grating is demonstrated in [135]. To provide random distributed feedback in the half-open cavity configuration, a 1-km-long single-mode fiber is used. The laser can be tuned continuously from 1040 to 1090 nm with maximum power fluctuations within 1 dB.

A random fiber laser with completely open cavity based on erbium-doped fiber without spectral filtration is realized in [136]. To obtain random lasing, two couples of pump lasers are used with wavelength of 1480 nm for 100-m-long Er-doped fiber and 1455 nm for two spoons of 50-km-long single-mode fibers (Fig. 7.42a). With increase of pump power, different transitions between chaotic and stable operation in the output spectrum were observed (Fig. 7.42b–e). In contrast to previously

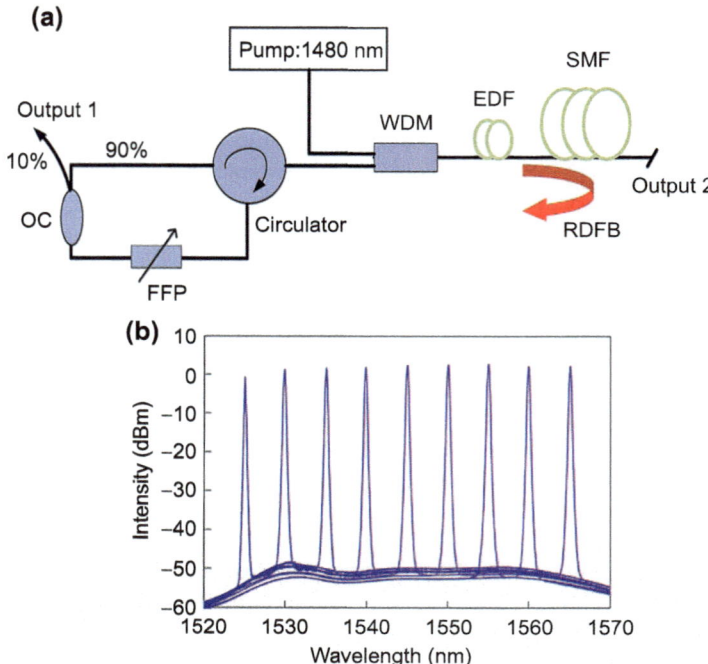

Fig. 7.41 (**a**) The scheme of tunable Random laser based on erbium-doped fiber. (**b**) Laser spectra at different wavelengths measured from output port 2 when pump power is 400 mW (From [134])

presented schemes, when only the Er-doped fiber is pumped (i.e., the active fiber is used as gain medium and the single-mode fiber performs a random distributed feedback), the instable chaotic random lasing takes place. The authors associate this instability with nonlinear effects, such as cascaded stimulated Brillouin scattering, and four-wave mixing, etc. Indeed, the cooperative Rayleigh-Brillouin scattering [31] is the main reason for such instability in random fiber lasers operating near the threshold, which could be suppressed far above the generation threshold. However, there is no switch from the unstable to the stable operation regime even at highest pump power for Er-doped fiber. Only when the second pair of pump lasers with wavelength of 1455 nm was used for creation of Raman amplification in the single-mode fiber the output spectrum is stabilized. As Raman pump increases, the chaotic output of Er doped is amplified in the single-mode fiber and experiences nonlinear effect that switches the laser output from chaotic (Fig. 7.42b) to stable state (Fig. 7.42c). In this case the Raman-pumped single-mode fibers perform a role of amplifiers for laser emission from the erbium-doped fiber where the spectrum is broadening due to nonlinear effects. When the Raman pump increases further, the threshold of random Raman lasing in the single-mode fibers is reached. In this case, the single-mode fiber begins to emit chaotic light as in random Raman fiber laser near the threshold. This chaotic emission from the single-mode fibers mixes with emission from the Er-doped, so the total output becomes chaotic again (Fig. 7.42d), which is the reason of transition from the stable to the chaotic states in the setup. The further increase of Raman pump leads to the output spectrum stabilization again (Fig. 7.42e). It should be noted that at high Raman pump power, the amplification is ensured by Raman gain rather than the erbium one. Also the authors checked the laser operation without Er-doped fiber. In addition to significant increase of lasing threshold in random pure RFL, they observed two-peak structure in spectrum (around 1555 and 1567 nm) which can be associated with two-peak structure of Raman gain, instead of single-peak spectrum in the scheme with additional Er-doped fiber.

Although the length of Er-doped fiber is rather long (100 m) in [136], the contribution to total random feedback from the active fiber is significantly smaller than that from 50-km-long single- mode fiber. So, this scheme as well as the previously presented ones can be called as hybrid schemes where amplification occurs in a short active fiber and the random feedback occurs in a separate long passive fiber. The scheme from [136] is close to the random RFL with in-cavity Er-doped amplifier. The hybrid schemes are closer to conventional lasers than to random ones. The only common feature with random lasers is modeless spectrum free of mode competition which results, e.g., in a flat tuning curve, just like in random RFLs. A realization of true random fiber laser based on an active fiber is complicated by short length leading to small value of random feedback. Indeed the concentration of dopants in commercially available active fibers is rather high aiming at short laser cavity design. Elongation of such fibers results in an increase of integral losses. This fact leads to undesirable effects such as signal reabsorption and generation instability (see, e.g., [137]).

The situation is quite different for recently developed bismuth-doped fibers. The typical concentration of active dopants in Bi-doped fibers is very low for the sake of elimination of bismuth (Bi) clustering [138]. The laser operation of random laser based on such active fiber in half-open cavity configuration is demonstrated in [139] (Fig. 7.43a) where distributed amplification and random feedback take place in the same piece of 150-m-long bismuth-doped fiber. The laser starts to operate at wavelength of 1420 nm selected by a FBG when pumped at wavelength of 1310 nm with rather low generation threshold for random lasers amounting to 150 mW. In contrast to the previously presented schemes, the contribution to total random feedback from bismuth-doped fiber was decisive due to higher Rayleigh scattering in comparison with ordinary telecommunication fiber. Despite this, the laser operation in the scheme with bismuth-doped fiber only was unstable. As mentioned above the main reason for such instability in random fiber lasers is cooperative Rayleigh-Brillouin scattering, which can be suppressed at high power due to spectral broadening. Indeed the stabilization of the generated spectrum was achieved by means of cavity lengthening with a passive single-mode fiber placed between FBG and the Bi-doped active fiber. The stability threshold decreases with increasing total cavity length: for 850-m-long phosphosilicate fiber, the stabilization occurs at pump powers above 1.7 W. It should be noted that in contrast to previously presented hybrid schemes where the single-mode fiber is added to the end of the

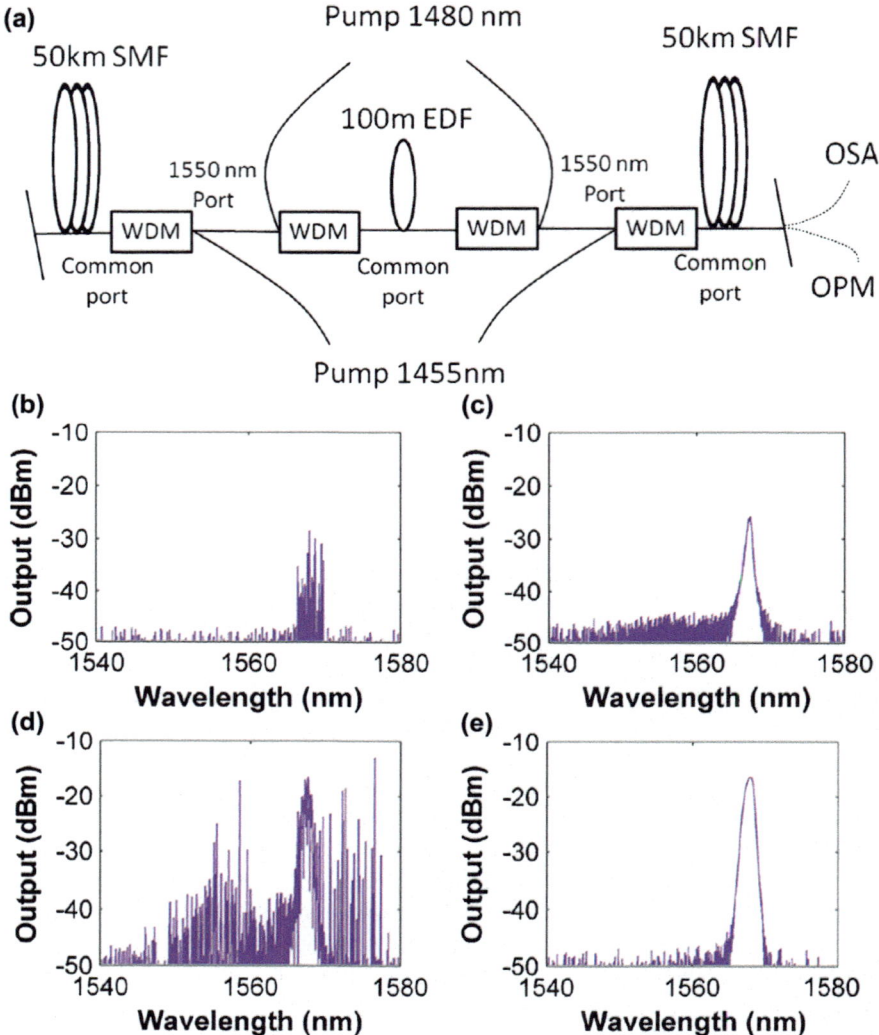

Fig. 7.42 (a) The scheme of random laser with opened cavity based on erbium-doped fiber with two different pumps. Output spectra for Raman pump power of (b) 0.25 W, (c) 0.79 W, (d) 2.2 W, and (e) 2.8 W at fixed power of erbium-doped fiber pump of 75 mW (From [136])

cavity with the purpose to increase feedback, in the scheme with Bi-doped fiber, the passive fiber is inserted between FBG and active fiber for the purpose of spectral broadening. Moreover, the Raman gain maximum in active and passive fibers at ∼1400 nm is well separated from generated Bi-related line at 1420 nm which means small effect of Raman process to the laser generation. To understand the features of developed random fiber laser, the authors compared it with the laser based

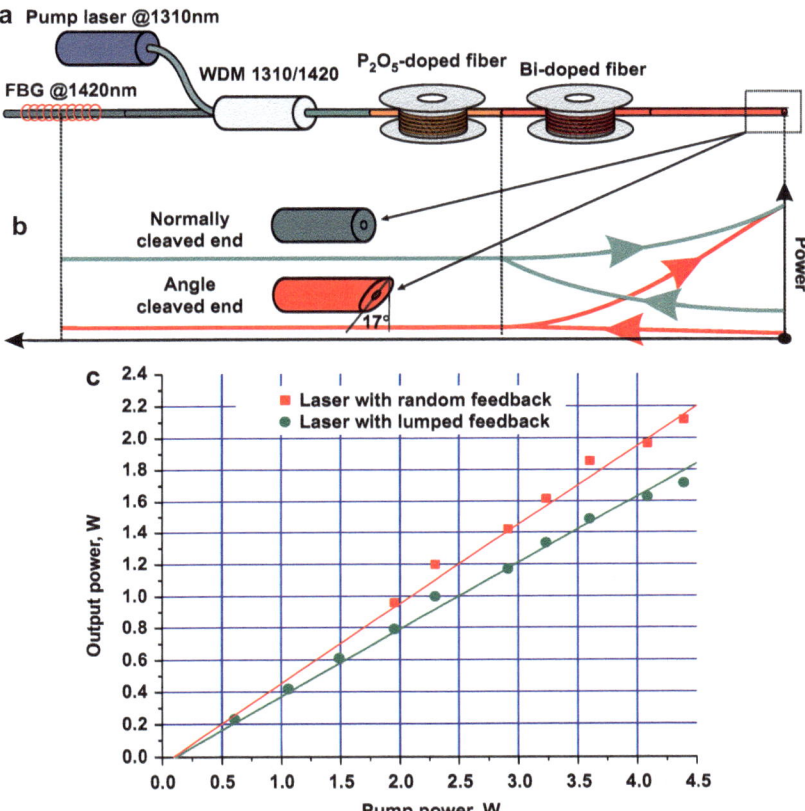

Fig. 7.43 (a) Scheme of bismuth-doped fiber lasers with random and lumped feedbacks and (b) corresponding qualitative longitudinal power distributions. (c) Output power as a function of the pump power for different cavity configurations (From [138])

on conventional cavity utilizing a lumped output reflector. For this purpose, the angle- cleaved fiber end was replaced by the normally cleaved one with the Fresnel reflection of $\sim$4% (Fig. 7.43a). In this case, the output power further decreases by 20% in comparison with the scheme with random feedback (Fig. 7.43c). Moreover, output spectra in two schemes are considerably different at the same level of output powers: for the random laser, the spectral linewidth is at least two times narrower than for traditional one (Fig. 7.44a). These facts can be explained in the terms of integral nonlinearity and longitudinal power distribution in the cavities $I(z)$. In random laser the most of power is concentrated in amplifying fiber due to the small value of random feedback in contrast to laser with lumped output reflector, where considerable value of power is present in non-amplifying fiber (Fig. 7.43b). Due to linear losses in passive fiber, the output power in traditional cavity is lower than that in the cavity with random feedback. Moreover, the authors theoretically and experimentally shown that the mechanisms of spectral broadening for random lasers are the same as for traditional ones based on active fibers, self- and cross-phase modulation, the last one can be usually neglected. In the case of self-phase modulation effect, the output spectral width is defined by nonlinear B-integral $B = \gamma A_{eff} \int I(z)dz$, where γ is the nonlinear Kerr coefficient, A_{eff} is the effective core area, and integration is performed over the cavity bypass. Thus, for the case of cavity with lumped mirror, where the power has more uniform distribution over all the cavity, the nonlinear effects appear to be stronger than those for random laser. For the laser with random distributed feedback, the main part of power is concentrated near the output end thus resulting in the lowest B-integral at the highest output power. For example, Fig. 7.44b shows the lasing spectra in the scheme with lumped reflector at output power of 2.2 W and with random feedback at output power of 0.2 W. Although the output power in these schemes differ by an order, the spectra are very similar due to similar value of B-integral.

In conclusion, random fiber lasers based on fiber with dopants have similar properties as for random RFLs: modeless spectrum generation, flat tuning curve, and reduced impact of nonlinear effects. But at the same time, the laser generation threshold in active fibers is sufficiently smaller than for Raman gain medium. The application of different dopants allows one to obtain new spectral ranges of laser generation that expands the range of possible applications of random lasers. For example, the demonstrated narrowband Bi-doped random fiber laser is attractive for applications such as sensing and telecommunications (e.g., as a modeless pump source for distributed Raman amplifiers) as well as bio-imaging taking into account a possibility of its efficient conversion into the visible range. Note that the lasers based on Bi-doped fibers with different host compositions and appropriate pumping can generate in an extremely broad range of wavelengths, 1150–1775 nm [139].

7.4 Advanced Applications of Random RFLs

7.4.1 Random Fiber Lasers for Optical Communications

Random fiber lasers have various applications in optical telecommunications. First of all, as conventional light sources, they can be used for generation of multiple ITU grid wavelengths [23, 39, 40, 140] (may be other refs to our and other papers on multiwavelengths), tunable sources of high-power radiation for Raman pumping or for optical signal across the telecom bands [22, 42, 43, 134, 141, 142], broadband or narrow linewidth light sources [24]. However, an ultra-long laser cavity implemented in optical fiber can also be seen as a transmission medium. In particular, ultra-long random DFB fiber laser generating in a 100 km scale fiber spans [15] is an interesting technique for enhancing distributed Raman amplification. Ultra-long quasi-lossless transmission systems [33, 143–148] may provide more uniform distribution of signal power along the fiber compared to standard

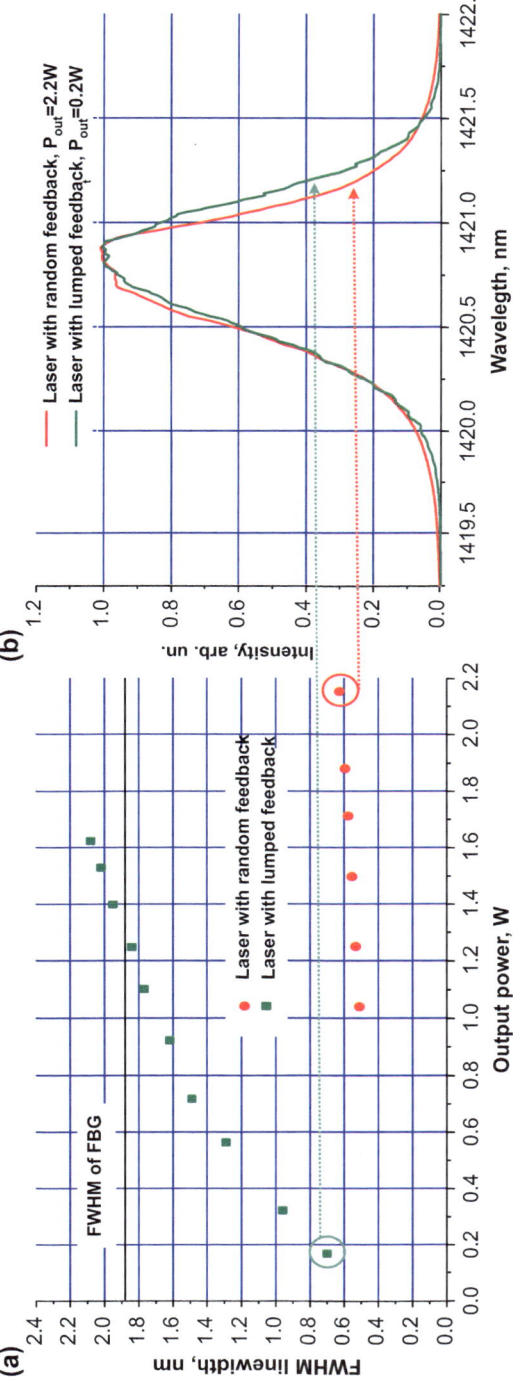

Fig. 7.44 (**a**) The comparison of power evolution of lasing linewidth (FWHM) in different schemes. (**b**) The spectra comparison in the scheme with lumped reflector at output power of 2.2 W and with random feedback at output power of 0.2 W (From [138])

backward Raman amplification. This technique can give a very low (± 0.8 dB) signal power excursion over an 80 km span. Minimization of the noise generation requires that the distributed Raman amplification would exactly counteract the fiber losses all along the transmission path, maintaining the signal power level at a near-constant value. Another important application of quasi-lossless or symmetric communication schemes with a constant and/or symmetric signal power distribution is in the emerging techniques of compensation or mitigation of nonlinear transmission impairments [149–154]. Ultra-long lasers allow to achieve spatio-spectral transparency over long fiber spans and over broad bandwidth even with only single-frequency laser pumps. Optimization of the pumping source wavelengths and powers can lead to distributed amplification schemes combining first- and second-order Raman pumping and providing reduced signal power excursion over a wide spatial-spectral range of 60 km $\times$ 80 nm in C + L bands [148].

Ultra-long fiber lasers are closely linked with the cascaded Raman amplification that is an attractive method for improving performance in long-span unrepeatered optical communications links [155–161]. Two main applications are typically considered: cascaded pump delivery for remotely pumped erbium-doped fiber amplifier (EDFA) [157, 158] and higher-order Raman amplification for improved optical noise figure performance [159]. High-order cascaded Raman pumping architectures present a natural and attractive mean to increase the pump power delivered to remote optically pumped amplifiers, by directly launching a high power at 1480 nm (first-order pumping scheme) or by utilizing high-order cascaded Raman processes for the generation of 1480 nm pump power inside the transmission fiber. An important issue affecting higher-order pumping is signal double Rayleigh scattering (DRS) that results in the noise enhancement. The longest demonstrated unrepeatered fiber links typically exploit a remotely pumped erbium-doped post-amplifier located in the interval of 30–70 km from the transmitter terminal. A remote optically pumped amplifier is typically placed at 100–140 km distance from the receiver terminal. One may anticipate that an even more advanced power delivery to remote optically pumped amplifiers solutions will be developed. Second- and third-order Raman pumping schemes have demonstrated significant margin improvements compared to traditional first-order pumping. Sixth-order cascaded Raman amplification has been demonstrated for the first time in [160] and has been shown to provide more than 10 dB improvement in receiver sensitivity, an increase of more than 1 dB compared to third-order pumping [159]. A simple Yb laser, operating at 1091 nm, may be used as the primary and the only active pump source. However, relatively high power required for pumping higher-order schemes limits their practical use. The performance improvement provided by higher-order cascaded counter-propagating Raman amplifiers is based on the fact that the final pump is effectively created through Raman cascades within the transmission fiber span itself, and as a result, the propagating signal is amplified further from the receiver terminal reducing signal power excursion.

One of the attractions of the distributed Raman amplification is an improvement of the transmission performance, compared with lumped amplification schemes based on Er-doped fiber amplifiers (EDFA). Conventional ultra-long Raman fiber laser based amplification technique (using second-order pump and two fiber Bragg gratings) [143–146] is based on exploiting a symmetrical bi-directional pumping with equal forward and backward pump powers to minimize the power excursion and amplified spontaneous emission noise. The relative disadvantage of such schemes is the use of forward pumping that is challenging at the moment due to relative intensity noise (RIN) being transferred from noisy pumps to the signal that dramatically degrades the performance improvement due to low-signal power variations and low ASE noise.

The RIN transfer is one of the main drawbacks in applications of the distributed Raman amplifiers in telecommunication. The RIN transfer leads to additional noise arising at signal wavelength. A numerical study of the RIN transfer function in a forward-pumped random fiber laser for telecom applications was performed in [35]. The power balance model is expanded with equations for the spectral density of the noise intensity for the pump and the Stokes waves. The RIN transfer function was calculated for different pump powers and fiber lengths. It was shown that the noise transfer decreases while the random laser cavity length or the pump power increases. The authors compared RIN transfer properties of a 100-km-long random fiber laser with those of the 100-km-long Raman fiber laser with the conventional FBG-based cavity. The averaged level of the RIN transfer function was found to be similar in both lasers. However, in the case of the modeless random fiber laser, there are no oscillations in the noise transfer function. Note that in conventional Raman fiber laser, the frequency noise originated not only from the beating of the distinct longitudinal modes of the laser cavity but also from the pump-to-Stokes wave noise transfer [162]. The dumping of the oscillations in random fiber laser could play a positive role in applications of the random fiber lasers for distributed amplification. Quite opposite to intuitive expectations, random fiber lasers could have noise performances which are, in some cases, superior to commercially available cavity-based solutions. Saxena et al. [163] recently showed that Rayleigh feedback actually suppresses noise at higher frequencies.

Using backward pumping it is possible to find a trade-off between an increase in the signal power excursion and ASE noise and reduction of RIN transfer. Distributed amplification using third-order backward pumping resulting into transmission performance improvement was demonstrated in [159]. Random lasing can be realized in a long-distance fiber span, making it intrinsically suitable for the distributed Raman amplification. Combination of backward pumping and a Rayleigh backscattering effect can lead to random distributed feedback (DFB) fiber lasing. In [164–169], Studies were performed on the transmission performance improvement and RIN mitigation using random DFB fiber laser based Raman amplification schemes of first- and second-order pumping. The second-order bi-directionally/backward-pumped Raman amplification schemes based on a random distributed feedback fiber laser configuration have been compared in [164] with conventional Raman amplification schemes. It was demonstrated in 10×116 Gb/s DP-QPSK WDM transmission that bi-directionally pumped random DFB fiber laser based amplification scheme can achieve low-signal power excursion improving transmission performance [164]. Using the random DFB laser based scheme with a signal power variation of about 4 dB, it was demonstrated in [164] that an extended reach of 7915 km in a recirculating loop experiment is feasible, compared to 5000 km using conventional backward first-order Raman pumping. No RIN increase was observed on the signal using random DFB laser transmission scheme even with bi-directional pumping. Unrepeated DP-QPSK transmission over 353 km SMF using random DFB fiber laser amplification was demonstrated in [169]. The random DFB fiber laser operating at 1455 nm was used as a pump in distributed Raman amplifier, and performance of the scheme was compared with performance of the conventional bi-directional first-order and second-order distributed Raman amplifiers in [170]. Random lasing was created in a 100 km fiber transmission fiber span. After comparison of various configurations of random DFB fiber laser, it was found that the transmission scheme based on the random laser exhibits $\sim$2.3 dB lower effective noise figure as compared to conventional bi-directional first-order and second-order distributed Raman amplifiers. Hybrid distributed Raman amplification combining random fiber laser based second-order pumping and low-noise LD based first-order pumping was used to improve system performance in [171].

Recent development in applications of random DFB fiber lasers in telecom includes the use of new types of pumping sources. Novel Bi-doped low-RIN fiber laser operating at 1450 nm in both CW and mode-locked regimes was used in [172] for forward pumping in 83 km telecom link using distributed Raman amplification with very low-signal power variations. Highly depolarized backward pump at 1366 nm with RIN of approximately -120 dB/Hz was used to create a random distributed feedback fiber laser at 1455 nm. Random fiber laser based distributed amplification played an important role in optimizing transmission link for the application on the mid-link optical phase conjugation [149, 151, 151, 173, 174] and the application on the multichannel digital back propagation [175, 176]. All the transmission schemes based on random DFB fiber laser amplification can be further optimized to satisfy the link requirement for specific nonlinearity compensation techniques. This makes random DFB fiber laser amplification schemes an important emerging technique in future telecommunication applications.

7.4.2 Random Fiber Lasers for Distributed Sensing

Due to possibility to obtain generation in very long fiber spans, the concept of random fiber laser based distributed amplification could be productively combined with a concept of distributed sensing. As a proof-of-principle experiment, Brillouin optical time-domain analysis (BOTDA) reflectometer was demonstrated in [177], with incorporated random fiber laser as an amplifier to increase the signal-to-noise ratio. These systems allow continuous sensing of temperature over 122 km fiber span with $\pm$2 m spatial resolution and $\pm$14 °C temperature accuracy.

Further extension of sensing distance is mainly restricted by the relatively high pump power requirement (multiple-Watt level typically) and detrimental nonlinear effects. To address this issue, hybrid pumping configuration was proposed in paper [171]. The system is pumped by a low-noise laser-diode based Raman fiber laser which acts as a first-order Raman pump for the random fiber laser. This allows to increase OSNR value on the far end of the sensing system and reach the sensing length of 154.4 km with 5 m spatial resolution and ±1.4 °C temperature accuracy. The demonstrated Brillouin optical time-domain reflectometer has a longest repeater-less sensing range.

Another configuration implementing only one pump source is described in [178]. In this system, a ring configuration is used to extend the system length: sensing over 142.2 km of fiber with 5 m spatial resolution and ±1.5 °C temperature accuracy was achieved. Although the sensing distance is 12 km less than that in paper [171], only one pump source is needed in [178].

The random fiber laser based distributed amplifier can also be incorporated in other types of distributed optical fiber sensing systems for extension of sensing distance. As an impressive demonstration, a phase-sensitive optical time-domain reflectometry (Φ-OTDR) with 175 km sensing range and ±12.5 m spatial resolution was demonstrated in paper [179]. In this work, authors utilize the combination of co-pumped random fiber laser based distributed amplifier with counter-pumped first-order Raman amplification and Brillouin amplification. The 1455 nm random lasing generated under the forward pumping provides distributed amplification for the probe pulse all the way along the first 88 km fiber span, making the probe pulse efficiently amplified via the Brillouin gain in the range from 88 to 138 km. The first-order Raman amplification takes over the signal boosting in the third segment of 37 km length. In such a configuration, the longest repeater-less distributed phase-sensitive time-domain reflectometer was demonstrated.

For the forward-pumped random fiber laser based distributed sensing systems, the sensing distance and OSNR could be further improved by using the high-order random fiber lasers [180]. Indeed, the position of maximum of longitudinal distribution in the random fiber laser shifts toward longer distances for the high-order Stokes waves [34]. Also, with the forward-pumped random fiber laser based distributed amplification, the lasing power distribution can be adjusted by tuning the reflectivity of the point-based mirrors located on the fiber span end [71] or by polarization control [62], making the manipulation of the signal power distribution viable. Various random fiber laser configurations have been utilized for remote ultra-long sensing in [181, 182].

More details on sensing applications of RRFL can be found in [183].

7.5 Conclusions and Perspectives

Thus, the performed analysis shows that the output characteristics of Raman fiber lasers with random distributed feedback reached to the moment are comparable and even outperform in many aspects those for conventional Raman fiber lasers, namely:

- High polarization extinction ratio (PER) has been achieved in the random RFLs based on polarization-maintaining fiber with linearly polarized pumping and the obtained PER values close to that for the pump ($\sim$25 dB) not only for the first Stokes but also for higher Stokes orders.
- The RRFL scheme based on short ($\leq$0.5 km) PM fiber exhibits also unique power/efficiency performance of cascaded generation of multiple Stokes orders: the absolute optical efficiency of pump-to-Stokes conversion exceeds 87% (corresponding to quantum efficiency of about 92%) for the first order, and it slightly decreases to 75% for the second order and to 70% for the third Stokes order, in correspondence with increasing attenuation in longer fiber (1 km) used. Using tunable pump source (within $\sim$50 nm) is possible to obtain 300 nm tuning range (1.1–1.4 nm) of the cascaded RRFL.
- The laser bandwidth grows with increasing Stokes order, but it is almost independent on power within the generation range of each order with variation 0.8–1.3 nm, 1.4–2.3 nm, and 2.4–3.3 nm for the first, second, and third Stokes waves, accordingly. At that, the generated spectrum remains to be sufficiently narrower than the Raman gain profile of corresponding order. An analytical model has been developed describing well the generated power and spectrum for all orders of the cascaded RRFL.
- Short-wavelength RFL operation (950–980 nm) has been achieved through the direct pumping of multimode graded-index fibers by high-power multimode LDs (915–940 nm) in both configurations (with linear cavity or random distributed feedback), at that the configuration with random distributed feedback demonstrates stronger beam cleaning effect with output beam close to diffraction limited.
- Short-wavelength operation via second harmonic generation (SHG) of RRFL radiation has been also explored. Direct comparison of the SHG efficiency in the whole power range for the Raman laser configurations with and without an output cavity reflector demonstrates a higher efficiency for the random fiber laser without the output reflector. The type of highly reflective mirror also influences the conversion efficiency, resulting in $\sim$2 times higher SHG power for the RDFB laser with narrowband FBG at laser powers <4 W. The generated quasi-continuous radiation with a nm-wide spectrum filled with random frequencies has great potential for speckle-free imaging and other applications that require low-coherence visible light.
- Actively Q-switched (by AOM) random Raman fiber laser has been also studied for the first time demonstrating efficient generation of $\sim$1 µs pulses, combination of AOM mode-locking with NPE-based self-amplitude modulation resulted in generation of 50 ps pulses, and other types of pulsed RRFLs have been also demonstrated.
- Random lasing appears possible not only in passive fibers with Raman gain but also in active fibers, the most interesting results have been obtained with bismuth-doped fibers characterized by relatively large length ($\sim$150 m), and narrowband ($\sim$0.4 nm) linewidth demonstrated at high output powers ($\sim$2 W), that is, several times narrower than that in the same fiber with conventional cavity at comparable

power. Random FBGs in active and passive fibers have been explored instead of Rayleigh backscattering to achieve single-frequency regime of random fiber lasers.

The demonstrated unique performance of random fiber lasers opens the door to their application in advanced technologies, such as long-distance transmission and remote sensing, low-coherence IR, and visible sources for speckle-free bio-imaging, laser displays, and others. Simple all-fiber scheme with direct LD pumping is one of the technological challenges for further improvement of their performance.

Overall, distributed feedback random fiber laser is an interesting and growing area of science and technology at the interface of the laser physics, theory of the disordered systems, optical engineering, fiber optics, theoretical physics, and various practical applications, and we anticipate new important developments in this field in the nearest future.

Acknowledgements The authors acknowledge financial support by the Russian Science Foundation (grant 14-22-00118).

References

1. Noginov, M.A.: Solid-State Random Lasers. Springer Series in Optical Sciences, vol. 105. Springer, New York (2005)
2. Cao, H.: J. Phys. A **38**(49), 10497 (2005). doi:10.1088/0305-4470/38/49/004, http://iopscience.iop.org/article/10.1088/0305-4470/38/49/004/meta
3. Wiersma, D.S.: Nat. Phys. **4**(5), 359 (2008). doi:10.1038/nphys971
4. Zhai, T., Zhang, X., Pang, Z., Su, X., Liu, H., Feng, S., Wang, L.: Nano Lett. **11**(10), 4295 (2011). doi:10.1021/nl2023096, PMID: 21879718
5. Cheng, S.H., Yeh, Y.C., Lu, M.L., Chen, C.W., Chen, Y.F.: Opt. Express **20**(S6), A799 (2012). doi:10.1364/OE.20.00A799, http://www.opticsexpress.org/abstract.cfm?URI=oe-20-106-A799
6. Ghofraniha, N., Viola, I., Maria, F.D., Barbarella, G., Gigli, G., Conti, C.: Laser Photon. Rev. **7**(3), 432 (2013). doi:10.1002/lpor.201200105
7. Baudouin, Q., Mercadier, N., Guarrera, V., Guerin, W., Kaiser, R.: Nat. Phys. **9**(6), 357 (2013). doi:10.1038/nphys2614, http://www.nature.com/nphys/journal/v9/n6/abs/nphys2614.html#supplementary-information
8. Song, Q., Xiao, S., Xu, Z., Liu, J., Sun, X., Drachev, V., Shalaev, V.M., Akkus, O., Kim, Y.L.: Opt. Lett. **35**(9), 1425 (2010). doi:10.1364/OL.35.001425, http://ol.osa.org/abstract.cfm?URI=ol-35-9-1425
9. Wang, C.S., Chang, T.Y., Lin, T.Y., Chen, Y.F.: Sci. Rep. **4**, 6736 (2014). doi:10.1038/srep06736
10. Gather, M.C., Yun, S.H.: Nat. Photon. **5**(7), 406 (2011). doi:10.1038/nphoton.2011.99, http://www.nature.com/nphoton/journal/v5/n7/abs/nphoton.2011.99.html#supplementary-information
11. Polson, R.C., Vardeny, Z.V.: J. Opt. **12**(2), 024010 (2010). doi:10.1088/2040-8978/12/2/024010, http://stacks.iop.org/2040-8986/12/i=2/a=024010
12. Folli, V., Ghofraniha, N., Puglisi, A., Leuzzi, L., Conti, C.: Sci. Rep. **3**, 2251 (2013). doi:10.1038/srep02251

13. Hokr, B.H., Bixler, J.N., Cone, M.T., Mason, J.D., Beier, H.T., Noojin, G.D., Petrov, G.I., Golovan, L.A., Thomas, R.J., Rockwell, B.A., Yakovlev, V.V.: Nat. Commun. **5**, 4356 (2014). doi:10.1038/ncomms5356, Supplementary information available for this article at http://www.nature.com/ncomms/2014/140711/ncomms5356/suppinfo/ncomms5356_S1.html

14. Redding, B., Choma, M.A., Cao, H.: Nat. Photon. **6**(6), 355–359 (2012). doi:10.1038/nphoton.2012.90, http://www.nature.com/nphoton/journal/v6/n6/abs/nphoton. 2012.90.html#supplementary-information

15. Turitsyn, S.K., Babin, S.A., El-Taher, A.E., Harper, P., Churkin, D.V., Kablukov, S.I., Ania-Castañón, J.D., Karalekas, V., Podivilov, E.V.: Nat. Photon. **4**(4), 231 (2010). doi:10.1038/nphoton.2010.4

16. Vatnik, I.D., Churkin, D.V., Podivilov, E.V., Babin, S.A.: Laser Phys. Lett. **11**(7), 075101 (2014). doi:10.1088/1612-2011/11/7/075101, http://stacks.iop.org/1612-202X/11/i=7/a= 075101

17. Zhang, H., Zhou, P., Xiao, H., Xu, X.: Laser Phys. Lett. **11**(7), 075104 (2014). doi:10.1088/1612-2011/11/7/075104, http://stacks.iop.org/1612-202X/11/i=7/a=075104

18. Wang, Z., Wu, H., Fan, M., Zhang, L., Rao, Y., Zhang, W., Jia, X.: IEEE J. Sel. Top. Quantum Electron. **21**(1), 10 (2015). doi:10.1109/JSTQE.2014.2344293

19. Babin, S.A., Vatnik, I.D., Laptev, A.Y., Bubnov, M.M., Dianov, E.M.: Opt. Express **22**(21), 24929 (2014). doi:10.1364/OE.22.024929, http://www.opticsexpress.org/abstract.cfm?URI= oe-22-21-24929

20. Du, X., Zhang, H., Wang, X., Zhou, P., Liu, Z.: Opt. Lett. **41**(3), 571 (2016). doi:10.1364/OL.41.000571, http://ol.osa.org/abstract.cfm?URI=ol-41-3-571

21. Churkin, D.V., Kolokolov, I.V., Podivilov, E.V., Vatnik, I.D., Nikulin, M.A., Vergeles, S.S., Terekhov, I.S., Lebedev, V.V., Falkovich, G., Babin, S.A., Turitsyn, S.K.: Nat. Commun. **2**, 6214 (2015). doi:10.1038/ncomms7214, Supplementary information available for this article at http://www.nature.com/ncomms/2015/150203/ncomms7214/suppinfo/ncomms7214_ S1.html

22. Babin, S.A., El-Taher, A.E., Harper, P., Podivilov, E.V., Turitsyn, S.K.: Phys. Rev. A **84**(2), 021805 (2011). doi:10.1103/PhysRevA.84.021805, http://link.aps.org/doi/10.1103/ PhysRevA.84.021805

23. El-Taher, A.E., Harper, P., Babin, S.A., Churkin, D.V., Podivilov, E.V., Ania-Castanon, J.D., Turitsyn, S.K.: Opt. Lett. **36**(2), 130 (2011). doi:10.1364/OL.36.000130, http://ol.osa.org/ abstract.cfm?URI=ol-36-2-130

24. Sugavanam, S., Tarasov, N., Shu, X., Churkin, D.V.: Opt. Express **21**(14), 16466 (2013). doi:10.1364/OE.21.016466, http://www.opticsexpress.org/abstract.cfm?URI=oe-21- 14-16466

25. Babin, S.A., Dontsova, E.I., Kablukov, S.I.: Opt. Lett. **38**(17), 3301 (2013). doi:10.1364/OL.38.003301, http://ol.osa.org/abstract.cfm?URI=ol-38-17-3301

26. Bravo, M., Fernandez-Vallejo, M., Lopez-Amo, M.: Opt. Lett. **38**(9), 1542 (2013). doi:10.1364/OL.38.001542, http://ol.osa.org/abstract.cfm?URI=ol-38-9-1542

27. Kuznetsov, A.G., Podivilov, E.V., Babin, S.A.: Laser Phys. Lett. **12**(3), 035102 (2015). doi:10.1088/1612-2011/12/3/035102, http://stacks.iop.org/1612-202X/12/i=3/a=035102

28. Tang, Y., Xu, J.: Sci. Rep. **5**, 9338 (2015). doi:10.1038/srep09338, http://www.nature.com/ articles/srep09338#supplementary-information

29. Nakazawa, M.: J. Opt. Soc. Am. **73**(9), 1175 (1983). doi:10.1364/JOSA.73.001175, http:// www.osapublishing.org/abstract.cfm?URI=josa-73-9-1175

30. Barnoski, M.K., Rourke, M.D., Jensen, S.M., Melville, R.T.: Appl. Opt. **16**(9), 2375 (1977). doi:10.1364/AO.16.002375, http://ao.osa.org/abstract.cfm?URI=ao-16-9-2375

31. Fotiadi, A.A., Kiyan, R.V.: Opt. Lett. **23**(23), 1805 (1998). doi:10.1364/OL.23.001805, http:// ol.osa.org/abstract.cfm?URI=ol-23-23-1805

32. Chen, C.J., Lee, H.K., Cheng, Y.J.: In: Optical Fiber Communication Conference, Technical Digest, Atlanta, 23 Mar, p. TuC2. OSA (2003). http://www.osapublishing.org/abstract.cfm? URI=OFC-2003-TuC2

33. Turitsyn, S.K., Ania-Castañón, J.D., Babin, S.A., Karalekas, V., Harper, P., Churkin, D., Kablukov, S.I., El-Taher, A.E., Podivilov, E.V., Mezentsev, V.K.: Phys. Rev. Lett. 103(13), 133901 (2009). doi:10.1103/PhysRevLett.103.133901, http://link.aps.org/doi/10.1103/PhysRevLett.103.133901
34. Turitsyn, S.K., Babin, S.A., Churkin, D.V., Vatnik, I.D., Nikulin, M., Podivilov, E.V.: Phys. Rep. 542(2), 133 (2014). doi:http://dx.doi.org/10.1016/j.physrep.2014.02.011, http://www.sciencedirect.com/science/article/pii/S0370157314001215
35. no, J.N., Alcon-Camas, M., Ania-Castañón, J.: Opt. Express 20(24), 27376 (2012). doi:10.1364/OE.20.027376, http://www.opticsexpress.org/abstract.cfm?URI=oe-20-24-27376
36. Vatnik, I.D., Churkin, D.V., Babin, S.A.: Opt. Express 20(27), 28033 (2012). doi:10.1364/OE.20.028033, http://www.opticsexpress.org/abstract.cfm?URI=oe-20-27-28033
37. Ambartsumyan, R.V., Basov, N.G., Kryukov, P.G., Letokhov, V.S.: Sov. Phys. JETP 24(3), 481 (1967). http://www.jetp.ac.ru/cgi-bin/r/index/e/24/3/p481?a=list
38. Babin, S.A., Karalekas, V., Podivilov, E.V., Mezentsev, V.K., Harper, P., Ania-Castañón, J.D., Turitsyn, S.K.: Phys. Rev. A 77, 033803 (2008). doi:10.1103/PhysRevA.77.033803, http://link.aps.org/doi/10.1103/PhysRevA.77.033803
39. Sugavanam, S., Yan, Z., Kamynin, V., Kurkov, A.S., Zhang, L., Churkin, D.V.: Opt. Express 22(3), 2839 (2014). doi:10.1364/OE.22.002839, http://www.opticsexpress.org/abstract.cfm?URI=oe-22-3-2839
40. Pinto, A.M.R., Frazão, O., Santos, J.L., Lopez-Amo, M.: Appl. Phys. B 99(3), 391 (2010). doi:10.1007/s00340-010-4037-3, http://link.springer.com/article/10.1007/s00340-010-4037-3
41. Pinto, A., Frazao, O., Santos, J., Lopez-Amo, M.: J. Lightwave Technol. 29(10), 1482 (2011). doi:10.1109/JLT.2011.2134833, https://www.osapublishing.org/jlt/abstract.cfm?uri=jlt-29-10-1482
42. Sarmani, A.R., Zamiri, R., Bakar, M.H.A., Azmi, B.Z., Zaidan, A.W., Mahdi, M.A.: J. Eur. Opt. Soc.-Rapid Publ. 6, 11043 (2011). doi:10.2971/jeos.2011.11043, https://www.jeos.org/index.php/jeos\relax$\@@underline\hbox{~}\mathsurround\z@$\relaxrp/article/view/11043
43. Zhu, Y.Y., Zhang, W.L., Jiang, Y.: IEEE Photon. Technol. Lett. 25(16), 1559 (2013). doi:10.1109/LPT.2013.2271044
44. Bednyakova, A.E., Gorbunov, O.A., Politko, M.O., Kablukov, S.I., Smirnov, S.V., Churkin, D.V., Fedoruk, M.P., Babin, S.A.: Opt. Express 21(7), 8177 (2013). doi:10.1364/OE.21.008177, http://www.opticsexpress.org/abstract.cfm?URI=oe-21-7-8177
45. Babin, S.A., Churkin, D.V., Ismagulov, A.E., Kablukov, S.I., Podivilov, E.V.: J. Opt. Soc. Am. B 24(8), 1729 (2007). doi:10.1364/JOSAB.24.001729, http://josab.osa.org/abstract.cfm?URI=josab-24-8-1729
46. Churkin, D., Smirnov, S.: Opt. Commun. 285(8), 2154 (2012). doi:http://dx.doi.org/10.1016/j.optcom.2011.12.082, http://www.sciencedirect.com/science/article/pii/S003040181101488X
47. Gorbunov, O.A., Sugavanam, S., Churkin, D.V.: Opt. Lett. 40(8), 1783 (2015). doi:10.1364/OL.40.001783, http://ol.osa.org/abstract.cfm?URI=ol-40-8-1783
48. Randoux, S., Suret, P.: Opt. Lett. 37(4), 500 (2012). doi:10.1364/OL.37.000500, http://ol.osa.org/abstract.cfm?URI=ol-37-4-500
49. Sugavanam, S., Tarasov, N., Wabnitz, S., Churkin, D.V.: Laser Photon. Rev. 9(6), L35 (2015). doi:10.1002/lpor.201500012, http://dx.doi.org/10.1002/lpor.201500012
50. Churkin, D.V., Smirnov, S.V., Podivilov, E.V.: Opt. Lett. 35(19), 3288 (2010). doi:10.1364/OL.35.003288, http://ol.osa.org/abstract.cfm?URI=ol-35-19-3288
51. Churkin, D.V., Gorbunov, O.A., Smirnov, S.V.: Opt. Lett. 36(18), 3617 (2011). doi:10.1364/OL.36.003617, http://ol.osa.org/abstract.cfm?URI=ol-36-18-3617
52. Randoux, S., Dalloz, N., Suret, P.: Opt. Lett. 36(6), 790 (2011). doi:10.1364/OL.36.000790, http://ol.osa.org/abstract.cfm?URI=ol-36-6-790

53. Walczak, P., Randoux, S., Suret, P.: Opt. Lett. **40**(13), 3101 (2015). doi:10.1364/OL.40.003101, http://ol.osa.org/abstract.cfm?URI=ol-40-13-3101
54. Liu, W., Kuang, W., Huang, L., Zhou, P.: Laser Phys. Lett. **12**(4), 045104 (2015). http://stacks.iop.org/1612-202X/12/i=4/a=045104
55. Smirnov, S.V., Churkin, D.V.: Opt. Express **21**(18), 21236 (2013). doi:10.1364/OE.21.021236, http://www.opticsexpress.org/abstract.cfm?URI=oe-21-18-21236
56. Suret, P., Koussaifi, R.E., Tikan, A., Evain, C., Randoux, S., Szwaj, C., Bielawski, S.: Nat. Commun. **7**, 13136 (2016). doi:10.1038/ncomms13136, http://www.nature.com/articles/ncomms13136#supplementary-information
57. Walczak, P., Randoux, S., Suret, P.: Phys. Rev. Lett. **114**, 143903 (2015). doi:10.1103/PhysRevLett.114.143903, http://journals.aps.org/prl/abstract/10.1103/PhysRevLett.114.143903
58. Randoux, S., Walczak, P., Onorato, M., Suret, P.: Phys. Rev. Lett. **113**, 113902 (2014). doi:10.1103/PhysRevLett.113.113902, http://journals.aps.org/prl/abstract/10.1103/PhysRevLett.113.113902
59. Gorbunov, O.A., Sugavanam, S., Churkin, D.: Opt. Express **22**(23), 28071 (2014). doi:10.1364/OE.22.028071, http://www.opticsexpress.org/abstract.cfm?URI=oe-22-23-28071
60. Gorbunov, O.A., Sugavanam, S., Vatnik, I.D., Churkin, D.V.: Opt. Express **24**(17), 19417 (2016). doi:10.1364/OE.24.019417, http://www.opticsexpress.org/abstract.cfm?URI=oe-24-17-19417
61. Sugavanam, S., Zulkifli, M.Z., Churkin, D.V.: Laser Phys. **26**(1), 015101 (2016). doi:10.1088/1054-660X/26/1/015101, http://stacks.iop.org/1555-6611/26/i=1/a=015101
62. Wu, H., Wang, Z.N., Churkin, D.V., Vatnik, I.D., Fan, M.Q., Rao, Y.J.: Laser Phys. Lett. **12**(1), 015101 (2015). doi:10.1088/1612-2011/12/1/015101, http://stacks.iop.org/1612-202X/12/i=1/a=015101
63. Du, X., Zhang, H., Wang, X., Zhou, P., Liu, Z.: Photon. Res. **3**(2), 28 (2015). doi:10.1364/PRJ.3.000028, http://www.osapublishing.org/prj/abstract.cfm?URI=prj-3-2-28
64. Zlobina, E.A., Kablukov, S.I., Babin, S.A.: Opt. Lett. **40**(17), 4074 (2015). doi:10.1364/OL.40.004074, http://ol.osa.org/abstract.cfm?URI=ol-40-17-4074
65. Fan, M., Wang, Z., Wu, H., Sun, W., Zhang, L.: IEEE Photon. Technol. Lett. **27**(3), 319 (2015). doi:10.1109/LPT.2014.2370644
66. Babin, S.A., Zlobina, E.A., Kablukov, S.I., Podivilov, E.V.: Sci. Rep. **6**, 22625 (2016). doi:10.1038/srep22625, http://www.nature.com/articles/srep22625
67. Feng, Y., Taylor, L.R., Calia, D.B.: Opt. Express **17**(26), 23678 (2009). doi:10.1364/OE.17.023678, http://www.opticsexpress.org/abstract.cfm?URI=oe-17-26-23678
68. Doutté, A., Suret, P., Randoux, S.: Opt. Lett. **28**(24), 2464 (2003). doi:10.1364/OL.28.002464, http://ol.osa.org/abstract.cfm?URI=ol-28-24-2464
69. Skubchenko, S.A., Vyatkin, M.Y., Gapontsev, D.V.: IEEE Photon. Technol. Lett. **16**(4), 1014 (2004). doi:10.1109/LPT.2004.823725
70. Stolen, R.H., Lee, C., Jain, R.K.: J. Opt. Soc. Am. B **1**(4), 652 (1984). doi:10.1364/JOSAB.1.000652, http://josab.osa.org/abstract.cfm?URI=josab-1-4-652
71. Wu, H., Wang, Z., Fan, M., Zhang, L., Zhang, W., Rao, Y.: Opt. Express **23**(2), 1421 (2015). doi:10.1364/OE.23.001421, http://www.opticsexpress.org/abstract.cfm?URI=oe-23-2-1421
72. Zhang, H., Zhou, P., Wang, X., Du, X., Xiao, H., Xu, X.: Opt. Express **23**(13), 17138 (2015). doi:10.1364/OE.23.017138, https://www.osapublishing.org/oe/abstract.cfm?&uri=oe-23-13-17138
73. Wang, X., Zhou, P., Miao, Y., Zhang, H., Xiao, H., Wang, X., Liu, Z.: J. Opt. Soc. Am. B **31**(10), 2476 (2014). doi:10.1364/JOSAB.31.002476, http://josab.osa.org/abstract.cfm?URI=josab-31-10-2476

74. Vatnik, I.D., Zlobina, E.A., Kablukov, S.I., Babin, S.A.: Opt. Express **25**(3), 2703 (2017). doi:10.1364/OE.25.002703, https://www.osapublishing.org/oe/abstract.cfm?uri=oe-25-3-2703

75. Zhang, L., Jiang, H., Yang, X., Pan, W., Feng, Y.: Opt. Lett. **41**(2), 215 (2016). doi:10.1364/OL.41.000215, http://ol.osa.org/abstract.cfm?URI=ol-41-2-215

76. Zhang, L., Jiang, H., Yang, X., Pan, W., Cui, S., Feng, Y.: Sci. Rep. **7**, 42611 (2017). doi:10.1038/srep42611, http://www.nature.com/articles/srep42611

77. Babin, S., Churkin, D., Podivilov, E.: Opt. Commun. **226**(1–6), 329 (2003). doi:10.1016/j.optcom.2003.09.003, http://www.sciencedirect.com/science/article/pii/S0030401803019230

78. Kolokolov, I.V., Lebedev, V.V., Podivilov, E.V., Vergeles, S.S.: J. Exp. Theor. Phys. **119**(6), 1134 (2014). doi:10.1134/S1063776114120061, http://dx.doi.org/10.1134/S1063776114120061

79. Dianov, E., Prokhorov, A.: IEEE J. Sel. Top. Quantum Electron. **6**(6), 1022 (2000). doi:10.1109/2944.902151

80. Archambault, J.L., Grubb, S.G.: J. Lightwave Technol. **15**(8), 1378 (1997). doi:10.1109/50.618351

81. Dianov, E.M., Bufetov, I.A., Bubnov, M.M., Grekov, M.V., Vasiliev, S.A., Medvedkov, O.I.: Opt. Lett. **25**(6), 402 (2000). doi:10.1364/OL.25.000402, http://ol.osa.org/abstract.cfm?URI=ol-25-6-402

82. Richardson, D.J., Nilsson, J., Clarkson, W.A.: J. Opt. Soc. Am. B **27**(11), B63 (2010). doi:10.1364/JOSAB.27.000B63, http://josab.osa.org/abstract.cfm?URI=josab-27-11-B63

83. Koponen, J., Söderlund, M., Hoffman, H.J., Kliner, D.A.V., Koplow, J.P., Hotoleanu, M.: Appl. Opt. **47**(9), 1247 (2008). doi:10.1364/AO.47.001247, http://ao.osa.org/abstract.cfm?URI=ao-47-9-1247

84. Baek, S.H., Roh, W.B.: Opt. Lett. **29**(2), 153 (2004). doi:10.1364/OL.29.000153, http://ol.osa.org/abstract.cfm?URI=ol-29-2-153

85. Terry, N.B., Alley, T.G., Russell, T.H.: Opt. Express **15**(26), 17509 (2007). doi:10.1364/OE.15.017509, https://www.osapublishing.org/oe/abstract.cfm?uri=oe-15-26-17509

86. Yao, T., Harish, A.V., Sahu, J.K., Nilsson, J.: Appl. Sci. **5**(4), 1323 (2015). doi:10.3390/app5041323, http://www.mdpi.com/2076-3417/5/4/1323

87. Glick, Y., Fromzel, V., Zhang, J., Dahan, A., Ter-Gabrielyan, N., Pattnaik, R.K., Dubinskii, M.: Laser Phys. Lett. **13**(6), 065101 (2016). doi:10.1088/1612-2011/13/6/065101, http://stacks.iop.org/1612-202X/13/i=6/a=065101

88. Codemard, C.A., Dupriez, P., Jeong, Y., Sahu, J.K., Ibsen, M., Nilsson, J.: Opt. Lett. **31**(15), 2290 (2006). doi:10.1364/OL.31.002290, http://ol.osa.org/abstract.cfm?URI=ol-31-15-2290

89. Codemard, C.A., Ji, J., Sahu, J.K., Nilsson, J.: In: Tankala, K. (ed.) Fiber Lasers VII: Technology, Systems, and Applications, San Francisco, 23 Jan. Proceedings of SPIE, vol. 7580, pp. 75801N–7 (2010). doi:10.1117/12.845606, http://dx.doi.org/10.1117/12.845606

90. Yao, T., Nilsson, J.: J. Opt. Soc. Am. B **31**(4), 882 (2014). doi:10.1364/JOSAB.31.000882, http://josab.osa.org/abstract.cfm?URI=josab-31-4-882

91. Kablukov, S.I., Dontsova, E.I., Zlobina, E.A., Nemov, I.N., Vlasov, A.A., Babin, S.A.: Laser Phys. Lett. **10**(8), 085103 (2013). doi:10.1088/1612-2011/10/8/085103, http://stacks.iop.org/1612-202X/10/i=8/a=085103

92. Abdullina, S.R., Babin, S.A., Vlasov, A.A., Kablukov, S.I.: Quantum Electron. **35**(9), 857 (2005). doi:10.1070/QE2005v035n09ABEH003440

93. AuYeung, J., Yariv, A.; J. Opt. Soc. Am. **69**(6), 803 (1979). doi:10.1364/JOSA.69.000803, http://www.opticsinfobase.org/abstract.cfm?URI=josa-69-6-803

94. Bufetov, I.A., Bubnov, M.M., Neustruev, V.B., Mashinsky, V.M., Shubin, A.V., Grekov, M.V., Guryanov, A.N., Khopin, V.F., Dianov, E.M., Prokhorov, A.M.: Laser Phys. **11**(1), 130 (2001). http://pleiades.online/cgi-perl/search.pl?type=abstract&name=lasphys&number=1&year=1&page=130

95. Kurukitkoson, N., Sugahara, H., Turitsyn, S.K., Egorova, O.N., Kurkov, A.S., Paramonov, V.M., Dianov, E.M.: Electron. Lett. **37**(21), 1281 (2001). doi:10.1049/el:20010848
96. Ravet, G., Fotiadi, A.A., Blondel, M., Mégret, P.: Electron. Lett. **40**(9), 528 (2004). doi:10.1049/el:20040390
97. Rodgers, B.C., Russell, T.H., Roh, W.B.: Opt. Lett. **24**(16), 1124 (1999). doi:10.1364/OL.24.001124, http://ol.osa.org/abstract.cfm?URI=ol-24-16-1124
98. Zlobina, E.A., Kablukov, S.I., Skvortsov, M.I., Nemov, I.N., Babin, S.A.: Laser Phys. Lett. **13**(3), 035102 (2016). doi:10.1088/1612-2011/13/3/035102, http://stacks.iop.org/1612-202X/13/i=3/a=035102
99. Gloge, D., Marcatili, E.A.J.: Bell Syst. Tech. J. **52**(9), 1563 (1973). doi:10.1002/j.1538-7305.1973.tb02033.x
100. Kawakami, S., Tanji, H.: Electron. Lett. **19**(3), 100 (1983). doi:10.1049/el:19830072
101. Mizunami, T., Djambova, T.V., Niiho, T., Gupta, S.: J. Lightwave Technol. **18**(2), 230 (2000). doi:10.1109/50.822797
102. Zlobina, E.A., Kablukov, S.I., Wolf, A.A., Dostovalov, A.V., Babin, S.A.: Opt. Lett. **42**(1), 9 (2017). doi:10.1364/OL.42.000009, https://www.osapublishing.org/ol/abstract.cfm?uri=ol-42-1-9
103. Zlobina, E.A., Kablukov, S.I., Wolf, A.A., Nemov, I.N., Dostovalov, A.V., Tyrtyshnyy, V.A., Myasnikov, D.V., Babin, S.A.: Opt. Express **25**(11), 12581 (2017). doi:10.1364/OE.25.012581, https://www.osapublishing.org/oe/abstract.cfm?uri=oe-25-11-12581
104. Feng, Y., Huang, S., Shirakawa, A., ichi Ueda, K.: Opt. Express **12**(9), 1843 (2004). doi:10.1364/OPEX.12.001843, http://www.opticsexpress.org/abstract.cfm?URI=oe-12-9-1843
105. Georgiev, D., Gapontsev, V.P., Dronov, A.G., Vyatkin, M.Y., Rulkov, A.B., Popov, S.V., Taylor, J.R.: Opt. Express **13**(18), 6772 (2005). doi:10.1364/OPEX.13.006772, http://www.opticsexpress.org/abstract.cfm?URI=oe-13-18-6772
106. Surin, A.A., Borisenko, T.E., Larin, S.V.: Opt. Lett. **41**(11), 2644 (2016). doi:10.1364/OL.41.002644, http://ol.osa.org/abstract.cfm?URI=ol-41-11-2644
107. Feng, Y., Taylor, L.R., Calia, D.B.: Opt. Express **17**(21), 19021 (2009). doi:10.1364/OE.17.019021, http://www.opticsexpress.org/abstract.cfm?URI=oe-17-21-19021
108. Kablukov, S.I., Babin, S.A., Churkin, D.V., Denisov, A.V., Kharenko, D.S.: Opt. Express **17**(8), 5980 (2009). doi:10.1364/OE.17.005980, http://www.opticsexpress.org/abstract.cfm?URI=oe-17-8-5980
109. Dontsova, E.I., Kablukov, S.I., Vatnik, I.D., Babin, S.A.: Opt. Lett. **41**(7), 1439 (2016). doi:10.1364/OL.41.001439, http://ol.osa.org/abstract.cfm?URI=ol-41-7-1439
110. Babin, S.A., Churkin, D.V., Kablukov, S.I., Rybakov, M.A., Vlasov, A.A.: Opt. Express **15**(13), 8438 (2007). doi:10.1364/OE.15.008438, http://www.opticsexpress.org/abstract.cfm?URI=oe-15-13-8438
111. Boyd, G.D., Kleinman, D.A.: J. Appl. Phys. **39**(8), 3597 (1968). doi:10.1063/1.1656831, http://link.aip.org/link/?JAP/39/3597/1
112. Ducuing, J., Bloembergen, N.: Phys. Rev. **133**(6A), A1493 (1964). doi:10.1103/PhysRev.133.A1493, http://link.aps.org/doi/10.1103/PhysRev.133.A1493
113. Baac, H.W., Uribe-Patarroyo, N., Bouma, B.E.: Opt. Express **22**(6), 7113 (2014). doi:10.1364/OE.22.007113, http://www.opticsexpress.org/abstract.cfm?URI=oe-22-6-7113
114. Kurkov, A.S., Dvoyrin, V.V., Paramonov, V.M., Medvedkov, O.I., Dianov, E.M.: Laser Phys. Lett. **4**(6), 449 (2007). doi:10.1002/lapl.200710009, http://stacks.iop.org/1612-202X/4/i=6/a=009
115. Grukh, D.A., Levchenko, A.E., Kurkov, A.S., Paramonov, V.M.: Quantum Electron. **35**(5), 442 (2005). doi:10.1070/QE2005v035n05ABEH003852, http://stacks.iop.org/1063-7818/35/i=5/a=A09
116. Zhao, Y., Jackson, S.D.: Opt. Lett. **31**(6), 751 (2006). doi:10.1364/OL.31.000751, http://ol.osa.org/abstract.cfm?URI=ol-31-6-751

117. Chernikov, S.V., Zhu, Y., Taylor, J.R., Gapontsev, V.P.: Opt. Lett. **22**(5), 298 (1997). doi:10.1364/OL.22.000298, http://ol.osa.org/abstract.cfm?URI=ol-22-5-298
118. Kuznetsov, A.G., Babin, S.A.: Laser Phys. **20**(5), 1266 (2010). doi:10.1134/S1054660X10100038
119. Yang, X., Zhang, L., Jiang, H., Fan, T., Feng, Y.: Opt. Express **23**(15), 19831 (2015). doi:10.1364/OE.23.019831, http://www.opticsexpress.org/abstract.cfm?URI=oe-23-15-19831
120. Kuznetsov, A.G., Kharenko, D.S., Podivilov, E.V., Babin, S.A.: Opt. Express **24**(15), 16280 (2016). doi:10.1364/OE.24.016280, http://www.opticsexpress.org/abstract.cfm?URI=oe-24-15-16280
121. Wu, H., Wang, Z., He, Q., Fan, M., Li, Y., Sun, W., Zhang, L., Li, Y., Rao, Y.: Laser Phys. Lett. **13**(5), 055101 (2016). doi:10.1088/1612-2011/13/5/055101, http://stacks.iop.org/1612-202X/13/i=5/a=055101
122. Wang, S., Lin, W., Chen, W., Li, C., Yang, C., Qiao, T., Yang, Z.: Appl. Phys. Express **9**(3), 032701 (2016). doi:10.7567/APEX.9.032701, http://stacks.iop.org/1882-0786/9/i=3/a=032701
123. Yao, B.C., Rao, Y.J., Wang, Z.N., Wu, Y., Zhou, J.H., Wu, H., Fan, M.Q., Cao, X.L., Zhang, W.L., Chen, Y.F., Li, Y.R., Churkin, D., Turitsyn, S., Wong, C.W.: Sci. Rep. **5**, 18526 (2015). doi:10.1038/srep18526, http://www.ncbi.nlm.nih.gov/pmc/articles/PMC4685245/
124. Lizárraga, N., Puente, N.P., Chaikina, E.I., Leskova, T.A., Méndez, E.R.: Opt. Express **17**(2), 395 (2009). doi:10.1364/OE.17.000395, http://www.opticsexpress.org/abstract.cfm?URI=oe-17-2-395
125. Gagné, M., Kashyap, R.: Opt. Express **17**(21), 19067 (2009). doi:10.1364/OE.17.019067, http://www.opticsexpress.org/abstract.cfm?URI=oe-17-21-19067
126. Li, Y., Lu, P., Bao, X., Ou, Z.: Opt. Lett. **39**(8), 2294 (2014). doi:10.1364/OL.39.002294, http://ol.osa.org/abstract.cfm?URI=ol-39-8-2294
127. Gagné, M., Kashyap, R.: Opt. Lett. **39**(9), 2755 (2014). doi:10.1364/OL.39.002755, http://ol.osa.org/abstract.cfm?URI=ol-39-9-2755
128. Abdullina, S.R., Vlasov, A.A., Lobach, I.A., Belai, O.V., Shapiro, D.A., Babin, S.A.: Laser Phys. Lett. **13**(7), 075104 (2016). doi:10.1088/1612-2011/13/7/075104, http://stacks.iop.org/1612-202X/13/i=7/a=075104
129. Vlasov, A.A., Churin, D.E., Babin, S.A.: Laser Phys. **20**(12), 2045 (2010). doi:10.1134/S1054660X10170159
130. Shapira, O., Fischer, B.: J. Opt. Soc. Am. B **22**(12), 2542 (2005). doi:10.1364/JOSAB.22.002542, http://josab.osa.org/abstract.cfm?URI=josab-22-12-2542
131. Belai, O., Podivilov, E., Shapiro, D.: Opt. Commun. **266**(2), 512 (2006). doi:10.1016/j.optcom.2006.05.032, http://www.sciencedirect.com/science/article/pii/S0030401806005499
132. Zhang, P., Wang, T., Jia, Q., Liu, X., Kong, M., Tong, S., Jiang, H.: In: Wilson, K.E., Ma, J., Liu, L., Jiang, H., Ke, X. (eds.) International Symposium on Photoelectronic Detection and Imaging 2013: Laser Communication Technologies and Systems, Beijing, 25 June. Proceedings of SPIE, vol. 8906, pp. 890617–6 (2013). doi:10.1117/12.2033151
133. Wang, L., Dong, X., Shum, P.P., Huang, C., Su, H.: Laser Phys. **24**(11), 115101 (2014). doi:10.1088/1054-660X/24/11/115101, http://stacks.iop.org/1555-6611/24/i=11/a=115101
134. Wang, L., Dong, X., Shum, P.P., Su, H.: IEEE Photon. J. **6**(5), 1 (2014). doi:10.1109/JPHOT.2014.2352623
135. Du, X., Zhang, H., Wang, X., Zhou, P.: Appl. Opt. **54**(4), 908 (2015). doi:10.1364/AO.54.000908, http://ao.osa.org/abstract.cfm?URI=ao-54-4-908
136. Zhang, W.L., Li, S.W., Ma, R., Rao, Y.J., Zhu, Y.Y., Wang, Z.N., Jia, X.H., Li, J.: IEEE J. Sel. Top. Quantum Electron. **21**(1), 44 (2015). doi:10.1109/JSTQE.2014.2344050
137. Li, J.L., Musha, M., Shirakawa, A., Ueda, K.I., Zhong, L.X.: Electron. Lett. **42**(8), 449 (2006). doi:10.1049/el:20060182

138. Lobach, I.A., Kablukov, S.I., Skvortsov, M.I., Podivilov, E.V., Melkumov, M.A., Babin, S.A., Dianov, E.M.: Sci. Rep. **6**, 30083 (2016). doi:10.1038/srep30083, http://www.nature.com/articles/srep30083
139. Dianov, E.M.: Light Sci. Appl. **1**, e12 (2012). doi:10.1038/lsa.2012.12, http://www.nature.com/lsa/journal/v1/n5/full/lsa201212a.html
140. El-Taher, A.E., Alcon-Camas, M., Babin, S.A., Harper, P., Ania-Castañón, J.D., Turitsyn, S.K.: Opt. Lett. **35**(7), 1100 (2010). doi:10.1364/OL.35.001100, http://ol.osa.org/abstract.cfm?URI=ol-35-7-1100
141. naki Aporta Litago, I., Pérez-Herrera, R.A., Ángeles Quintela, M., López-Amo, M., López-Higuera, J.M.: J. Lightwave Technol. **34**(17), 4148 (2016). doi:10.1109/JLT.2016.2585742, http://jlt.osa.org/abstract.cfm?URI=jlt-34-17-4148
142. DeMiguel-Soto, V., Bravo, M., Lopez-Amo, M.: Opt. Lett. **39**(7), 2020 (2014). doi:10.1364/OL.39.002020, http://ol.osa.org/abstract.cfm?URI=ol-39-7-2020
143. Ania-Castañón, J.D.: Opt. Express **12**(19), 4372 (2004). doi:10.1364/OPEX.12.004372, http://www.opticsexpress.org/abstract.cfm?URI=oe-12-19-4372
144. Ania-Castañón, J.D., Ellingham, T.J., Ibbotson, R., Chen, X., Zhang, L., Turitsyn, S.K.: Phys. Rev. Lett. **96**, 023902 (2006). doi:10.1103/PhysRevLett.96.023902, http://link.aps.org/doi/10.1103/PhysRevLett.96.023902
145. Ellingham, T.J., Ania-Castanon, J.D., Ibbotson, R., Chen, X., Zhang, L., Turitsyn, S.K.: IEEE Photon. Technol. Lett. **18**(1), 268 (2006). doi:10.1109/LPT.2005.862001
146. Ania-Castañón, J.D., Karalekas, V., Harper, P., Turitsyn, S.K.: Phys. Rev. Lett. **101**, 123903 (2008). doi:10.1103/PhysRevLett.101.123903, http://link.aps.org/doi/10.1103/PhysRevLett.101.123903
147. Ania-Castañón, J., Turitsyn, S.: Opt. Commun. **281**(23), 5760 (2008). doi:10.1016/j.optcom.2008.08.021, http://www.sciencedirect.com/science/article/pii/S0030401808008110
148. Bednyakova, A.E., Fedoruk, M., Harper, P., Turitsyn, S.: Opt. Express **21**(24), 29140 (2013). doi:10.1364/OE.21.029140, http://www.opticsexpress.org/abstract.cfm?URI=oe-21-24-29140
149. Phillips, I., Tan, M., Stephens, M.F., McCarthy, M., Giacoumidis, E., Sygletos, S., Rosa, P., Fabbri, S., Le, S.T., Kanesan, T., Turitsyn, S.K., Doran, N.J., Harper, P., Ellis, A.D.: In: Optical Fiber Communication Conference. OSA (2014), p. M3C.1. doi:10.1364/OFC.2014.M3C.1, http://www.osapublishing.org/abstract.cfm?URI=OFC-2014-M3C.1
150. Rosa, P., Le, S.T., Rizzelli, G., Tan, M., Ania-Castañón, J.D.: Opt. Express **23**(25), 31772 (2015). doi:10.1364/OE.23.031772, http://www.opticsexpress.org/abstract.cfm?URI=oe-23-25-31772
151. Ellis, A.D., Tan, M., Iqbal, M.A., Al-Khateeb, M.A.Z., Gordienko, V., Mondaca, G.S., Fabbri, S., Stephens, M.F.C., McCarthy, M.E., Perentos, A., Phillips, I.D., Lavery, D., Liga, G., Maher, R., Harper, P., Doran, N., Turitsyn, S.K., Sygletos, S., Bayvel, P.: J. Lightwave Technol. **34**(8), 1717 (2016). doi:10.1109/JLT.2016.2521430, http://jlt.osa.org/abstract.cfm?URI=jlt-34-8-1717
152. Prilepsky, J.E., Derevyanko, S.A., Blow, K.J., Gabitov, I., Turitsyn, S.K.: Phys. Rev. Lett. **113**, 013901 (2014). doi:10.1103/PhysRevLett.113.013901, http://link.aps.org/doi/10.1103/PhysRevLett.113.013901
153. Le, S.T., Philips, I.D., Prilepsky, J.E., Harper, P., Ellis, A.D., Turitsyn, S.K.: J. Lightwave Technol. **34**(10), 2459 (2016). doi:10.1109/JLT.2016.2536780, http://jlt.osa.org/abstract.cfm?URI=jlt-34-10-2459
154. Aref, V., Le, S., Buelow, H.: In: 2016 European Conference on Optical Communication (ECOC) (2016), p. Th.3.B.2. Postdeadline paper
155. Okuno, T., Tsuzaki, T., Nishimura, M.: IEEE Photon. Technol. Lett. **13**(8), 806 (2001). doi:10.1109/68.935810
156. Bouteiller, J.C., Brar, K., Headley, C.: In: 28th European Conference on Optical Communication, ECOC 2002, vol. 3, pp. 1–2 (2002). http://ieeexplore.ieee.org/document/1601058/

157. Karpov, V., Papernyi, S., Ivanov, V., Clements, W., Araki, T., Koyano, Y.: In: SubOptic 2004: International Submarine Convention: Proceedings (2004), p. We 8.8. http://www.mpbcommunications.com/images/papers/SubOptic\relax$\@@underline\hbox{~}\mathsurround\z@$\relax2004-Poster.pdf. Poster paper March 29th to April 1st
158. Labrunie, L., Boubal, F., Roux, P.L., Brandon, E.: Electron. Lett. **39**(19), 1394 (2003). doi:10.1049/el:20030876
159. Papernyi, S., Karpov, V., Clements, W.: In: Optical Fiber Communications Conference, p. FB4. OSA (2002). http://www.osapublishing.org/abstract.cfm?URI=OFC-2002-FB4. Postdeadline
160. Papernyi, S., Ivanov, V., Koyano, Y., Yamamoto, H.: In: Optical Fiber Communication Conference and Exposition and the National Fiber Optic Engineers Conference, p. OThF4. OSA (2005). http://www.osapublishing.org/abstract.cfm?URI=OFC-2005-OThF4
161. Faralli, S., Sugliani, S., Sacchi, G., Pasquale, F.D., Papernyi, S.: In: Optical Fiber Communication Conference and Exposition and the National Fiber Optic Engineers Conference, p. OThF3. OSA (2005). http://www.osapublishing.org/abstract.cfm?URI=OFC-2005-OThF3
162. Babin, S., Churkin, D., Fotiadi, A., Kablukov, S., Medvedkov, O., Podivilov, E.: IEEE Photon. Technol. Lett. **17**(12), 2553 (2005). doi:10.1109/LPT.2005.859547
163. Saxena, B., Bao, X., Chen, L.: Opt. Lett. **39**(4), 1038 (2014). doi:10.1364/OL.39.001038, http://ol.osa.org/abstract.cfm?URI=ol-39-4-1038
164. Tan, M., Rosa, P., Le, S.T., Iqbal, M.A., Phillips, I.D., Harper, P.: Opt. Express **24**(3), 2215 (2016). doi:10.1364/OE.24.002215, http://www.opticsexpress.org/abstract.cfm?URI=oe-24-3-2215
165. Tan, M., Rosa, P., Le, S.T., Phillips, I.D., Harper, P.: Opt. Express **23**(17), 22181 (2015). doi:10.1364/OE.23.022181, http://www.opticsexpress.org/abstract.cfm?URI=oe-23-17-22181
166. Tan, M., Rosa, P., Iqbal, M.A., Phillips, I., no, J.N., Ania-Castanon, J.D., Harper, P.: In: Asia Communications and Photonics Conference 2015, p. AM2E.6. OSA (2015). doi:10.1364/ACPC.2015.AM2E.6, http://www.osapublishing.org/abstract.cfm?URI=ACPC-2015-AM2E.6
167. Tan, M., Rosa, P., Phillips, I., Harper, P.: In: Optical Fiber Communication Conference, p. W4E.1. OSA (2015). doi:10.1364/OFC.2015.W4E.1, http://www.osapublishing.org/abstract.cfm?URI=OFC-2015-W4E.1
168. Tan, M., Rosa, P., Phillips, I.D., Harper, P.: In: Asia Communications and Photonics Conference 2014, p. ATh1E.4. OSA (2014). doi:10.1364/ACPC.2014.ATh1E.4. http://www.osapublishing.org/abstract.cfm?URI=ACPC-2014-ATh1E.4
169. Rosa, P., Tan, M., Le, S.T., Philips, I.D., Ania-Castañón, J.D., Sygletos, S., Harper, P.: IEEE Photon. Technol. Lett. **27**(11), 1189 (2015). doi:10.1109/LPT.2015.2414712
170. Jia, X.H., Rao, Y.J., Peng, F., Wang, Z.N., Zhang, W.L., Wu, H.J., Jiang, Y.: Opt. Express **21**(5), 6572 (2013). doi:10.1364/OE.21.006572, http://www.opticsexpress.org/abstract.cfm?URI=oe-21-5-6572
171. Jia, X.H., Rao, Y.J., Yuan, C.X., Li, J., Yan, X.D., Wang, Z.N., Zhang, W.L., Wu, H., Zhu, Y.Y., Peng, F.: Opt. Express **21**(21), 24611 (2013). doi:10.1364/OE.21.024611, http://www.opticsexpress.org/abstract.cfm?URI=oe-21-21-24611
172. Dvoyrin, V., Tan, M., Zlenko, A., Mashinsky, V., Melkumov, M., Dianov, E., Harper, P., Turitsyn, S.K.: In: Conference on Lasers and Electro-optics, p. STu1F.4. Optical Society of America (2016). doi:10.1364/CLEO_SI.2016.STu1F.4, http://www.osapublishing.org/abstract.cfm?URI=CLEO\relax$\@@underline\hbox{~}\mathsurround\z@$\relaxSI-2016-STu1F.4
173. Rosa, P., Rizzelli, G., Tan, M., Harper, P., Ania-Castañón, J.D.: Opt. Express **23**(22), 28634 (2015). doi:10.1364/OE.23.028634, http://www.opticsexpress.org/abstract.cfm?URI=oe-23-22-28634

174. Ellis, A.D., Phillips, I.D., Tan, M., Stephens, M.F.C., McCarthy, M.E., Kahteeb, M.A.Z.A., Iqbal, M.A., Perentos, A., Fabbri, S., Gordienko, V., Lavery, D., Liga, G., Saavedra, M.G., Maher, R., Sygletos, S., Harper, P., Doran, N.J., Bayvel, P., Turitsyn, S.K.: In: European Conference on Optical Communication (ECOC 2015), pp. 1–3 (2015). doi:10.1109/ECOC.2015.7341993
175. Galdino, L., Tan, M., Alvarado, A., Lavery, D., Rosa, P., Maher, R., Ania-Castañón, J.D., Harper, P., Makovejs, S., Thomsen, B.C., Bayvel, P.: J. Lightwave Technol. **34**(9), 2221 (2016). doi:10.1109/JLT.2016.2521002, http://jlt.osa.org/abstract.cfm?URI=jlt-34-9-2221
176. Galdino, L., Tan, M., Lavery, D., Rosa, P., Maher, R., Phillips, I.D., Ania-Castañón, J.D., Harper, P., Killey, R.I., Thomsen, B.C., Makovejs, S., Bayvel, P.: Opt. Lett. **40**(13), 3025 (2015). doi:10.1364/OL.40.003025, http://ol.osa.org/abstract.cfm?URI=ol-40-13-3025
177. Jia, X.H., Rao, Y.J., Wang, Z.N., Zhang, W.L., Jiang, Y., Zhu, J.M., Yang, Z.X.: In: Liao, Y., Jin, W., Sampson, D.D., Yamauchi, R., Chung, Y., Nakamura, K., Rao, Y. (eds.) OFS2012 22nd International Conference on Optical Fiber Sensors, Beijing, 14 Oct. Proceedings of SPIE, vol. 8421, pp. 842127–4 (SPIE, 2012). doi:10.1117/12.975168
178. Jia, X.H., Rao, Y.J., Wang, Z.N., Zhang, W.L., Yuan, C.X., Yan, X.D., Li, J., Wu, H., Zhu, Y.Y., Peng, F.: Opt. Express **21**(18), 21208 (2013). doi:10.1364/OE.21.021208, http://www.opticsexpress.org/abstract.cfm?URI=oe-21-18-21208
179. Wang, Z.N., Zeng, J.J., Li, J., Fan, M.Q., Wu, H., Peng, F., Zhang, L., Zhou, Y., Rao, Y.J.: Opt. Lett. **39**(20), 5866 (2014). doi:10.1364/OL.39.005866, http://ol.osa.org/abstract.cfm?URI=ol-39-20-5866
180. Wang, Z., Wu, H., Fan, M., Rao, Y., Jia, X., Zhang, W.: Opt. Express **21**(17), 20090 (2013). doi:10.1364/OE.21.020090, http://www.opticsexpress.org/abstract.cfm?URI=oe-21-17-20090
181. Leandro, D., deMiguel-Soto, V., López-Amo, M.: J. Lightwave Technol. **34**(19), 4596 (2016). doi:10.1109/JLT.2016.2536650, http://jlt.osa.org/abstract.cfm?URI=jlt-34-19-4596
182. Leandro, D., deMiguel Soto, V., Perez-Herrera, R.A., Acha, M.B., López-Amo, M.: J. Lightwave Technol. **34**(19), 4430 (2016). doi:10.1109/JLT.2016.2547868, http://jlt.osa.org/abstract.cfm?URI=jlt-34-19-4430
183. Churkin, D.V., Sugavanam, S., Vatnik, I.D., Wang, Z., Podivilov, E.V., Babin, S.A., Rao, Y., Turitsyn, S.K.: Adv. Opt. Photon. **7**(3), 516 (2015). doi:10.1364/AOP.7.000516, http://aop.osa.org/abstract.cfm?URI=aop-7-3-516

Erratum to: Distributed Feedback Raman and Brillouin Fiber Lasers

Paul S. Westbrook, Kazi S. Abedin, and Tristan Kremp

Erratum to:
Chapter 6 in: Y. Feng (ed.), *Raman Fiber Lasers*,
Springer Series in Optical Sciences 207,
https://doi.org/10.1007/978-3-319-65277-1_6

On page 249, the labels in figures 6.5(a) and (b), should have read "FOM $= \infty$" and not "FOM $= \alpha$".

These labels have been updated.

The updated online version of this chapter can be found at
https://doi.org/10.1007/978-3-319-65277-1_6

© Springer International Publishing AG 2017
Y. Feng (ed.), *Raman Fiber Lasers*, Springer Series in Optical Sciences 207,
DOI 10.1007/978-3-319-65277-1_8

Index

© Springer International Publishing AG 2017
Y. Feng (ed.), *Raman Fiber Lasers*, Springer Series in Optical Sciences 207,
DOI 10.1007/978-3-319-65277-1

Printed by Printforce, the Netherlands